D1127140

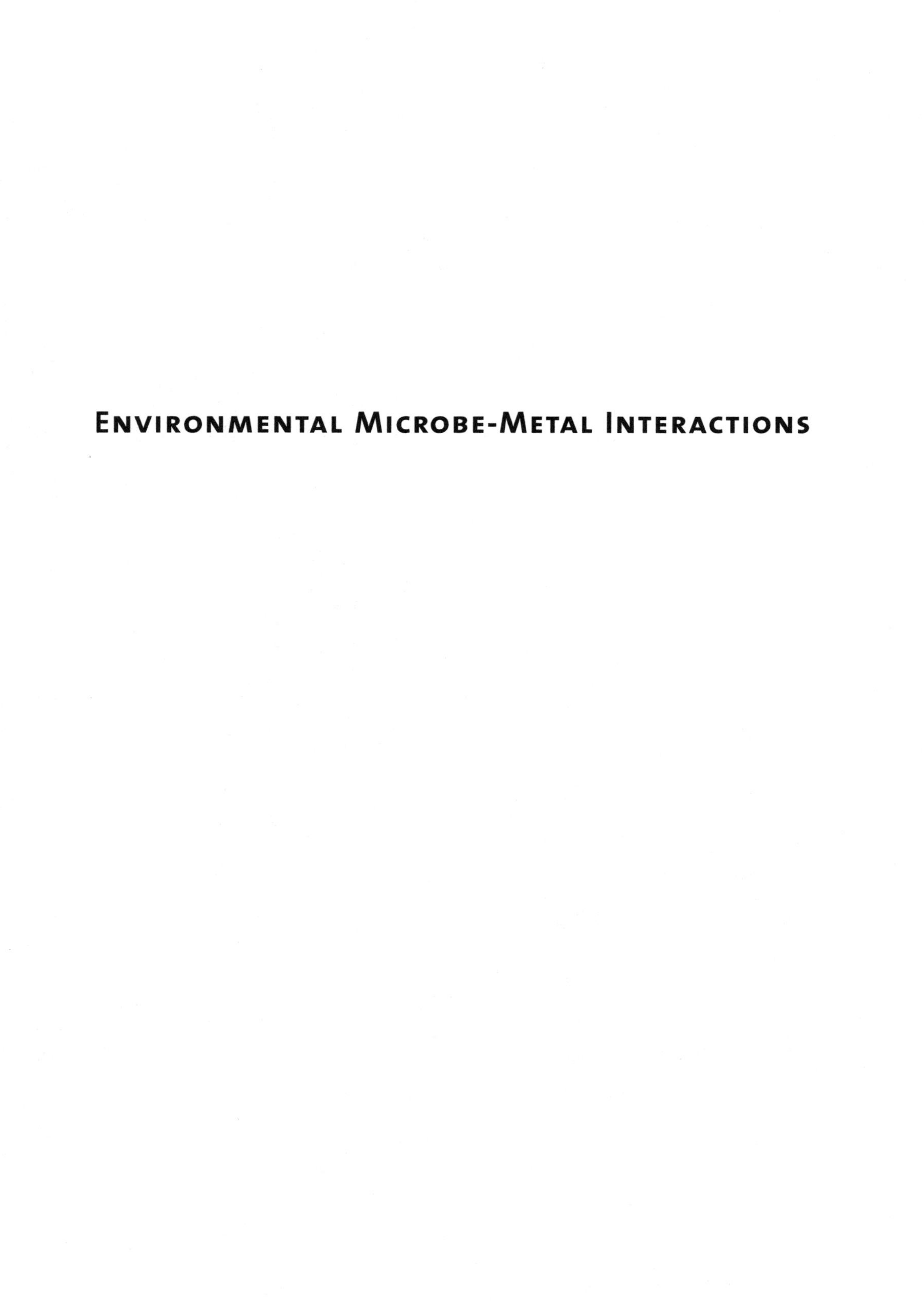

Environmental Microbe-Metal Interactions

Environmental Microbe-Metal Interactions

EDITED BY

Derek R. Lovley

Department of Microbiology
University of Massachusetts
Amherst, Massachusetts

ASM PRESS WASHINGTON, D.C.

American Society for Microbiology
1752 N St. NW
Washington, DC 20036-2804

Library of Congress Cataloging-in-Publication Data

Environmental microbe-metal interactions / edited by Derek R. Lovley.
p. cm.
Includes bibliographical references and index.
ISBN 1-55581-195-7
1. Microorganisms—Effect of metals on. 2. Metals—Metabolism. 3. Biogeochemical cycles. 4. Bioremediation. 5. Biomineralization. I. Lovley, Derek R.
QR92.M45 E586 2000
579—dc21

00-025073

Printed in the United States of America

Cover photo: Scanning electron micrograph of the dissimilatory metal-reducing microorganism *Geobacter metallireducens* and the uraninite precipitate that results from the reduction of soluble uranium.

CONTENTS

CONTRIBUTORS

Dennis A. Bazylinski • Department of Microbiology, Iowa State University, Ames, Iowa 50011

Robert Blake II • College of Pharmacy, Xavier University, New Orleans, Louisiana 70125

Harvey Bolton, Jr. • Pacific Northwest National Laboratory, Richland, Washington 99352

Nigel L. Brown • School of Biological Sciences, University of Birmingham, Edgbaston, Birmingham B15 2TT, United Kingdom

Ralf Cord-Ruwisch • Biotechnology, Murdoch University, Perth, Western Australia 6150, Australia

David Emerson • American Type Culture Collection, 10801 University Boulevard, Manassas, Virginia 20110, and Institute for Biosciences, Bioinformatics, and Biotechnology, George Mason University, Manassas, Virginia 20110

Richard B. Frankel • Department of Physics, California Polytechnic State University, San Luis Obispo, California 93407

Geoffrey M. Gadd • Department of Biological Sciences, University of Dundee, Dundee DD1 4HN, Scotland

Don C. Girvin • Pacific Northwest National Laboratory, Richland, Washington 99352

Larry E. Hersman • Environmental Molecular Biology Group, Los Alamos National Laboratory, Los Alamos, New Mexico 87545

Jon L. Hobman • School of Biological Sciences, University of Birmingham, Edgbaston, Birmingham B15 2TT, United Kingdom

D. Barrie Johnson • School of Biological Sciences, University of Wales, Bangor, United Kingdom

Jon R. Lloyd • Department of Microbiology, University of Massachusetts, Amherst, Massachusetts 01003

Derek R. Lovley • Department of Microbiology, University of Massachusetts, Amherst, Massachusetts 01003

Lynne E. Macaskie • School of Biological Sciences, University of Birmingham, Edgbaston, Birmingham B15 2TT, United Kingdom

Ronald S. Oremland • U.S. Geological Survey, ms 480, 345 Middlefield Road, Menlo Park, California 94025

Jacqueline A. Sayer • Department of Biological Sciences, University of Dundee, Dundee DD1 4HN, Scotland

Silke Schiewer • Department of Chemical Engineering, McGill University, M. H. Wong Building, 3610 University Street, Montreal, Quebec H3A 2B2, Canada

Gordon Southam • Department of Biological Sciences, Northern Arizona University, Flagstaff, Arizona 86011-5640

John Stolz • Department of Biological Sciences, Duquesne University, Pittsburgh, Pennsylvania 15282

William G. Sunda • Beaufort Laboratory, National Oceanic and Atmospheric Administration, Beaufort, North Carolina 28516

Bohumil Volesky • Department of Chemical Engineering, McGill University, M. H. Wong Building, 3610 University Street, Montreal, Quebec H3A 2B2, Canada

Yi-Tin Wang • Department of Civil Engineering, University of Kentucky, Lexington, Kentucky 40506

Jon R. Wilson • School of Biological Sciences, University of Birmingham, Edgbaston, Birmingham B15 2TT, United Kingdom

Luying Xun • Washington State University, Pullman, Washington 99164

PREFACE

The importance of metals in the life of microorganisms is well known. Iron and other metals are key components in many proteins that are necessary for microbial respiration and metabolism. Less appreciated is the major impact that microorganisms can have on the fate of metals in the environment. Environmental science textbooks typically discuss the geochemical cycles of metals in terms of equilibrium thermodynamics and Eh-pH diagrams. However, research on microbe-metal interactions, much of it conducted within the last decade, has demonstrated that new models that take into account nonequilibrium biochemical processes are required in order to truly understand metal geochemistry. The purpose of this volume is to give an overview of the current understanding of environmental microbe-metal interactions and to provide the basis for improved models of metal cycling.

This book deals first with biogeochemical cycling of iron and manganese. These are the two most abundant redox-active metals in the Earth's crust. As outlined in chapter 1, the origin of life may have been intimately connected to the ability of iron to readily cycle between the ferric [Fe(III)] and ferrous [Fe(II)] states. Some of the earliest geochemical signals of life on Earth are the conversion of the Fe(II) dissolved in the archaean seas to massive Fe(III) oxide deposits, possibly as the result of activity of Fe(II)-oxidizing phototrophs (chapter 2), and the conversion of this Fe(III) oxide to magnetite by Fe(III)-reducing microorganisms (chapter 1). In modern environments, microbial oxidation of organic matter to carbon dioxide, coupled to the reduction of Fe(III) or Mn(IV), is an important process for the degradation of both naturally occurring and contaminant organic compounds in a variety of sedimentary environments and the subsurface (chapter 1). Microbial oxidation of the Fe(II) produced from Fe(III) reduction provides a "ferrous wheel" to complete the iron cycle (chapter 2). Microbially catalyzed Fe(III) reduction and Fe(II) oxidation are also important processes in acidic environments, such as those that result from mine drainage. Recent studies have begun to elucidate which microorganisms might catalyze these reactions at low pH and the biochemical mechanisms for these reactions (chapter 3).

Iron is also an important nutrient for environmentally relevant microbes. Nowhere is this more apparent than in the dramatic effect that iron and other trace metals can have on primary productivity in the ocean (chapter 4). The study of this phenomenon is essential to understanding global carbon cycling and nutrient dynamics in marine systems.

An amazing impact of microorganisms on the iron cycle is the ability of magnetotactic bacteria to concentrate iron from the environment into intracellular chains of single-domain magnetite crystals (chapter 5). Magnetite from magnetotactic bacteria provides one of the best-characterized geological signatures of microbial activity on Earth and possibly other planets.

One of the most-studied forms of microbe-metal interactions is microbial acquisition of iron through the use of siderophores. Nearly all of the research on microbial siderophores has focused on iron uptake by microorganisms of medical interest. However, as described in chapter 6, siderophores may also have an impact on iron cycling and microbial metabolism in aerobic soils. Studies in this area are in their infancy, but it seems likely that this form of metal solubilization plays an important role in metal cycling.

Microbially influenced corrosion of metallic iron is of considerable economic concern and has been studied intensively for many years. Chapter 7 reviews the diverse concepts that have been developed in this area and presents a new model for the mechanisms by which microorganisms enhance iron corrosion.

A major factor driving recent increased interest in environmental microbe-metal interactions is the need to remediate extensive metal contamination of water and soils. Microorganisms are not alchemists and cannot change a toxic metal to a less-toxic element. However, microbially catalyzed precipitation or volatilization of metals can remove them from polluted environments. The most-studied form of microbial metal reduction is reduction of soluble Hg(II) to volatile Hg(0) (chapter 8). Hg(II) reduction is not linked to respiration but rather is a detoxification strategy which removes mercury from the cell and may also promote mercury volatilization from contaminated environments. The detailed information that is available on Hg(II) reduction serves as an excellent model for the study of microbial reduction of other metals.

Reductive precipitation of metal and metalloid contaminants from waters and waste streams also shows promise for environmental restoration. For example, microbes can use oxidized forms of the metalloids selenium and arsenic as terminal electron acceptors to support anaerobic growth (chapter 9). Reduction of soluble selenium in insoluble elemental selenium naturally removes this toxic metalloid from agricultural drainage waters and can be stimulated in order to promote selenium removal (chapter 9). The metalloid arsenic can also be microbially reduced, which, depending upon environmental conditions, can lead to solubilization or precipitation of arsenic (chapter 9). Microbial reduction of soluble, toxic Cr(VI) to less soluble, less toxic Cr(III) provides a potential mechanism for remediation of chromium-contaminated waters and soils (chapter 10). Microorganisms can even conserve energy to support growth from the reduction of uranium and other radioactive metals and in the process immobilize these metals in the environment (chapter 13).

Although most investigations of microbe-metal interactions have focused on prokaryotes, fungi have similar abilities to affect the fate of metals (chapter 11). Considering the vast biomass of fungi in many soils, fungus-metal interactions are certain to be a major area of future study on the biogeochemistry of metals.

A major environmental fate of toxic metals may be adsorption to microbial biomass. As outlined in chapters 12 and 14, there are numerous mechanisms by which metals can bind to cell surfaces. Microorganisms can also promote cell-associated precipitation of toxic metals by the release of phosphate, which can form insoluble metal phosphates (chapter 13). These various adsorption processes can lead to mineral formation in the environment (chapter 12), and they have practical application for the removal of metals from contaminated waters and waste streams (chapters 13 and 14).

One of the most important factors promoting the solubility of contaminant metals, and hence their mobility in subsurface environments, is complexation of the metals to synthetic chelators (chapter 15). The discovery that microbes can degrade the organic portion of metal chelates, thus diminishing the mobility of the metals, has been a major advancement in the field of microbe-metal interactions.

A take-home message of many of the chapters in this book is the need for better understanding of many facets of environmental microbe-metal interactions. The level of inquiry into microbe-metal interactions has been modest in comparison with the intensive study of microbial carbon metabolism and microbial interactions with major, nonmetallic inorganic species such as oxygen, nitrogen, phosphorous, sulfur, and hydrogen. Hopefully, this book makes clear the significance of environmental microbe-metal interactions and provides some clues for fruitful areas of further investigation.

Derek R. Lovley

I. BIOGEOCHEMICAL CYCLING OF IRON AND MANGANESE

Environmental Microbe-Metal Interactions
Edited by Derek R. Lovley

Chapter 1

Fe(III) and Mn(IV) Reduction

Derek R. Lovley

The abundance of iron in the Earth's crust and the ability of iron to readily transition between the Fe(III) and Fe(II) states has resulted in iron becoming a key metal in environmental microbe-metal interactions. One of the most recently recognized environmental roles of iron is that Fe(III) serves as an electron acceptor in microbial respiration. Respiration with Fe(III) oxides as the electron acceptor is commonly referred to as dissimilatory Fe(III) reduction. However, dissimilatory Fe(III) reduction need not necessarily be coupled to energy conservation (62, 64). Therefore, the preferred term for energy conservation via Fe(III) reduction is Fe(III) respiration. In Fe(III) respiration, as well as other forms of dissimilatory Fe(III) reduction, Fe(III) is reduced to Fe(II). As discussed below, a wide phylogenetic diversity of microorganisms is capable of Fe(III) respiration. Many of these organisms are also capable of using Mn(IV) as an electron acceptor, reducing Mn(IV) to Mn(II) (69).

Fe(III) respiration is an important process primarily in anoxic environments. Although some microorganisms may reduce Fe(III) in the presence of oxygen (8), in aerobic environments the Fe(II) produced will react with oxygen (see chapter 2) and be converted back to Fe(III). Generally, there is also no net Fe(III) reduction in the presence of nitrate due to a combination of preferential reduction of nitrate by Fe(III)-reducing microorganisms (30), biological oxidation of Fe(II) with nitrate (130, 131), and possible abiological oxidation of Fe(II) by nitrite (64). Fe(II) is also rapidly oxidized abiotically with Mn(IV) (64). Thus, in a typical aquatic sediment or subsurface environment, Fe(III) reduction is localized in zones in which oxygen, nitrate, and Mn(IV) have been depleted (64, 74). Fe(III) reduction typically takes place prior to sulfate reduction and methane production, because Fe(III)-reducing microorganisms can outcompete sulfate-reducing and methanogenic microorganisms for electron donors (21, 87).

Fe(III) respiration impacts not only on the speciation of iron in anoxic sedimentary environments but also on the carbon cycle and the fate of a variety trace metals and nutrients, as well as the physical structure of the sediments themselves.

Derek R. Lovley • Department of Microbiology, University of Massachusetts, Amherst, MA 01003.

Microbiological and geological evidence suggests that in addition to its importance in modern environments, Fe(III) respiration may have been one of the earliest important forms of respiration as life was first evolving on Earth. The purpose of this chapter is to summarize the known environmental consequences of Fe(III) respiration as well as related forms of respiration such as reduction of Mn(IV) and humic substances.

BRIEF HISTORY OF THE CONCEPTS OF ENVIRONMENTAL Fe(III) REDUCTION

The finding that specialized microorganisms are capable of Fe(III)-based respiration substantially changed our understanding of iron geochemistry in soils and sediments. Until recently, even microbially oriented texts considered that Fe(III) reduction in sedimentary environments was primarily an abiotic process (36, 41, 141). As previously reviewed in detail (64), this abiotic concept dates back to some of the earliest studies on microbial Fe(III) reduction (128). In these studies it was concluded that Fe(III) reduction occurred when microorganisms generated chemical conditions which would promote the nonenzymatic conversion of Fe(III) to Fe(II) and that Fe(III) reduction was merely a function of changes in pH and/or redox potential. However, no direct evidence to support this nonenzymatic model for Fe(III) reduction was provided (64).

Over time, numerous microorganisms that reduce Fe(III) during anaerobic growth have been described (see reference 62 for a review), but these are fermentative microorganisms that do not require Fe(III) as an electron acceptor in order to grow. Typically, less than 5% of the reducing equivalents in the substrates fermented are transferred to Fe(III) (64). Thus, it was considered that "reduction of iron is not a vital process for these bacteria; the presence of ferric iron cannot be considered a necessary condition for their development" (6).

Ottow and coworkers (references 103, 104, 112, and 114 and references therein) championed the concept that Fe(III) reduction during anaerobic fermentation was the result of enzymatic rather than nonenzymatic processes. Support for this came from studies that indirectly implicated the nitrate reductase of some organisms as an Fe(III) reductase (113). However, the strongest argument for enzymatic Fe(III) reduction was provided in studies demonstrating that Fe(III) reduction was inhibited when direct contact between the Fe(III)-reducing microorganisms and Fe(III) oxides was prevented by placing the Fe(III) within a semipermeable membrane (104). The assumption in these studies was that the semipermeable membrane would permit the passage of soluble reductants and would provide for equivalent pH and redox potential on either side of the membrane. Thus, the Fe(III) oxide should still be reduced when it was within the semipermeable membrane if Fe(III) reduction was the result of changes in pH, redox potential, or the generation of soluble reductants. The validity of this assumption has recently been questioned (K. P. Nevin and D. R. Lovley, submitted for publication), but such studies nevertheless significantly promoted the concept of Fe(III) reduction as an enzymatic process.

Studies on Fe(III) reduction in the organism *Pseudomonas* strain 200 (now considered a strain of *Shewanella putrefaciens*) suggested that Fe(III) reduction could be linked to an electron transport chain involving cytochromes and other electron carriers (7, 9, 31, 110, 111). Although this organism is now known to conserve energy from Fe(III) reduction (33), this was not apparent from earlier studies, and thus the metabolic and environmental significance of the enzymatic Fe(III) reduction in *Pseudomonas* strain 200 was not initially clear.

The first unequivocal demonstration of anaerobic growth with Fe(III) serving as the sole electron acceptor was the finding that a "pseudomonad" could grow via hydrogen oxidation coupled to Fe(III) reduction (10). The stoichiometries of hydrogen uptake and Fe(II) production were consistent with Fe(III) serving as the sole electron acceptor for hydrogen oxidation. This suggested that Fe(III) reduction was linked to energy-conserving electron transport. However, hydrogen oxidation coupled to Fe(III) reduction could not be considered to account for the majority of Fe(III) reduction in sedimentary environments, and Fe(III) reduction was still discussed as a primarily abiotic process (141).

The finding that microorganisms could completely oxidize organic compounds to carbon dioxide with Fe(III) serving as the sole electron acceptor (95) provided the first model to explain how Fe(III) respiration could be responsible for most of the Fe(III) reduction in sedimentary environments. Strain GS-15 (later named *Geobacter metallireducens* [81]) was found to oxidize acetate to carbon dioxide with reduction of Fe(III) (88, 95) according to the following equation:

$$CH_3COO^- + 8Fe(III) + 4H_2O \rightarrow 2HCO_3^- + 8Fe(II) + 9H^+$$

This suggested that fermentable compounds could be completely oxidized to carbon dioxide through the cooperative activity of fermentative microorganisms and the Fe(III)-respiring microorganisms (FRM) which could oxidize fermentation products (88). Analysis of glucose metabolism by microbial communities (89) suggested that this is how fermentable compounds are processed in environments in which Fe(III) reduction is the predominant terminal electron-accepting process (TEAP). As expected from previous studies on carbon metabolism under methanogenic and sulfate-reducing conditions, acetate was the principal fermentation acid produced in sediments in which Fe(III) reduction was the TEAP (89). Subsequent studies have demonstrated that FRM can also completely oxidize other electron donors such as short-chain fatty acids other than acetate, long-chain fatty acids, alcohols, and a variety of monoaromatic compounds (69). This suggests that complex organic matter can be oxidized to carbon dioxide by the cooperative activity of various fermentative microorganisms and FRM (Fig. 1). This model is analogous to the pathways proposed to account for organic-matter oxidation in sediments in which sulfate reduction is the TEAP (74).

Comparative studies have indicated that Fe(III) reduction by FRM is likely to be much more important than completely abiotic processes for Fe(III) reduction, especially under conditions in which sulfate reduction is minimal. Few organic compounds react directly with Fe(III), and the few organics that can reduce Fe(III) will reduce only a minor amount of Fe(III) compared to the amount of Fe(III) reduction when FRM oxidize the same organics to carbon dioxide (93). For example, hydroquinones are among the most reactive organic compounds in reducing

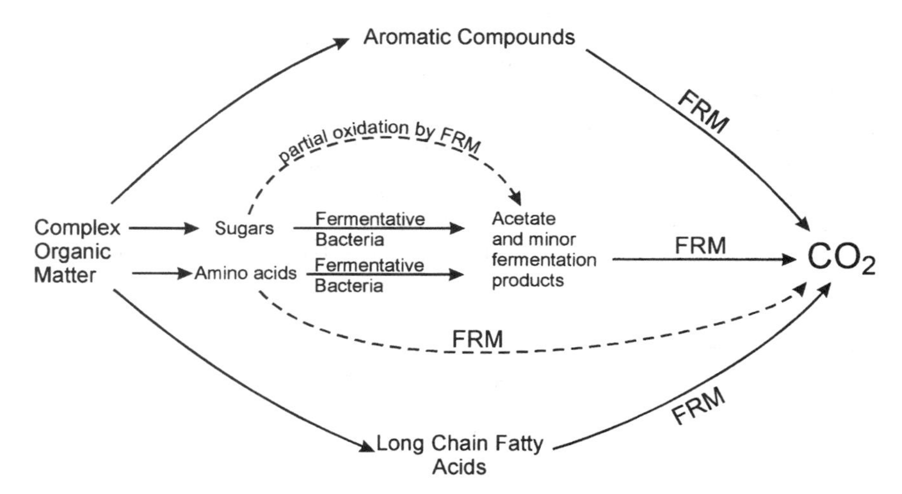

Figure 1. Model for degradation of organic matter in anoxic environments in which Fe(III) reduction is the TEAP. Dashed lines represent minor pathways. A similar model is likely to apply to sediments in which Mn(IV) reduction predominates.

Fe(III) (56), but hydroquinones only abiotically reduce 2 mol of Fe(III) per mol of hydroquinone whereas over 20 mol of Fe(III) is typically reduced when FRM enzymatically oxidize these compounds to carbon dioxide with the reduction of Fe(III) (93). Furthermore, it has been demonstrated that the development of a low redox potential through the consumption of oxygen and the production of hydrogen and organic acids is not a sufficient condition to bring about Fe(III) reduction in the absence of FRM (93).

Sulfide can nonenzymatically reduce Fe(III) and is often suggested to be the prime reductant for Fe(III), especially in marine sediments, where there are large supplies of sulfate for dissimilatory sulfate reduction. However, such a reduction of Fe(III) by sulfide is not expected in the vast majority of sediments, in which Fe(III) is reduced prior to significant sulfate reduction (64). As mentioned above, there is generally little sulfate reduction in the Fe(III) reduction zone of sediments because FRM can outcompete sulfate-reducing microorganisms for important electron donors such as acetate and hydrogen. FRM maintain the concentration of these electron donors too low for sulfate reduction to be thermodynamically favorable. Even in marine sediments, the inhibition of sulfate reduction has no effect on the rates of Fe(III) reduction (127). This suggests that sulfide production from sulfate reduction is not an important mechanism for Fe(III) reduction.

Another possible mechanism for nonenzymatic Fe(III) reduction has arisen with the discovery that FRM can reduce humic substances (humics) and other extracellular quinones (77, 79, 83, 124; K. P. Nevin and D. R. Lovley, submitted for publication). As detailed below, once these extracellular quinones are reduced, they in turn can reduce Fe(III) oxides via a nonenzymatic reaction. It has yet to be determined if this form of electron shuttling between FRM and Fe(III) oxides is an important environmental mechanism for Fe(III) reduction. However, if electron shuttling via extracellular quinones does account for a substantial amount of Fe(III) reduction, then even though metabolism of FRM would be the source of electrons for Fe(III) reduction, the actual reduction of the Fe(III) oxides would proceed through an abiotic process.

The concept of metal-based respiration has been expanded from Fe(III) to a wide variety of redox-active metals and metalloids that have now been found to serve as terminal electron acceptors in microbial respiration. For example, microbial reduction of Mn(IV) to Mn(II) is discussed here and the potentials for reduction of other metals and metalloids are discussed in subsequent chapters.

MICROORGANISMS INVOLVED IN DISSIMILATORY Fe(III) REDUCTION

A substantial number of microorganisms capable of conserving energy to support growth via Fe(III) reduction are known (Fig. 2 and 3). The physiology and phylogeny of these organisms have been recently reviewed in detail (69, 78). A brief overview of the major groups of Fe(III)-reducing microorganisms that grow at circumneutral pH follows. Acidophilic Fe(III)-reducing microorganisms are discussed in chapter 3.

Mesophilic Fe(III)-Respiring Microorganisms Available in Pure Culture

The largest known group of FRM is the *Geobacteraceae* family in the delta subclass of the *Proteobacteria* (Fig. 2). All of the organisms within this family are

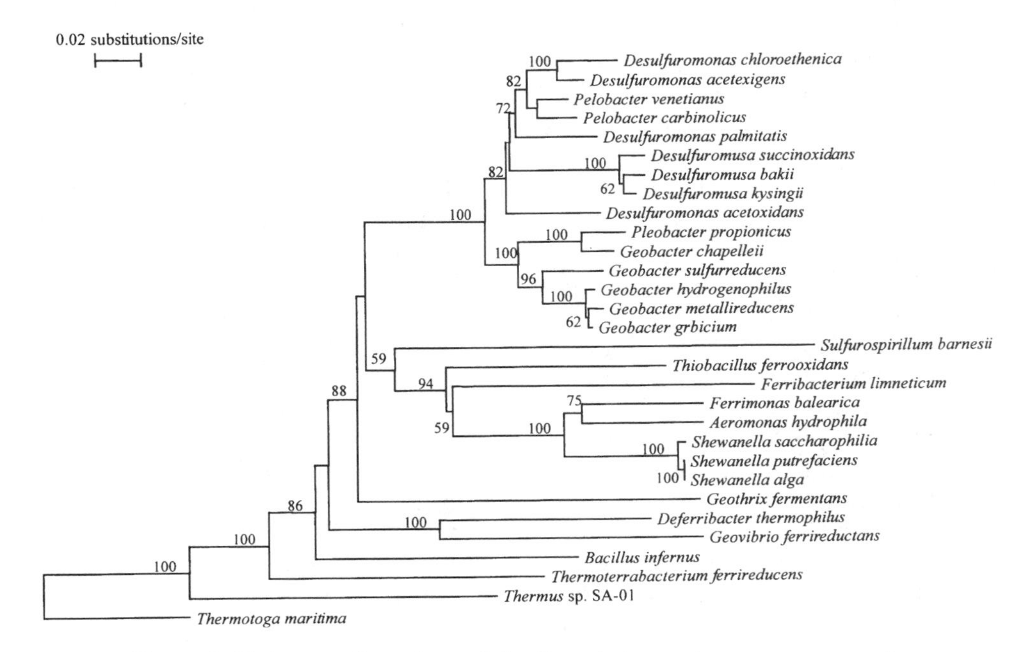

Figure 2. Phylogeny of members of the *Bacteria* known to conserve energy to support growth via Fe(III) reduction based on 16S rDNA sequences.

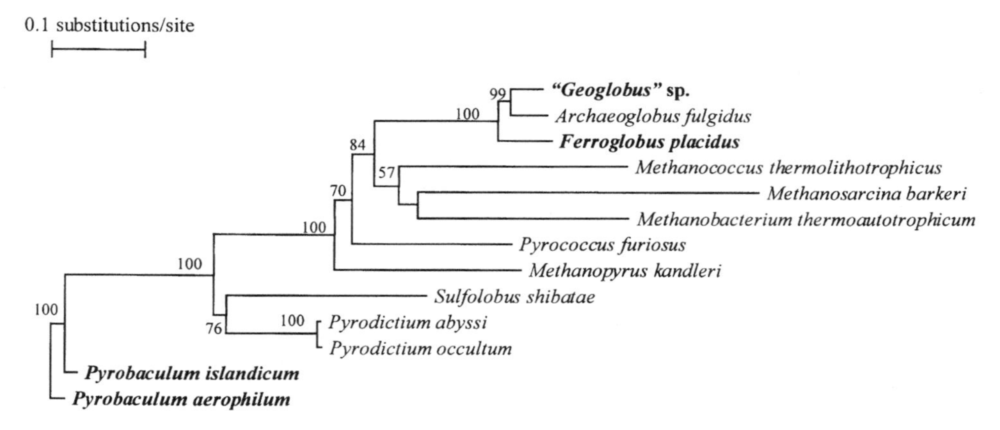

Figure 3. Phylogeny of members of the *Archaea* known to reduce Fe(III) based on 16S rDNA sequences. Those highlighted in bold conserve energy to support growth from Fe(III) reduction. All others, with the exception of *Pyrobaculum occultum* and *Sulfolobus shibatae*, which have not yet been evaluated, have the ability to reduce Fe(III) in cell suspensions.

capable of conserving energy to support growth from Fe(III) reduction (69). With the exception of the *Pelobacter* species, these organisms are capable of completely oxidizing acetate to carbon dioxide with the reduction of Fe(III). Most of these organisms can completely oxidize other organic acids (69). *Geobacter metallireducens* and *G. sulfurreducens* are capable of oxidizing monoaromatic compounds, including the aromatic hydrocarbon toluene, to carbon dioxide. Many of the members of the *Geobacteraceae* can use hydrogen as an electron donor for Fe(III) reduction (69). Members of the *Geobacteraceae* are the Fe(III) reducers most commonly recovered from a variety of aquatic sediments and subsurface environments when the culture medium contains acetate as the electron donor and Fe(III) oxide or the humic acid analog anthraquinone-2,6-disulfonate (AQDS) as the electron acceptor (18, 24, 25, 95, 132).

Other organisms capable of completely oxidizing multicarbon organic acids to carbon dioxide with Fe(III) as the electron acceptor include *Geothrix fermentans* (23), *Geovibrio ferrireducens* (17), and *Ferribacter limneticum* (29). These organisms are phylogenetically distinct from the *Geobacteraceae* and from each other. To date, each of these organisms has been isolated from only a single type of environment. *Geothrix fermentans* was isolated from the Fe(III)-reducing zone of shallow contaminated aquifers whereas *Geovibrio ferrireducens* came from contaminated soil and *F. limneticum* was isolated from freshwater aquatic sediments.

Another intensely studied group of FRM are those in the gamma subclass of the *Proteobacteria*, which, in contrast to the organisms described above, can also use oxygen as an electron acceptor. Organisms in this group include species of *Shewanella* (16, 22, 92, 108), *Ferrimonas* (122), and *Aeromonas* (57). In general, these organisms can use hydrogen and only a limited range of organic acids as electron donors for Fe(III) and Mn(IV) reduction (69). The organic acids that do serve as electron donors, such as lactate and pyruvate, are not expected to be important extracellular intermediates in anaerobic metabolism in sediments (74, 89). *Shewanella saccharophila* also has the ability to use glucose as an electron donor for Fe(III) reduction (22). The *Shewanella* species which have been studied in detail, including *S. saccharophila*, only incompletely oxidize multicarbon organic electron donors to acetate. *Sulfurospirillum barnesii*, which also incompletely oxidizes lactate and pyruvate to acetate with the reduction of Fe(III), is in the epsilon subclass of the *Proteobacteria* (58, 129). The FMR in the gamma *Proteobacteria* have been recovered from a variety of sedimentary environments including various aquatic sediments (16, 22, 58, 108) and the subsurface (38, 116).

Mesophilic Fe(III) Reducers That Predominate in Sediments

The isolation and characterization of many pure cultures of FRM has expanded the known diversity of these organisms. Much less is known about which of these organisms may be important in Fe(III) reduction in the various environments in which Fe(III) reduction takes place. Evaluations of the potential contribution of the various known FRM to Fe(III) reduction in soils and sediments should involve molecular techniques in order to avoid many of the biases that are associated cultivation-based approaches.

The community structure of Fe(III)-reducing environments has been studied most intensively in sandy aquifer sediments. The practical reason for this is that, as described below, oxidation of organic contaminants is an important natural mechanism for the removal of contaminants from the subsurface, and this metabolism can be stimulated to enhance contaminant degradation. Furthermore, stimulating the activity of FRM in the subsurface may be a useful strategy for immobilizing toxic metals and metalloids in situ (65, 76).

Evaluation of the microbial community structure in a petroleum-contaminated aquifer demonstrated that FRM in the genus *Geobacter* were highly enriched in the Fe(III)-reducing zone in which aromatic hydrocarbons were being most rapidly degraded (5, 121). When 16S rDNA sequences in the sediments were amplified with PCR primers expected to amplify the 16S rDNA sequences of most members of the *Bacteria*, sequences most closely related to known species of *Geobacter* become prominent in the Fe(III)-reducing sediments. Enumeration of *Geobacter* sequences by a most-probable-number PCR approach demonstrated that the Fe(III)-reducing sediments in which contaminant degradation was most active had orders of magnitude more *Geobacter* organisms than did the uncontaminated aerobic sediments.

Geobacter species also became the predominant FRM when Fe(III) reduction in the uncontaminated aquifer sediments was stimulated by incubating the sediments under anaerobic conditions and adding electron donors such as glucose, acetate, lactate, benzoate, or formate (132a). In addition to the large enrichment of *Geobacter* associated with the development of Fe(III)-reducing conditions, there was a slight enrichment of *Geothrix* species. However, the number of *Geothrix* species was several orders of magnitude lower than the number of *Geobacter* species in all treatments. Field studies in which Fe(III) reduction was artificially stimulated by the addition of benzoate or aromatic hydrocarbons also resulted in an enrichment for *Geobacter* species (132a). Similar *Geobacter* sequences were recovered from the Fe(III)-reducing zone of three geographically distinct aquifers (132a).

These results indicate that in the Fe(III) reduction zone of sandy aquifers, *Geobacter* species are important members of the microbial community and by far the most numerous organisms that can clearly be identified as FRM. It is of interest that *Geobacter* species with 16S rDNA sequences similar to those that predominate in the sediments can be recovered in culture (121, 132a). This provides a rare opportunity to study the physiology of microorganisms known to be important in the environment in pure culture. Studies on these pure cultures may eventually lead to identification of gene sequences that can provide probes for estimating in situ metabolic activities.

The primary enrichment of *Geobacter* species and secondary enrichment of *Geothrix* species associated with Fe(III)-reducing conditions is consistent with the metabolic capabilities of these organisms. All *Geobacter* strains that have been studied have the ability to oxidize acetate to carbon dioxide with the reduction of Fe(III). As noted above, acetate is likely to be the prime electron donor for Fe(III) reduction in many sedimentary environments. Furthermore, *Geobacter* species are the only FRM known to be able to oxidize aromatic compounds, including aromatic hydrocarbons (69). Thus, in petroleum-contaminated aquifers, the ability to oxidize aromatic compounds should give *Geobacter* species a competitive edge over other FRM such as *Geothrix* species, which are known to be able to oxidize fatty acids, including acetate, but do not oxidize aromatic compounds (23).

Shewanella species are frequently recovered in pure culture from various sedimentary environments (16, 32, 108), and they accounted for ca. 2% of the microbial population in some surficial aquatic sediments (32). However, *Shewanella* 16S rDNA sequences could not be recovered from aquifer sediments in which Fe(III) reduction was the predominant terminal electron-accepting process (132a). This was the case even when electron donors such as lactate and formate, which are preferred by *Shewanella* species, were added to stimulate Fe(III) reduction. The factors likely to limit the contribution of *Shewanella* and related species to Fe(III) reduction in sedimentary environments are as follows: (i) as discussed above, the organic electron donors utilized by these organisms are not major intermediates in carbon flow; and (ii) these organisms only incompletely oxidize their electron donors to acetate, and thus most of the electrons available in the electron donors are eventually consumed by acetate-oxidizing organisms such as *Geobacter* and *Geothrix* species.

Efforts to determine which FRM are most numerous and active in other sedimentary environments are under way, since it would not necessarily be expected that *Geobacter* species would be the predominant FRM in all anaerobic environments. This is especially true for environments with extremes of pH or temperature.

Fermentative and Sulfate-Reducing Microorganisms

In addition to FRM, other organisms may contribute to Fe(III) reduction in sedimentary environments. These include the fermentative microorganisms discussed above, which can divert a small amount of their electron flow during a primarily fermentative metabolism. The large diversity of microorganisms that are capable of such metabolism have been previously reviewed (62, 69) and are not discussed further here. According to the current understanding of carbon and electron flow in sedimentary environments in which Fe(III) reduction is the TEAP (Fig. 1), the proportion of Fe(III) reduction that can be attributed to the activity of fermentative microorganisms is expected to be small.

Many sulfate-reducing microorganisms also have the ability to reduce Fe(III) and may preferentially reduce Fe(III) at the low electron donor concentrations typically found in sedimentary environments (26, 94). Analysis of microbial lipids suggested that sulfate-reducing microorganisms in the genus *Desulfovibrio* were enriched in zones of Fe(III) reduction in salt marsh sediments, suggesting that they might be involved in Fe(III) reduction in this environment (26). However, a full microbial analysis of Fe(III) reduction in such environments has not yet been completed.

Thermophilic and Hyperthermophilic Fe(III)-Respiring Microorganisms

Thermophilic and hyperthermophilic microorganisms are expected to be most important in Fe(III) reduction in environments at elevated temperatures such as the deep subsurface and near hydrothermal vents. Although the microbial community responsible for Fe(III) reduction in such environments has not been studied in detail, thermophilic and hyperthermophilic microorganisms capable of Fe(III) reduction have been recovered in pure culture. The first of these was the thermophile *Bacillus infernus*, which has a temperature optimum of 60°C (14). *Deferribacter thermophilus*, which has a temperature optimum of 65°C, was isolated from the production waters of a petroleum reservoir (45); *Thermoterrabacterium ferrireducens* was isolated from hot springs (126), as were Fe(III)-reducing *Thermus* species, which were also recovered from a South African gold mine (52).

A wide diversity of hyperthermophilic microorganisms which have been recovered from a variety of hot environments can reduce Fe(III). In fact, all of the hyperthermophiles in the domains *Archaea* and *Bacteria* that have been evaluated can reduce Fe(III) with hydrogen as the electron donor (133). Most of the organisms that have been evaluated to date are members of the *Archaea* (Fig. 3). In addition to the hyperthermophilic *Archaea* members, some mesophilic and thermophilic methanogens have been found to reduce Fe(III) (Fig. 3). The capacity of most of these members of the *Archaea* to reduce Fe(III) has been evaluated only in cell suspensions, and data on their potential for growth with Fe(III) serving as the electron acceptor are not available. The hyperthermophiles that have been examined in more detail, *Pyrobaculum islandicum* and *Thermotoga maritima*, conserve energy to support growth from hydrogen oxidation coupled to Fe(III) reduction (51a, 133).

Highly conserved characteristics of hyperthermophilic microorganisms are often considered to represent the characteristics of the last common ancestor of extant life because phylogenies based on 16S rDNA sequences suggest that these organisms are most closely related to the last common ancestor (115). If so, it seems likely that the last common ancestor was an FRM (133). The ability to reduce Fe(III) is more widespread among deeply branching members of the *Archaea* and *Bacteria* than is the capacity to reduce any other commonly considered electron acceptor. This and the fact that *T. maritima* can conserve energy to support growth via Fe(III) respiration but grows fermentatively in the presence of other potential electron acceptors suggest that Fe(III) reduction was a core metabolic function of the last common ancestor (133).

ENVIRONMENTAL SIGNIFICANCE OF MICROBIAL Fe(III) AND Mn(IV) REDUCTION

Respiration on Early Earth

A variety of studies involving a diversity of approaches have suggested that Fe(III) reduction was one of the first forms, if not the first, of microbial respiration (28, 29, 64, 123, 133, 138). As discussed in the previous section, it has been suggested that the last common ancestor of extant life was likely to have been an FRM. However, even earlier life forms may have been involved in Fe(III) reduction (28). Numerous studies (see reference 19 for a review) have suggested that hydrogen and Fe(III) accumulated on prebiotic Earth as the result of high-intensity UV radiation impinging on Archaean seas that were high in dissolved Fe(II) (Fig. 4A). Hydrothermal fluids may have also contributed Fe(III) (19), and geotectonic sources of hydrogen were also probably available (136). Thus, prior to the evolution of life, environmental conditions may have been ideal to support the development of a hydrogen-oxidizing, Fe(III)-reducing biological system.

In fact, it has been proposed that hydrogen oxidation coupled to Fe(III) reduction was the first protometabolism that eventually led to the evolution of life (123). In this model, membrane-like accumulations of the iron-sulfur mineral mackinawite catalyzed the oxidation of hydrogen and transferred the electrons to Fe(III) adsorbed on the exterior of the mineral surface (123). If this is so, UV radiation from the Sun could have provided the energy to support a protometabolism based upon an iron cycle ("ferrous wheel") on prebiotic Earth (Fig. 4A). As inorganic membranous structures were replaced with organic membranes and life forms that were more similar to present-day microorganisms developed (19, 28, 123), the continued availability of hydrogen and Fe(III) could have supported the growth of these organisms (Fig. 4B).

With further evolution of microbial life and the development of either Fe(II)-based (34, 46, 140) and/or oxygenic photosynthesis, even more Fe(III) would have been generated and, along with it, organic matter (Fig. 4C). Fe(II)-based photosynthesis would have directly generated Fe(III) whereas oxygenic photosynthesis would have produced Fe(III) as the result of oxygen reacting with Fe(II). In either event, sedimentation of the organic matter and Fe(III) oxides would have provided ideal conditions for the activity of heterotrophic FRM. Thus, the ferrous wheel driven by solar radiation could have continued, but with increasingly more of the Fe(II) oxidation resulting from microorganisms harnessing longer-wavelength light sources.

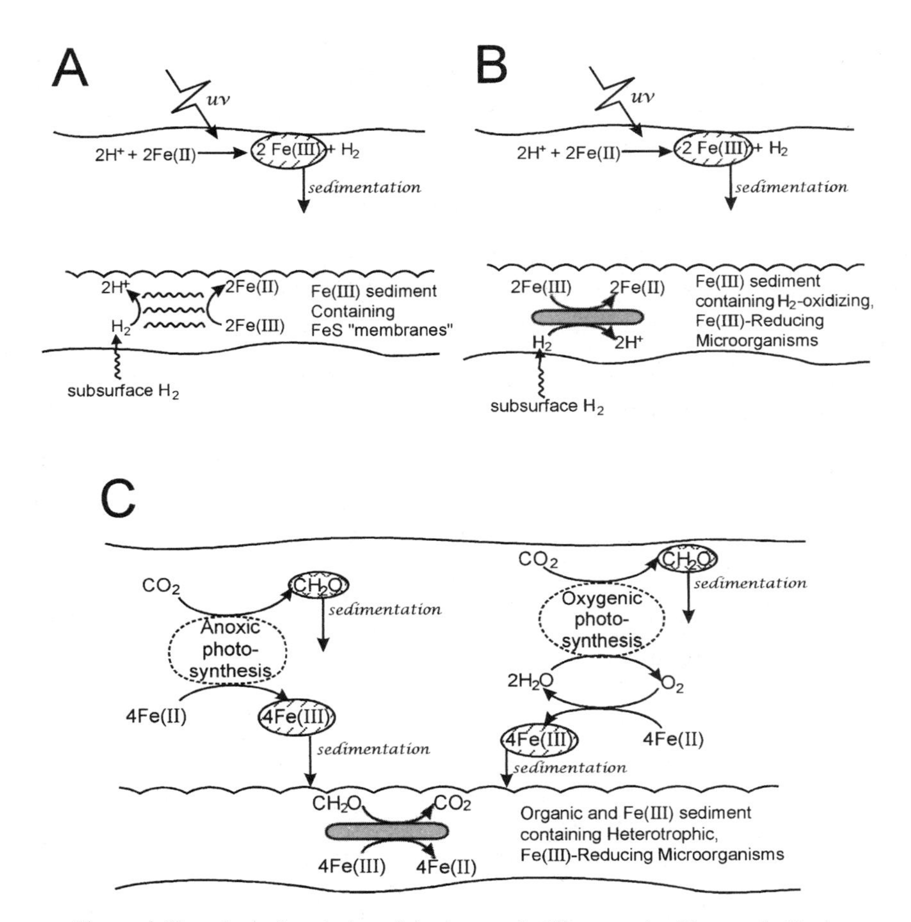

Figure 4. Hypothetical evolution of the iron cycle ('ferrous wheel') on early Earth. (A) Generation of Fe(III) and hydrogen from UV-mediated oxidation of dissolved Fe(II) as well as Fe(III) and hydrogen from other sources may have provided hydrogen and Fe(III) for protometabolism catalyzed by iron-sulfur nanocrystal membranes. (B) Iron cycle with hydrogen oxidation coupled to Fe(III) reduction by early microorganisms. (C) Enhanced precipitation of Fe(III) with the evolution of Fe(II)-based and/or oxygenic photosynthesis resulting in the coprecipitation of organic matter and Fe(III) oxides which supported heterotrophic Fe(III) reduction.

Evidence for extensive oxidation of organic matter coupled to Fe(III) reduction in sediments deposited during the early Precambrian period is provided by the accumulation of massive amounts of magnetite in the Precambrian banded iron formations. Carbon isotopic signatures suggest that the reduction of Fe(III) to magnetite in the banded iron formations was linked to the oxidation of organic matter to carbon dioxide (11, 137, 138). All FRM that have been examined, including the deeply branching hyperthermophiles, produce magnetite as an end product of Fe(III) reduction (51a, 63, 95, 133). Since there is no known abiotic mechanism to account for organic-matter oxidation coupled to Fe(III) reduction to magnetite, this suggests that the Fe(III) reduction was the result of the activity of FRM. The geological record indicates that this oxidation of organic matter coupled to Fe(III) reduction was globally significant before other mechanisms for anaerobic or aerobic oxidation of organic matter to carbon dioxide (138).

Fe(III) Reduction in Modern Hot Environments

The hydrogen-based Fe(III) reduction hypothesized to have taken place on early Earth may have an analog on modern Earth. Hydrothermal vents release large quantities of Fe(II), which, once mixed in cold, oxygen-containing waters, is oxidized to Fe(III) (50, 51). This Fe(III) precipitates to the sediments near the vents. Hot, hydrogen-rich water seeps through these sediments. Thus, conditions may be ideal for modern hyperthermophilic, hydrogen-oxidizing FRM. Conditions favoring the activity of thermophilic FRM may also be found around terrestrial hot springs (15). The recovery of large quantities of ultrafine-grained magnetite from the deep hot subsurface has been suggested to provide evidence for microbial Fe(III) reduction in this hot environment (43). However, the environmental significance of microbial Fe(III) reduction in any of these hot environments has not been quantitatively evaluated. Furthermore, it is not yet known whether the thermophilic and hyperthermophilic FRM that are available in culture are the organisms responsible for the Fe(III) reduction in hot environments.

Modern-Day Carbon Cycle

Fe(III) reduction and Mn(IV) reduction are important processes for the oxidation of organic matter in a diversity of sedimentary environments including freshwater aquatic sediments, submerged soils, marine sediments, deep pristine aquifers, and shallow aquifers contaminated with organic compounds (20, 64, 109). The concentrations of Fe(III) and Mn(IV) in the deposited sediments and the availability of mechanisms for the regeneration of Fe(III) and Mn(IV) from Fe(II) and Mn(II) are important factors which influence the relative importance of Fe(III) and Mn(IV) reduction in organic matter oxidation. For example, high concentrations of Fe(III) in sedimenting particles, resulting from iron release from a waste-treatment plant (42) or acid-mine drainage (27), increased the significance of Fe(III) reduction in the sediments. Sedimentation of soils naturally high in Fe(III) or Mn(IV) may also enhance Fe(III) and Mn(IV) reduction in aquatic sediments (2, 59).

Reoxidation of Fe(II) and Mn(II) produced as the result of Fe(III) and Mn(IV) reduction can also be an important source of Fe(III) and Mn(IV) for microbial reduction. Some of the Fe(II) and Mn(II) generated within the zones of Fe(III) and Mn(IV) reduction will diffuse to the surface, where they will be oxidized and redeposited as Fe(III) and Mn(IV) oxides. However, most of the Fe(II) and Mn(II) produced enters solid phases in sediments (64). Therefore, mechanisms that expose solid-phase Fe(II) and Mn(II) to oxygen can greatly increase the extent of Fe(III) and Mn(IV) reduction. In aquatic environments that contain rooted plants, the oxidation of Fe(II) in the rhizosphere may be an important factor promoting Fe(III) reduction (120). Physical mixing such as that seen in the periodic resuspension of large amounts of sediments of the Amazon (2) can help reoxidize Fe(II) and Mn(II) solids to Fe(III) and Mn(IV) oxides, as can bioturbation by benthic organisms in marine sediments (1). Draining submerged soils as part of the management of rice paddies introduces air into the soils and oxidizes Fe(II), which accounts for the importance of Fe(III) reduction when the paddy soils are subsequently flooded. Similar oxidation events as a result of fluctuating water levels are likely to play a role in regenerating Fe(III) and Mn(IV) oxides in natural wetlands as well.

Periodic oxidation of Fe(II) may also contribute to the importance of Fe(III) reduction in shallow aquifers. During recharge events, dissolved oxygen that is introduced into zones in which methane production or sulfate reduction is the TEAP can oxidize Fe(II) in the sediments and the increased Fe(III) concentrations can switch the TEAP to Fe(III) reduction (135).

A major factor accounting for the importance of Fe(III) reduction in subsurface environments, other than high concentrations of Fe(III) in many subsurface soils, is the movement of the Fe(III)-reducing zone over time (68). For example, shortly after the contamination of subsurface environments with organic contaminants such as petroleum or landfill leachate, a zone in which Fe(III) reduction predominates is often found directly downgradient of the zone of contamination. As Fe(III) is depleted from these sediments closest to the source of contamination, the zone of Fe(III) reduction is displaced further downgradient and is replaced by another terminal electron-accepting process, typically methanogenesis. Since many organic contaminants are only slowly degraded under methanogenic conditions, significant organic inputs are still available when the groundwater enters the zone where Fe(III) is present. Thus, by continually moving further downgradient as Fe(III) is consumed closest to the source of contamination, Fe(III) continues to be an important process for organic matter degradation.

Effects on Water Quality and Soil Chemistry

As previously reviewed in detail (64, 66), reduction of insoluble Fe(III) oxides has a major impact on several aspects of pore water chemistry. For example, the release of dissolved Fe(II) into groundwater as a result of microbial Fe(III) reduction may be one of the most prevalent groundwater problems worldwide (3). Fe(II) in groundwater causes plumbing and staining problems and can be expensive to remediate. Fe(II) released from microbial Fe(III) reduction may also influence the growth of plants, including rice, in submerged soils (66). Phosphate and trace metals adsorbed on Fe(III) and Mn(IV) oxides may be released into solution when the oxides are reduced (57, 64). This is environmentally significant because phosphate is often a nutrient limiting the development of plant growth in aquatic systems and because many of the trace metals bound to Fe(III) and Mn(IV) oxides can be toxic. Fe(III) and Mn(IV) reduction typically result in an increase in the pH, the ionic strength of the pore water, and the concentration of a variety of cations (118, 119). These changes may impact on plant growth in soils and influence the composition and thus the quality of groundwater.

Formation and Dissolution of Minerals

Microbial Fe(III) reduction can lead to the formation of geologically significant minerals (26, 64). These minerals can sometimes be used to infer the prior activity of FRM such as the magnetite formed in the banded iron formations and the ultra-fine-grained magnetite recovered from the deep subsurface discussed above. It has been stated that "if it were not for the bacterium GS-15 [i.e., *Geobacter metallireducens*] we would not have radio and television today" (134). The rationale for this is that the study of magnetite from the banded iron formations led to the discovery of important principles of magnetism and electricity. Magnetite formed from microbial Fe(III) reduction is a likely explanation for the magnetic anomalies found near petroleum deposits and may contribute to the magnetic remanence of sediments (42, 63, 95). It was speculated that some of the ultrafine-grained magnetite recovered in a Martian meteorite was the result of the activity of FRM on Mars (99).

Siderite ($FeCO_3$) is another geologically significant mineral that was thought to be formed primarily under methanogenic conditions until studies in salt marsh sediments demonstrated that siderite could also be rapidly formed under conditions in which Fe(III) reduction was the TEAP (26). Siderite formed as the result of microbial Fe(III) reduction may differ in isotopic and trace metal composition from what would be expected from purely abiological mechanisms of formation (101, 102). The accumulation of rhodochrosite ($MnCO_3$) may provide a geological signature of past microbial Mn(IV) reduction (64). Less-well-defined Fe(III)-Fe(II) minerals, including "green rusts," may be active in the abiological reduction of contaminants such as chlorinated solvents (37) and nitroaromatics (47–49).

As discussed in chapter 13, many FRM can also reduce U(VI) to U(IV). The formation of the insoluble U(IV) mineral, uraninite, as the result of microbial U(VI) reduction may explain the reductive sequestration of uranium in marine sediments, the concentration of uranium in reduction spots in rocks, as well as the formation of some forms of uranium deposits (90, 91). FRM can also reduce soluble gold in the oxidized form, Au(III), to gold metal, Au(0) (K. Kashefi and D. R. Lovley, unpublished data). This precipitates gold from solution and may provide an explanation for some gold deposits.

Microbial Fe(III) reduction can also lead to the dissolution of minerals. Obviously, the reduction of Fe(III) and Mn(IV) oxides destroys these minerals. Culture studies have suggested that FRM can also dissolve magnetite via Fe(III) reduction (54), but magnetite accumulates in most FRM cultures and the role of FRM in magnetite dissolution in sediments has yet to be evaluated. Culture studies have demonstrated that FRM can reduce the structural Fe(III) in clays and, in the process, influence the swelling and size of clay minerals, which could influence soil porosity and other factors (35, 40, 55). These culture studies emphasize the importance of further studies on the potential for FRM to reduce structural Fe(III) in clays under environmentally relevant conditions.

Degradation of Organic Contaminants in Subsurface Environments

The potential for FRM to remove organic contaminants from polluted groundwater has been clearly established (3, 68). Extensive contamination of groundwater with common pollutants such as petroleum and landfill leachate results in the development of anaerobic conditions near the source of contamination (68, 71). As oxygen is depleted from groundwater, Fe(III) is generally the most abundant potential electron acceptor for organic matter oxidation in typical sandy aquifers. FRM have the ability to oxidize a variety of aromatic compounds, which are often the groundwater contaminants of greatest concern (61, 72, 84, 97), and geochemical studies have suggested that a wide variety of organic contaminants are removed within the Fe(III) reduction zone of contaminated aquifers (71). The increased activity of FRM as the result of the introduction of organic contaminants into aquifers is often visually apparent, since the sediment color changes significantly when Fe(III) oxides in the sediments are reduced (Fig. 5). The reddish-brown color characteristic of Fe(III) oxides is lost, and the sediments turn grey, characteristic of the formation of Fe(II) minerals.

The role of microbial Fe(III) reduction in the removal of organic contaminants from groundwater has been studied most extensively in petroleum-contaminated aquifers, in which aromatic hydrocarbons such as toluene and benzene are the primary contaminants of concern. Evaluation of several petroleum-contaminated aquifers demonstrated that the Fe(III)-reducing community could effectively degrade toluene in all of the aquifers studied (4, 5). However, benzene, the aromatic hydrocarbon of greater concern because of its higher toxicity, persisted under Fe(III)-reducing conditions in all but one of the aquifers evaluated. In the one petroleum-contaminated aquifer in which benzene was degraded, benzene degradation took place only within a limited portion of the Fe(III)-reducing zone. Evaluation of a variety of geochemical factors failed to provide a parameter which could be used to predict when a particular aquifer sediment would have the ability to degrade benzene (4). However, molecular analysis of the sediments did indicate

Figure 5. Appearance of sandy aquifer material from an uncontaminated portion of a shallow aquifer (left) and of aquifer material from within a petroleum-contaminated portion of the aquifer in which Fe(III) has been extensively reduced as the result of microbial oxidation of contaminants coupled to Fe(III) reduction (right).

that sediments within the Fe(III) reduction zone that had the capacity for benzene degradation were greatly enriched in *Geobacter* species compared with other sediments in which Fe(III) reduction was the TEAP but in which benzene persisted (121). This and the fact that a *Geobacter* species was recovered in a benzene-oxidizing, Fe(III)-reducing enrichment culture established with sediments from the aquifer suggested that *Geobacter* species were associated with benzene degradation and that the presence of certain *Geobacter* species might be indicative of the potential for benzene degradation in contaminated aquifers (121).

When benzene persists in the Fe(III) reduction zone of contaminated aquifers, it may be possible to enhance benzene degradation by stimulating Fe(III) reduction. For example, when chelators such as nitrilotriacetic acid were added to sediments from the Fe(III) reduction zone of a petroleum-contaminated aquifer, benzene degradation was greatly stimulated after an adaptation period (97, 98). Stimulation of benzene degradation was associated with higher rates of Fe(III) reduction. Nitrilotriacetic acid and other chelators stimulated Fe(III) reduction by solubilizing Fe(III) from the Fe(III) oxides in the aquifer sediments (96). The soluble Fe(III) is much more accessible for reduction than are the insoluble Fe(III) oxides, and alleviating the need for Fe(III)-reducing microorganisms to establish direct contact with the Fe(III) oxides (67).

Although the addition of chelators alone is sufficient to stimulate anaerobic benzene degradation in contaminated aquifer sediments that contain Fe(III), remediation of benzene in aquifers in which the Fe(III) has been depleted due to intensive Fe(III) reduction may require the addition of Fe(III) coupled to a chelator (Fig. 6). Under these conditions, chelated Fe(III) pumped into the subsurface would serve

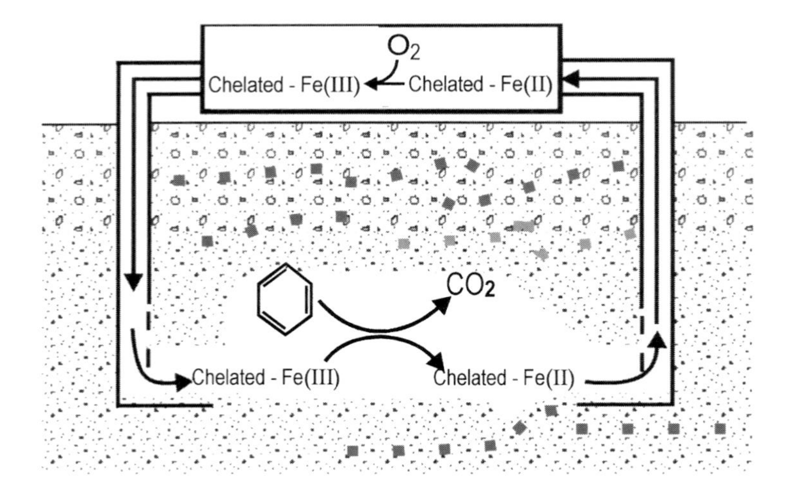

Figure 6. Schematic of remediation of petroleum-contaminated aquifers with soluble Fe(III). Chelated Fe(III) is introduced in an injection well. The Fe(III) serves as an electron acceptor for the oxidation of benzene and other contaminants. The Fe(II) produced can be recovered in a downgradient well, exposed to air to reoxidize the Fe(II) to Fe(III), and then reintroduced into the aquifer.

as the electron acceptor for benzene oxidation; the Fe(II) produced from Fe(III) reduction, as well as the chelator, could be recovered in a downgradient recovery well; the Fe(II) could be oxidized to Fe(III) by exposure to air; and the chelated Fe(III) that is produced could be reintroduced into the aquifer.

Addition of humics and other quinone-containing compounds to aquifer sediments can also stimulate anaerobic benzene oxidation and Fe(III) reduction in aquifer sediments (4, 98; Nevin and Lovley, submitted). As discussed in detail below, extracellular quinones, such as humics, serve as electron shuttles between FRM and Fe(III) oxides. This also accelerates Fe(III) reduction by eliminating the need for the FRM to contact insoluble Fe(III) oxides, but by a fundamentally different mechanism from that with Fe(III) chelators.

Reduction and Oxidation of Humic Substances

Given the ability of FRM to transfer electrons to insoluble, extracellular electron acceptors such as Fe(III) and Mn(IV) oxides, it may not be surprising to find that these organisms can transfer electrons to another class of extracellular electron acceptors, humics. Humics are one of the most abundant forms of organic matter in many soils and sediments. Microorganisms may slowly degrade humics, especially under aerobic conditions, and microbial biomass may contribute to humic formation (100). However, until the discovery of microbial electron transport to humics (77), humics were generally considered to be nearly biochemically inert under anaerobic conditions.

All of the FRM that have been evaluated, including the hyperthermophiles, have the ability to transfer electrons to humics or the humic acid analog AQDS (24, 77, 79, 83). Electron transport to humics can yield energy to support the growth of pure cultures (77, 79). Furthermore, the addition of humics or AQDS to aquifer sediments resulted in the growth of *Geobacter* species in the sediments (132a).

Current evidence suggests that the primary electron-accepting moieties in humics are quinones. There is a direct correlation between the quinone content of humics, as determined by electron spin resonance measurements, and the electron-accepting capacity of the humics (124). Microbial transfer of electrons to humics results in an increase in the number of quinone free radicals that is directly proportional to the number of electrons transferred to the humics. All microorganisms that have the ability to transfer electrons to extracellular quinones such as AQDS also have the ability to transfer electrons to humics, whereas those that do not reduce AQDS also do not reduce humics (24, 77, 79, 83). Enrichment and isolation of microorganisms with AQDS invariably recovers microorganisms which can reduce humics (24). It has also been suggested that iron bound in humics might be an important electron acceptor in humics respiration (13). However, subsequent studies have indicated that the iron content of most humics is much too low to account for a significant proportion of the electron accepting capacity of the humics (73).

Electron transfer to humics is likely to be most important in environments which contain Fe(III). This is because, once microbially reduced, humics can abiotically transfer electrons to Fe(III) (77). Studies with the humic acid analog AQDS have demonstrated that the abiotic oxidation of extracellular quinones by Fe(III) can regenerate the quinone form of the molecule, which can then undergo another cycle of microbial reduction and oxidation by Fe(III). In this manner, humics and other extracellular quinones have the potential to accept a significant number of electrons from the microbial oxidation of organic matter and hydrogen because each quinone molecule may be recycled numerous times.

The electron shuttling between Fe(III)-reducing microorganisms can greatly accelerate the rate of Fe(III) reduction (77, 79, 83; Nevin and Lovley submitted). This has been found with pure Fe(III) phases such as poorly crystalline Fe(III) oxide, goethite, and hematite, as well as Fe(III) in clays and the poorly characterized Fe(III) in aquifer sediments. It is likely that the FRM can access the soluble quinones more readily than they can access insoluble Fe(III) oxides and that the reduced quinones can establish contact with insoluble Fe(III) oxides more easily than can the Fe(III) reductases of FRM. This latter factor was evident in studies in which poorly crystalline Fe(III) oxide was incorporated into microporous beads with pore diameters too small for the Fe(III)-reducing microorganism *G. metallireducens* to enter (Nevin and Lovley, submitted). *G. metallireducens* could not reduce the Fe(III) within the beads. However, when AQDS was added to the cultures, the Fe(III) was reduced, since *G. metallireducens* reduced AQDS, which then entered the beads and reduced the Fe(III). This serves as a model for how extracellular quinones may facilitate the reduction of Fe(III) oxides that are occluded in pore spaces within soils that are too small for FRM to penetrate.

Although the potential for humics and other extracellular quinones to serve as electron acceptors and electron shuttles in soils and sediments has been demonstrated, the amount of electron flow through humics has yet to be quantified in any environment. The addition of humics and/or AQDS accelerates the oxidation of organic contaminants in Fe(III)-reducing sediments from petroleum-contaminated aquifers (4, 98, 139) and stimulates Fe(III) reduction in aquifer sediments (Nevin and Lovley, submitted). It is assumed that in these instances, the increased rate of electron flow to Fe(III) can be attributed to electron shuttling via humics.

Some FRM, as well as other microorganisms, can use microbially reduced humics as an electron donor for the reduction of electron acceptors such as nitrate, selenate, arsenate, and fumarate (80). The environmental significance of this respiration has yet to be evaluated in detail. It has been proposed (70) that microbial humics oxidation could be important in organic-rich soil aggregates, which would not be expected to contain high concentrations of Fe(III). Humics reduced in the anaerobic center of the aggregates might diffuse out and serve as an electron donor for the reduction of nitrate or other electron acceptors near the surface of the aggregate, where conditions are less reducing (Fig. 7).

MECHANISMS FOR Fe(III) REDUCTION

Mechanisms for Enzymatic Fe(III) Reduction in Fe(III)-Respiring Microoganisms

If, as generally believed, FRM directly use insoluble Fe(III) oxides as an electron acceptor, then FRM are faced with the dilemma of how to transfer electrons to an

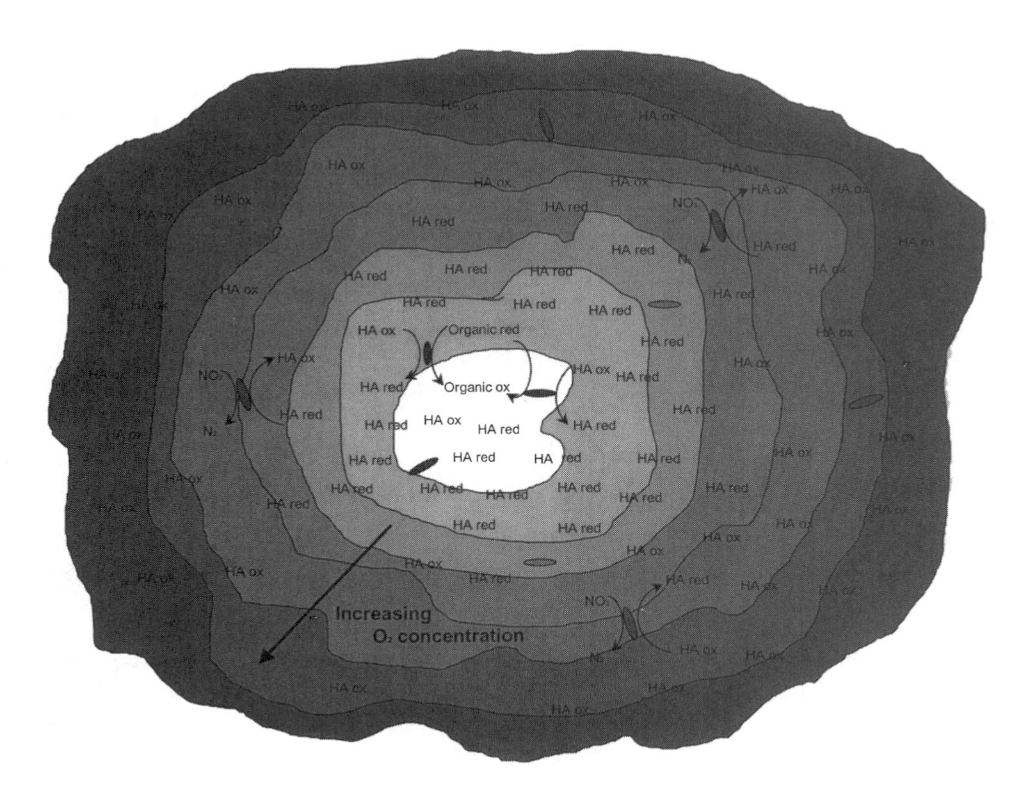

Figure 7. Model for the potential significance of microbially reduced humics serving as an electron donor for denitrification in organic-rich, iron-depleted soil aggregates.

extracellular electron acceptor. A detailed review of proposed mechanisms by which FRM might transfer electrons to Fe(III) is beyond the scope of this chapter. Briefly, it has been shown that FRM such as *S. putrefaciens* and several *Geobacter* species contain membrane-bound Fe(III) reductase activity and that electron carriers such as cytochromes, capable of transferring electrons to Fe(III), are located in the membranes of these organisms (12, 44, 81, 105–107; T. S. Magnuson, A. L. Hodges-Myerson, and D. R. Lovley, submitted for publication). It has been suggested, but not proven, that such membrane-bound systems might transfer electrons to extracellular insoluble Fe(III) oxides.

Recent studies have led to a preliminary model for electron transfer to insoluble Fe(III) oxides by *Geobacter* species (Fig. 8). *Geobacter* species contain a membrane-bound NADH-dependent Fe(III) reductase activity (39, 44; Magnuson et al., submitted). The NADH hydrogenase appears to be localized in the inner membrane and is part of a respiratory complex that contains a 89-kDa *c*-type cytochrome (Magnuson et al., submitted). Although the 89 kDa *c*-type cytochrome is capable of transferring electrons to Fe(III), its apparent localization on the inner

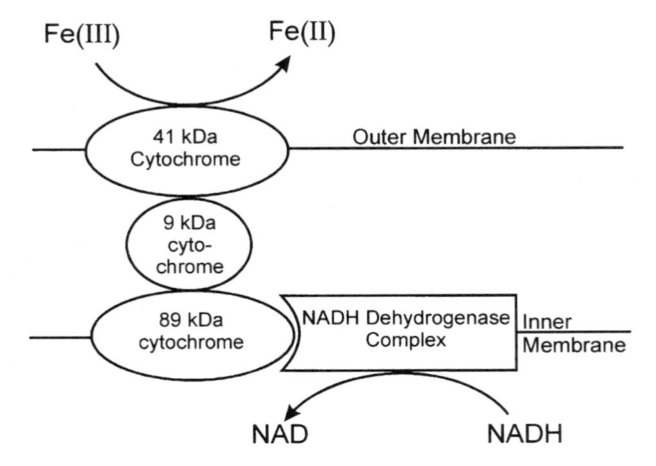

Figure 8. Model for electron transport to extracellular Fe(III) in *G. sulfurreducens.*

membrane would preclude directly accessing Fe(III). However, the midpoint potential of the 89-kDa cytochrome is such that it would be capable of transferring electrons to the periplasmic 9-kDa cytochrome (Magnuson et al., submitted).

There is yet another membrane-bound *c*-type cytochrome in *Geobacter* species, which has a molecular mass of 41 kDa (J. R. Lloyd and D. R. Lovley, unpublished data). Protease treatment of whole cells of *G. sulfurreducens* resulted in digestion of the 41 kDa cytochrome, but not the periplasmic 9 kDa cytochrome or the 89 kDa cytochrome (Lloyd and Lovley, submitted). The 41-kDa cytochrome is capable of donating electrons to Fe(III). Protease digestion of the 41-kDa cytochrome in whole cells inhibited Fe(III) reductase activity but had no effect on fumarate reduction which is assumed to be an intracellular electron-accepting reaction (Lloyd and Lovley, submitted). These results suggest that electrons derived from NADH oxidation are passed to the 89-kDa cytochrome, then to the periplasmic 9-kDa cytochrome, and then to the 41-kDa cytochrome, which can transfer electrons to Fe(III) (Fig. 8). The soon to be completed sequencing of the genome of *G. sulfurreducens* and the development of a genetic system for this organism should aid in further assessing the validity of this model.

An alternative model, in which *G. sulfurreducens* does not reduce Fe(III) with a membrane-bound Fe(III) reductase but, rather, releases a cytochrome that acts as a soluble electron shuttle between *G. sulfurreducens* and Fe(III) oxides, has also been proposed (125), but more in-depth analysis has indicated that this mechanism is highly unlikely (60). This is because the cytochrome is not consistently released from healthy cells. Furthermore, even when the cytochrome is added at high levels, it does not serve as an electron shuttle between *G. sulfurreducens* and Fe(III) oxides.

Mechanisms for Microbial Fe(III) Reduction in Soils and Sediments

It is apparent that more research is required to better understand the mechanisms for Fe(III) oxide reduction by FRM in pure culture, but elucidating the pathways for Fe(III) reduction in soils and sediments may be even more difficult. Reduction of Fe(III) oxides in soils and sediments could potentially take place as a direct enzymatic reduction by FRM, as discussed above for pure cultures of microorganisms in culture media. However, several other mechanisms are potentially possible.

There is considerable uncertainty over which types of Fe(III) oxides are available for microbial reduction. Studies with aquatic sediments indicated that poorly crystalline Fe(III) oxides, which typically exist as coatings on clays and other surfaces in sediments, are readily available for microbial reduction (85, 86). When poorly crystalline Fe(III) oxides are present, FRM can outcompete sulfate-reducing and methanogenic microorganisms for electron donors, and Fe(III) reduction functions as the predominant terminal electron accepting mechanism (87).

However, much of the Fe(III) in aquatic sediments was not available for microbial reduction and persisted with depth in the sediments (85). Most of this unavailable Fe(III) was probably in the form of crystalline Fe(III) forms (117). When only crystalline Fe(III) forms are available in sediments, FRM are not able to outcompete sulfate-reducing or methanogenic microorganisms for electron donors (86, 87; R. T. Anderson, M. E. Housewright, and D. R. Lovley, submitted for publication). Comparable studies of aquifer sediments have provided similar results and have demonstrated that crystalline Fe(III) oxides persist even in subsurface sediments that have high rates of organic-carbon degradation (Anderson et al., submitted). Thus, although some pure cultures of FRM can partially reduce some crystalline Fe(III) forms under artificial laboratory conditions, the evidence currently available suggests that crystalline Fe(III) forms are not important electron acceptors for FRM in sedimentary environments.

Although most recent studies on microbial reduction of Fe(III) have emphasized direct enzymatic reduction of Fe(III) oxides, the finding that humics and other extracellular quinones may serve as electron shuttles to promote Fe(III) reduction has indicated that, at least in some instances, much of the Fe(III) oxide reduction may actually result from the nonenzymatic reduction of Fe(III) oxides by microbially reduced humics. However, ascertaining how much of the electron flow to Fe(III) proceeds through humics is technically difficult because of the rapid transition between oxidized and reduced states of the humics. It is clear that extremely low concentrations of humics could have a major impact on the rate of Fe(III) reduction. For example, concentrations as low as 10 μM AQDS (the lowest concentration evaluated) stimulated the reduction of Fe(III) oxides in aquifer sediments (Nevin and Lovley, submitted) and as little as 1 μM AQDS appeared to promote electron shuttling in aquatic sediments (R. T. Anderson, S. Walker, M. E. Housewright, and D. R. Lovley, submitted for publication).

Hydrogen measurements may provide a tool for determining whether electron shuttling via humics is an important mechanism for Fe(III) reduction (Anderson et al., submitted). Previous studies have demonstrated that under steady-state conditions, there is a distinct range of hydrogen concentrations associated with each of the major anaerobic electron-accepting processes, such as methanogenesis, sulfate reduction, and Fe(III) reduction (75, 82). Studies with aquatic sediments and sediments from a sandy aquifer have demonstrated that when AQDS is provided as an electron shuttle to promote Fe(III) reduction, the steady-state concentration of hydrogen is consistently lower than in sediments in which an electron shuttle is not present (Anderson et al., submitted).

A relatively unexplored area is the possibility that soluble Fe(III) is a major source of the Fe(III) that FRM reduce. Although Fe(III) is highly insoluble at neutral pH, organic compounds in soils and sediments may act as Fe(III) chelators. As discussed above, the solubilization of Fe(III) with synthetic chelators greatly stimulates the activity of FRM in sediments. The degree to which naturally occurring organic compounds may provide soluble chelated Fe(III) to FRM has yet to be examined in detail.

SUMMARY

Studies to date suggest that the microorganisms that conserve energy to support growth from Fe(III) and Mn(IV) reduction can influence important biogeochemical cycles in anaerobic sedimentary environments. FRM may be useful agents for the remediation of both organic and metal contaminants in the subsurface. Much has been learned about the diversity of microorganisms that might be responsible for Fe(III) and Mn(IV) reduction through the isolation and characterization of pure cultures. However, more information is required about which microorganisms are responsible for Fe(III) and Mn(IV) reduction in environments in which Fe(III) and Mn(IV) reduction are significant. In a similar manner, studies with pure cultures have demonstrated much about the physiological capabilities of FRM, but there is little information on the activity of these organisms in their native environments. Hydrothermal environments represent an underexplored habitat for FRM. Studies in this area may not only provide better insights into the biogeochemistry of these environments as they exist today but also provide hints on the functioning of microbial ecosystems on early Earth. Further investigation into the biochemical and biogeochemical mechanisms of Fe(III) and Mn(IV) reduction is required in order to fully realize the potential of these processes for environmental remediation.

Acknowledgments. Research from my laboratory summarized in this chapter was supported by grants from the Department of Energy, the Office of Naval Research, the National Science Foundation, and the American Petroleum Institute.

REFERENCES

1. **Aller, R. C.** 1990. Bioturbation and manganese cycling in hemipelagic sediments. *Philos. Trans. R. Soc. London Ser. A* **331:**51–68.
2. **Aller, R. C., J. E. Macklin, and R. T. J. Cox.** 1986. Diagenesis of Fe and S in Amazon inner shelf muds: apparent dominance of Fe reduction and implications for the genesis of ironstones. *Cont. Shelf Res.* **6:**263–289.
3. **Anderson, R. T., and D. R. Lovley.** 1997. Ecology and biogeochemistry of in situ groundwater bioremediation. *Adv. Microb. Ecol.* **15:**289–350.
4. **Anderson, R. T., and D. R. Lovley.** 1999. Naphthalene and benzene degradation under Fe(III)-reducing conditions in petroleum-contaminated aquifers. *Bioremediation J.* **3:**121–135.
5. **Anderson, R. T., J. Rooney-Varga, C. V. Gaw, and D. R. Lovley.** 1998. Anaerobic benzene oxidation in the Fe(III)-reduction zone of petroleum-contaminated aquifers. *Environ. Sci. Technol.* **32:**1222–1229.
6. **Aristovskaya, T. V., and G. A. Zavarzin.** 1971. Biochemistry of iron in soil, p. 385–408. *In* A. D. McLaren and J. Skujins (ed.), *Soil Biochemistry*. Marcel Dekker, Inc., New York, N.Y.

7. **Arnold, R. G., T. J. DiChristina, and M. R. Hoffmann.** 1986. Inhibitor studies of dissimilative Fe(III) reduction by *Pseudomonas* sp. strain 200 ("*Pseudomonas ferrireductans*"). *Appl. Environ. Microbiol.* **52:**281–289.
8. **Arnold, R. G., M. R. Hoffman, T. J. DiChristina, and F. W. Picardal.** 1990. Regulation of dissimilatory Fe(III) reduction activity in *Shewanella putrefaciens*. *Appl. Environ. Microbiol.* **56:** 2811–2817.
9. **Arnold, R. G., T. M. Olson, and M. R. Hoffmann.** 1986. Kinetics and mechanism of dissimilative Fe(III) reduction by *Pseudomonas* sp. 200. *Biotechnol. Bioeng.* **28:**1657–1671.
10. **Balashova, V. V., and G. A. Zavarzin.** 1980. Anaerobic reduction of ferric iron by hydrogen bacteria. *Microbiology* **48:**635–639.
11. **Baur, M. E., J. M. Hayes, S. A. Studley, and M. R. Walter.** 1985. Millimeter-scale variations of stable isotope abundances in carbonates from banded iron-formations in the Hamersley Group of Western Australia. *Econ. Geol.* **80:**270–282.
12. **Beliaev, A. S., and D. A. Saffarini.** 1998. *Shewanella putrefaciens mtrB* encodes an outer membrane protein required for Fe(III) and Mn(IV) reduction. *J. Bacteriol.* **180:**6292–6297.
13. **Benz, M., B. Schink, and A. Brune.** 1998. Humic acid reduction by *Propionibacterium freudenreichii* and other fermenting bacteria. *Appl. Environ. Microbiol.* **64:**4507–4512.
14. **Boone, D. R., Y. Liu, Z.-J. Zhao, D. L. Balkwill, G. T. Drake, T. O. Stevens, and H. C. Aldrich.** 1995. *Bacillus infernus* sp. nov., an Fe(III)- and Mn(IV)-reducing anaerobe from the deep terrestrial subsurface. *Int. J. Syst. Bacteriol.* **45:**441–448.
15. **Brock, T. D., C. S., S. Petersen, and J. L. Mosser.** 1976. Biogeochemistry and bacteriology of ferrous iron oxidation in geothermal habitats. *Geochim. Cosmochim. Acta* **40:**493–500.
16. **Caccavo, F., Jr., R. P. Blakemore, and D. R. Lovley.** 1992. A hydrogen-oxidizing, Fe(III)-reducing microorganism from the Great Bay Estuary, New Hampshire. *Appl. Environ. Microbiol.* **58:**3211–3216.
17. **Caccavo, F., Jr., J. D. Coates, R. A. Rossello-Mora, W. Ludwig, K. H. Schleifer, D. R. Lovley, and M. J. McInerney.** 1996. *Geovibrio ferrireducens*, a phylogenetically distinct dissimilatory Fe(III)-reducing bacterium. *Arch. Microbiol.* **165:**370–376.
18. **Caccavo, F., D. J. Lonergan, D. R. Lovley, M. Davis, J. F. Stolz, and M. J. McInerney.** 1994. *Geobacter sulfurreducens* sp. nov., a hydrogen- and acetate-oxidizing dissimilatory metal-reducing microorganism. *Appl. Environ. Microbiol.* **60:**3752–3759.
19. **Cairns-Smith, A. G., A. J. Hall, and M. J. Russell.** 1992. Mineral theories of the origin of life and an iron sulfide example. *Origins Life Evol. Biosph.* **22:**161–180.
20. **Canfield, D. E., B. B. Jørgensen, H. Fossing, R. Glud, J. Gundersen, N. B. Ramsing, B. Thamdrup, J. W. Hansen, L. P. Nielsen, and P. O. J. Hall.** 1993. Pathways of organic carbon oxidation in three continental margin sediments. *Mar. Geol.* **113:**27–40.
21. **Chapelle, F. H., and D. R. Lovley.** 1992. Competitive exclusion of sulfate reduction by Fe(III)-reducing bacteria: a mechanism for producing discrete zones of high-iron ground water. *Ground Water* **30:**29–36.
22. **Coates, J. D., T. B. Councell, D. J. Ellis, and D. R. Lovley.** 1998. Carbohydrate-oxidation coupled to Fe(III) reduction, a novel form of anaerobic metabolism. *Anaerobe* **4:**277–282.
23. **Coates, J. D., D. J. Ellis, and D. R. Lovley.** 1999. *Geothrix fermentans* gen. nov. sp. nov., an acetate-oxidizing Fe(III) reducer capable of growth via fermentation. *Int. J. Syst. Bacteriol.* **49:**1615–1622.
24. **Coates, J. D., D. J. Ellis, E. Roden, K. Gaw, E. L. Blunt-Harris, and D. R. Lovley.** 1998. Recovery of humics-reducing bacteria from a diversity of sedimentary environments. *Appl. Environ. Microbiol.* **64:**1504–1509.
25. **Coates, J. D., D. J. Lonergan, H. Jenter, and D. R. Lovley.** 1996. Isolation of *Geobacter* species from diverse sedimentary environments. *Appl. Environ. Microbiol.* **62:**1531–1536.
26. **Coleman, M. L., D. B. Hedrick, D. R. Lovley, D. C. White, and K. Pye.** 1993. Reduction of Fe(III) in sediments by sulphate-reducing bacteria. *Nature* **361:**436–438.
27. **Cummings, D. E., F. Caccavo Jr., S. Spring, and R. F. Rosenzweig.** 1999. *Ferribacter limneticum*, gen. nov., sp. nov., an Fe(III)-reducing microorganism isolated from mining-impacted freshwater lake sediments. *Arch. Microbiol.* **171:**183–188.
28. **de Duve, C.** 1995. *Vital Dust.* Basic Books, New York, N.Y.

29. **de Duve, C.** 1998. Clues from present-day biology: the thioester world, p. 219–236. *In* A. Brack (ed.), *The Molecular Origins of Life*. Cambridge University Press, Cambridge, United Kingdom.
30. **DiChristina, T. J.** 1992. Effects of nitrate and nitrite on dissimilatory iron reduction by *Shewanella putrefaciens* 200. *J. Bacteriol.* **174:**1891–1896.
31. **DiChristina, T. J., R. G. Arnold, M. E. Lidstrom, and M. R. Hoffmann.** 1988. Dissimilative iron reduction by the marine eubacterium *Alteromonas putrefaciens* strain 200. *Water Sci. Technol.* **20:** 69–79.
32. **DiChristina, T. J., and E. F. DeLong.** 1993. Design and application of rRNA-targeted oligonulceotide probes for the dissimilatory iron- and manganese-reducing bacterium *Shewanella putrefaciens*. *Appl. Environ. Microbiol.* **59:**4152–4160.
33. **DiChristina, T. J., and E. F. DeLong.** 1994. Isolation of anaerobic respiratory mutants of *Shewanella putrefaciens* and genetic analysis of mutants deficient in anaerobic growth on Fe^{3+}. *J. Bacteriol.* **176:**1468–1474.
34. **Ehrenreich, A., and F. Widdel.** 1994. Anaerobic oxidation of ferrous iron by purple bacteria, new type of phototrophic metabolism. *Appl. Environ. Microbiol.* **60:**4517–4526.
35. **Ernstsen, V., W. P. Gates, and J. W. Stucki.** 1998. Microbial reduction of structural iron in clays—a renewable source of reduction capacity. *J. Environ. Qual.* **27:**761–766.
36. **Fenchel, T., and T. H. Blackburn.** 1979. *Bacteria and Mineral Cycling*. Academic Press, Ltd., London, United Kingdom.
37. **Fredrickson, J. K., and Y. A. Gorby.** 1996. Environmental processes mediated by iron-reducing bacteria. *Curr. Opin. Biotechnol.* **7:**287–294.
38. **Fredrickson, J. K., J. M. Zachara, D. W. Kennedy, H. Dong, T. C. Onstott, N. W. Hinman, and S.-M. Li.** 1998. Biogenic iron mineralization accompanying the dissimilatory reduction of hydrous ferric oxide by a groundwater bacterium. *Geochim. Cosmochim. Acta* **62:**3239–3257.
39. **Gaspard, S., F. Vazquez, and C. Holliger.** 1998. Localization and solubilization of the iron(III) reductase of *Geobacter sulfurreducens*. *Appl. Environ. Microbiol.* **64:**3188–3194.
40. **Gates, W. P., A. M. Jaunet, D. Tessier, M. A. Cole, H. T. Wilkinson, and J. W. Stucki.** 1998. Swelling and texture of iron-bearing smectites reduced by bacteria. *Clays Clay Miner.* **46:**487–497.
41. **Ghiorse, W. C.** 1988. Microbial reduction of manganese and iron, p. 305–331. *In* A. J. B. Zehnder (ed.), *Biology of Anaerobic Microorganisms*. John Wiley & Sons Inc., New York, N.Y.
42. **Gibbs-Eggar, Z., B. Jude, J. Dominik, J.-L. Loizeau, and F. Oldfield.** 1999. Possible evidence for dissimilatory bacterial magnetite dominating the magnetic properties of recent lake sediments. *Earth Planet. Sci. Lett.* **168:**1–6.
43. **Gold, T.** 1992. The deep, hot biosphere. *Proc. Natl. Acad. Sci. USA* **89:**6045–6049.
44. **Gorby, Y., and D. R. Lovley.** 1991. Electron transport in the dissimilatory iron-reducer, GS-15. *Appl. Environ. Microbiol.* **57:**867–870.
45. **Greene, A. C., B. K. C. Patel, and A. J. Sheehy.** 1997. *Deferribacter thermophilus* gen. nov., sp. nov., a novel thermophilic manganese- and iron-reducing bacterium isolated from a petroleum reservoir. *Int. J. Syst. Bacteriol.* **47:**505–509.
46. **Hartman, H.** 1984. The evolution of photosynthesis and microbial mats: a speculation on the banded iron formations, p. 449–453. *In* B. Crawford (ed.), *Microbial Mats: Stromatolites*. Alan R. Liss, Inc., New York, N.Y.
47. **Heijman, C. G., E. Grieder, C. Holliger, and R. P. Schwarzenbach.** 1995. Reduction of nitroaromatic compounds coupled to microbial iron reduction in laboratory aquifer columns. *Environ. Sci. Technol.* **29:**775–783.
48. **Heijman, C. G., C. Holliger, M. A. Glaus, R. P. Schwarzenbach, and J. Zeyer.** 1993. Abiotic reduction of 4-chloronitrobenzene to 4-chloroaniline in a dissimilatory iron-reducing enrichment culture. *Appl. Environ. Microbiol.* **59:**4350–4353.
49. **Hofstetter, T. B., C. G. Heijman, S. B. Haderlein, C. Holliger, and R. P. Schwarzenbach.** 1999. Complete reduction of TNT and other (poly)nitroaromatic compounds under iron-reducing subsurface conditions. *Environ. Sci. Technol.* **33:**1479–1487.
50. **Jannasch, H. W.** 1995. Microbial interactions with hydrothermal fluids. Seafloor hyrothermal Systems: Physical, Chemical, Biological, and Geological Interactions. *Geophys. Monogr.* **91:**273–296.
51. **Karl, D. M.** 1995. Ecology of free-living hydrothermal vent microbial communities, p. 35–124. *In* D. M. Karl (ed.), *The Microbiology of Deep-Sea Hydrothermal Vents*. CRC Press, Inc., New York, N.Y.

51a. **Kashefi, K., and D. R. Lovley.** 2000. Reduction of Fe(III), Mn(IV), and toxic metals at 100°C by *Pyrobaculum islandicum*. *Appl. Environ. Microbiol.* **66:**1050–1056.
52. **Kieft, T. L., J. K. Fredrickson, T. C. Onstott, Y. A. Gorby, H. M. Kostandarithes, T. J. Bailey, D. W. Kennedy, W. Li, A. E. Plymale, C. M. Spadoni, and M. S. Gray.** 1999. Dissimilatory reduction of Fe(III) and other electron acceptors by a *Thermus* isolate. *Appl. Environ. Microbiol.* **65:**1214–1221.
53. **Knight, V., and R. Blakemore.** 1998. Reduction of diverse electron acceptors by Aeromonas hydrophila. *Arch. Microbiol.* **169:**239–248.
54. **Kostka, J. E., and K. H. Nealson.** 1995. Dissolution and reduction of magnetite by bacteria. *Environ. Sci. Technol.* **29:**2535–2540.
55. **Kostka, J. E., J. W. Stucki, K. H. Nealson, and J. Wu.** 1996. Reduction of structural Fe(III) in smectite by a pure culture of *Shewanella putrefaciens* strain MR-1. *Clays Clay Miner.* **44:**522–529.
56. **LaKind, J. S., and A. T. Stone.** 1989. Reductive dissolution of goethite by phenolic reductants. *Geochim. Cosmochim. Acta* **53:**961–971.
57. **Landa, E. R., E. J. P. Phillips, and D. R. Lovley.** 1991. Release of 226-Ra from uranium mill tailings by microbial Fe(III) reduction. *Appl. Geochem.* **6:**647–652.
58. **Laverman, A. M., J. Switzer Blum, J. K. Schaefer, E. J. P. Phillips, D. R. Lovley, and R. S. Oremland.** 1995. Growth of strain SES-3 with arsenate and other diverse electron acceptors. *Appl. Environ. Microbiol.* **61:**3556–3561.
59. **Litaor, M. I., and R. B. Keigley.** 1991. Geochemical equilibria of iron in sediments of the rorarine river alluvial fan, Rocky Mountain National Park, Colorado. *Earth Surf. Proc. Landforms* **16:**533–546.
60. **Lloyd, J. R., E. L. Blunt-Harris, and D. R. Lovley.** 1999. The periplasmic 9.6 kDa *c*-type cytochrome is not an electron shuttle to Fe(III). *J. Bacteriol.* **181:**7647–7649.
61. **Lonergan, D. J., and D. R. Lovley.** 1991. Microbial oxidation of natural and anthropogenic aromatic compounds coupled to Fe(III) reduction, p. 327–338. *In* R. A. Baker (ed.), *Organic Substances and Sediments in Water*. Lewis Publishers, Inc., Chelsea, Mich.
62. **Lovley, D. R.** 1987. Organic matter mineralization with the reduction of ferric iron: a review. *Geomicrobiol. J.* **5:**375–399.
63. **Lovley, D. R.** 1990. Magnetite formation during microbial dissimilatory iron reduction. p. 151–166. *In* R. B. Frankel, and R. P. Blakemore (ed.), *Iron Biominerals*. Plenum Press, New York, N.Y.
64. **Lovley, D. R.** 1991. Dissimilatory Fe(III) and Mn(IV) reduction. *Microbiol. Rev.* **55:**259–287.
65. **Lovley, D. R.** 1995. Bioremediation of organic and metal contaminants with dissimilatory metal reduction. *J. Ind. Microbiol.* **14:**85–93.
66. **Lovley, D. R.** 1995. Microbial reduction of iron, manganese, and other metals. *Adv. Agron.* **54:**175–231.
67. **Lovley, D. R.** 1997. Microbial Fe(III) reduction in subsurface environments. *FEMS Microbiol. Rev.* **20:**305–315.
68. **Lovley, D. R.** 1997. Potential for anaerobic bioremediation of BTEX in petroleum-contaminated aquifers. *J. Ind. Microbiol.* **18:**75–81.
69. **Lovley, D. R.** Fe(III)- and Mn(IV)-reducing prokaryotes. *In* M. Dworkin, S. Falkow, E. Rosenberg, K.-H. Schleifer, and E. Stackebrandt (ed.), *The Prokaryotes*, in press. Springer-Verlag, New York, N.Y.
70. **Lovley, D. R.** Reduction of iron and humics in subsurface environments. in press. *In* J. Fredrickson and M. Fletcher (ed.), *Subsurface Microbiology and Biogeochemistry*, John Wiley & Sons Inc., New York, N.Y..
71. **Lovley, D. R., and R. T. Anderson.** 2000. The influence of dissimilatory metal reduction on the fate of organic and metal contaminants in the subsurface. *J. Hydrol.* in press.
72. **Lovley, D. R., M. J. Baedecker, D. J. Lonergan, I. M. Cozzarelli, E. J. P. Phillips, and D. I. Siegel.** 1989. Oxidation of aromatic contaminants coupled to microbial iron reduction. *Nature* **339:** 297–299.
73. **Lovley, D. R., and E. L. Blunt-Harris.** 1999. Role of humics-bound iron as an electron transfer agent in dissimilatory Fe(III) reduction. *Appl. Environ. Microbiol.* **65:**4252–4254.
74. **Lovley, D. R., and F. H. Chapelle.** 1995. Deep subsurface microbial processes. *Rev. Geophys.* **33:** 365–381.

75. **Lovley, D. R., F. H. Chapelle, and J. C. Woodward.** 1994. Use of dissolved H_2 concentrations to determine the distribution of microbially catalyzed redox reactions in anoxic ground water. *Environ. Sci. Technol.* **28:**1205–1210.
76. **Lovley, D. R., and J. D. Coates.** 1997. Bioremediation of metal contamination. *Curr. Opin. Biotechnol.* **8:**285–289.
77. **Lovley, D. R., J. D. Coates, E. L. Blunt-Harris, E. J. P. Phillips, and J. C. Woodward.** 1996. Humic substances as electron acceptors for microbial respiration. *Nature* **382:**445–448.
78. **Lovley, D. R., J. D. Coates, D. A. Saffarini, and D. J. Lonergan.** 1997. Dissimilatory iron reduction, p. 187–215. *In* G. Winkelman and C. J. Carrano (ed.), *Iron and Related Transition Metals in Microbial Metabolism*. Harwood Academic Publishers, Chur, Switzerland.
79. **Lovley, D. R., J. L. Fraga, E. L. Blunt-Harris, L. A. Hayes, E. J. P. Phillips, and J. D. Coates.** 1998. Humic substances as a mediator for microbially catalyzed metal reduction. *Acta Hydrochim. Hydrobiol.* **26:**152–157.
80. **Lovley, D. R., J. L. Fraga, J. D. Coates, and E. L. Blunt-Harris.** 1999. Humics as an electron donor for anaerobic respiration. *Environ. Microbiol.* **1:**89–98.
81. **Lovley, D. R., S. J. Giovannoni, D. C. White, J. E. Champine, E. J. P. Phillips, Y. A. Gorby, and S. Goodwin.** 1993. *Geobacter metallireducens* gen. nov. sp. nov., a microorganism capable of coupling the complete oxidation of organic compounds to the reduction of iron and other metals. *Arch. Microbiol.* **159:**336–344.
82. **Lovley, D. R., and S. Goodwin.** 1988. Hydrogen concentrations as an indicator of the predominant terminal electron accepting reactions in aquatic sediments. *Geochim. Cosmochim. Acta* **52:**2993–3003.
83. **Lovley, D. R., K. Kashefi, M. Vargas, J. M. Tor, and E. L. Blunt-Harris.** Reduction of humic substances and Fe(III) by hyperthermophilic microorganisms. *Chem. Geol.* in press.
84. **Lovley, D. R., and D. J. Lonergan.** 1990. Anaerobic oxidation of toluene, phenol, and *p*-cresol by the dissimilatory iron-reducing organism, GS-15. *Appl. Environ. Microbiol.* **56:**1858–1864.
85. **Lovley, D. R., and E. J. P. Phillips.** 1986. Availability of ferric iron for microbial reduction in bottom sediments of the freshwater tidal Potomac River. *Appl. Environ. Microbiol.* **52:**751–757.
86. **Lovley, D. R., and E. J. P. Phillips.** 1986. Organic matter mineralization with reduction of ferric iron in anaerobic sediments. *Appl. Environ. Microbiol.* **51:**683–689.
87. **Lovley, D. R., and E. J. P. Phillips.** 1987. Competitive mechanisms for inhibition of sulfate reduction and methane production in the zone of ferric iron reduction in sediments. *Appl. Environ. Microbiol.* **53:**2636–2641.
88. **Lovley, D. R., and E. J. P. Phillips.** 1988. Novel mode of microbial energy metabolism: organic carbon oxidation coupled to dissimilatory reduction of iron or manganese. *Appl. Environ. Microbiol.* **54:**1472–1480.
89. **Lovley, D. R., and E. J. P. Phillips.** 1989. Requirement for a microbial consortium to completely oxidize glucose in Fe(III)-reducing sediments. *Appl. Environ. Microbiol.* **55:**3234–3236.
90. **Lovley, D. R., and E. J. P. Phillips.** 1992. Reduction of uranium by *Desulfovibrio desulfuricans*. *Appl. Environ. Microbiol.* **58:**850–856.
91. **Lovley, D. R., E. J. P. Phillips, Y. A. Gorby, and E. R. Landa.** 1991. Microbial reduction of uranium. *Nature* **350:**413–416.
92. **Lovley, D. R., E. J. P. Phillips, and D. J. Lonergan.** 1989. Hydrogen and formate oxidation coupled to dissimilatory reduction of iron or manganese by *Alteromonas putrefaciens*. *Appl. Environ. Microbiol.* **55:**700–706.
93. **Lovley, D. R., E. J. P. Phillips, and D. J. Lonergan.** 1991. Enzymatic versus nonenzymatic mechanisms for Fe(III) reduction in aquatic sediments. *Environ. Sci. Technol.* **25:**1062–1067.
94. **Lovley, D. R., E. E. Roden, E. J. P. Phillips, and J. C. Woodward.** 1993. Enzymatic iron and uranium reduction by sulfate-reducing bacteria. *Mar. Geol.* **113:**41–53.
95. **Lovley, D. R., J. F. Stolz, G. L. Nord, and E. J. P. Phillips.** 1987. Anaerobic production of magnetite by a dissimilatory iron-reducing microorganism. *Nature* **330:**252–254.
96. **Lovley, D. R., and J. C. Woodward.** 1996. Mechanisms for chelator stimulation of microbial Fe(III)-oxide reduction. *Chem. Geol.* **132:**19–24.
97. **Lovley, D. R., J. C. Woodward, and F. H. Chapelle.** 1994. Stimulated anoxic biodegradation of aromatic hydrocarbons using Fe(III) ligands. *Nature* **370:**128–131.

98. **Lovley, D. R., J. C. Woodward, and F. H. Chapelle.** 1996. Rapid anaerobic benzene oxidation with a variety of chelated Fe(III) forms. *Appl. Environ. Microbiol.* **62:**288–291.
99. **McKay, D. S., E. K. Gibson, Jr., K. L. Thomas-Deprta, H. Vali, C. S. Romanek, S. J. Clement, X. D. F. Chillier, C. R. Maechling, and R. N. Zare.** 1996. Search for past life on Mars: possible relic biogenic activity in Martian meteorite ALH84001. *Science* **273:**924–930.
100. **McKnight, D. M., P. Behmel, D. A. Francko, E. T. Gjessing, U. Munster, R. C. Petersen Jr., O. M. Skulberg, C. E. W. Steinberg, E. Tipping, S. A. Visser, P. W. Werner, and R. G. Wetzel.** 1990. Group report how do organic acids interact with solutes, surfaces, and organisms, p. 223–243. *In* E. M. Perdue and E. T. Gjessing (ed.), *Organic Acids in Aquatic Ecosystems*. John Wiley & Sons, Inc., New York, N.Y.
101. **Mortimer, R. J. G., and M. L. Coleman.** 1997. Microbial influence on the oxygen isotopic composition of diagenetic siderite. *Geochim. Cosmochim. Acta* **61:**1705–1711.
102. **Mortimer, R. J. G., M. L. Coleman, and J. E. Rae.** 1997. Effect of bacteria on the elemental composition of early diagenetic siderite: implications for paleoenvironmental interpretations. *Sedimentology* **44:**759–765.
103. **Munch, J. C., and J. C. G. Ottow.** 1977. Modelluntersuchungen zum Mechanismus der bakteriellen Eisenreduktion in Hydromorphen Boden. *Z. Pflanzenernaehr. Bodenkd.* **140:**549–562.
104. **Munch, J. C., and J. C. G. Ottow.** 1983. Reductive transformation mechanism of ferric oxides in hydromorphic soils. *Environ. Biogeochem. Ecol. Bull. (Stockholm)* **35:**383–394.
105. **Myers, C. R., and J. M. Myers.** 1992. Localization of cytochromes to the outer membrane of anaerobically grown *Shewanella putrefaciens* MR-1. *J. Bacteriol.* **174:**3429–3438.
106. **Myers, C. R., and J. M. Myers.** 1993. Ferric reductase is associated with the membranes of anaerobically grown *Shewanella putrefaciens* MR-1. *FEMS Microbiol. Lett.* **108:**15–22.
107. **Myers, C. R., and J. M. Myers.** 1997. Cloning and sequencing of *cymA*, a gene encoding a tetraheme cytochrome c required for reduction of iron(III), fumarate, and nitrate by *Shewanella putrefaciens* MR-1. *J. Bacteriol.* **179:**1143–1152.
108. **Myers, C. R., and K. H. Nealson.** 1988. Bacterial manganese reduction and growth with manganese oxide as the sole electron acceptor. *Science* **240:**1319–1321.
109. **Nealson, K. H., and D. Saffarini.** 1994. Iron and manganese in anaerobic respiration: environmental significance, physiology, and regulation. *Annu. Rev. Microbiol.* **48:**311–343.
110. **Obuekwe, C. O., and D. W. S. Westlake.** 1982. Effects of medium composition on cell pigmentation, cytochrome content, and ferric iron reduction in a *Pseudomonas* sp. isolated from crude oil. *Can. J. Microbiol.* **28:**989–992.
111. **Obuekwe, C. O., D. W. S. Westlake, and F. D. Cook.** 1981. Effect of nitrate on reduction of ferric iron by a bacterium isolated from crude oil. *Can. J. Microbiol.* **27:**692–697.
112. **Ottow, J. C. G.** 1970. Bacterial mechanism of gley formation in artificially submerged soil. *Nature* **225:**103.
113. **Ottow, J. C. G.** 1970. Selection, characterization and iron-reducing capacity of nitrate reductaseless (nit^-) mutants of iron-reducing bacteria. *Z. Allg. Mikrobiol.* **10:**55–62.
114. **Ottow, J. C. G., and H. Glathe.** 1971. Isolation and identification of iron-reducing bacteria from gley soils. *Soil Biol. Biochem.* **3:**43–55.
115. **Pace, N. R.** 1991. Origin of life-facing up to the physical setting. *Cell* **65:**531–533.
116. **Pedersen, K., J. Arlinger, S. Ekendahl, and L. Hallbeck.** 1996. 16S rRNA gene diversity of attached and unattached bacteria in boreholes along the access tunnel to the Aspo hard rock laboratory, Sweden. *FEMS Microbiol. Ecol.* **19:**249–262.
117. **Phillips, E., D. R. Lovley, and E. E. Roden.** 1993. Composition of non-microbially reducible Fe(III) in aquatic sediments. *Appl. Environ. Microbiol.* **59:**2727–2729.
118. **Ponnamperuma, F. N.** 1972. The chemistry of submerged soils. *Adv. Agron.* **24:**29–96.
119. **Ponnamperuma, F. N.** 1984. Effects of flooding on soils, p. 9–45. *In* T. T. Kozlowski (ed.), *Flooding and Plant Growth*. Academic Press, Inc., New York, N.Y.
120. **Roden, E. E., and R. G. Wetzel.** 1996. Organic carbon oxidation and suppression of methane production by microbial Fe(III) oxide reduction in vegetated and unvegetated freshwater wetland sediments. *Limnol. Oceanogr.* **41:**1733–1748.
121. **Rooney-Varga, J. N., R. T. Anderson, J. L. Fraga, D. Ringelberg, and D. R. Lovley.** 1999. Microbial communities associated with anaerobic benzene mineralization in a petroleum-contaminated aquifer. *Appl. Environ. Microbiol.* **65:**3056–3063.

122. **Rossello-Mora, R. A., W. Ludwig, P. Kampfer, R. Amann, and K.-H. Schleifer.** 1995. *Ferrimonas balearica* gen. nov. spec. nov., a new marine facultative Fe(III)-reducing bacterium. *Syst. Appl. Microbiol.* **18:**196–202.
123. **Russell, M. J., D. E. Daia, and A. J. Hall.** 1998. The emergence of life from FeS bubbles at alkaline hot springs in an acid ocean, p. 77–126. *In* J. Wiegel and M. W. W. Adams (ed.), *Thermophiles: the Keys to Molecular Evolution and the Origin of Life*? Taylor & Francis, Inc., Philadelphia, Pa.
124. **Scott, D. T., D. M. McKnight, E. L. Blunt-Harris, S. E. Kolesar, and D. R. Lovley.** 1998. Quinone moieties act as electron acceptors in the reduction of humic substances by humics-reducing microorganisms. *Environ. Sci. Technol.* **32:**2984–2989.
125. **Seeliger, S., R. Cord-Ruwisch, and B. Schink.** 1998. A periplasmic and extracellular *c*-type cytochrome of *Geobacter sulfurreducens* acts as a ferric iron reductase and as an electron carrier to other acceptors or to partner bacteria. *J. Bacteriol.* **180:**3686–3691.
126. **Slobodkin, A., A.-L. Reysenbach, N. Strutz, M. Dreier, and J. Wiegel.** 1997. *Thermoterrabacterium ferrireducens* gen. nov., sp. nov., a thermophilic anaerobic dissimilatory Fe(III)-reducing bacterium from a continental hot spring. *Int. J. Syst. Bacteriol.* **47:**541–547.
127. **Sørensen, J.** 1982. Reduction of ferric iron in anaerobic, marine sediment and interaction with reduction of nitrate and sulfate. *Appl. Environ. Microbiol.* **43:**319–324.
128. **Starkey, R. L., and H. O. Halvorson.** 1927. Studies on the transformations of iron in nature. II. Concerning the importance of microorganisms in the solution and precipitation of iron. *Soil Sci.* **24:**381–402.
129. **Stolz, J. F., D. J. Ellils, J. Switzer Blum, D. Ahmann, D. R. Lovley, and R. S. Oremland.** 1999. *Sulfurospirillum barnesii* sp. nov., *Sulfurospirillum arsenophilum* sp. nov., new members of the Sulfurospirillum clade of the Epsilon Proteobacteria. *Int. J. Syst. Bacteriol.* **49:**1177–1180.
130. **Straub, K. L., M. Benz, B. Schink, and F. Widdel.** 1996. Anaerobic, nitrate-dependent microbial oxidation of ferrous iron. *Appl. Environ. Microbiol.* **62:**1458–1460.
131. **Straub, K. L., and B. E. E. Cuchholz-Cleven.** 1998. Enumeration and detection of anaerobic ferrous iron-oxidizing, nitrate-reducing bacteria from European sediments. *Appl. Environ. Microbiol.* **64:**4846–4856.
132. **Straub, K. L., M. Hanzlik, and B. E. E. Buchholz-Cleven.** 1998. The use of biologically produced ferrihydrite for the isolation of novel iron-reducing bacteria. *Syst. Appl. Microbiol.* **21:**442–449.
132a. **Synoeyenbos-West, O. L., K. P. Nevin, and D. R. Lovley.** Stimulation of dissimilatory Fe(III) reduction results in a predominance of *Geobacter* species in a variety of sandy aquifers. *Microb. Ecol.*, in press.
133. **Vargas, M., K. Kashefi, E. L. Blunt-Harris, and D. R. Lovley.** 1998. Microbiological evidence for Fe(III) reduction on early Earth. *Nature* **395:**65–67.
134. **Verschuur, G. L.** 1993. *The History and Mystery of Magnetism.* Oxford University Press, New York, N.Y.
135. **Vroblesky, D. A., and F. H. Chapelle.** 1994. Temporal and spatial changes of terminal electron accepting processes in a petroleum hydrocarbon contaminated aquifer and the significance for contaminant biodegradation. *Water Resour. Res.* **30:**1561–1570.
136. **Walker, J. C. G.** 1980. Atmospheric constraints on the evolution of metabolism. *Origins Life* **10:**93–104.
137. **Walker, J. C. G.** 1984. Suboxic diagenesis in banded iron formations. *Nature* **309:**340–342.
138. **Walker, J. C. G.** 1987. Was the Archaean biosphere upside down? *Nature* **329:**710–712.
139. **Weiner, J. M., T. S. Lauck, and D. R. Lovley.** 1998. Enhanced anaerobic benzene degradation with the addition of sulfate. *Bioremediation J.* **2:**159–173.
140. **Widdel, F., S. Schnell, S. Heising, A. Ehrenreich, B. Assmus, and B. Schink.** 1993. Ferrous iron oxidation by anoxygenic phototrophic bacteria. *Nature* **362:**834–835.
141. **Zehnder, A. J. B., and W. Stumm.** 1988. Geochemistry and biogeochemistry of anaerobic habitats, p. 1–38. *In* A. J. B. Zehnder (ed.), *Biology of Anaerobic Microorganisms.* John Wiley & Sons, Inc., New York, N.Y.

Environmental Microbe-Metal Interactions
Edited by Derek R. Lovley

Chapter 2

Microbial Oxidation of Fe(II) and Mn(II) at Circumneutral pH

David Emerson

Iron and manganese represent the 4th and 12th most abundant elements, respectively, in the Earth's crust (45). Both of these transition metals are essential trace elements in biological systems. Over the past decade it has become clear that they can be important electron acceptors for anaerobic respiration carried out by a diverse array of prokaryotes (see chapter 1). The role played by microbes in the oxidation of Fe at circumneutral pH is poorly understood, and while it is widely recognized that Mn oxidation is often microbially mediated, the mechanisms of how this oxidation occurs are only beginning to be elucidated. In comparison to other biogeochemically important inorganic electron donors such as reduced S compounds, methane, or ammonia, the reduced forms of Fe and Mn have received much less study in terms of their capacity to sustain lithotrophic microbial growth. Two of the early "fathers" of general microbiology, Sergei Winogradsky and Martinus Beijerinck, both devoted studies to Fe- and Mn-oxidizing bacteria and demonstrated an undeniable association between biological activity and metal oxidation (4, 100); however, few of their contemporaries or scientific "progeny" took up the gauntlet laid down by the often enigmatic organisms thought to be responsible for these reactions. For reasons cited below, the biological oxidation of these metals has proven inherently difficult to study. As a result, there are few good experimental systems to investigate the mechanisms of microbial Mn and Fe oxidation; it has therefore been hard to unequivocally link lithotrophic growth to the oxidation of these metals. Nonetheless, the goal of science is to understand the unknown, and in the last decade steady progress has been made in our understanding of microbial mediation of Fe and Mn oxidation.

This chapter is biased toward iron, in part because a number of recent findings concerning the role of microbes in iron oxidation make this area of particular interest at the organismal level and in part because an excellent review on Mn

David Emerson • American Type Culture Collection, 10801 University Blvd., Manassas, VA 20110, and Institute for Biosciences, Bioinformatics, and Biotechnology, George Mason University, Manassas, VA 20110.

oxidation was recently published by Tebo et al. (90). It is also biased toward organisms rather than molecules, since virtually nothing is know about the molecules, involved in Fe oxidation at neutral pH, although there is an emerging story in this regard concerning Mn. For the sake of brevity, after a short historical aside, the focus is on the work of the last decade. The reader is referred to prior reviews that have dealt with earlier work (21, 32, 42, 48, 68–70, 75, 85, 89).

IRON OXIDATION

Historical Overview

The notion that microbes are involved in Fe oxidation dates back to the first half of the 19th century, when Ehrenberg suggested that bog iron may form as a result of biological action (19). He also went on to describe the first bacterium directly associated with Fe oxides, *Gallionella ferruginea* (18). In the latter part of the 19th century, Winogradsky used laboratory microcosms to demonstrate that bacteria could play a role in Fe oxidation at near-neutral pH (100). He described *Leptothrix ochracea* and provided circumstantial evidence that this organism was a lithoautotroph; however, this has never been directly proven. In 1919, Harder, a geologist by training, published a treatise (42) on what had by then become known as the iron bacteria. This work firmly established the importance of these organisms as geochemical agents. Harder described in detail several freshwater habitats in which these bacteria occurred, the natural history of some of the organisms, and the results of enrichment cultures. His enrichments also provided circumstantial evidence for lithotrophic growth by Fe-oxidizing bacteria; however, again the inability to obtain pure cultures frustrated attempts to provide definitive proof for lithotrophy. In the subsequent 75 years, a number of studies have added to our understanding of the circumneutral pH habitats that Fe-oxidizing microbes prefer, their importance in geochemical reactions, and their role in biofouling and biocorrosion. Still, with the exception of *G. ferruginea* (53), few breakthroughs have been made in culturing these organisms, and so their diversity and taxonomy have largely been restricted to morphological descriptions. Understandably, this has led to considerable confusion with regard to both describing the organisms and understanding their physiology. A 1984 review by Ghiorse (32), which remains relevant today, discusses the many problems with the nomenclature and taxonomy of this intriguing group of organisms.

Bioenergetics and Mineralogy

In terms of biological reactivity, the two most relevant oxidation states of iron are Fe(II), the reduced ferrous form, and Fe(III), the oxidized ferric form. If a prokaryote is going to gain energy by capturing electrons from this oxidation, an abundant supply of Fe(II) must be available, and herein lies the first problem. At neutral pH and under fully aerated conditions, there is a rapid chemical oxidation of Fe(II). Thus, while it is not uncommon for anaerobic sediments to generate significant concentrations of Fe(II), once that Fe(II) enters a fully aerobic zone, its half-life may be only a matter of minutes. The kinetics that describe chemical Fe oxidation are

$$\frac{-d[\mathrm{Fe(II)}]}{dt} = k[\mathrm{Fe(II)}]\,[\mathrm{OH^-}]^2 p_{\mathrm{O_2}}$$

where $k = (8 \pm 2.5) \times 10^{13}$ min^{-1} atm^{-1} mol^{-2} liter^{-2} (94). As is evident from this equation, the pH has a strong influence on the reaction rate, which is why under very acidic conditions (pH < 4), Fe(II) is quite stable in the presence of air. It should also be pointed out that the half-life of Fe(II) can be up to 100 times longer in seawater than in freshwater (88). The most efficient way for a microbe to overcome the Fe(II) stability problem is to either grow at very low pH, which is what *Thiobacillus ferrooxidans* and other acidophilic Fe oxidizers do (76), or to grow at a very low O_2 concentration at circumneutral pH, where the half-life of Fe(II) may be much longer (54). This is what *G. ferruginea* and other iron oxidizers appear to do.

A second problem faced by these organisms is that the product of the oxidation is an insoluble ferric hydroxide. To avoid mineralizing the cytoplasm of the cell iron oxidation must occur at the exterior of the cell surface. This requires that cells possess a chemical mechanism for transporting electrons across the periplasm to the cytoplasmic membrane, where a chemiosmotic potential is established. Thus, one would expect to find the initial Fe oxidase, as well as soluble electron transport components, exterior to the cytoplasmic membrane. This seems to be the case in *T. ferrooxidans* (76). Related to these problems is the determination of the free energy available from Fe oxidation at neutral pH, which is not straightforward. The $\Delta G^{\circ\prime}$ of the process, as generally stated,

$$\mathrm{Fe^{2+}} + 0.25\ \mathrm{O_2} + \mathrm{H^+} \rightarrow \mathrm{Fe^{3+}} + \tfrac{1}{2}\mathrm{H_2O}$$

yields only about 29 kJ mol of iron^{-1} (21). However, in situ the process is more complicated, especially in circumneutral environments. At a pH of 6 or 7, the Fe^{2+} will often be in the form of $FeCO_3$ and the Fe^{3+} will be in the form of an insoluble amorphous hydroxide, e.g., $Fe(OH)_3$ or FeOOH; thus, the product of the oxidation is continuously removed from solution (99). This is not the case under the acidic conditions where *T. ferrooxidans* grows, and the Fe^{3+} formed may remain soluble (76). At pH 7, the redox potential of the couple $Fe(OH)_3 + HCO^{-1}/FeCO_3$ ($E'_0 =$ 0.200 V) or $Fe(OH)_3/Fe^{2+}$ ($E'_0 = -0.236$ V) is substantial (99) compared to the pH 7 redox couple of O_2/H_2O ($E'_0 = 0.810$ V); this indicates that Fe^{2+} oxidation coupled to O_2 respiration can generate a substantial redox potential for establishing a chemiosmotic gradient at the cell membrane. Indeed, it is this interpretation of the bioenergetics that explains why newly discovered anoxic phototrophic bacteria are able to couple the oxidation of Fe^{2+} to Fe^{3+} as an electron donor for anoxygenic photosynthesis (20, 99; see below).

Habitats

The most common habitat of prokaryotes associated with Fe oxides occurs at sites where anoxic water containing significant concentrations of Fe(II) flows into an oxic region. In the oxic-anoxic mixing zone, copious quantities of rust-colored flocculent material, composed of amorphous Fe hydroxides, will form. These hydroxides are often deposited on the remains of biogenic structures such as sheaths and stalks. While initial microscopic observations of such samples often indicate a general paucity of either prokaryotes or eukaryotes, staining of the flocculent material with a fluorescent dye for nucleic acid (e.g., acridine orange) or protein [e.g., 5-(4,6-dichlorotiazinyl)aminofluorescein (DTAF)] often reveals high densities of prokaryotic cells tightly associated with the Fe oxides (24, 34). These types of environments are found in streams (82), groundwater springs (24, 34), wetlands (95; E. I. Robbins and A. W. Norden, Coal-Energy Environ., 1994), caves (7, 73) and at both freshwater (17) and marine (47, 51) hydrothermal vents. They are also common in irrigation ditches (46), boreholes (W. C. Ghiorse, Int. Symp. Biofouled Aquifers: Prevention and Restoration, 1986), and municipal and industrial water distribution systems (64), where they can cause major economic problems due to biofouling and corrosion of waterways and pipelines. For a more thorough inventory of these habitats, see reference 34. To illustrate the commonalities and differences that are manifested by these sites, four quite different examples are discussed below.

Marselisborg

The Marselisborg iron seep is located in a forest park in Denmark and is fed by groundwater with an average temperature of 10°C and an average Fe(II) concentration of 100 μM (24). Copious quantities of iron oxides formed where the groundwater flowed out of a rock wall and formed a loose but integral microbial mat. This Fe mat was constructed largely on a matrix of abandoned *L. ochracea*-like sheaths that were encrusted with amorphous Fe hydroxides. Both the Fe(III) concentration and total cell numbers increased with depth in the mat. Microscopic counts revealed that up to 10^8 bacterial cells/ml of mat were closely associated with these amorphous oxides. The entire system appeared to be driven by the flow of Fe(II)-rich water. Laboratory microcosm experiments done using fresh mat material collected from this site suggested that up to 80% of the Fe oxidation in the mat was microbially mediated (25).

Loihi

The Loihi seamount is an undersea volcano located 40 km off the coast of Hawaii. At its summit is a hydrothermal vent field, discovered in 1987, at a water depth of approximately 1,200 m (51). The vent fluid was highly charged with CO_2, resulting in a pH of 5.5, and it contained up to 1,000 μM Fe^{2+} and 20 μM Mn^{2+}. The temperature of fluid at the vent sites was 20 to 30°C. Unlike many of the axial hydrothermal vents that have been described, Loihi was largely devoid of macrofaunal life; however, extensive fluffy mats of Fe oxides had developed on the sea floor surrounding the vents. Microscopic observations indicated that these Fe deposits often consisted of sheaths and filaments reminiscent of *L. ochracea*, suggesting that the oxides may be of biogenic origin (50; D. Emerson, unpublished results). A phylogenetic analysis based on the DNA sequences of small subunit (SSU) rRNA genes extracted from the iron deposits surrounding the vents revealed two dominant phylotypes (65, 66). One of these (OTU1) formed a lineage within the xanthomonad group in the gamma proteobacteria. This lineage is essentially identical to novel Fe oxidizers recently isolated from groundwater in Michigan (see below), suggesting the possibility that a related group of Fe oxidizers is involved in Fe oxidation at Loihi. Large masses of similar type Fe hydroxide mats have been observed at other hydrothermal vent sites (22) (Fig 1B).

Plant Rhizosphere

Another variation on the microaerobic habitats where iron-oxidizing prokaryotes may occur is in association with the rhizospheres of some plants. For years, scientists studying wetland plants have described iron plaques associated with the roots of certain species. It has been hypothesized that these iron hydroxide deposits form as a result of radial oxygen loss from the plant root (61). The diffusion of O_2 from the root into the surrounding anoxic soil, which may contain significant quantities of Fe(II), results in the formation of iron oxides or iron plaque on the plant roots. While there have been suggestions that microbes could be involved in iron plaque formation, there has been no direct evidence (92). However, recent observations indicate that large numbers of bacteria, up to 10^5 or 10^6 cells/cm^2 of root, are specifically associated with the iron plaque (P. Megonigal, D. Emerson, and J. Weiss, Biogeomon, 1997). Iron-oxidizing bacteria can also be easily enriched from Fe-plaque encrusted roots. These results warrant further investigation to determine the specific role that iron-oxidizing prokaryotes may be playing in the formation of iron plaques.

Anaerobic Environments

Until recently, it was thought that microbial iron oxidation was limited to oxic or microaerobic zones. However, the finding of anoxic phototrophic microbes that can utilize Fe(II) for anoxygenic photosynthesis and the finding of microbes capable of anaerobic respiration by coupling nitrate reduction to Fe oxidation (see below for details) now indicate that microbes can play a role in iron oxidation in anoxic environments. This has added a novel twist to the habitats where Fe oxidation can occur, and although the extent of this activity remains to be shown, these processes will probably have important biogeochemical implications.

Representative Organisms

Representatives of some of the Fe oxidizers that are known to occur in the habitats described above are discussed in more detail below.

L. ochracea

L. ochracea is perhaps the most common visible inhabitant of many freshwater Fe seeps (95), where it may form brown fluffy filamentous layers that are visible to the naked eye. Microscopic inspection of this material reveals a tangled matrix of tubular sheaths encrusted with iron (Fig 1A). The sheaths are straight tubes that appear quite robust (Fig. 2A) and often appear refractile by phase-contrast micro-

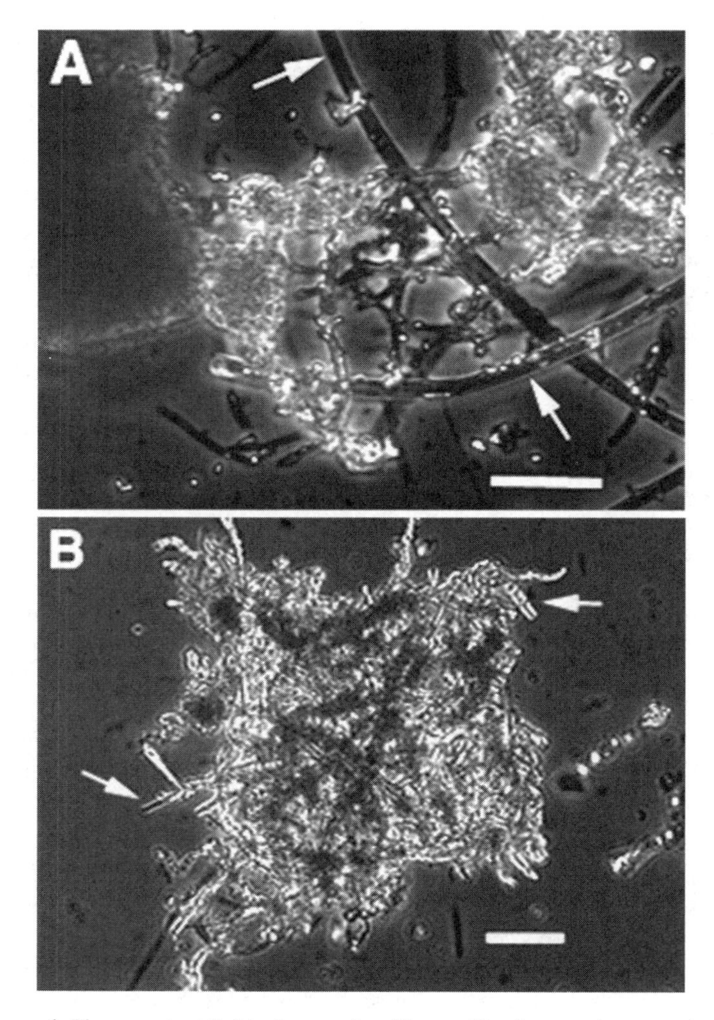

Figure 1. Phase contrast light micrographs of iron oxides from environmental samples. (A) Sample from a circumneutral iron spring in Northern Virginia. The arrows denote *L. ochracea*-like sheaths. Note also the presence of finer filaments of Fe oxides that are of unknown origin, as well as the larger amorphous particles of Fe oxides. Bar, 10 μm. (B) Sample collected from an iron-rich hydrothermal vent site on the North Gorda Ridge in the Pacific Ocean. The arrows again denote the remains of *L. ochracea*-like sheath structures. Bar, 20 μm.

scopy. It appears that as the sheaths age they continue to accumulate Fe oxides. Due to the lack of pure cultures, the organism(s) is identified solely by morphotype, which leaves open all questions on systematics and diversity. For example, we do not know if the morphotype that is referred to as *L. ochracea* is one organism or a group of organisms that produce sheaths of similar appearance. One consistent observation of *L. ochracea* morphotypes is that it is rare to see filaments of cells

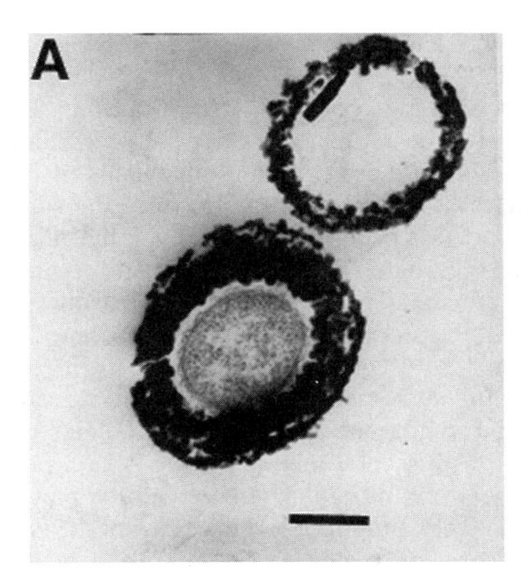

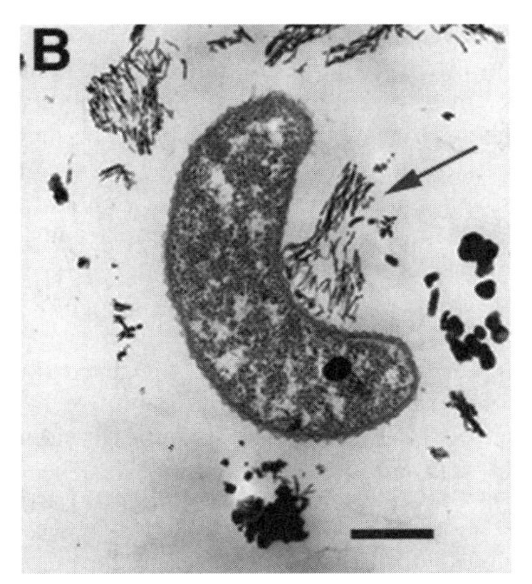

Figure 2. Transmission electron micrographs of Fe-oxidizing bacteria from the Marselisborg iron seep. (A) *L. ochracea* ensheathed cell and empty sheath in cross-section. Note the thick Fe oxide crust on the sheath. Bar, 0.5 μm. (B) *G. ferruginea*. The arrow points to a portion of the stalk that is attached to the cell. Bar, 0.5 μm.

inside the sheaths (24, 34, 95). The organism is capable of accumulating large amounts of empty sheath material in a short time, at the Marselisborg iron mat, dominated by *L. ochracea* (see above), accretion rates of up to 3 mm/day were measured (24). Depth profiles taken at this site and analyzed microscopically indicated that sheaths containing filaments of cells were almost all in the top 1 mm of the mat and that even then only about 9% of the sheaths actually contained cells. It has been speculated that as this organism oxidizes iron, it deposits the ferric hydroxides on the sheath and in this way prevents itself from becoming encrusted in the Fe oxide precipitate (95). Again, because of the lack of pure cultures, little is known about the physiology of this organism. Winogradsky was the first to propose that *L. ochracea* might be a lithotroph (100); however, this has never been confirmed, although others have been able to obtain enrichments in systems where Fe was the only major electron donor (25, 42, 67). Oxygen microsensor studies done in situ at the Marselisborg Fe seep indicated that the organism could grow in water that was at least 50% air saturated with respect to O_2, suggesting that it was not a strict microaerophile (24). Despite its appearance, the sheath does not appear to consist of a robust extracellular matrix, since phase-contrast microscopy shows that it virtually disappears when the Fe is reduced (Emerson, unpublished); this is in contrast to the more visible organic sheaths produced by *Leptothrix discophora* and *Sphaerotilus natans* (see below).

Gallionella spp.

Gallionella was first described by Ehrenberg in the 1800s (18) and is immediately recognizable due to the helical stalk that it forms (Fig. 2B). The cells are bean-shaped and grow at the termini of the stalks; thus, one cell is capable of producing a large amount of stalk material. The stalk material is composed largely of iron hydroxides; presumably there is an underlying organic matrix, but this not been characterized (40). The stalks continue to accumulate Fe after the producing cells have gone (44). Like *L. ochracea*, *Gallionella* are common inhabitants of Fe springs, and where they are abundant, the stalk material appears to form the substratum upon which much subsequent Fe oxidation occurs. Interestingly, it has been observed that actively growing *Gallionella* spp. and *L. ochracea* appear to occupy separate microniches within the same environment (24); one possible reason for this is that *Gallionella* spp., which appear to be strict microaerophiles (40), prefer lower O_2 tensions than *L. ochracea* does.

A gradient enrichment technique for *Gallionella* spp. was developed by Kucera and Wolfe (53) that used tubes containing FeS to generate opposing gradients of O_2 and Fe(II). This technique has been used by several laboratories to obtain purified cultures of *G. ferruginea* (39, 40); therefore, more is known about the physiology of this organism than of other putatively lithotrophic Fe oxidizers. There has been one report that *G. ferruginea* is capable of fixing CO_2 by the Calvin cycle, although this work has yet to be followed up (40). Studies by Hallbeck and coworkers (37, 38) demonstrated that *G. ferruginea* was capable of lithotrophic growth in a mineral medium with FeS or $FeCO_3$ as the energy source. It was shown that the cells could fix all their carbon from CO_2, demonstrating that they are true lithoautotrophs. These workers also showed that *G. ferruginea* was capable of mixotrophic growth. In the presence of glucose, the cells could obtain all their cell carbon from the glucose, although the addition of glucose to the medium did not appreciably stimulate the growth rate or yield of the cells over those in the presence of Fe(II) alone. There has also been a report that *G. ferruginea* can utilize sulfide and thiosulfate as energy sources and grow to cell densities of 10^8 cells/ml (56); however, Hallbeck and Pederson were unable to repeat these results with their strain (37). Another interesting physiological finding reported by these workers was that *G. ferruginea* does not form a stalk at a pH below 6 or under very microaerobic conditions where O_2 is present but the redox potential is -40 mV (39); their interpretation was that these are the most favorable growth conditions for *G. ferruginea* and that the stalk represents a survival structure, which the organisms produce as a means of protecting themselves until more favorable growth conditions occur. These findings do not seem entirely consistent with the numerous observations of stalked *Gallionella* cultures occurring in nature, where, at least judging from the sheer quantity of stalks and the rate at which they accumulate, the cells would appear to be quite active. Careful field studies using molecular probes to identify stalked and unstalked cells might sort out the microscale spatial distributions of these cells and provide a more definitive answer.

Novel Anaerobic Fe Oxidizers

Widdel et al. noticed that when both freshwater and marine anaerobic sediments were stored anoxically in the light for several weeks, rust colored spots appeared on the walls of the containment vessels (99). They then set up sulfide-free, anoxic enrichments containing sediment samples amended with $FeSO_4$, $NaHCO_3$, and vitamins that were incubated under low light levels. The Fe(II) was oxidized, with concomitant increases in cell number. From such enrichments, several isolates were obtained through dilution series, and two of these have been characterized in detail (20). Strain L7 is a member of the alpha subclass of the proteobacteria, and its closest known relative is *Chromatium vinosum.* Strain SW-2 is in the gamma subclass of the proteobacteria, and its closest relative is *Rhodobacter capsulatus*. Both strains are capable of photoautotrophic growth on Fe(II) and CO_2 alone in the presence of light. In the absence of light, there is no growth. Both strains can also utilize H_2 for autotrophic growth, as well as a limited number of organic carbon sources including glucose and acetate. Neither strain can utilize sulfide, thiosulfate, or Mn(II) as electron donors for growth. Both strains have growth yields of 7.6 g of dry mass (as CH_2O)/mol of Fe oxidized. This is in good agreement with the theoretical value of 7.5 g/mol derived from the equation

$$4Fe^{2+} + CO_2 + 11H_2O \rightarrow 4Fe(OH)_3 + (CH_2O) + 8H^+$$

It is presumed that these organisms utilize photosystem I for Fe oxidation, since this is the photosystem used by all known anoxygenic phototrophs and since inhibitors of photosystem II did not appear to inhibit growth on iron by some of these strains (20, 99). Recently, two new species of marine phototrophic Fe(II)-oxidizing bacteria in the genus *Rhodovulum* have been isolated (87); these strains were also able to oxidize sulfide and thiosulfate. Interestingly, none of the related anoxygenic phototrophs from culture collections oxidize iron, suggesting that iron oxidation is a specialized adaptation for these organisms.

Another interesting development in the story of anaerobic iron oxidation is the discovery that some nitrate-reducing organisms are capable of utilizing Fe(II) as the electron donor and NO_3 as the electron acceptor (86). This process was initially discovered using freshwater sediments to inoculate anoxic, bicarbonate buffered medium containing NO_3 and $FeSO_4$. Nitrate reduction occurred, with the concomitant accumulation of rust colored Fe oxides; uninoculated or heat-inactivated controls failed to oxidize Fe. Quantitative estimates from the enrichment culture demonstrated that the accumulation of Fe(III) relative to the reduction of NO_3 was in the ratio 1:0.22, which was in agreement with the proposed stoichiometry for the process:

$$10FeCO_3 + 2NO_3^- + 24H_2O \rightarrow 10Fe(OH)_3 + N_2 + 10HCO_3^- + 8H^+$$

Three strains, HidR2, BrG1, and BrG2, were subsequently isolated using dilution series and shown to couple Fe oxidation with nitrate reduction. HidR2 and BrG1 required acetate to carry out this process, but BrG2 was capable of Fe oxidation with no organic substrate present. It was not clear from these studies that this form of Fe oxidation actually stimulated growth of the organisms, since direct cell yields were not determined, but they were reported to be very low. When other known nitrate reducers were tested for their capacity to carry out Fe(II) oxidation, several of them, including *Thiobacillus denitrificans* and *Pseudomonas stutzeri*, were capable of it. These results suggest that this form of Fe oxidation has the potential to be widespread in anoxic habitats.

A phylogenetic analysis of five of the new isolates of NO_3-reducing Fe oxidizers indicated that four strains fell within the Rubrivax group of the beta-subclass of the proteobacteria (6). It is interesting that *G. ferruginea* (38), *L. discophora* (84), and *S. natans* (84) all fall within this group as well. Strain BrG3 formed a novel lineage in the *Xanthomonas* group of the gamma proteobacteria. Its closest relatives, with a sequence similarity of 92.8%, were the environmental clones PVB OTU 1 (for operational taxonomic unit 1) and PVB OTU 11, these two clones are essentially identical to the lithotrophic, microaerophilic Fe oxidizing strains ES-1 and ES-2, which are described below.

Perhaps the most interesting newly discovered microbe capable of coupling Fe oxidation to NO_3 reduction is *Ferroglobus placidus*. This obligate anaerobe was isolated from a shallow marine hydrothermal vent system in Italy and was the first thermophilic archaeon to grow by oxidizing Fe(II) at neutral pH (36). *F. placidus* grew rapidly and with good cell yields in a mineral medium (Fig. 3) and has a temperature optimum of 85°C. In addition to Fe(II), H_2 and sulfide served as electron donors for NO_3 reduction. The organism could also reduce thiosulfate by using H_2 as an electron donor and was even reported to grow in the presence of $S_2O_3^{2-}$ and Fe(II), producing FeS and SO_3^{2-}, although this reaction was not further char-

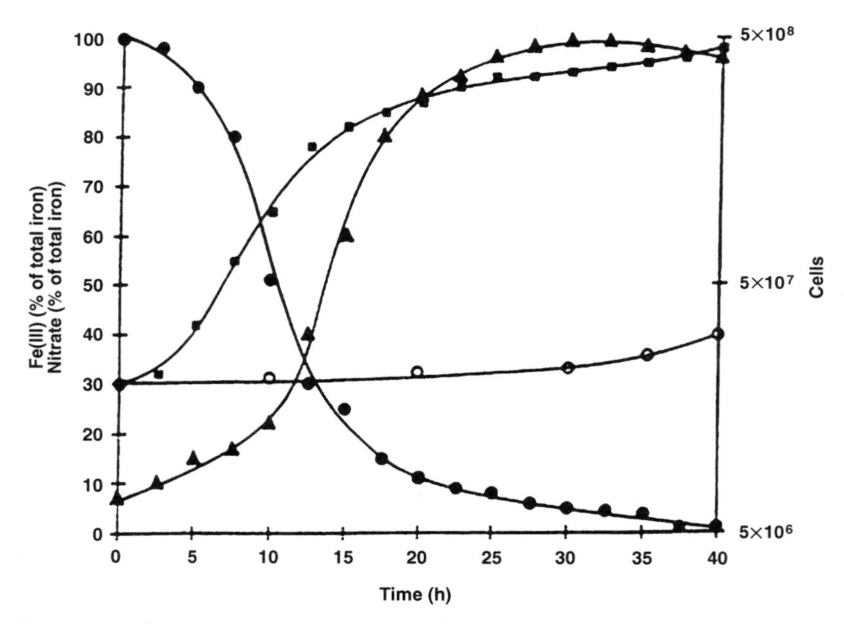

Figure 3. Anaerobic growth of *F. placidus* on iron and nitrate under chemolithoautotrophic conditions. Symbols: ■, ferric iron in growing culture, ○, autoxidation in uninoculated culture medium at 85°C, ●, nitrate in growing culture, ▲, number of cells per milliliter. Reprinted from reference 36 with permission of the publisher.

acterized. The presence of organic matter stimulated growth but was not required. The closest phylogenetic relative of *F. ferroglobus* is *Archaeoglobus fulgidus*.

Novel Microaerobic Iron Oxidizers

Several new microaerobic iron-oxidizing bacteria have been isolated using a gel-stabilized gradient enrichment technique. This technique is a modification of the method used for culturing *G. ferruginea*, in which opposing gradients of Fe(II) and O_2 are established in an agarose-stabilized gel rather than a liquid (30). This is also similar to the techniques devised for growing the lithotrophic S oxidizer *Beggiatoa* (72). When samples from an iron-rich groundwater source were inoculated into these gradients, a rust-colored band formed at the oxic-anoxic interface within a few days (30). Epifluorescence microscopy combined with acridine orange staining revealed that a large population of bacterial cells was tightly associated with the Fe hydroxides. Using this technique to perform successive dilution series, it was possible to isolate two strains, ES-1 and ES-2, from Fe floc-containing groundwater in Michigan. A third isolate, RL-1, came from a high-dilution (10^{-7}) enrichment obtained from a wetland in Michigan that had 100 to 200 μM Fe^{2+} in the water (Emerson, unpublished). Each of these strains is morphologically unique, but all are unicellular microbes that do not produce sheaths or stalks. Strains ES-1 and ES-2 will utilize Fe^{2+} from either $FeCO_3$ or FeS but will not grow on the organic substrates glucose, succinate, pyruvate, or acetate; furthermore, acetate did not stimulate the growth of ES-1 when it was grown on FeS (30). Neither sulfide or Mn^{2+} appeared to stimulate growth of these two strains. Oxygen microsensor studies revealed that both ES-1 and ES-2 preferred to grow right at the oxic-anoxic boundary and that they would track this boundary if it was shifted deeper into the agarose gel layer; these results indicate that these organisms are microaerophiles.

Although there were clear morphological and physiological differences between isolates ES-1, ES-2, and RL-1, a phylogenetic analysis of the 16S SSU rRNA molecule of isolates ES-1 and ES-2 indicated that they are closely related (30). They formed a novel lineage within the *Xanthomonas* group of the gamma proteobacteria. As mentioned above, they also had an equally high phylogenetic similarity to PVB OTU 1 and PVB OTU 11, two SSU rRNA phylotypes that were cloned from DNA isolated from the Loihi Seamount (66). This same phylotype has been recovered from an iron-rich permafrost soil in Siberia that is up to 2 million years old, and it also has been recovered from a deep aquifer in Washington State (C. L. Moyer, N. K. Hobbs, and D. Emerson, Abstr. 97th Gen. Meet. Am. Soc. Microbiol. 1997, abstr. N017). These results would imply that this Fe-oxidizing lineage is surprisingly monophyletic. Significantly more information needs to be gathered to confirm these initial findings, however.

Heterotrophic Iron Oxidizers

Although putatively lithotrophic iron-oxidizing bacteria tend to predominate in environments where there is a low concentration of organic matter and a high concentration of Fe^{2+}, there are also obligately heterotrophic bacteria that are known to either oxidize Fe(II) or preferentially accumulate Fe oxides in association with the cells or exopolymers produced by the cells. The filamentous, sheath-forming organism *Sphaerotilus natans* is the most commonly recognized iron oxidizer among this group (67). This organism prefers eutrophic environments and is often implicated in causing bulking problems in activated sludge. While it is recognized for its capacity to deposit iron oxides on its sheath, very few studies have attempted to elucidate the mechanism of the oxidation or to even prove that it is an enzymatically mediated oxidation rather than a passive precipitation (78). More recent studies have suggested that *L. discophora*, better known as an Mn oxidizer (see below), possesses an Fe-oxidizing protein (9, 13). Another morphotypic group of "iron bacteria" contains members of the *Siderocapsaceae*. This group includes four genera, *Siderocapsa*, *Naumanniella*, *Siderococcus*, and *Ochrobium*, consisting of 18 species (41). The species descriptions are based primarily on the morphology of the oxide coatings that form around the cells. This group is thought to be predominantly heterotrophic; however, it has been difficult to obtain pure cultures that are definitively *Siderocapsaceae*. A recent study of metal-coated biofilms that formed in the Elbe River in Germany suggested that the causative organisms might be in the genus *Siderocapsa* (55). In light of recent findings of unicellular Fe oxidizers, it is possible that some of the descriptions of the *Siderocapsaceae* have identified the morphological indicators of these organisms (30).

MANGANESE OXIDATION

Mineralogy and Bioenergetics

The oxidation of the manganous ion, Mn(II), to the manganic form, Mn(IV), is a two-electron transfer, which can proceed via one-electron steps through an unstable intermediate, Mn(III). As is the case for Fe(II), Mn(II) is quite soluble but the oxidation product Mn(IV) is highly insoluble. Unlike Fe(II), however, Mn(II) is quite stable under fully aerobic conditions at neutral pH, and only at pH $\geq$ 8 does chemical oxidation of Mn begin to become appreciable (88). These properties have made it much easier to unequivocally demonstrate that Mn oxidation is mediated biologically. In terms of energetics, for the two-electron transfer to oxygen the reaction is

$$Mn^{2+} + 0.5O_2 + H_2O \rightarrow MnO_2 + 2H^+$$

The standard free energy at pH 7 ($\Delta G°$) of this reaction is −70.9 kJ/mol (90). Compared to the free energy required to form an ATP molecule (−30 kJ/mol), this would suggest that Mn oxidation could quite easily support lithotrophic energy metabolism. However, the bioenergetics can be complicated by the different mineral forms that Mn oxides may take; these include hausmannite (Mn_3O_4) and manganite (MnOOH) which are Mn(III) forms, as well as Mn(IV) minerals such as buserite (90). Measurements of the oxidation state of the Mn oxides from pure cultures of Mn-oxidizing bacteria have reported values of between 3 and 4 (90). Recent evidence suggests that microbes do oxidize Mn(II) directly to Mn(IV) (58). In any event, the precise form of the oxide may be affected by the initial Mn(II) concentration, as well as by the pH and ionic strength of the medium (58). This has real bioenergetic implications, in that the redox potentials of the different couples between Mn(II) and mixed Mn(III)/Mn(IV) can be quite variable, which will in turn affect how much chemiosmotic potential an organism can harness to make ATP from this reaction. See reference 90 for more discussion of the bioenergetics of growth on Mn. What makes these bioenergetic arguments especially cogent is that despite the presence of a relatively abundant electron donor, it remains to be conclusively demonstrated that any prokaryote can grow lithotrophically using Mn.

Habitats

Mn often occurs in the same types of oxic-anoxic interfacial environments where Fe is found, although it is generally 5 to 10 times less abundant, reflecting the relative total abundance of these two metals (80). In high-iron environments, similar to those described above for Fe, it is common for ferromanganic oxides to form. However, detailed studies examining the composition of biogenic ferromanganic oxides or the way in which different microbes may influence the relative abundances of the oxides of the two metals have yet to be done. Since it is not subject to as rapid a chemical oxidation, Mn(II) may accumulate to greater concentrations in more aerobic regions of the water column of marine waters and freshwaters than Fe(II) does, and it can be found in oxic soils. Perhaps as a result, heterotrophic Mn-oxidizing bacteria have been isolated from or found in lakes (81), wetlands (33), soils (16), particles of marine snow (11), and hydrothermal vents (49); for a more detailed description of these habitats, see reference 34.

Organisms

Given the range of habitats where appreciable concentrations of Mn(II) can occur, it is perhaps not surprising that a diversity of heterotrophic microbes capable of Mn oxidation have been described (32, 69, 90). One common trait that is emerging among microbes that oxidize and precipitate Mn is that the oxides often seem to be associated with extracellular polymers produced by the organisms (70). Fungi are also known to oxidize Mn (90). The important lignin-degrading white rot fungus *Phanerochaete chrysosporium* is known to produce an Mn peroxidase that plays an important role in lignin breakdown. Examples of three quite different Mn-oxidizing organisms are described below.

L. discophora

L. discophora grows for a portion of its life as filaments of cells encased in a fibrillar sheath layer. The sheath is a complex heteropolymer consisting of protein (20%), lipid (8%), and carbohydrate (40%), the latter is made up a mix of amino sugars and uronic acids (29). The fibrils that compose the organized sheath are held together by disulfide bonds, and there is an excess of free sulfhydryl groups within the sheath (27). The availability of COOH and SH groups provides potential ligands for sorption of metal ions to the sheath. In sheathed cells, the protein(s) involved in Mn oxidation is excreted and become associated with the sheath layer, where Mn oxidation occurs (26). Sheathless variants of *L. discophora* also excrete Mn-oxidizing proteins, but in this case the Mn oxides form as disorganized amorphous particulates that are free in the medium (2). The specific reasons why *L.*

discophora (or any other organism) oxidizes Mn are the subject of speculation, but no hard evidence has been obtained (28). The oxide-encrusted sheath could serve a number of protective functions, including protection from grazing by protozoa, viral attack, UV irradiation, or detoxification of O_2 radicals; it could also aid in sequestering nutrients. One intriguing hypothesis (89) follows from the demonstration that the reaction of Mn oxides with humic acids resulted in degradation of the humic materials into low-molecular-weigh organic acids, including pyruvate, which are the type of C sources that *L. discophora* prefers. Thus, Mn-oxidizing organisms could potentially use the Mn oxides as a catalyst for converting a recalcitrant C source into a food supply. To date, no empirical evidence to support this hypothesis has been found.

Several groups have isolated an Mn-oxidizing protein from both sheathed (strain SP-6) and nonsheathed (strain SS-1) *L. discophora* (1, 5, 26). In cultures of SS-1, the Mn-oxidizing factors are isolated from spent culture medium (1, 5); in cultures of SP-6, the factor can be concentrated from preparations of the sheath material (26). This protein has a molecular mass of approximately 110 kDa, as determined by sodium dodecyl sulfate-polyacrylamide gel electrophoresis (SDS-PAGE) (1, 5, 26). The Mn-oxidizing activity is inactivated by heat, sodium azide, and potassium cyanide, although it is stable in the presence of detergents such as SDS and *n*-lauroylsarcosine. Attempts to purify Mn-oxidizing factors from *L. discophora* have been hindered by apparent inactivation of the proteins during purification (1, 5, 26). These results suggest that either unknown cofactors or accessory proteins are involved in the oxidation or that the proteins require the polymeric sheath material to enhance their activity in some way. A recent breakthrough was made by Corstjens et al. (10), who were able to isolate enough of the Mn-oxidizing protein from *L. discophora* SS-1 to raise polyclonal antibodies. They used the antibodies to screen expression libraries of cloned SS-1 DNA and obtained several recombinants from which they were ultimately able to construct a 4,986-bp open reading frame, dubbed *mofA* (for "Mn-oxidizing factor"). This coded for a putative protein of 174,292 kDa. This protein had a signal sequence at the N terminus, and it had similarity to several known copper oxidase proteins (see below). The discrepancy between the molecular *mofA* masses of the product (174 kDa) and the Mn-oxidizing factor in *L. discophora* cultures (110 kDa) was speculated to be due to either posttranslational modifications or breakdown of the protein during SDS-PAGE. Unfortunately, the cloned *mofA* was lethal to its *Escherichia coli* host; however, it was possible to express the protein using an in vitro transcription-translation system. The authors did not report if the translated product had Mn-oxidizing activity. An environmental study of *mofA* revealed that it was present in a wetland habitat typical of the habitat of *L. discophora* (83).

Marine *Bacillus* sp. strain SG-1

The gram-positive, spore-forming strain SG-1 was isolated from a marine sediment during selection for Mn-oxidizing phenotypes (71). It is actually the mature spore, free of the vegetative cell which catalyzes the oxidation of Mn (79). The spore coats become encrusted with amorphous Mn oxides. Again, it is not clear what benefit accrues to the bacterium from Mn oxidation. Nevertheless, a good deal of work has been undertaken to elucidate the mechanism by which Mn oxidation occurs on the spores. Inhibitor studies including heat inactivation and treatment with mercuric chloride, potassium cyanide, or azide all resulted in loss of Mn-oxidizing activity, suggesting a proteinaceous factor (90). Attempts to purify an Mn-oxidizing protein from the spore coat have met with limited success, in part due to the inherent difficulties of working with spore proteins. A genetic approach taken by van Waasbergen et al. (96, 97) used transposon mutagenesis to probe the chromosome of the vegetative cell and met with more success. Mutants were obtained that had lost their capacity to oxidize Mn but otherwise behaved as wild-type spores in their ability to germinate and be resistant to conventional heat and chemical treatments (97). Characterization and sequencing of regions identified by insertion mutagenesis revealed a gene cluster of seven open reading frames, designated *mnxA* to *mnxG*, that appeared to form a single operon (97). Six of these genes were speculated to play a structural role in the spore coat formation or to be of indeterminate function. *mnx*G, however, coded for a polypeptide with a molecular mass of 138 kDa, which had significant similarity to a family of multicopper oxidase proteins. These workers were also able to show that low levels of Cu^{2+} (1 μM) increased the Mn-oxidizing activity of spores although higher concentrations were inhibitory. The multicopper oxidases are known to be involved in accepting electrons from a substrate and transferring them to O_2 (90). They are capable of catalyzing the four electron reduction of O_2 to H_2O. Most multicopper oxidases utilize organic compounds as substrates, although some oxidize Fe^{2+} in conjunction with iron uptake systems. While it remains to be biochemically confirmed that MnxG codes for a Cu-containing enzyme that has Mn-oxidizing enzyme activity, these studies, along with those done on *L. discophora*, suggest that a multicopper oxidase is a good potential candidate for the Mn-oxidizing enzyme(s).

"Metallogenium"

Perhaps, no "organism" demonstrates how enigmatic biological Fe and Mn oxidation can be. "*Metallogenium*" was first described by Perfil'ev, a Russian microbiologist, who used specially constructed microscope slides or "peloscopes" to directly observe microbial growth in sediments (74). Perfil'ev made detailed observations on the development of filamentous structures that formed in the sediment and deposited copious quantities of ferromanganic oxides. Later work described "*Metallogenium symbioticum*," which could grow only in conjunction with a fungus; the prokaryotic partner was proposed to be of mycoplasmal origin (101). Since then, "*Metallogenium*"-like particles have been described in a number of environments including desert varnish (15), sand filters (12), and the water columns of a number of lakes (35, 52, 57, 62, 81, 91). A number of these subsequent studies, however, have been unable to show any evidence that these "*Metallogenium*"-like particles possess the molecules and ultrastructure required for life, including nucleic acids or membranes (35, 52, 57, 81). In one such study, a basidiomycetous fungus was described that formed "*Metallogenium*"-like particles when grown with Mn (23). When slide cultures were used to closely observe the formation of the particles, distinct cell-like structures were never visible, nor were any cells visible in the absence of Mn. Nucleic acid dyes did not bind to the particles, and serial thin sections showed no evidence for membranes or any definitive subcellular structure. The particles could be stained with Coomassie blue, suggesting that they may contain protein. Given the fact that microorganisms may excrete Mn- and Fe-oxidizing factors and the propensity for the binding of Fe(II) and Mn(II) to extracellular polymers, an alternative explanation for "*Metallogenium*" formation is that the binding of these metals by the extracellular polymers and subsequent oxidation causes them to fold into the distinctive shapes characteristic of "*Metallogenium.*" Unequivocal demonstration that "*Metallogenium*" is a living organism will require first an unambiguous generic description and then evidence that the organism possesses nucleic acids capable of replication and has a cellular ultrastructure that is consistent with prokaryotic cell structure.

BIOGEOCHEMICAL, PALEOBIOLOGICAL, AND EXTRATERRESTRIAL IMPLICATIONS

Because of many of the problems alluded to above, the biogeochemical cycles of Mn and Fe are less well understood than are those of other important inorganic electron donors and acceptors, such as S, N, and C. Virtually nothing is known about the in situ kinetics of microbially mediated iron oxidation at circumneutral pH. Furthermore, aside from the numerous field observations of iron bacteria, there are few quantitative data on the extent of these processes in nature or the abundance and diversity of the microbes responsible. For Mn there are more data. The importance of microbes directly mediating Mn oxidation in a number of environments is convincing (49, 90) and in some cases rate measurements have been made. Still, compared to other important cycles, the number of studies is small, and our knowledge of the organisms is far from complete. Simply from the point of view of these being important trace metals for the functioning of the Earth's ecosystems, more knowledge in these areas is desirable. For example, a recent microscopic study of water samples collected from fjords in Norway revealed bacteria with a variety of appendages that were encrusted with ferromanganic oxides (43). While it was estimated that these organisms made up only 0.5% of the total cell population, their ability to concentrate Fe and Mn suggested that they could deplete these essential trace nutrients in the water column. If similar organisms exist in open oceanic waters, they could have a bearing on the recent finding that large portions of the world's open oceans appear to be iron limited (59). Finally, it is understood that the different oxidation and mineral states of Fe and Mn can interact with one another, as well as with other important elements such as S and P (88); however, the biological and/or geochemical implications of these interactions in nature are barely recognized.

The Isua rock formations on the coast of Greenland are the oldest known rocks on Earth, dating back 3.85 billion years. Banded iron formations (BIFs) are clearly evident in these rocks, and recent carbon isotope evidence indicates that life was present when these BIFs formed (63). The role played by microorganisms in the formation of BIFs (which are a major source of mined iron) has been a source of ongoing debate. It has implications for the evolution of Earth's atmosphere, since it is assumed that the great quantities of free Fe(II) in the early anoxic Earth's ocean consumed any O_2 that was produced as a result of photosynthesis and maintained a reducing atmosphere (8). This, in turn, obviously influenced the evolution of multicellular life. Some arguments support a completely abiological deposition of BIFs involving UV photooxidation of Fe(II) in the anoxic ocean (31). Others suggest that life played an indirect role, in that the iron was oxidized as a result of oxygenic photosynthesis by cyanobacteria and subsequent Fe oxidation occurred spontaneously (98). The finding that anoxygenic photosynthesis can be coupled to Fe oxidation also presents a plausible scenario whereby the Fe oxidation could have occurred in a completely anoxic atmosphere (20, 98). Although this reaction requires light, the bacteria could have grown near the sea surface, precipitated Fe, and then rained down on the seafloor, resulting in large accumulations of Fe oxides. Of course, the presence of abundant microaerophilic, lithotrophic iron oxidizers could have also played a role in the deposition of BIFs, utilizing low concentrations of O_2 to oxidize Fe(II). These types of organisms are certainly capable of depositing significant quantities of iron in very localized environments on the present-day Earth (24), and their microfossils appear to be present in some BIFs (3). Clearly, the debate on BIF formation will remain active, and the recent microbiological discoveries clearly strengthen the hand of those who favor biogenically influenced BIF deposition.

In terms of extraterrestrial implications, it has been speculated that if life did exist at one time on Mars (the Red Planet) (60), iron oxidation might have been an important energetic mechanism for growth (77). In terms of finding hard evidence for ancient Martian life, this is especially intriguing in view of the signature morphologies that many Fe-oxidizing prokaryotes leave behind as mineralized remains.

PRACTICAL CONSIDERATIONS

Although this review has focused on the basic science behind the study of Fe- and Mn-oxidizing microbes, they are of significant economic importance. Iron and manganese oxidizers have long been recognized as important agents in biofouling. Their growth and, more importantly, their deposition of metals can cause restricted flow and even complete clogging of water distribution pipes (94; Ghiorse, Int. Symp. Biofouled Aquifers Prevention Restoration). They have also been implicated in corrosion of metals. One recent study showed that *L. discophora* promoted ennoblement of stainless steel, which is the initial process that results in pitting and then corrosion of the steel (14). On the beneficial side, it has been recognized in Europe that when water treatment plants promote the growth of Fe and Mn oxidizers, they can very efficiently remove Fe and Mn from municipal water supplies, eliminating both biofouling of distribution pipelines and discoloration of water used by the consumer (64). In addition, Fe oxides tend to be quite efficient at binding cells and so may aid in reducing the number of microbes in a given system (12).

CONCLUSIONS

The last decade of research on Fe and Mn oxidation has led to a number of significant discoveries. Real progress has been made in beginning to elucidate the molecules that are involved in Mn oxidation. The hints that both *Bacillus* sp. strain SG-1 and *L. discophora* may utilize multicopper oxidase-type enzymes as the mechanistic proteins for Mn oxidation presents an exciting possibility for further studies aimed at understanding the molecular basis for Mn oxidation. Ecologically, the importance of both Mn and Fe oxidation continue to emerge. The discovery of anaerobic and thermophilic Fe oxidizers has significantly altered our view of the biogeochemistry of iron cycling and provides a fascinating new aspect of microbial physiology to study. The isolation of novel microaerobic Fe oxidizers helps to reinforce the notion that microbes do play an important role in Fe oxidation at circumneutral pH. Many open questions and exciting research opportunities remain. These include a continued search for lithotrophic Mn oxidizers, more studies on the diversity and distributions of lithotrophic iron oxidizers, physiological and biochemical studies that elucidate the pathways of iron oxidation, and process studies that determine the cycling of iron and manganese between reductive and oxidative organisms.

REFERENCES

1. **Adams, L. F., and W. C. Ghiorse.** 1987. Characterization of extracellular Mn^{2+}-oxidizing activity and isolation of an Mn^{2+}-oxidizing protein from *Leptothrix discophora* SS-1. *J. Bacteriol.* **169:** 1279–1285.
2. **Adams, L. F., and W. C. Ghiorse.** 1986. Physiology and ultrastructure of *Leptothrix discophora* SS-1. *Arch. Microbiol.* **145:**126–135.
3. **Barghoorn, E. S., and S. A. Tyler.** 1965. Microorganisms from Gunflint chert. *Science* **147:**563–577.
4. **Beijerinck, M.** 1913. Oxydation des Mangancarbonates durch Bakterien. *Folia Microbiol. Delft* **2:** 123–134.
5. **Boogerd, F. C., and J. P. M. De Vrind.** 1987. Manganese oxidation by *Leptothrix discophora*. *J. Bacteriol.* **169:**489–494.
6. **Buchholz-Cleven, B. E. E., B. Rattunde, and K. L. Straub.** 1997. Screening for genetic diversity of isolates of anaerobic Fe(II)-oxidizing bacteria using DGGE and whole-cell hybridization. *Syst. Appl. Microbiol.* **20:**301–309.
7. **Caldwell, D. E., and S. J. Caldwell.** 1980. Fine structure of in situ microbial iron deposits. *Geomicrobiol. J.* **2:**39–53.
8. **Cloud, P.** 1973. Paleoecological significance of the banded iron-formation. *Econ. Geol.* **68:**1135–1143.
9. **Corstjens, P. L. A. M., J. P. M. De Vrind, P. Westbroek, and E. W. De Vrind-De Jong.** 1992. Enzymatic iron oxidation by *Leptothrix discophora*: identification of an iron-oxidizing protein. *Appl. Environ. Microbiol.* **58:**450–454.
10. **Corstjens, P. L. A. M., J. P. M. de Vrind, T. Goosen, and E. W. deVrind-de Jong.** 1997. Identification and molecular analysis of the *Leptothrix discophora* SS-1 *mofA* gene, a gene putatively encoding a manganese-oxidizing protein with copper domains. *Geomicrobiol. J.* **14:**91–108.
11. **Cowen, J. P., and M. W. Silver.** 1984. The association of iron and manganese with bacteria on marine macroparticulate material. *Science* **224:**1340–1342.
12. **Czekalla, C., W. Mevius, and H. Hanert.** 1985. Quantitative removal of iron and manganese by microorganisms in rapid sand filters (in situ investigations). *Water Supply* **3:**111–123.
13. **De Vrind-De Jong, E. W., P. L. A. M. Corstjens, E. S. Kempers, P. Westbroek, and J. P. M. De Vrind.** 1990. Oxidation of manganese and iron by *Leptothrix discophora*: use of N,N,N′,N′-tetramethyl-*p*-phenylenediamine as an indicator of metal oxidation. *Appl. Environ. Microbiol.* **56:** 3458–3462.

14. **Dickinson, W. H., F. Caccavo, Jr., B. Olesen, and Z. Lewandowski.** 1997. Ennoblement of stainless steel by the manganese-depositing bacterium *Leptothrix discophora. Appl. Environ. Microbiol.* **63:**2502–2506.
15. **Dorn, R. I., and T. M. Oberlander.** 1981. Microbial origin of desert varnish. *Science* **213:**1245–1247.
16. **Douka, C.** 1980. Kinetics of manganese oxidation by cell-free extracts of bacteria isolated from manganese concretions from soil. *Appl. Environ. Microbiol.* **39:**74–80.
17. **Dymond, J., R. W. Collier, and M. E. Watwood.** 1989. Bacterial mats from Crater Lake, Oregon and their relationship to possible deep-lake hydrothermal venting. *Nature* **342:**673–675.
18. **Ehrenberg, C. G.** 1838. *Gallionella ferruginea. Taylor's Scientific Mem.* **1:**402.
19. **Ehrenberg, C. G.** 1836. Vorlage mettheilungen ueber das wirklige vorkommen fossiler infusorien und ihre grosse verbreitung. *Poggendorfs Ann. Phys. Chem.* **38:**213–227.
20. **Ehrenreich, A., and F. Widdel.** 1994. Anaerobic oxidation of ferrous iron by purple bacteria, a new type of phototrophic metabolism. *Appl Environ. Microbiol.* **60:**4517–4526.
21. **Ehrlich, H. L., W. J. Ingledew, and J. C. Salerno.** 1991. Iron- and manganese-oxidizing bacteria, p. 147–170. *In* J. M. Shively and L. L. Barton (ed.), *Variations in Autotrophic Life.* Academic Press Inc., San Diego, Calif.
22. **Embley, R. W., W. W. Chadwick, I. R. Jonasson, D. A. Butterfield, and E. T. Baker.** 1995. Initial results of the rapid response to the 1993 CoAxial event: relationships between hydrothermal and volcanic processes. *Geophys. Res. Lett.* **22:**143–146.
23. **Emerson, D., R. E. Garen, and W. C. Ghiorse.** 1989. Formation of *Metallogenium*-like structures by a manganese-oxidizing fungus. *Arch. Microbiol.* **151:**223–231.
24. **Emerson, D., and N. P. Revsbech.** 1994. Investigation of an iron-oxidizing microbial mat community located near Aarhus, Denmark: field studies. *Appl. Environ. Microbiol.* **60:**4022–4031.
25. **Emerson, D., and N. P. Revsbech.** 1994. Investigation of an iron-oxidizing microbial mat community located near Aarhus, Denmark: laboratory studies. *Appl. Environ. Microbiol.* **60:**4032–4038.
26. **Emerson, D., and W. C. Ghiorse.** 1992. Isolation, cultural maintenance, and taxonomy of a sheath-forming strain of *Leptothrix discophora* and characterization of manganese-oxidizing activity associated with the sheath. *Appl. Environ. Microbiol.* **58:**4001–4010.
27. **Emerson, D., and W. C. Ghiorse.** 1993. Role of disulfide bonds in maintaining the structural integrity of the sheath of *Leptothrix discophora* SP-6. *J. Bacteriol.* **175:**7819–7827.
28. **Emerson, D.** 1989. Ph.D. thesis. Cornell University, Ithaca, N.Y.
29. **Emerson, D., and W. C. Ghiorse.** 1993. Ultrastructure and chemical composition of the sheath of *Leptothrix discophora* SP-6. *J. Bacteriol.* **175:**7808–7818.
30. **Emerson, D., and. C. L. Moyer.** 1997. Isolation and characterization of novel iron-oxidizing bacteria that grow at circumneutral pH. *Appl. Environ. Microbiol.* **63:**4784–4792.
31. **Francois, L. M.** 1986. Extensive deposition of banded iron formations was possible without photosynthesis. *Nature* **320:**352–354.
32. **Ghiorse, W. C.** 1984. Biology of iron- and manganese-depositing bacteria. *Annu. Rev. Microbiol.* **38:**515–550.
33. **Ghiorse, W. C., and S. C. Chapnick.** 1983. Metal-depositing bacteria and the distribution of manganese and iron in swamp waters, p. 367–376. *In* R. Hallberg (ed.), *Environmental Biogeochemistry*, vol. 35. Ecological Bulletin, Stockholm, Sweden.
34. **Ghiorse, W. C., and H. L. Ehrlich.** 1993. Microbial biomineralization of iron and manganese, p. 75–107. *In* R. W. Fitzpatrick and H. C. W. Skinner (ed.), *Iron and Manganese Biomineralization Processes in Modern and Ancient Environments.* Catena, Cremlingen-Destedt, Germany.
35. **Gregory, E., R. S. Perry, and J. T. Staley.** 1980. Characterization, distribution, and significance of *Metallogenium* in Lake Washington. *Microb. Ecol.* **6:**125–140.
36. **Hafenbrandl, D., M. Keller, R. Dirmeier, R. Rachel, P. Roβnagel, S. Burggraf, H. Huber, and K. O. Stetter.** 1996. *Ferroglobus placidus* gen. nov., sp. nov. a novel hyperthermophilic archaeum that oxidizes Fe^{2+} at neutral pH under anoxic conditions. *Arch. Microbiol.* **166:**308–314.
37. **Hallbeck, L., and K. Pederson.** 1991. Autotrophic and mixotrophic growth of *Gallionella ferruginea. J. Gen. Microbiol.* **137:**2657–2661.

38. **Hallbeck, L., F. Ståhl, and K. Pedersen.** 1993. Phylogeny and phenotypic characterization of the stalk-forming and iron-oxidizing bacterium *Gallionella ferruginea*. *J. Gen. Microbiol.* **139:**1531–1535.
39. **Hallbeck, L., and. K. Pedersen.** 1990. Culture parameters regulating stalk formation and growth rate of *Gallionella ferruginea*. *J. Gen. Microbiol.* **136:**1675–1680.
40. **Hanert, H. H.** 1992. The Genus *Gallionella*, p. 4082–4088. *In* H. G. Trüper, A. Balows, M. Dworkin, W. Harder, and K. H. Schleifer (ed.), *The Prokaryotes*, 2nd ed., vol. 4. Springer-Verlag, New York, N.Y.
41. **Hanert, H. H.** 1992. The genus *Siderocapsa* (and other iron- or manganese-oxidizing eubacteria), p. 4102–4113. *In* H. G. Trüper, A. Balows, M. Dworkin, W. Harder, and K. H. Schleifer (ed.), *The Prokaryotes*, 2nd ed., vol. 4. Springer-Verlag, New York, N.Y.
42. **Harder, E. C.** 1919. Iron-depositing bacteria and their geologic relations. *U.S. Geol. Surv. Prof. Pap.* **113:**7–89.
43. **Heldal, M., K. M. Fagerbakke, P. Tuomi, and G. Bratbak.** 1996. Abundant populations of iron and manganese sequestering bacteria in coastal water. *Aquat. Microb. Ecol.* **11:**127–133.
44. **Heldal, M., and O. Tumyr.** 1983. Gallionella from metaliminion in an eutrophic lake: morphology and X-ray energy-dispersive microanalysis of apical cells and stalks. *Can. J. Microbiol.* **29:**303–308.
45. **Horne, R. A.** 1978. *The Chemistry of Our Environment.* John Wiley & Sons, New York, N.Y.
46. **Ivarson, K. C., and M. Sojak.** 1978. Microorganisms and ochre deposits in field drains of Ontario. *Can. J. Soil Sci.* **58:**1–17.
47. **Jannasch, H. W., and M. J. Mottl.** 1985. Geomicrobiology of deep-sea hydrothermal vents. *Science* **229:**717–725.
48. **Jones, J. G.** 1986. Iron transformations by freshwater bacteria. *Adv. Microb. Ecol.* **9:**149–185.
49. **Juniper, S. K., and. B. M. Tebo.** 1995. Microbe-metal interactions and mineral deposition at hydrothermal vents, p. 219–253. *In* D. M. Karl (ed.), *The Microbiology of Deep-Sea Hydrothermal Vents.* CRC Press Inc., Boca Raton, Fla.
50. **Karl, D. M., A. M. Brittain, and B. D. Tilbrook.** 1989. Hydrothermal and microbial processes at Loihi Seamount, a mid-plate hot-spot volcano. *Deep-Sea Res.* **36:**1655–1673.
51. **Karl, D. M., G. M. McMurtry, G. M. Malahoff, and M. O. Garcia.** 1988. Loihi seamount, Hawaii: a mid-plate volcano with a distinctive hydrothermal system. *Nature* **335:**532–535.
52. **Klaveness.** 1977. Morphology, distribution and significance of the manganese-accumulating microorganism *Metallogenium* in lakes. *Hydrobiologia* **56:**25–33.
53. **Kucera, S., and R. S. Wolfe.** 1957. A selective enrichment method for *Gallionella ferruginea*. *J. Bacteriol.* **74:**344–349.
54. **Liang, L., J. A. McNabb, J. M. Paulk, B. Gu, and J. F. McCarthy.** 1993. Kinetics of Fe(II) oxygenation at low partial pressure of oxygen in the presence of natural organic matter. *Environ. Sci. Technol.* **27:**1864–1870.
55. **Lünsdorf, H., I. Brümmer, K. N. Timmis, and I. Wagner-Dobler.** 1997. Metal selectivity of in situ microcolonies in biofilms of the Elbe River. *J. Bacteriol.* **179:**31–40.
56. **Lutters-Czekalla.** 1990. Lithoautotrophic growth of the iron bacterium *Gallionella ferruginea* with thiosulfate or sulfide as energy source. *Arch. Microbiol.* **154:**417–421.
57. **Maki, J. S., B. M. Tebo, F. E. Palmer, K. H. Nealson, and J. T. Staley.** 1987. The abundance and biological activity of manganese-oxidizing bacteria and *Metallogenium*-like morphotypes in Lake Washington, USA. *FEMS Microbiol. Ecol.* **45:**21–29.
58. **Mandernack, K. W., J. Post, and B. M. Tebo.** 1995. Manganese mineral formation by bacterial spores of the marine *Bacillus*, strain SG-1: evidence for the direct oxidation of Mn (II) to Mn (IV). *Geochim. Cosmochim. Acta* **59:**4393–4408.
59. **Martin, J. H., and 42 others.** 1994. Testing the iron hypothesis in ecosystems of the equatorial Pacific Ocean. *Nature* **371:**1973–1802.
60. **McKay, D. S., E. K. Gibson Jr, K. L. Thomas-Keprta, H. Vali, C. S. Romanek, S. J. Clemett, X. D. F. Chillier, C. R. Maechling, and R. N. Zare.** 1996. Search for past life on Mars: possible relic biogenic activity in Martian meteorite ALH84001. *Science* **273:**924–930.
61. **Mendelssohn, I. A., B. A. Kleiss, and J. S. Wakeley.** 1995. Factors controlling the formation of oxidized root channels: a review. *Wetlands* **15:**37–46.

62. **Miyajima, T.** 1992. Production of *Metallogenium*-like particles by heterotrophic bacteria collected from a lake. *Arch. Microbiol.* **158:**100–106.
63. **Mojzsiz, S. J., G. Arrhenius, K. D. McKeegan, T. M. Harrison, A. P. Nutman, and C. R. L. Friend.** 1996. Evidence for life on Earth before 3,800 million years ago. *Nature* **384:**55–59.
64. **Mouchet, P.** 1992. From conventional fo biological removal of iron and manganese in France. *J. Am. Water Works Assoc.* **84:**158–167.
65. **Moyer, C. L., F. C. Dobbs, and D. M. Karl.** 1994. Estimation of diversity and community structure through restriction fragment polymorphism distribution analysis of bacterial 16S rRNA genes from a microbial mat at an active, hydrothermal vent system, Loihi Seamount, Hawaii. *Appl. Environ. Microbiol.* **60:**871–879.
66. **Moyer, C. L., F. C. Dobbs, and D. M. Karl.** 1995. Phylogenetic diversity of the bacterial community from a microbial mat at an active, hydrothermal vent system, Loihi Seamount, Hawaii. *Appl. Environ. Microbiol.* **61:**1555–1562.
67. **Mulder, E. G., and M. H. Deinema.** 1992. The sheathed bacteria, p. 2612–2624. *In* H. G. Trüper, A. Balows, M. Dworkin, W. Harder, and K. H. Schleifer (ed.), *The Prokaryotes*, vol. 2. Springer-Verlag, New York, N.Y.
68. **Nealson, K. H.** 1983. The microbial iron cycle, p. 159–190. *In* W. Krumbein (ed.), *Microbial geochemistry*. Blackwell Scientific, Boston, Mass.
69. **Nealson, K. H.** 1983. The microbial manganese cycle, p. 191–221. *In* W. Krumbein (ed.), *Microbial Geochemistry*. Blackwell Scientific, Boston, Mass.
70. **Nealson, K. H., B. M. Tebo, and R. A. Rosson.** 1988. Occurrence and mechanisms of microbial oxidation of manganese. *Adv. Appl. Microbiol.* **33:**279–318.
71. **Nealson, K. H., and J. Ford.** 1980. Surface enhancement of bacterial manganese oxidation: implications for aquatic environments. *Geomicrobiol. J.* **2:**21–37.
72. **Nelson, D. C., and H. W. Jannasch.** 1983. Chemoautotrophic growth of a marine *Beggiatoa* in sulfide-gradient cultures. *Arch. Microbiol.* **136:**262–269.
73. **Peck, S. B.** 1986. Bacterial deposition of iron and manganese oxides in North American caves. *Natl. Speleol. Soc. Bull.* **44:**26–30.
74. **Perfil'ev, B. V., and D. R. Gabe.** 1961. *Capillary Methods of Investigating Microorganisms.* Oliver & Boyd, Edinburgh, United Kingdom.
75. **Pringsheim, E. G.** 1949. Iron bacteria. *Biol. Rev.* **24:**200–245.
76. **Rawlings, D. E., and T. Kusano.** 1994. Molecular genetics of *Thiobacillus ferrooxidans. Microbiol. Rev.* **58:**39–55.
77. **Robbins, E. I., and A. S. Iberall.** 1991. Mineral remains of early life on Earth? On Mars? *Geomicrobiol. J.* **9:**51–66.
78. **Rogers, S. R., and J.J. Anderson.** 1976. Measurement of growth and iron deposition in *Spaerotilus discophorus. J. Bacteriol.* **126:**257–263.
79. **Rosson, R. A., and K. H. Nealson.** 1982. Manganese binding and oxidation by spores of a marine bacillus. *J. Bacteriol.* **151:**1027–1034.
80. **Schlesinger, W. H.** 1997. *Biogeochemistry: an Analysis of Global Change*, 2nd ed. Academic Press Inc., New York, N.Y.
81. **Schmidt, W. D., and J. Overbeck.** 1984. Studies of 'iron bacteria' from Lake Pluss. *Z. Allg. Mikrobiol.* **24:**329–339.
82. **Sheldon, S. P., and D. K. Skelly.** 1990. Differential colonization and growth of algae and ferromanganese-depositing bacteria in a mountain stream. *J. Fresh Water Ecol.* **5:**475–485.
83. **Siering, P. L., and W. C. Ghiorse.** 1997. PCR detection of a putative manganese oxidation gene (*mofA*) in environmental samples and assessement of *mofA* homology among diverse manganese-oxidizing bacteria. *Geomicrobiol. J.* **14:**109–125.
84. **Siering, P. L., and W. C. Ghiorse.** 1996. Phylogeny of the *Sphaerotilus-Leptothrix* group inferred from morphological comparisons, genomic fingerprinting, and 16S ribosomal DNA sequence analysis. *Int. J. Syst. Bacteriol.* **46:**173–182.
85. **Starkey, R. L.** 1945. Transformations of iron by bacteria in water. *J. Am. Water Works Assoc.* **37:** 963–984.
86. **Straub, K. L., M. Benz, B. Schink, and F. Widdel.** 1996. Anaerobic, nitrate-dependent microbial oxidation of ferrous iron. *Appl. Environ. Microbiol.* **62:**1458–1460.

87. **Straub, K. L., F. A. Rainey, and F. Widdel.** 1999. Isolation and characterization of marine phototrophic ferrous iron-oxidizing purple bacteria, *Rhodovulum iodosum* sp. nov. and *Rhodovulum robiginosum* sp. nov. *Int. J. Syst. Bacteriol.* **49**:729–735.
88. **Stumm, W., and J. J. Morgan.** 1981. *Aquatic Chemistry*, 2nd ed. John Wiley & Sons, Inc., New York, N.Y.
89. **Sunda, W. G., and D. J. Kieber.** 1994. Oxidation of humic substances by manganese oxides yields low-molecular-weight organic substrates. *Nature* **367:**62–64.
90. **Tebo, B. M., W. C. Ghiorse, L. G. van Waasbergen, P. L. Siering, and R. Caspi.** 1997. Bacterially-mediated mineral formation: insights into manganese (II) oxidation from molecular genetic and biochemical studies. *Rev. Mineral.* **35**:225–266.
91. **Tipping, E. A., J. G. Jones, and C. Woof.** 1985. Lacustrine manganese oxides: Mn oxidation states and relationships to "Mn depositing bacteria." *Arch. Hydrobiol.* **105:**161–175.
92. **Trolldenier, G.** 1988. Visualization of oxidizing power of rice roots and of possible participation of bacteria in iron deposition. *Z. Pflanzeneraehr. Bodenkd.* **151:**117–121.
93. **Tuhela, L., L. Carlson, and O. H. Tuovinen.** 1997. Biogeochemical transformations of Fe and Mn in oxic groundwater and well water environments. *J. Environ. Sci. Health* **A32:**407–426.
94. **Tyrrel, S. F., and P. Howsam.** 1994. Field observations of iron biofouling in water supply boreholes. *Biofouling* **8:**65–69.
95. **van Veen, W. L., E. G. Mulder, and M. H. Deinema.** 1978. The *Sphaerotilus-Leptothrix* group of bacteria. *Microbiol. Rev.* **42:**329–356.
96. **Van Waasbergen, L. G., J. A. Hoch, and B. M. Tebo.** 1993. Genetic analysis of the marine manganese-oxidizing *Bacillus* sp. strain SG-1L protoplast transformation, Tn*917* mutagenesis, and identification of chromosomal loci involved in manganese oxidation. *J. Bacteriol.* **175:**7594–7603.
97. **Van Waasbergen, L. G., M. Hildebrand, and B. M. Tebo.** 1996. Identification and characterization of a gene cluster involved in manganese oxidation by spores of the marine *Bacillus* sp. strain SG-1. *J. Bacteriol.* **178:**3517–3530.
98. **Walker, J. C. G., C. Klein, M. Schidlowski, J. W. Schopf, D. J. Stevenson, and M. R. Walter.** 1983. Environmental evolution of the Archaean-Early Proterozoic biosphere, p. 260–290. *In* J. W. Schopf (ed.), *Earth's Earliest Biosphere.* Princeton University Press, Princeton, N.J.
99. **Widdel, F., S. Schnell, S. Heising, A. Ehrenreich, B. Assmus, and B. Schink.** 1993. Ferrous iron oxidation by anoxygenic phototrophic bacteria. *Nature* **362:**834–836.
100. **Winogradsky, S.** 1888. Ueber Eisenbakterien. *Bot. Z.* **46:**262–270.
101. **Zavarzin.** 1992. The genus *Metallogenium*, p. 524–528. *In* H. G. Trüper, A. Balows, M. Dworkin, W. Harder, and K. H. Schleifer (ed.), *The Prokaryotes*, 2nd ed., vol. 2. Springer-Verlag, New York, N.Y.

Environmental Microbe-Metal Interactions
Edited by Derek R. Lovley

Chapter 3

Phylogenetic and Biochemical Diversity among Acidophilic Bacteria That Respire on Iron

Robert Blake II and D. Barrie Johnson

Iron is the fourth most abundant element in the lithosphere and the most abundant in the planet as a whole. Besides its highly reactive, pure metallic form (Fe^0), iron exists in two ionic states, ferrous (Fe^{2+}) and ferric (Fe^{3+}). The two ionic forms of iron differ greatly in their stabilities and solubilities. In oxidizing environments, free uncomplexed ferrous iron is thermodynamically unstable and oxidizes spontaneously to the ferric state; the rate at which this occurs is, however, mediated by pH and is far lower in acidic than in circumneutral environments. Under reducing conditions ($E_h < 0$ to +500 mV, depending on pH), ferrous iron is the more thermodynamically stable form, particularly in environments at pH $\leq$ 8 (12). Another significant difference is that whereas ferrous iron is relatively stable under reducing conditions, ferric iron is highly insoluble in most environments and hydrolyzes to form a variety of amorphous and crystalline products. The exceptions to this generalization are environments that are extremely acidic (typically, pH < 2.5), where ferric iron forms complexes such as $Fe(SO_4)^+$ that are relatively soluble. (55). For these reasons, acidophiles (defined in this chapter as microorganisms that show optimal growth at pH $\leq$ 3) inhabit environments in which the physicochemical behavior of iron is markedly different from that encountered by the majority of other life-forms on our planet. Soluble ionic iron is particularly abundant in extremely acidic environments that result from the oxidation of sulfidic minerals such as pyrite (FeS_2), the most abundant sulfide mineral. Many of these acidophilic environments are associated with human activities, particularly the mining of coal and metal ores, and they typically contain concentrations of soluble iron that far exceed those found in more "normal" environments.

The ferric/ferrous couple has a standard reduction potential (E_0') of +770 mV at pH 2 to 3, indicating that it could, in principle, be utilized by biological systems as both a source and a sink of electrons. This value is sufficiently greater than that of the nitrate/nitrite couple (E_0' = +420 mV) to ensure that acidophilic micro-

Robert Blake II • College of Pharmacy, Xavier University, New Orleans, LA 70125. ***D. Barrie Johnson*** • School of Biological Sciences, University of Wales, Bangor, United Kingdom.

organisms that use ferrous iron as an electron donor in energy-transducing reactions must necessarily use molecular oxygen (E_0' for the oxygen/water couple = +820 mV) as the electron sink:

$$2Fe^{2+} + 0.5O_2 + 2H^+ \rightarrow 2Fe^{3+} + H_2O \quad (1)$$

In higher-pH environments, there is a marked shift in the redox potential of the ferrous/ferric couple, so that values close to +200 mV are more appropriate at pH values close to neutrality (22). Anaerobic bacteria that couple ferrous ion oxidation to nitrate reduction have been isolated from such environments (69).

The free energy associated with the reaction in equation 1, −30 kJ mol^{-1} at 30°C, is sufficiently low, even by chemolithotrophic standards (48), that an energy metabolism based on the oxidation of ferrous iron might not seem the most attractive of microbial life-styles. On the other hand, ferrous iron (and associated reduced sulfur compounds) is often a more abundant and accessible energy resource in extremely acidic environments, where organic substrates are often present in very low concentrations. In addition, ferrous iron is a potentially renewable energy source in environments that contain sulfide (mineral waste heaps and biomining operations), since reduced iron may be regenerated abiotically as a result of ferric iron oxidation of minerals, as in equation 2:

$$FeS_2 + 6Fe^{3+} + 3H_2O \rightarrow 7Fe^{2+} + S_2O_3^{2-} + 6H^+ \quad (2)$$

In contrast, the relatively high reduction potential of the ferric/ferrous couple suggests that ferric iron could be an attractive alternative to molecular oxygen as an electron acceptor, particularly in acidic environments, where, as noted, Fe^{3+} is much more soluble and therefore more readily available to biological systems. The juxtaposition of aerobic and anaerobic microsites in acidic environments can promote a rapid cycling of iron between the ferrous and ferric forms (45).

The first isolate from an extremely acidic environment that was shown to catalyze the dissimilatory oxidation of ferrous iron was the chemolithotrophic bacterium *Thiobacillus ferrooxidans* (16). *T. ferrooxidans* has traditionally been the major focus of research in pure and applied aspects of microbial iron oxidation, although in recent times its presumed preeminent role in catalyzing sulfide mineral dissolution in industrial and environmental situations has been questioned, since other metal-mobilizing bacteria have often been shown to be more numerous and more active in many such situations (63). It is likely that the bulk of investigative research on acidophilic microorganisms will remain focused on acidophilic bacteria and archaea that perform the dissimilatory oxidation of ferrous iron rather than the dissimilatory reduction of ferric iron although reports of the latter have occasionally appeared in the literature (44). This trend reflects the environmental and economic importance of iron oxidizing acidophiles in accelerating the oxidative dissolution of sulfidic minerals. On the one hand, this bacterial activity is largely responsible for the chronic acid pollution of streams that issue from surface and underground mining operations. Acid mine drainage is formed by the weathering and oxidation of pyritic materials exposed during the mining of coal and base metal deposits and is a major source of water pollution wherever minerals are extracted for commercial gain. On the other hand, this bacterial activity solubilizes thousands of tons of commercially valuable metals from low-grade sulfide ores each year. The possibility of exploiting this bacterial activity to extract metals for commercial gain, a process known as bacterial leaching, has received increasing attention in recent years (62). One widespread application of bioleaching is its use in the pretreatment of refractory gold-bearing ores (54). It is estimated that gold worth $10 million to $50 million was recovered using biooxidation in 1988, rising to $2 billion to $3 billion in 1998 (64). Further, the global market for the bioleaching of base metals (mainly copper) and uranium amounted to $2 billion in 1988 and reached an estimated $8 billion by 1998.

Despite the widespread interest in both tempering the harmful effects of these bacteria and extending and improving the range of purposeful bioleaching activities, there is still much uncertainty regarding the bacteria involved and the complex biological events that result in aerobic respiration on iron. It is now clear that iron-metabolizing acidophilic microorganisms comprise a diverse range of prokaryotes that vary considerably in aspects of their physiologies, phylogenies, and biochemistries, as described below.

PHYLOGENETIC DIVERSITY AMONG BACTERIA THAT RESPIRE ON IRON

Acidophilic ferrous iron-oxidizing prokaryotes have frequently been categorized on the basis of their temperature optima for growth (59). Three groups have been described: mesophilic iron oxidizers, which have temperature optima of ca. 25 to 37°C, thermotolerant (or moderately thermophilic) iron oxidizers, which have temperature optima of ca. 40 to 60°C, and thermophilic iron oxidizers, which have temperature optima of >60°C (Table 1). Quite coincidentally, this approach also delineates microbial groups that have very different phylogenies and physiological characteristics (Table 1; Fig. 1), although this correlation is becoming less distinct with the discovery of new species and genera of iron-oxidizing acidophiles. Mesophilic iron-oxidizing acidophiles are principally gram-negative eubacteria, moderate thermophiles are gram-positive eubacteria, and thermophilic iron oxidizers are exclusively archaea. No truly psychrophilic iron-oxidizing acidophiles have been isolated, despite numerous efforts. Psychrotolerant iron-oxidizing acidophiles that have been isolated from low-temperature environments appear to be *Thiobacillus*-like gram-negative bacteria (53). A variety of obligately acidophilic eukaryotic life-forms (e.g., fungi, protozoa, and rotifers) often coexist with prokaryotes in metalliferous acidic environments (41) although there is no evidence that any eukaryotic acidophile can catalyze the oxidoreduction of iron.

Mesophilic Iron Oxidizers

Among the mesophilic iron-oxidizing acidophiles, by far the best-known bacterium is *T. ferrooxidans*, which has been the subject of numerous reviews (53). This organism has traditionally been considered to be an obligately iron- and sulfur-

Table 1. Acidophilic microorganisms that respire on iron

Species	Cell morphology	pH optimum for growth	Mol% G+C	Iron oxidation	Iron reduction
Mesophilic bacteria					
Thiobacillus ferrooxidans	Straight rods, variable motility	2.5	58–59	+	+
"*Leptospirillum ferrooxidans*"	Curved rods, highly motile	1.5–2.0	51–56	+	–
Thiobacillus prosperus	Straight rods, motile	2.0	63–64	+	ND[c]
Acidiphilium acidophilum	Straight rods, motile	2.5–3.0	63–64	–	+
Acidiphilium spp.	Straight rods, motile	ca. 3–4	60–70	–	+[a]
"*Ferrimicrobium acidophilus*"	Straight rods, motile	2.0–2.5	51–55	+	+
Sulfobacillus spp.	Sporulating rods, motile	1.5	ND	+	+
Thermotolerant bacteria					
Sulfobacillus thermosulfidooxidans	Sporulating rods, nonmotile	ca. 2	48–50	+	+
Sulfobacillus acidophilus	Sporulating rods, limited motility	ca. 2	55–57	+	+
Acidimicrobium ferrooxidans	Nonsporulating rods, motile	ca. 2	67–68.5	+	+
"*Leptospirillum thermoferrooxidans*"	Curved rods, highly motile	1.6–1.9	56	+	ND
Thermophilic archaea					
Sulfolobus metallicus	Cocci (irregular)	ca. 2–3	38	+	ND
Acidianus brierleyi	Cocci (irregular)	1.5–2.0	31	+	ND
Metallosphaera sedula	Cocci	ca. 2–3	45	+	ND
Sulfurococcus yellowstonii	Cocci (chains/ aggregates)	2.0–2.5	44.6	+	ND
Sulfolobus acidocaldarius[b]	Cocci (irregular)	3.0–3.5	37	–	+

[a]Not all strains reduce ferric iron.
[b]Optimal temperature, ca. 80°C.
[c]ND, not determined.

oxidizing chemolithotrophic aerobe, although it is now known that it can also grow by oxidation of hydrogen or formic acid, using oxygen as an electron acceptor, or as an anaerobe, using sulfur or formic acid as an electron donor and ferric iron as an electron acceptor (61). *T. ferrooxidans* has, however, a very limited capacity for mixotrophic or heterotrophic metabolism; addition of yeast extract to ferrous iron-oxidizing cultures typically results in no increase in growth rates or cell yields (D. B. Johnson, unpublished data). However, considerable care has to be exercised when drawing general conclusions from experiments that involve different strains of *T. ferrooxidans*, since these may display considerable physiological and phylogenetic variation. Also, difficulties that have been encountered in past work with culturing this iron-oxidizing acidophile (and others) have meant that the purity of cultures used in experimental work has sometimes been open to question, although advances in solid-medium design have essentially eliminated this potential problem (40).

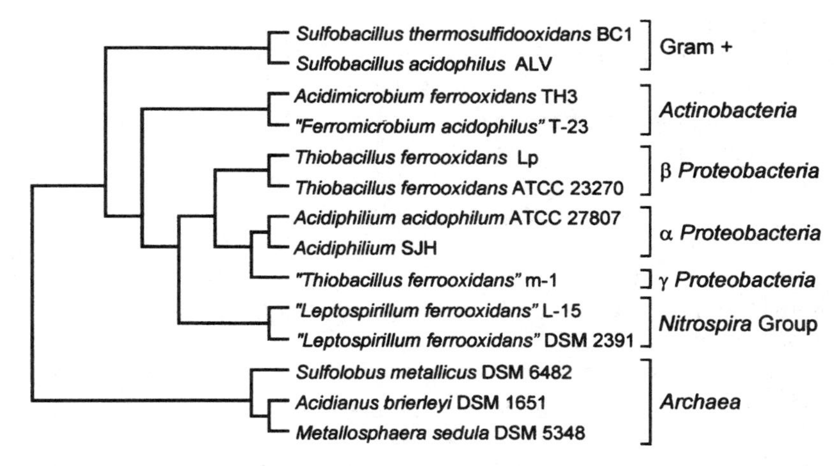

Figure 1. Phylogenetic tree based on a comparison of 16S rRNA gene sequences, showing the distribution on iron-metabolizing prokaryotes in the domains *Archaea* and *Bacteria*. With the exceptions of *Acidiphilium acidophilum* and *Acidiphilium* strain SJH, which catalyze the dissimilatory reduction of ferric iron, all the microorganisms shown are iron-oxidizing prokaryotes.

The phylogeny of bacteria labeled as *T. ferrooxidans* has been, and continues to be, a subject of considerable conjecture and debate. Apparent differences among the abilities of rod-shaped iron-oxidizing bacteria to oxidize reduced sulfur compounds resulted in the early designation of three bacterial species: *T. ferrooxidans*, *Ferrobacillus ferrooxidans*, and *Ferrobacillus sulfooxidans*. The nutritional basis for this differentiation was later shown to be not valid, and the three isolates reverted to a single species name, *T. ferrooxidans* (49). Heterogeneity among strains of *T. ferrooxidans* was once again noted by Harrison (31), who subdivided 23 strains of iron-oxidizing bacteria into seven groups on the basis of DNA-DNA homologies, although four of these groups contained single isolates only and a fifth (group 1) comprised strains of a quite separate iron-oxidizing acidophile, "*Leptospirillum ferrooxidans*" (described below). The ability of bacteria in some of the other groups to oxidize reduced sulfur (an obvious prerequisite for their inclusion in the genus *Thiobacillus*) was also apparently lacking, and chromosomal G+C base contents were also often well outside that (58 to 59 mol%) of "mainstream" *T. ferrooxidans* isolates. One such anomalous bacterium was *T. ferrooxidans* strain m-1, a rod-shaped iron-oxidizing acidophile isolated from coal mine waste, which apparently could not oxidize elemental sulfur and which had a chromosomal G+C base content of 65 mol%. The phylogenetic relationships of different strains of *T. ferrooxidans* and other acidophilic prokarotes were resolved by comparison of 16S rRNA base sequences (52). Six of the seven strains compared were found to form a cluster in the beta subdivision of the *Proteobacteria*, branching near to the beta-gamma division along with the sulfur-oxidizing acidophile *Thiobacillus thiooxidans*. The single exception to this trend was strain m-1, which occurred as a deep branch in the gamma subdivision of the *Proteobacteria*, confirming that isolate m-1 is incorrectly classified as a strain of *T. ferrooxidans*. The extent to which iron-oxidizing bacteria like strain m-1 are present in the environment and bioleaching operations is unknown, since strain m-1 is currently the only known isolate of its type. Other phylogenetic trees, based on more complete sequences of 16S rDNA (20, 27, 68), indicate that considerable phylogenetic heterogeneity exists among strains of *T. ferrooxidans* that group in the beta division of the *Proteobacteria*. For example, isolates have been noted to branch frequently on either side of the sulfur oxidizer *T. thiooxidans*. One strain that shows significant sequence disparity with most others is strain Lp (IFO 14245), an isolate that had also been found by Harrison (31) to have limited DNA homology to the type strain of *T. ferrooxidans*. This iron-oxidizing acidophile, isolated by W. W. Leathen from a coal mine in Pennsylvania, appears to be the organism originally labelled *Ferrobacillus ferrooxidans*. It is clear that the question of phylogenetic heterogeneity among bacteria labeled *T. ferrooxidans* needs to be reassessed.

A mesophilic halotolerant acidophile, *Thiobacillus prosperus*, was isolated from a geothermal marine environment by Huber and Stetter (34). Like *T. ferrooxidans*, it can use ferrous iron or reduced sulfur compounds as electron donors, although its G+C content (63 to 64 mol%) is significantly higher than that of *T. ferrooxidans* and isolates of *T. prosperus* have very little (6 to 14%) DNA-DNA homology to the type strain of *T. ferrooxidans*. *T. prosperus* is located in the gamma subdivision of the *Proteobacteria*.

"*Leptospirillum ferrooxidans*" is another chemolithophic iron-oxidizing acidophile, readily differentiated from *T. ferrooxidans* by its cell morphology (curved rods, occasionally forming spirilla [Fig. 2]) and its more noticeable motility; isolates may also form copious amounts of exopolymer, giving rise to floc formation in liquid media. Although it was first isolated and described in the early 1970s, the description and nomenclature of this acidophile have yet to be formalized. In the light of recent results from 16S rRNA gene sequencing (described below), this might be somewhat fortuitous and could avoid similar problems to those existing with strains of "*T. ferrooxidans*." Physiologically, "*L. ferrooxidans*" differs from *T. ferrooxidans* in being unable to oxidize reduced sulfur or to use ferric iron as an electron sink; ferrous iron appears to be the sole electron donor used by this acidophile, making it the archetypal iron bacterium. Growth of "*L. ferrooxidans*" is restricted to aerobic environments, since only oxygen can serve as a thermodynamically viable electron acceptor, for reasons discussed above. "*L. ferrooxidans*" is, however, capable of accelerating the oxidative dissolution of a number of sulfide minerals, presumably as a consequence of its production of ferric iron, which oxidizes the minerals abiotically and results in the regeneration of ferrous iron (equation 2). Other ways in which "*L. ferrooxidans*" contrasts with *T. ferrooxidans* include its greater tolerance of acidity, higher temperature optimum (of most strains), higher affinity for ferrous iron, and lesser sensitivity to ferric iron (Tables 1 and 2). These characteristics probably explain why "*L. ferrooxidans*" is often the more dominant iron-oxidizing mesophile in acid mine drainage waters and mineral-processing bioreactors and heaps, an observation made by research groups using both molecular and more traditional approaches (20, 28, 65, 72).

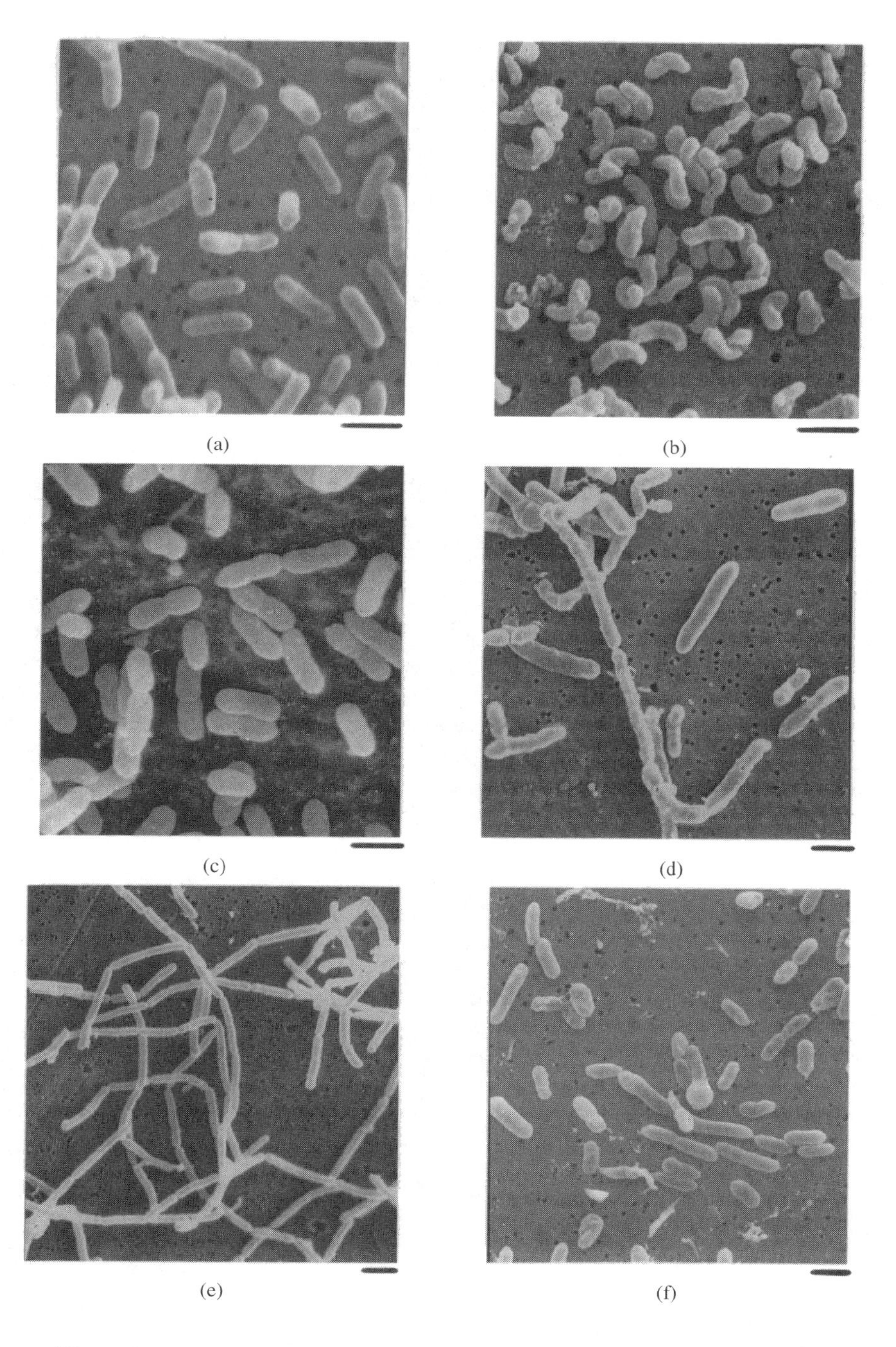

Figure 2. Scanning electron micrographs of some iron-oxidizing acidophilic bacteria. (a) *T. ferrooxidans*; (b) "*L. ferrooxidans*"; (c) "*F. acidophilus*"; (d) *S. acidophilus*; (e) *A. ferrooxidans*; and (f) a mesophilic *Sulfobacillus* sp. (geothermal spring, Montserrat). Bars, 1 μm.

Table 2. Kinetic data relating to ferrous iron oxidation by acidophilic microorganisms[a]

Strain	Temp (°C)	Specific rate of Fe^{2+} oxidation[b]	K_m (mM Fe^{2+})	K_i (mM Fe^{3+})
T. ferrooxidans				
(DSM 538)	30		1.34 ± 0.16	3.10 ± 0.18
(ATCC 23270)	30	559.6 ± 34.6		
"*L. ferrooxidans*"				
BC	30		0.25 ± 0.08	42.8
CF12	30	422.0 ± 19.9		
"*F. acidophilus*" T23	30	214.7 ± 14.1	ND[c]	ND
S. thermosulfido-oxidans TH1	50	ND	1.04 ± 0.08	2.74 ± 0.13
S. acidophilus				
ALV	50	ND	2.96 ± 0.68	1.13 ± 0.34
YTF1	50	276		
A. ferrooxidans TH3	50	ND	0.47 ± 0.10	1.89 ± 0.39
S. metallicus BC	65	ND	0.56 ± 0.03	1.65

[a]Data from references 1, 9, and 57.
[b]Micrograms of Fe^{2+} oxidized per minute per milligram of protein.
[c]ND, not determined.

"*L. ferrooxidans*" is, phylogenetically, very distant from all other acidophilic bacteria. The closest relationship appears to be with the magnetotactic bacterium "*Magnetobacterium bavaricum*," the sulfate-reducing bacterium *Thermodesulfovibrio yellowstonii*, and the nitrite-oxidizing *Nitrospira* spp. (20). The "*Leptospirillum*" group is related only moderately to all other members of the domain *Bacteria* and is thought to represent a distinct phylum. Comparison of the 16S rRNA genes of the relatively few strains and clones of "*L. ferrooxidans*" that have been sequenced indicates that the original Armenian isolate (strain L-15) may be an atypical strain (Fig. 1) and that the "*L. ferrooxidans*" cluster includes at least two species (or even two genera) of bacteria (20, 65).

Other iron-oxidizing mesophilic acidophiles with distinct physiological and/or biochemical properties, such as strains Funis (8) and T3.2 (19), have been described. One interesting group of novel iron-oxidizing acidophiles differs from both *T. ferrooxidans* and "*L. ferrooxidans*" in being obligately heterotrophic. One such isolate (CCH7) was a filamentous bacterium that formed streamerlike growths in liquid media (43); the phylogeny of this bacterium has not yet been established. More recently, unicellular iron-oxidizing heterotrophic bacteria have been isolated from a wide range of acidic environments (42, 46, 73). Rates of ferrous iron oxidation by these novel bacteria are comparable to those by the chemolithotrophic mesophiles (Table 2). The name "*Ferromicrobium acidophilus*" has been proposed for these bacteria, although 16S rDNA analysis has found significant phylogenetic diversity among isolates, indicating that bacteria with these physiological characteristics encompass more than a single bacterial species. The type strain of "*F. acidophilus*" (strain T-23) branches close to the cusp between gram-negative and gram-positive bacteria within the *Actinobacteria*. The closest known association is with the thermotolerant iron-oxidizing acidophile *Acidimicrobium ferrooxidans* (Fig. 1), although the overall sequence similarity (67%) indicates that the two are related only distantly (46). It appears, therefore, that the biodiversity of mesophilic iron-oxidizing acidophiles is considerably greater than was once supposed. Many novel isolates, including gram-positive spore-forming bacteria (Fig. 2), that have been isolated from weathering regoliths in the United States and acidic sites on the island of Montserrat (73) have yet to be fully characterized, although some isolates have been noted to oxidize sulfide minerals in extremely acidic ($pH < 1$) laboratory media.

Thermotolerant Iron Oxidizers

Thermotolerant (or moderately thermophilic) iron-oxidizing acidophiles differ from the majority of mesophiles in several important respects in addition to their temperature response characteristics (Table 1). First, they are gram-positive eubacteria, many of which form endospores. Second, isolates often display considerable metabolic versatility and may grow autotrophically (on ferrous iron or reduced sulfur), mixotrophically (e.g., in glucose-ferrous iron medium, where both glucose and CO_2 act as carbon sources), heterotrophically (e.g., in media containing glucose and ferric iron), and chemolithoheterotrophically (e.g., in ferrous iron-yeast extract medium, where ferrous iron acts as an energy source and yeast extract acts as a carbon source). However, different strains of thermotolerant iron oxidizers vary in their preponderance for autotrophy and heterotrophy (25).

Thermotolerant iron-oxidizing acidophiles have been isolated from a variety of sites, including geothermal springs and pools, bioleaching heaps, and self-heating coal spoils. Two genera (*Sulfobacillus* and *Acidimicrobium*) and three species (*Sulfobacillus thermosulfidooxidans*, *Sulfobacillus acidophilus*, and *Acidimicrobium ferrooxidans*) are currently recognized. The type strain (VKM B-1269; DSM 9293) of *S. thermosulfidooxidans* was isolated and described by Golovacheva and Karavaiko (30). Other isolates (strains TH1 and BC1) have been isolated from a geothermal spring in Iceland and a self-heating coal spoil in England, respectively. *S. thermosulfidooxidans* differs from *S. acidophilus* by its lower G+C content and its poor autotrophic growth on elemental sulfur (30). *S. acidophilus* appears to be widely distributed in geothermal environments and is readily isolated on agarose-gelled media; designated strains include NAL (the type strain) ALV, THWX, and N (58). The thermotolerant acidophile *A. ferrooxidans* is also a gram-positive eubacterium, although endospore production has not been observed in the two designated strains, ICP (the type strain), isolated from a geothermal pool in Iceland, and TH3 (the original isolate of this species), isolated from a copper leaching dump (34). Strain TH3 (and another *A. ferrooxidans*-like isolate, strain HPTH) (25) shows characteristic filamentous growth, whereas strain ICP occurs typically as single or paired rods. *A. ferrooxidans* is also readily differentiated from *Sulfobacillus* spp. by its greater chromosomal G+C content (Table 1). Strain TH3 was once thought to be obligately heterotrophic, although it is now known that both *A. ferrooxidans* strains are capable of autotrophic growth on ferrous iron. They appear to possess inducible, high-affinity mechanisms for uptake of carbon dioxide, in contrast to *Sulfobacillus* spp., which tend to show poor autotrophic growth when cultured under atmospheric concentrations of carbon dioxide (58). Mixed cultures of *A.*

ferroooxidans and either *Sulfobacillus* sp. are more efficient at oxidizing ferrous iron than are pure cultures of any of the three moderate thermophiles (14).

All three species of iron-oxidizing thermotolerant acidophiles are positioned in the gram-positive eubacterial division, *A. ferrooxidans* (strain TH3), as a deep-branching member of the high G+C division (within the class *Actinobacteria* and *Sulfobacillus* spp., strains BC1 and ALV) close to the root of the high-G+C/low-G+C division (52). There has been some debate over the phylogeny of the type strain of *S. thermosulfidooxidans*. A sequence published by Tourova et al. (70) clustered, somewhat surprisingly, with those of *Alicyclobacillus* spp. (obligately heterotrophic, non-iron-oxidizing thermotolerant acidophiles) rather than with other *Sulfobacillus* spp., including *S. thermosulfidooxidans* strain BC1. In contrast, the 16S rRNA gene from the type strain of *S. thermosulfidooxidans* sequenced by Durand (21) showed a high degree of similarity to those of other *Sulfobacillus* spp. It has now been acknowledged that the earlier sequence (70) was inaccurate, possibly as a result of strain misidentification or contamination of the laboratory culture.

A quite distinct thermotolerant iron-oxidizing acidophile, isolated and described by Golovacheva et al. (29), was a *Leptospirillum*-like bacterium, which differed from the mesophilic "*L. ferrooxidans*" in its higher temperature optimum and greater chromosomal G+C content (Table 1); DNA-DNA homology between the thermotolerant and mesophilic *Leptospirillum* spp. was 26.7%. The novel isolate, with the proposed name "*Leptospirillum thermoferrooxidans*," has apparently been lost, and other strains of this type await isolation.

Thermophilic Iron Oxidizers

All thermophilic iron-oxidizing acidophiles that have been described are archaea rather than eubacteria. They also differ from their mesophilic and thermotolerant counterparts in their cell morphologies; isolates are small (1 to 2 μm) cocci rather than rods, although irregular spheres are characteristic of some strains. While a number of different species of thermophilic archaea are known to catalyze the oxidoreduction of sulfur, a more restricted range is capable of the dissimilatory oxidation of ferrous iron (Table 1). Those that have this ability can promote rapid and efficient extraction of metals from sulfide minerals although there is as yet no commercial industrial bioreactor that makes use of thermophilic metal-mobilizing archaea.

Some confusion has arisen regarding the mineral-oxidizing ability of *Sulfolobus acidocaldarius*, possibly resulting from the inadvertent use of mixed cultures. The type strain of *S. acidocaldarius* (ATCC 33909) has apparently no ability to oxidize iron or reduced sulfur and is an obligate heterotroph. The isolate named *S. acidocaldarius* strain BC (which was isolated from same self-heating coal spoil as *Sulfobacillus thermosulfidooxidans* strain BC1, and which has been used extensively in mineral leaching studies) and *Sulfolobus* strain LM (an Icelandic isolate) are both probably strains of a separate species, *Sulfolobus metallicus* (Paul Norris, University of Warwick, personal communication). The type strain of *S. metallicus* was also isolated from a solfataric field in Iceland (35). This archaeon is an obligate chemolithotroph, able to oxidize ferrous iron and reduced sulfur and to oxidize sulfidic ores, and it is also described as being obligately aerobic. In contrast, *Acidianus brierleyi* displays considerably greater metabolic versatility. It can grow chemolithotrophically on ferrous iron, as well as mixotrophically or heterotrophically in the presence of organic substrates (67). *A. brierleyi* may also grow either aerobically or anaerobically by using elemental sulfur as an electron acceptor, although ferrous iron oxidation only occurs under aerobic conditions. Two other iron-oxidizing thermophilic archaea are known: *Metallosphaera sedula* (and the closely related *Metallosphaera prunae*), which was isolated from a solfataric field in Italy (33), and *Sulfurococcus yellowstonensis*, which was isolated from a geothermal spring in Yellowstone National Park. Both are facultative chemolithotrophs that have negligible DNA-DNA homologies to each other or to other iron-oxidizing archaea, and they may also be differentiated from *Sulfolobus and Acidianus* spp. by their higher chromosomal G+C contents (Table 1).

From analysis of 16S rDNA base sequences, the thermoacidophilic iron-oxidizing archaea have been assigned to the order *Sulfolobales*, which occurs as a monophyletic group within the *Crenarchaeota* branch of the domain *Archaea* (23). *Metallosphaera* spp. occur as a phylogenetically well-defined cluster within the *Sulfolobales*, whereas *A. brierleyi* displays a surprisingly large phylogenetic distance from other (non-iron-oxidizing) *Acidianus* spp. and branches more closely to *Metallosphaera* (23). A significant phylogenetic distance also occurs between *Sulfolobus* spp., possibly reflecting differences in their metabolisms noted above, and this has prompted the suggestion that this particular genus should be reclassified (23). No sequence data have been published for *Sulfurococcus yellowstonensis*.

Iron-Reducing Acidophiles

The ability to reduce ferric iron to ferrous ion appears to be fairly widespread among acidophilic bacteria. Thermoacidophilic archaea have been less well studied in this respect, although an early report (11) showed that *S. acidocaldarius* could reduce ferric iron when growing on organic substrates. Both chemolithotrophic and heterotrophic acidophilic bacteria catalyze the dissimilatory reduction of ferric iron, often in reactions that are coupled to energy conservation (61). Strictly anoxic conditions are generally not required, and iron reduction may be most rapid under microaerobic conditions (i.e., reduced oxygen tensions). Mixed cultures of iron-oxidizing and iron-reducing acidophiles that experience fluctuations in concentrations of dissolved oxygen will cycle iron between the ferrous and ferric states (45).

Both *T. ferrooxidans* and *T. thiooxidans* can couple the oxidation of reduced sulfur compounds to the reduction of ferric iron, although only *T. ferrooxidans* is capable of growth under strictly anoxic conditions (11, 18, 61). The mixotrophic acidophile *Acidiphilium acidophilum* (formerly *Thiobacillus acidophilus)* can reduce ferric iron when using organic electron donors (such as glucose) during microaerobic growth, a trait that has also been noted in a large number of obligately heterotrophic acidophilic bacteria. Thermotolerant iron-oxidizing bacteria (both *Sulfobacillus* and *Acidimicrobium* spp.) may also reduce ferric iron to ferrous iron under conditions of low oxygen tension. These bacteria are facultative anaerobes that are capable of growth under anoxic conditions with either an organic or (in the case of *Sulfobacillus* spp.) an organic or inorganic electron donor and ferric iron as the sole electron acceptor (10).

BIOCHEMICAL DIVERSITY AMONG IRON RESPIRATORY ELECTRON TRANSPORT CHAINS

In principle, aerobic respiration on ferrous ions should represent a simple biological system. Thermodynamic and chemiosmotic constraints dictate that the electron transport chain responsible for the transfer of electrons from iron to molecular oxygen be relatively short (36). Furthermore, studies on these respiratory chains should be facilitated by the unusually high concentrations of redox components present (37, 71). Disadvantages arise from low growth yields and hence large volumes of expensive and corrosive media that require processing. Nonetheless, aerobic respiration on ferrous ions represents an ideal experimental system with which to seek new insights on energy coupling in biological systems.

An immediate and striking feature of cell extracts derived from bacteria grown by aerobic respiration on iron is the rich and varied color of the extracted material. Thus, cell-free extracts of *T. ferrooxidans*, "*L. ferrooxidans*," and *S. thermosulfidooxidans* are deep blue, red, and bright yellow, respectively. These colors correspond to the conspicuous redox-active biomolecules expressed by each organism as it respires aerobically on ferrous ions. As part of a comparative study to identify and characterize the prominent redox-active biomolecules that might participate in the initial reactions of the respiratory chain in each organism, a standard protocol was developed and adopted to prepare cell extracts from iron-grown bacteria under strongly acidic conditions (7). This protocol was designed with two factors in mind. First, like all chemolithotrophs that utilize substrates of high reduction potential, each iron-oxidizing organism requires a high throughput of Fe(II) to obtain sufficient energy for growth. This requirement necessitates that the relevant electron transfer components be expressed in high yield during growth on Fe(II). Second, there is no electron microscopic or other evidence that bulk ferrous ions must enter the cytoplasm of the bacterial cell during the respiratory process. It is therefore assumed that the initial electron transfer reaction from a soluble ferrous ion to a cell-associated biomolecule must occur outside the cell in the acidic milieu required by these obligate acidophiles. Attention must thus be focused on acid-stable, redox-active biomolecules that appear to be present in relative abundance in iron-grown cells.

Thiobacillus ferrooxidans

Any description of the iron respiratory chain of *T. ferrooxidans* must account for the presence of rusticyanin. Rusticyanin is an acid-stable, acid-soluble type I copper protein that may constitute as much as 5% of the total soluble protein synthesized by this bacterium when cultured autotrophically on ferrous ions. Cell extracts of iron-grown *T. ferrooxidans* prepared at pH 2.0 contain sufficient rusticyanin to give the extract an unmistakable blue color even before centrifugation and removal of the cell fragments. The blue protein is readily purified to electrophoretic homogeneity using standard biochemical techniques (4, 17). The visible absorbance spectrum of the purified, oxidized rusticyanin is shown by the solid line in Fig. 3. The absorbance peak at 595 nm is accompanied by two less intense peaks on either side, one at about 445 nm and one at about 740 nm. All three visible absorbance peaks in the oxidized rusticyanin spectrum disappear upon one-electron reduction of the protein, as shown by the dashed line in Fig. 3.

T. ferrooxidans is representative of a large number of mesophilic iron-oxidizing bacteria that express rusticyanin and/or *c*-type cytochromes in conspicuous quantities when cultured on ferrous ions (2). The absorbance spectra in Fig. 4 depict the relative amounts of each type of biomolecule expressed by representative members of well-defined subgroups within this category. Figure 4A shows the absorbance spectra of a cell extract derived from *T. ferrooxidans* ATCC 23270. A principal and striking feature of cell extracts derived from bacteria in this subclass (which includes all strains of *T. ferrooxidans* currently available from the American Type Culture Collection, as well as numerous other strains) is the deep blue color due to the presence of abundant levels of oxidized rusticyanin. The loss in absorbance at 600 nm upon the addition of ferrous ions to the oxidized extract can be taken as a reliable measure of the amount of rusticyanin present in the extract, as illustrated by the large absorbance difference shown in the inset of Fig. 4A.

Figure 4B shows the absorbance spectra of a cell extract derived from strain JWC. Strain JWC is a nonmotile, gram-negative rod that is morphologically indistinguishable from the type strain of *T. ferrooxidans*. Heterologous DNA hybridization experiments indicated that strain JWC showed only 40% homology to the type strain of *T. ferrooxidans* (G. T. Howard, and R. C. Blake, unpublished data). Cell extracts derived from strain JWC and other bacteria within this subclass (which included numerous strains isolated from the southwestern United States and northern Mexico) were bright yellow and thus readily distinguishable from those derived

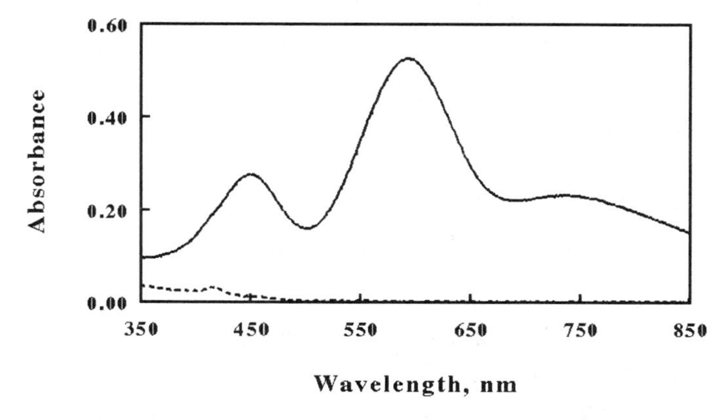

Figure 3. Absorbance spectra of oxidized (solid line) and reduced (dashed line) rusticyanin from *T. ferrooxidans*. The absorbance spectrum of the reduced rusticyanin was determined 10 min after the sample of oxidized rusticyanin was mixed with excess ferrous sulfate.

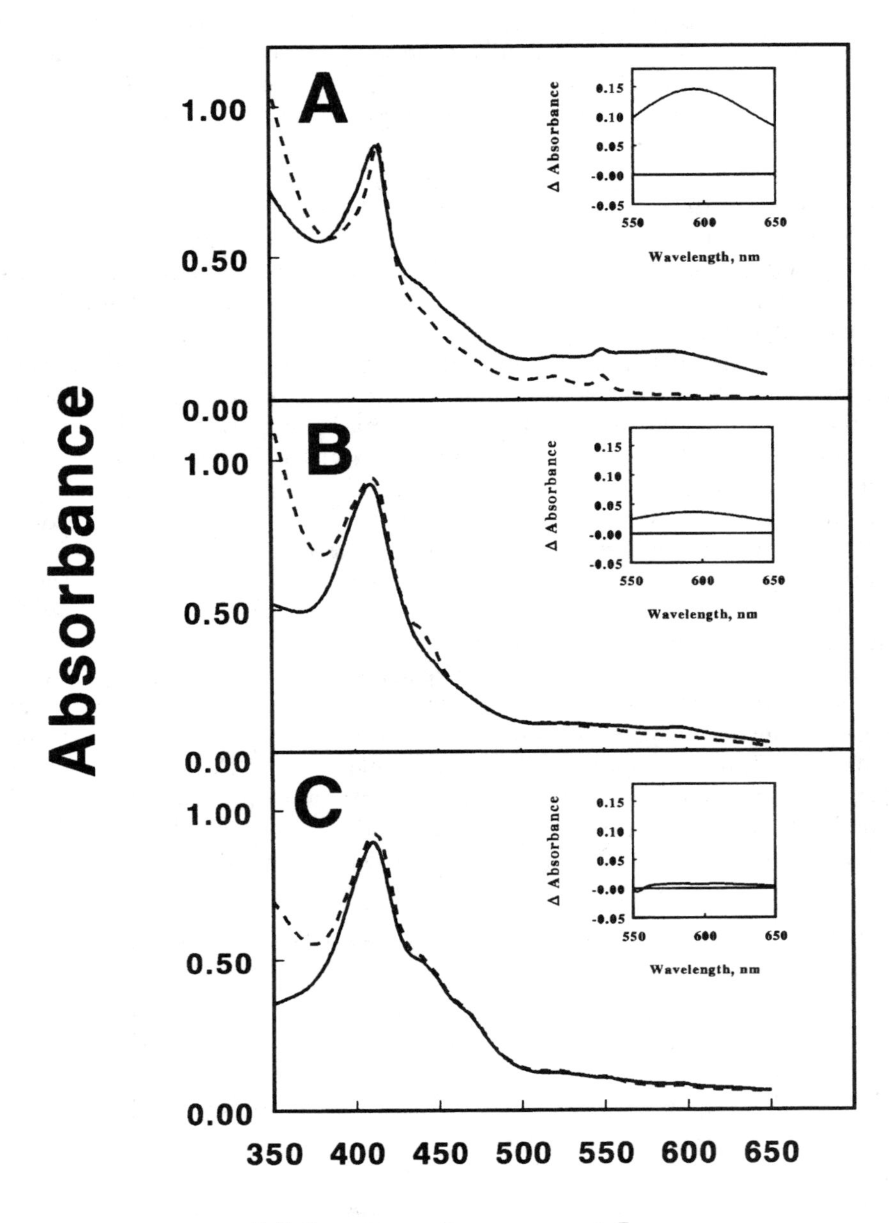

Figure 4. Absorbance spectra of oxidized (solid line) and reduced (dashed line) cell extracts from *T. ferrooxidans* ATCC 23270 (A), strain JWC (B), and *T. prosperus* DSM 5130 (C). Each inset shows a difference spectrum in the far-visible region that represents the absolute spectrum of the oxidized extract minus that of the Fe(II)-reduced extract.

from the type strain of *T. ferrooxidans*. Spectral analyses of cell extracts from bacteria represented by strain JWC revealed that far less blue copper protein is expressed by these organisms. The inset of Fig. 4B shows that the amount of holorusticyanin detected in cell extracts of strain JWC is far smaller than that detected in *T. ferrooxidans*.

Figure 4C shows the absorbance spectra of a cell extract derived from *T. prosperus*. *T. prosperus* is a motile, mesophilic iron-oxidizing bacterium that exhibits very little DNA homology to *T. ferrooxidans* (34). The extract derived from iron-grown *T. prosperus* was a deep golden yellow, which was distinguishable from the bright yellow extracts from bacteria represented by strain JWC. As indicated by the lack of absorbance in the difference spectrum in Fig. 4C, no spectroscopic evidence for a blue copper protein could be detected in cell extracts of *T. prosperus*. Similar observations were made with cell extracts derived from strain CF27, a gram-negative mesophile isolated from an abandoned cobalt mine in Idaho. It is possible that both strains do express a rusticyanin-like protein but at levels insufficient for detection by the spectroscopic methods used thus far.

Comparisons such as these among the organisms represented in Fig. 4 underscore the necessity for caution when evaluating published investigations of the molecular mechanism(s) of respiratory iron oxidation. In view of the diversity that is apparent in the levels of rusticyanin expressed by physiologically and morphologically similar bacteria (bacteria that nonetheless exhibit little or no homology in heterologous DNA hybridization experiments), it would perhaps not be surprising were we to learn that slightly different electron transfer mechanisms and pathways exist within each distinct group of bacteria that express a common conspicuous redox-active biomolecule such as the rusticyanin. Indeed, different investigators who work with different strains of *T. ferrooxidans* support molecular models of the iron respiratory chain that differ considerably in certain details.

As an example, consider the putative role of rusticyanin in the electron transport chain. If one focuses on organisms such as ATCC 23270, the evidence in support of the hypothesis that rusticyanin must be an important component of the iron respiratory chain is quite compelling. Not only does rusticyanin constitute as much as 5% of the total soluble protein synthesized by iron-grown cells (15, 17), but also the synthesis of the protein is repressed when *T. ferrooxidans* is grown solely on reduced sulfur compounds and induced when such sulfur-grown cells are subsequently exposed to soluble iron (39, 50). The remarkable acid stability of the protein is consistent with the evidence that the rusticyanin is expressed into the periplasmic space of these gram-negative bacteria, where the initial electron transfer from iron to a cellular component is thought to occur (38). The standard reduction potential of the purified protein, 680 mV (37), is compatible with a role in a respiratory chain where the physiological electron donor, sulfatoiron(II), has a potential no lower than 650 mV (36). Furthermore, the rate of the Fe(II)-dependent reduction of rusticyanin in crude cell extracts is sufficiently rapid to be of physiological significance (2).

Possible physiological electron transfer partners for the rusticyanin have been proposed (5). An acid-stable, highly aggregated *c*-type cytochrome that served to catalyze the iron-dependent reduction of the purified rusticyanin was partially purified from cell extracts of *T. ferrooxidans*. Catalysis by this iron:rusticyanin oxidoreductase occurred only in the presence of sulfate or selenate out of a total of 14 different anions investigated, an anion specificity identical to that exhibited by the intact bacterium for the Fe(II)-dependent reduction of molecular oxygen. Furthermore, a kinetic analysis of the cytochrome *c*-catalyzed electron transfer reaction produced values of V_{max} and a K_m for Fe(II) consistent with those obtained with intact cells for the overall aerobic respiratory process. On the basis of these favorable comparisons between the behavior patterns of isolated biomolecules and those of whole cells, this iron:rusticyanin oxidoreductase was postulated to be the physiological donor for electrons to the rusticyanin.

The physiological electron acceptor for electrons from rusticyanin was postulated to be an acid-stable c_4-type cytochrome isolated from cell extracts of *T. ferrooxidans* (13, 26). The purified cytochrome contained two *c*-type cytochromes per single polypeptide chain of 20,000 Da. Rusticyanin was required for the Fe(II)-dependent reduction of the cytochrome c_4. Furthermore, the standard reduction potential of the rusticyanin was lowered 150 mV to +530 mV in the presence of this cytochrome c_4, indicating that the two proteins were involved in a specific interaction. It is clearly of interest to determine whether there is a direct interaction between this cytochrome c_4 and a terminal cytochrome oxidase in *T. ferrooxidans*.

Another group of investigators, working with an entirely different strain of *T. ferrooxidans*, has offered experimental evidence for a different working hypothesis that provides no significant role for the rusticyanin (74). Instead, the electron transport chain from Fe(II) to molecular oxygen is seen to be composed of, in order of electron flow, an iron-sulfur protein, an acid-stable *c*-type cytochrome, and an *a*-type cytochrome oxidase. All three electron transfer components were purified to electrophoretic homogeneity (24, 47, 66, 74). The gene for the iron-sulfur protein was cloned, and the primary structure of the 6,400-Da protein was deduced from the DNA sequence of the open reading frame in the cloned DNA (51).

The different working hypotheses noted above may represent investigator-dependent differences in data interpretation that arise as an inevitable consequence of the revisionary nature of hypothesis-driven investigations, or they may represent actual differences in the respective biological systems. One other point that may bear on the role of rusticyanin in the energy metabolism of *T. ferrooxidans* concerns the hypothesis that aporusticyanin mediates the specific adhesion of the organism to pyrite (3, 60). Purified recombinant aporusticyanin was shown to adhere to different sulfide minerals with a pattern of reactivity identical to that exhibited by intact cells of *T. ferrooxidans*. Furthermore, preincubation of the mineral with aporusticyanin served to inhibit the adhesion of intact cells to the mineral. Holorusticyanin did not adhere to any of the minerals, nor did it influence the adhesion of intact cells. These observations are consistent with a model where aporusticyanin located on the outer surface of the bacterial cell acts as a mineral-specific receptor for the initial adhesion of *T. ferrooxidans* to the solid.

The latter observation raises a number of interesting questions. Is the true physiological role of rusticyanin that of a cellular adhesion protein rather than an electron transfer protein? That seems unlikely, given the large quantities of holorusticyanin present in some strains of *T. ferrooxidans*. What about those organisms such as strain JWC and *T. prosperus*, where the concentration of holorusticyanin is considerably lower? Do those organisms simply produce proportionately greater quantities of the spectrophotometrically invisible aporusticyanin? Is the iron respiratory chain actually composed of multiple electron transfer pathways where holorusticyanin is but one component in one branch of the pathway? What about the iron-oxidizing bacteria that appear to express no holorusticyanin whatsoever (see below)? Do they still express an aporusticyanin for binding to iron-bearing minerals? A reevaluation of the occurrence and quantity of both holo- and aporusticyanin in many strains of iron-oxidizing bacteria is clearly warranted.

"*Leptospirillum ferrooxidans*"

Neither cytochromes *c* nor rusticyanin-like molecules could be detected in cell-free extracts of "*L. ferrooxidans*" grown autotrophically on ferrous ions. Instead, iron-grown "*L. ferrooxidans*" expressed abundant levels of an acid-stable, acid-soluble novel red cytochrome, whose absorbance properties are illustrated in Fig. 5. The oxidized cytochrome exhibited a broad absorbance peak at 426 nm (dashed line). Electrochemical reduction of the cytochrome by Fe(II) at pH 2.0 produced new absorbance peaks at 441 and 579 nm, respectively, as shown by the solid line in Fig. 5. The inset in Fig. 5 is a difference spectrum of the absolute spectrum of the Fe(II)-reduced cytochrome minus that of the oxidized cytochrome. While the peak at 442 nm in the difference spectrum is reminiscent of that of an *a*-type cytochrome, the peak at 579 nm is considerably blue shifted from that anticipated for a typical cytochrome *a*. Preliminary structural experiments indicated that this novel heme is a derivative of heme *a* that is covalently bonded to the polypeptide

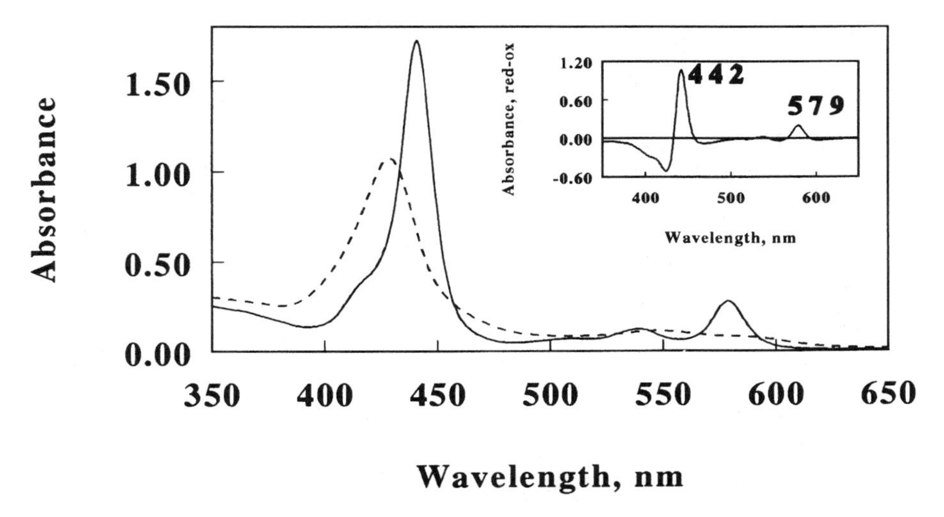

Figure 5. Absorbance spectra of oxidized (dashed line) and reduced (solid line) red cytochrome from "*L. ferrooxidans*." The inset shows a difference spectrum representing the absolute spectrum of the Fe(II)-reduced red cytochrome minus that of the oxidized red cytochrome.

with a sulfhydryl bond (R. Timkovich, University of Alabama, Tuscaloosa, unpublished data). Since the peak at 579 nm in the difference spectrum is unique to this particular respiratory component, this cytochrome was designated cytochrome$_{579}$ (6).

Cytochrome$_{579}$ was purified to electrophoretic homogeneity from cell extracts of "*L. ferrooxidans*" strain P3A (ATCC 49881) (6) and strain DSM 2705 (32). The subunit molecular mass of the former protein was determined to be 16,140 Da by liquid chromatography coupled with electrospray mass spectrometry, while that of the latter protein was estimated to be 17,900 Da by sodium dodecyl sulfate-polyacrylamide gel electrophoresis (SDS-PAGE). Both purified proteins were redox active with soluble iron at acid pH. The dependence of the pseudo-first order rate constant for the Fe(II)-dependent reduction of the cytochrome$_{579}$ from strain P3A was clearly nonlinear (6); the reaction order with respect to the concentration of ferrous ions was 5. At concentrations of ferrous ions in excess of 5.0 to 10 mM, the Fe(II)-dependent reduction of cytochrome$_{579}$ was sufficiently rapid to be of physiological significance. Given the abundant levels of this red cytochrome that are present and the rapidity of its reduction by Fe(II) in cell extracts, a rigorous investigation of the role of this protein in the iron-respiratory chain of "*L. ferrooxidans*" is clearly warranted.

Sulfobacillus thermosulfidooxidans

No evidence for either rusticyanin-like molecules or acid-soluble novel red cytochromes could be detected in cell extracts of iron-grown *S. thermosulfidooxidans* strain BC (7). However, strain BC did produce copious amounts of a novel acid-stable yellow material, whose visible absorbance properties are featured in Fig. 6. The yellow color of the oxidized chromophore represented by the solid line in Fig. 6 was slowly bleached upon addition of Fe(II) at acidic pH to yield the stable spectrum of the reduced chromophore represented by the accompanying dashed line. The inset in Fig. 6 shows the difference spectrum obtained when the spectrum of the Fe(II)-reduced chromophore was subtracted from that of the yellow oxidized form. The shape of the difference spectrum and the peak at 458 nm are reminiscent of those of a flavin, but there are no known flavoproteins that exhibit reduction potentials compatible with facile reduction by sulfatoiron(II) at a standard reduction potential in excess of +650 mV. Efforts to identify the chemical structure of this novel yellow material are clearly warranted.

Substantial purification of the protein(s) responsible for the yellow color in cell extracts of strain BC was accomplished by the simple addition of ammonium sulfate to 95% saturation. The majority of the proteins in the cell extract precipitated, while the yellow chromophore remained soluble. Only two bands of equal staining intensity at around 8,000 and 12,000 Da were revealed when the yellow ammonium sulfate-saturated supernatant was subjected to SDS-PAGE.

Abundant levels of these proteins are expressed by strain BC during growth on iron; the wet cell paste is actually canary yellow in appearance. A reasonable working hypothesis is that this yellow chromophore may represent a new redox-active enzyme that is unique to respiratory iron oxidation in these thermotolerant,

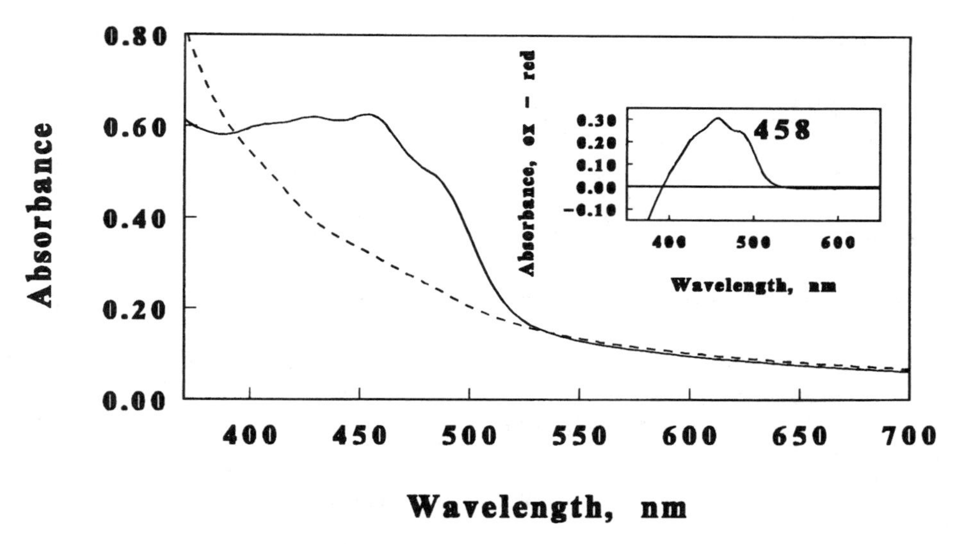

Figure 6. Absorbance spectra of oxidized (solid line) and reduced (dashed line) yellow chromophore from *S. thermosulfidooxidans*. The inset shows a difference spectrum representing the absolute spectrum of the oxidized yellow chromophore minus that of the Fe(II)-reduced yellow chromophore.

gram-positive organisms. Descriptions of respiratory components expressed in abundance in other members of the gram-positive, thermotolerant iron oxidizers have not been reported. Given the phenotypic and genotypic diversity observed in this functional class of bacteria, it would be very interesting to determine whether all members express identical respiratory components when grown on reduced iron.

Metallosphaera sedula

Soluble cell extracts derived from iron-grown *M. sedula* or *A. brierleyi* contained no apparent redox-active biomolecules. There was no evidence whatsoever for a rusticyanin, a $cytochrome_{579}$, or a BC-like yellow chromophore. Instead, iron-grown cells of both *M. sedula* (7) and *A. brierleyi* (56) expressed large quantities of a novel membrane-associated cytochrome that could be reduced directly with Fe(II) at acidic pH. Absorbance spectra of the detergent-solubilized and partially purified cytochrome from *M. sedula* are shown in Fig. 7. The oxidized cytochrome (dashed line) had an absorbance peak at 422 nm. The Fe(II)-reduced cytochrome (solid line) had peaks at 433 and 572 nm. The inset in Fig. 7 shows a difference spectrum of the Fe(II)-reduced form of the cytochrome minus that of the oxidized form. The positions of the two peaks in the difference spectrum are quite unlike those anticipated for the cytochromes of the classical *a*, *b*, or *c* types. These observations suggest that this yellow cytochrome may represent a new redox-active protein that is unique to respiratory iron oxidation in these extremely thermoacidophilic archaea.

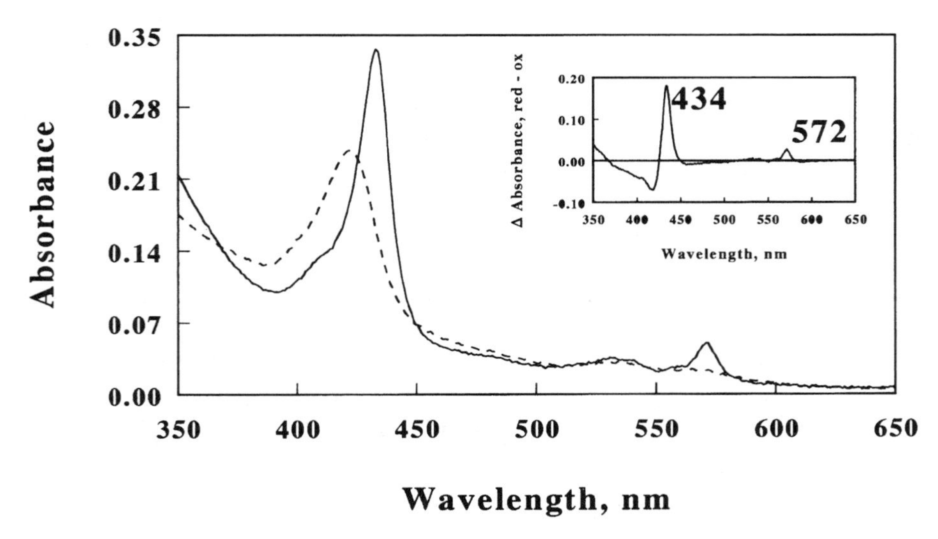

Figure 7. Absorbance spectra of oxidized (dashed line) and reduced (solid line) yellow cytochrome from *M. sedula*. The inset shows a difference spectrum representing the absolute spectrum of the Fe(II)-reduced yellow cytochrome minus that of the oxidized yellow cytochrome.

Even though the cytochrome featured in Fig. 7 and cytochrome_{579} are yellow and red, respectively, in their oxidized forms, one cannot help noticing at least a superficial resemblance in the respective Fe(II)-reduced forms of the two cytochromes. This resemblance is more evident in the comparison in Fig. 8, which shows the spectra of just the Fe(II)-reduced forms of the cytochromes produced in

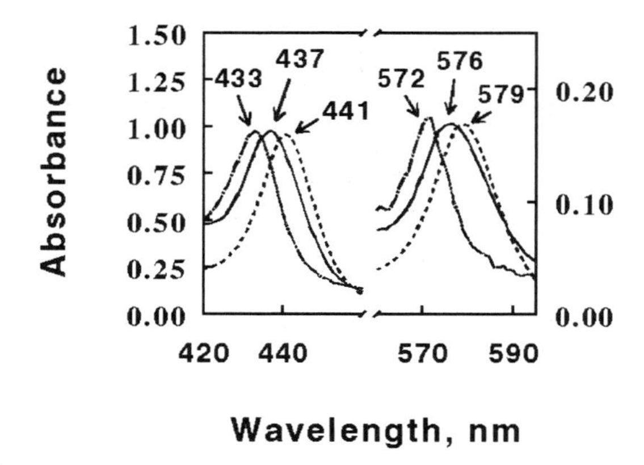

Figure 8. Absorbance spectra of the reduced cell extracts from strain JW14C (solid line), "*L. ferrooxidans*" P3A (dashed line), and *M. sedula* DSM 5348 (dot-dashed line). The left and right halves represent the Soret and alpha absorbance bands, respectively, of each cytochrome preparation, showing the wavelengths of maximum absorbance of each spectrum.

conspicuous quantities by "*L. ferrooxidans*" (dot-dashed line), strain JW14c (a gram-negative, nonmotile mesophilic straight rod) (solid line), and *M. sedula* (dashed line). While each cytochrome is easily distinguished on the basis of the precise wavelengths of its maximum absorbance in both the alpha and Soret bands, the similarities in the Fe(II)-reduced spectra are nonetheless evident. It is tempting to hypothesize that all three spectra arise from the same novel heme group and that the observed spectral differences are simply due to differences in the way that each apoprotein interacts with the same chromophore. If that were true, it would suggest that the iron-respiratory chains of selected mesophilic eubacteria and those of the thermophilic archaebacteria were perhaps descended from a common ancestor. One could then argue that the iron respiratory chains of *T. ferrooxidans* and *S. thermosulfidooxidans* each arose independently in evolution.

Other Acidophilic Iron-Oxidizing Organisms

The extent of the apparent diversity in iron respiratory chains and their components is not limited to just the novel chromophores described above. Other acidophilic iron-oxidizing organisms express respiratory chains that contain none of the conspicuous redox-active biomolecules described above. A case in point is the respiratory chain elaborated by "*F. acidophilus*." When grown either mixotrophically in the presence of Fe(II) or heterotrophically, "*F. acidophilus*" expressed large quantities of a *c*-type cytochrome (R. C. Blake II, and N. Ohmura, unpublished data). This cytochrome *c* was not reduced by Fe(II) either in crude cell extracts or when purified to electrophoretic homogeneity. The current hypothesis is that this cytochrome *c* is involved in the heterotrophic metabolism of the cell. Although intact cells of "*F. acidophilus*" exhibit a ferrous iron oxidase capacity that is severalfold lower than that of *T. ferroxidans* or "*L. ferrooxidans*" (Table 2), the organism does not appear to express conspicuous or even easily detected quantities of any other redox-active biopolymer beyond a membrane-associated terminal oxidase. How, then, does this organism accomplish aerobic respiration on iron?

The same question applies to a series of mesophilic iron oxidizing strains obtained from mining sites in the Malay peninsula and neighboring islands (2). When grown autotrophically on ferrous ions, these strains appeared to express no water- or detergent-soluble redox-active biomolecules with absorbance properties in the visible region. Indeed, cell pellets of these iron-grown cells were themselves white with no visible hint of coloration whatsoever. Cell extracts of these organisms were observed to contain conspicuous quantities of redox-active biomolecules that absorbed light in the ultraviolet, but these molecules were not reduced by ferrous ions under any conditions.

SUMMARY

If there is any truth in the widely held belief that as progress is achieved in understanding nature, simplicity rather than intricacy will be found to prevail, this must indicate that our goal to understand and describe iron respiratory systems in simple terms is far in the future. First of all, it is likely that we do not yet possess a complete inventory of all the principal microorganisms that participate in respiratory iron oxidation in acidic environments. Although a handful of microorganisms have been cultivated and isolated from biomining or acid mine drainage ecosystems, there is general agreement that many bacteria in these environments remain uncultivated and unrecognized, at least by current methods. Second, the simple picture of a highly conserved universal mechanism for respiratory iron oxidation is clearly inaccurate, and we have to prepare ourselves to face a picture of unforeseen and almost discouraging complexity.

The comparative spectroscopic analyses summarized above are intended to provide an overview of the most conspicuous components of the respiratory chains involved in iron oxidation as the first step in a more detailed investigation of the oxidation process in the different bacteria. It is evident that different iron oxidation pathways, electron transport mechanisms, and modes of energy conservation exist in different species and genera of acidophilic bacteria that respire on iron. Probably the most productive hypothesis for current and future work is that each new organism must be analyzed on the assumption that it will exhibit at least some unique features in the molecular details of its iron respiratory chain. In this way, the stages of the oxidative process that are common, if such exist, can eventually be identified. The good news is that the existence of multiple biochemical strategies to extract energy from the aerobic oxidation of iron provides more opportunities to investigate the molecular basis of energy conservation in these bacteria.

Acknowledgments. We thank Kevin Hallberg for the preparation of the phylogenetic tree in Fig. 1.
This work was partially supported by grant DE-FG02-96ER20228 from the U.S. Department of Energy (R.C.B.)

REFERENCES

1. **Bacelar-Nicolau, P.** 1996. Novel iron-oxidising acidophilic heterotrophic bacteria from mineral leaching environments. Ph.D. thesis. University of Wales, Bangor, United Kingdom.
2. **Blake, R. C., II, and S. McGinness.** 1993. Electron-transfer proteins of bacteria that respire on iron, p. 615–628. *In* A. E. Torma, M. L. Apel, and C. L. Brierley (ed.), *Biohydrometallurgical Technologies*, vol. II. *Minerals*. Metals & Materials Society, Warrendale, Pa.
3. **Blake, R. C., II, and N. Ohmura.** 1999. *Thiobacillus ferrooxidans* binds specifically to iron atoms at the exposed edge of the pyrite crystal lattice, p. 663–672. *In* R. Amils and A. Ballester (ed.), *Biohydrometallurgy and the Environment toward the Mining of the 21st Century.* Elsevier, New York, N.Y.
4. **Blake, R. C., II, and E. A. Shute.** 1987. Respiratory enzymes of *Thiobacillus ferrooxidans*. A kinetic study of electron transfer between iron and rusticyanin in sulfate media. *J. Biol. Chem.* **262:** 14983–14989.
5. **Blake, R. C., II, and E. A. Shute.** 1994. Respiratory enzymes of *Thiobacillus ferrooxidans*. Kinetic properties of an acid-stable iron:rusticyanin oxidoreductase. *Biochemistry* **33:**9220–9228.
6. **Blake, R. C., II, and E. A. Shute.** 1997. Purification and characterization of a novel cytochrome from *Leptospirillum ferrooxidans*, p. PB3.1–PB3.10. *In International Biohydrometallurgy Symposium '97.* Australian Mineral Foundation, Glenside, Australia.
7. **Blake, R. C., II, E. A. Shute, M. M. Greenwood, G. H. Spencer, and W. J. Ingledew.** 1993. Enzymes of aerobic respiration on iron. *FEMS Microbiol. Rev.* **11:**9–18.
8. **Blake, R. C., II, E. A. Shute, J. Waskovsky, and A. P. Harrison, Jr.** 1993. Respiratory components in acidophilic bacteria that respire on iron. *Geomicrobiol. J.* **10:**173–192.
9. **Bridge, T. A. M.** 1995. Iron reduction by acidophilic bacteria. Ph.D. thesis. University of Wales, Bangor, United Kingdom.

10. **Bridge, T. A. M., and D. B. Johnson.** 1998. Reduction of soluble iron and reductive dissolution of ferric iron-containing minerals by moderately thermophilic iron-oxidizing bacteria. *Appl. Environ. Microbiol.* **64:**2181–2186.
11. **Brock, T. D., and J. Gustafson.** 1976. Ferric iron reduction by sulfur- and iron-oxidizing bacteria. *Appl. Environ. Microbiol.* **32:**567–571.
12. **Brookins, D. G.** 1988. *Eh-pH Diagrams for Geochemistry.* Springer-Verlag KG, Berlin, Germany.
13. **Cavazza, C., M. T. Giudici-Orticoni, W. Nitschke, C. Appia, V. Bonnefoy, and M. Bruschi.** 1996. Characterization of a soluble cytochrome C_4 isolated from *Thiobacillus ferrooxidans. Eur. J. Biochem.* **242:**308–314.
14. **Clark, D. A., and P. R. Norris.** 1996. *Acidimicrobium ferrooxidans* gen. nov., sp. nov.: mixed-culture ferrous iron oxidation with Sulfobacillus species. *Microbiology* **141:**785–790.
15. **Cobley, J. G., and B. A. Haddock.** 1975. The respiratory chain of *Thiobacillus ferrooxidans*: the reduction of cytochromes by Fe^{2+} and the preliminary characterization of rusticyanin a novel 'blue' copper protein. *FEBS Lett.* **60:**29–33.
16. **Colmer, A. R., K. L. Temple, and M. E. Hinkle.** 1950. An iron-oxidizing bacterium from the acid drainage of some butiminous coal mines. *J. Bacteriol.* **59:**317–328.
17. **Cox, J. C., and D. H. Boxer.** 1978. The purification and some properties of rusticyanin, a blue copper protein involved in iron(II) oxidation from *Thiobacillus ferrooxidans. Biochem. J.* **174:**497–502.
18. **Das, A., A. K. Mishra, and P. Roy.** 1992. Anaerobic growth on elemental sulfur using dissimilar iron reduction by *Thiobacillus ferrooxidans. FEMS Microbiol. Lett.* **97:**167–172.
19. **De Siloniz, M. A., P. Lorenzo, M. Murua, and J. Perera.** 1993. Characterization of a new metal-mobilizing *Thiobacillus* isolate. *Arch. Microbiol.* **159:**237–243.
20. **De Wulf-Durand, P., L. J. Bryant, and L. I. Sly.** 1997. PCR-mediated detection of acidophilic, bioleaching-associated bacteria. *Appl. Environ. Microbiol.* **63:**2944–2948.
21. **Durand, P.** 1996. Primary structure of the 16S rRNA gene *of Sulfobacillus thermosulfidooxidans* by direct sequncing of PCR amplified gene and its similarity with that of other moderately thermophilic chemolithotrophic bacteria. *Syst. Appl. Microbiol.* **19:**360–364.
22. **Ehrenreich, A., and F. Widdel.** 1994. Anaerobic oxidation of ferrous iron by purple bacteria, a new type of phototrophic metabolism. *Appl. Environ. Microbiol* **60:**4517–4526.
23. **Fuchs, T., H. Huber, S. Burggraf, and K. O. Stetter.** 1996. 16S rDNA-based phylogeny of the archaeal order *Sulfolobales* and reclassification of *Desulfurolobus ambivalens* as *Acidianus ambivalens* comb. nov. *Syst. Appl. Microbiol.* **19:**56–60.
24. **Fukumori, Y., T. Yano, A. Sato, and T. Yamanaka.** 1988. Fe(II)-oxidizing enzyme purified from *Thiobacillus ferrooxidans. FEMS Microbiol. Lett.* **50:**169–172.
25. **Ghauri, M. A., and D. B. Johnson.** 1991. Physiological diversity amongst some moderately thermophilic iron-oxidising bacteria. *FEMS Microbiol. Ecol.* **85:**327–334.
26. **Giudici-Orticoni, M. T., W. Nitschke, C. Cavazza, and M. Bruschi.** 1997. Characterization and functional role of a cytochrome C_4 involved in the iron respiratory electron transport chain of *Thiobacillus ferrooxidans*, p. PB4.1–PB4.10. *In International Biohydrometallurgy Symposium '97.* Australian Mineral Foundation, Glenside, Australia.
27. **Goebel, B. M., and E. Stackebrandt.** 1994. Cultural and phylogenetic analysis of mixed microbial populations found in natural and commercial bioleaching environments. *Appl. Environ. Microbiol.* **60:**1614–1621.
28. **Goebel, B. M., and E. Stackebrandt.** 1995. Molecular analysis of the microbial biodiversity in a natural acidic environment, p. 43–52. *In* T. Vargas, C. A. Jerez, J. V. Wiertz, and H. Toledo (ed.), *Biohydrometallurgical Processing II.* University of Chile, Santiago.
29. **Golovacheva, R. S., O. V. Golyshina, G. I. Karavaiko, A. G. Dorofeev, T. A. Pivovarova, and N. A. Chernykh.** 1992. A new iron-oxidizing bacterium, *Leptospirillum thermoferrooxidans*, sp. nov. *Microbiology* **61:**1056–1065.
30. **Golovacheva, R. S., and G. I. Karavaiko.** 1979. *Sulfobacillus*—a new genus of spore-forming thermophilic bacteria. *Microbiology* **48:**658–665.
31. **Harrison, A. P., Jr.** 1982. Genomic and physiological diversity amongst strains of *Thiobacillus ferrooxidans*, and genomic comparison with *Thiobacillus thiooxidans. Arch. Microbiol.* **131:**68–76

32. **Hart, A., J. C. Murrell, R. K. Poole, and P. R. Norris.** 1991. An acid-stable cytochrome in iron-oxidizing *Leptospirillum ferrooxidans. FEMS Microbiol. Lett.* **81:**89–94.
33. **Huber, H., C. Spinnler, A. Gambacorta, and K. O. Stetter.** 1989. *Metallosphaera sedula* gen. and sp. nov. represents a new genus of aerobic, metal-mobilizing, thermoacidophilic archae bacteria. *Syst. Appl. Microbiol.* **12:**38–47.
34. **Huber, H., and K. O. Stetter.** 1989. *Thiobacillus prosperus*, sp. nov., represents a new group of halotolerant metal-mobilizing bacteria isolated from a marine geothermal field. *Arch. Microbiol.* **151:** 479–485.
35. **Huber, G., and K. O. Stetter.** 1991. *Sulfolobus metallicus*, sp. nov., a novel strictly chemolithoautotrophic thermophilic archaeal species of metal-mobilizers. *Syst. Appl. Microbiol.* **14:**372–378.
36. **Ingledew, W. J.** 1982. The bioenergetics of an acidophilic chemolithotroph. *Biochim. Biophys. Acta* **683:**89–117.
37. **Ingledew, W. J., and J. C. Cobley.** 1980. A potentiometric and kinetic study on the respiratory chain of ferrous-iron-grown *Thiobacillus ferrooxidans. Biochim. Biophys. Acta* **590:**141–158.
38. **Ingledew, W. J., and A. Houston.** 1986. The organization of the respiratory chain of *Thiobacillus ferrooxidans. Biotechnol. Appl. Biochem.* **8:**242–248.
39. **Jedlicki, E., R. Reyes, X. Jordana, O. Mercereau-Puijalon, and J. E. Allende.** 1986. Rusticyanin: initial studies on the regulation of its synthesis and gene isolation. *Biotechnol. Appl. Biochem.* **8:** 342–350.
40. **Johnson, D. B.** 1995. Selective solid media for isolating and enumerating acidophilic bacteria. *J. Microbiol. Methods* **23:**205–218.
41. **Johnson, D. B.** 1998. Biodiversity and ecology of acidophilic microorganisms. *FEMS Microbiol. Ecol.* **27:**307–317.
42. **Johnson, D. B., P. Bacelar-Nicolau, D. F. Bruhn, and F. F. Roberto.** 1995. Iron-oxidising heterotrophic acidophiles: ubiquitous novel bacteria in leaching environments, p. 47–56. *In* T. Vargas, C. A. Jerez, J. V. Wiertz, and H. Toledo (ed.), *Biohydrometallurgical processing I.* University of Chile, Santiago.
43. **Johnson, D. B., M. A. Ghauri, and M. F. Said.** 1992. Isolation and characterization of an acidophilic heterotrophic bacterium capable of oxidizing ferrous iron. *Appl. Environ. Microbiol.* **58:**1423–1428.
44. **Johnson, D. B., and S. McGinness.** 1991. Ferric iron reduction by acidophilic heterotrophic bacteria. *Appl. Environ. Microbiol.* **57:**207–211.
45. **Johnson, D. B., S. McGinness, and M. A. Ghauri.** 1993. Biogeochemical cycling of iron and sulfur in leaching environments. *FEMS Microbiol. Rev.* **11:**63–70.
46. **Johnson, D. B., and F. F. Roberto.** 1997. Heterotrophic acidophiles and their roles in the bioleaching of sulfide minerals, p. 259–279. *In* D. W. Rawlings (ed.), *Biomining: Theory, Microbes and Industrial Processes.* Landes Bioscience, Austin, Tex.
47. **Kai, M., T. Yano, H. Tamegai, Y. Fukumori, and T. Yamanaka.** 1992. *Thiobacillus ferrooxidans* cytochrome *c* oxidase: purification, and molecular and enzymatic features. *J. Biochem.* **112:**816–821.
48. **Kelly, D. P.** 1978. Bioenergetics of chemolithotrophic bacteria, p. 363–386. *In* A. T. Bull and P. M. Meadows (ed.), *Companion to Microbiology.* Longman, London, United Kingdom.
49. **Kelly, D. P., and O. H. Tuovinen.** 1972. Recommendation that the names *Ferrobacillus ferrooxidans* Leathen and Braley and *Ferrobacillus sulfooxidans* Kinsel be recognized as synonyms of *Thiobacillus ferrooxidans* Temple and Colmer. *Int. J. Syst. Bacteriol.* **22:**170–172.
50. **Kulpa, C. F., Jr., N. Mjoli, and M. T. Roskey.** 1986. Comparison of iron and sulfur oxidation in *Thiobacillus ferrooxidans*: inhibition of iron oxidation by growth on sulfur. *Biotechnol. Bioeng. Symp.* **16:**289–295.
51. **Kusano, T., T. Takeshima, K. Sugawara, C. Inoue, T. Shiratori, T. Yano, Y. Fukumori, and T. Yamanaka.** 1992. Molecular cloning of the gene encoding *Thiobacillus ferrooxidans* Fe(II) oxidase. *J. Biol. Chem.* **267:**11242–11247.
52. **Lane, D. J., A. P. Harrison, Jr., D. Stahl, B. Pace, S. J. Giovannoni, G. J. Olsen, and N. R. Pace.** 1992. Evolutionary relationships among sulfur- and iron-oxidizing eubacteria. *J. Bacteriol.* **174:**269–278.

53. **Leduc, L. G., and G. D. Ferroni.** 1994. The chemolithotrophic bacterium *Thiobacillus ferrooxidans. FEMS Microbiol. Rev.* **14:**103–120.
54. **Lindstrom, E. B., E. Gunneriusson, and O. H. Tuovinen.** 1992. Bacterial oxidation of refractory sulfide ores for gold recovery. *Crit. Rev. Biotechnol.* **12:**133–155.
55. **Nordstrom, D. K., E. A. Jenne, and J. W. Ball.** 1979. Redox equilibria of iron in acid mine waters. *ACS Symp. Ser.* **93:**51–79.
56. **Norris, P. R.** 1990. Acidophilic bacteria and their activity in mineral sulfide oxidation, p. 3–27. *In* H. L. Ehrlich and C. L. Brierley (ed.), *Microbial Mineral Recovery.* McGraw-Hill, New York, N.Y.
57. **Norris, P. R., D. W. Barr, and D. Hinson.** 1988. Iron and mineral oxidation by acidophilic bacteria: affinities for iron and attachment to pyrite, p. 43–59. *In* P. R. Norris and D. P. Kelly (ed.), *Biohydrometallurgy: Proceedings of the 1987 International Symposium.* Science and Technology Letters, Kew, United Kingdom.
58. **Norris, P. R., D. A. Clark, J. P. Owen, and S. Waterhouse.** 1996. Characteristics of *Sulfobacillus acidophilus* so. nov. and other moderately thermophilic mineral sulphide-oxidizing bacteria. *Microbiology* **141:**775–783.
59. **Norris, P. R., and D. B. Johnson.** 1998. Acidophilic microorganisms, p. 133–154. *In* K. Horikoshi and W. D. Grant (ed.), *Extremophiles: Microbial Life in Extreme Environments.* John Wiley & Sons, Inc., New York, N.Y.
60. **Ohmura, N., and R. C. Blake II.** 1997. Aporusticyanin mediates the adherence *of Thiobacillus ferrooxidans* to pyrite, p. PB1.1–PB1.10. *In International Biohydrometallurgy Symposium '97.* Australian Mineral Foundation, Glenside, Australia.
61. **Pronk, J. T., and D. B. Johnson.** 1992. Oxidation and reduction of iron by acidophilic bacteria. *Geomicrobiol. J.* **10:**153–171.
62. **Rawlings, D. E., and S. Silver.** 1995. Mining with microbes. *Bio/Technology* **13:**773–778.
63. **Rawlings, D. E., H. Tributsch, and G. S. Hansford.** 1999. Reasons why '*Leptospirillum*'-like species rather than *Thiobacillus ferrooxidans* are the dominant iron-oxidizing bacteria in many commercial processes for the biooxidation of pyrite and related ores. *Microbiology* **145:**5–13.
64. **Rawlings, D. E., D. R. Woods, and N. P. Mjoli.** 1991. The cloning and structure of genes from the autotrophic biomining bacterium *Thiobacillus ferrooxidans*, p. 215–237. *In* P. J. Greenaway (ed.), *Advances in Gene Technology.* JAI Press, Ltd., London, United Kingdom.
65. **Sand, W., K. Rohde, B. Sobotke, and C. Zenneck.** 1992. Evaluation of *Leptospirillum ferrooxidans* for leaching. *Appl. Environ. Microbiol.* **58:**85–92.
66. **Sato, A., Y. Fukumori, T. Yano, M. Kai, and T. Yamanaka.** 1989. *Thiobacillus ferrooxidans* cytochrome *c*-552: purification and some of its molecular features. *Biochim. Biophys. Acta* **976:**129–134.
67. **Segerer, A., A. Neuner, J. K. Kristjansson, and K. O. Stetter.** 1986. *Acidianus infernus* gen. nov., sp. nov., and *Acidianus brierleyi* comb. nov.: facultatively aerobic, extremely acidophilic, thermophilic sulfur-metabolizing archaebacteria. *Int. J. Syst. Bacteriol.* **36:**559–564.
68. **Shooner, F., J. Bousquet, and R. D. Tyagi.** 1996. Isolation, phenotypic characterization, and phylogenetic position of a novel, facultatively autotrophic, moderately thermophilic bacterium, *Thiobacillus thermosulfatus* sp. nov. *Int. J. Syst. Bacteriol.* **46:**409–415.
69. **Straub, K. L., and B. E. E. Buchholz-Cleven.** 1998. Enumeration and detection of anaerobic ferrous iron-oxidizing, nitrate-reducing bacteria from diverse European sediments. *Appl. Environ. Microbiol.* **64:**4846–4856.
70. **Tourova, T. P., A. B. Poltoraus, I. A. Lebedeva, I. A. Tsaplina, T. I. Bogdanova, and G. I. Karavaiko.** 1994. 16S ribosomal RNA (rDNA) sequence analysis and phylogenetic position of *Sulfobacillus thermosulfidooxidans. Syst. Appl. Microbiol.* **17:**509–512.
71. **Vernon, L. P., J. H. Mangum, J. V. Beck, and F. M. Shafia.** 1960. Studies on a ferrous-ion-oxidizing bacterium. II. Cytochrome composition. *Arch. Biochem. Biophys.* **88:**227–231.
72. **Walton, K. C., and D. B. Johnson.** 1992. Microbiological and chemical characteristics of an acidic stream draining a disused copper mine. *Environ. Pollut.* **76:**169–175.
73. **Yahya, A., F. F. Roberto, and D. B. Johnson.** 1999. Novel mineral-oxidizing bacteria from Montserrat (W.I.): physiological and phylogenetic characteristics, p. 729–741. *In* R. Amils and A. Ballester

(ed.), *Biohydrometallurgy and the Environment toward the Mining of the 21st Century*. Elsevier, New York, N.Y.

74. **Yamanaka, T., T. Yano, M. Kai, H. Tamegai, A. Sato, and Y. Fukumori.** 1991. The electron transfer system in an acidophilic iron-oxidizing bacterium, p. 223–246. *In* Y. Mukohata (ed.), *New Era of Bioenergetics*. Academic Press, Inc., Tokyo, Japan.

Environmental Microbe-Metal Interactions
Edited by Derek R. Lovley

Chapter 4

Trace Metal-Phytoplankton Interactions in Aquatic Systems

William G. Sunda

Trace metals influence phytoplankton communities both as limiting nutrients and as toxicants at elevated concentrations (Fig. 1). Until recently, phytoplankton productivity was thought to be limited primarily by major nutrients: phosphorus in freshwater and nitrogen in the ocean (102, 103). However, results of in situ iron enrichment experiments indicate that iron limits phytoplankton growth in large regions of the ocean (21, 64). Moreover, there is evidence that the supply of fixed nitrogen to aquatic algae may be limited by Fe availability, via Fe controls on N_2 fixation (36, 100, 136). Algal species have large differences in their growth requirements for Fe and other micronutrient metals (Zn, Mn, and Co) (10, 107), indicating that in addition to controlling productivity, trace metals influence the species composition of algal communities.

This chapter examines the effects of trace metals on the growth, composition, and trophic dynamics of phytoplankton communities. To do this we must consider metal interactions at several levels, including the molecular, cellular, community, and ecosystem levels, of organization. Specifically, we must know the chemistry of bioactive metals in various aquatic systems, the interaction of that chemistry with cellular metal uptake systems, the metabolic roles of metals within cells, metal controls on cellular growth and metabolic rates, and overall effects on different members of the phytoplankton community. We also need to consider the effect of phytoplankton communities on the concentration, complexation, and redox cycling of metals, which results in feedback relationships between plankton dynamics and trace metal chemistry (Fig. 1). This chapter emphasizes interactions in marine systems due to the recent discovery of iron limitation in the ocean as well as the growing body of knowledge about trace metal chemistry, biogeochemical cycling, and controls on phytoplankton dynamics in these systems.

William G. Sunda • Beaufort Laboratory, National Oceanic and Atmospheric Administration, Beaufort, NC 28516.

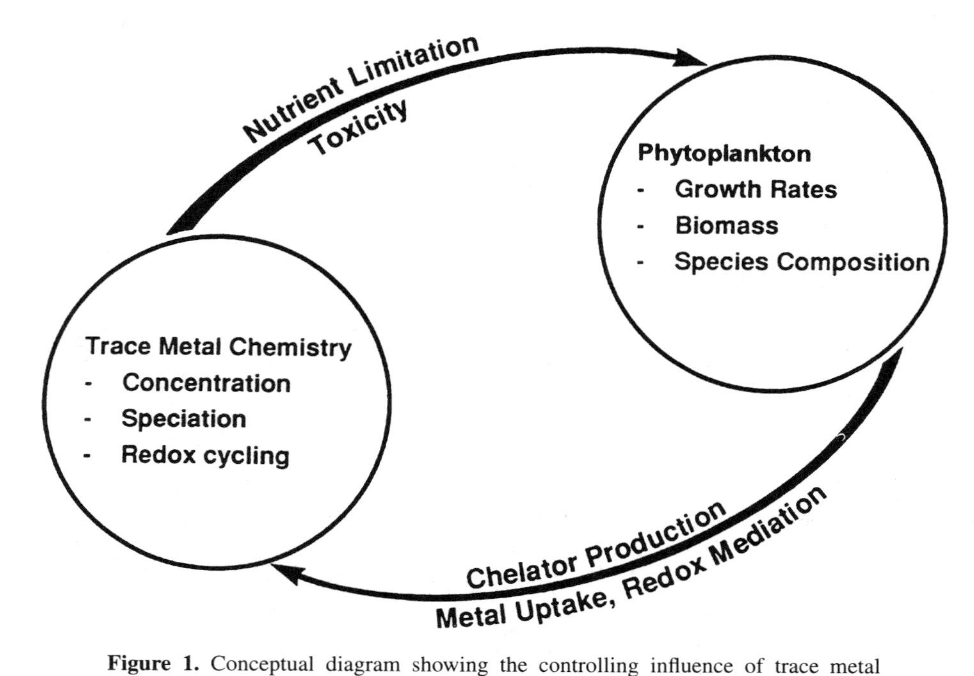

Figure 1. Conceptual diagram showing the controlling influence of trace metal chemistry on the productivity and species composition of phytoplankton communities and the reciprocal effects of phytoplankton on trace metal concentrations, complexation, and redox cycling. Adapted from reference 125.

TRACE METAL CHEMISTRY

Metal Distributions

Trace metal concentrations vary widely in aquatic systems due to differences in rates of input, loss, and internal cycling. Surface water concentrations in the ocean (8, 13–15) and lakes (105) are often lower than those at depth due to uptake of bioactive metals by phytoplankton in the euphotic zone followed by plankton settling and microbial remineralization (Fig. 2). Major nutrients (N, P, and Si) have long been known to be controlled by such cycles (92, 93), and algal uptake and regeneration of both sets of constituents is thought to explain the close correlations between concentrations of most bioactive metals (Zn, Cd, Fe, Ni, and Cu) and major nutrients in the ocean (8, 13, 53, 78, 108). In the North Pacific, efficient removal by phytoplankton has led to exceedingly low concentrations of dissolved Fe (0.02 to 0.1 nM), Zn (~0.07 nM), and Cd (~0.002 nM) at the surface relative to those at 1,000 m (Fig. 2). The concentrations of these and other trace metals increase substantially along surface transects from oceanic to coastal and estuarine waters due to inputs from continental sources, such as rivers, groundwater, eolian dust, and shelf sediments (15, 53, 135). Filterable iron concentrations are even higher (up to 20 μM) in rivers, where iron occurs largely as iron hydroxide-organic colloids. This colloidal iron is rapidly lost from the water column in estuaries, due

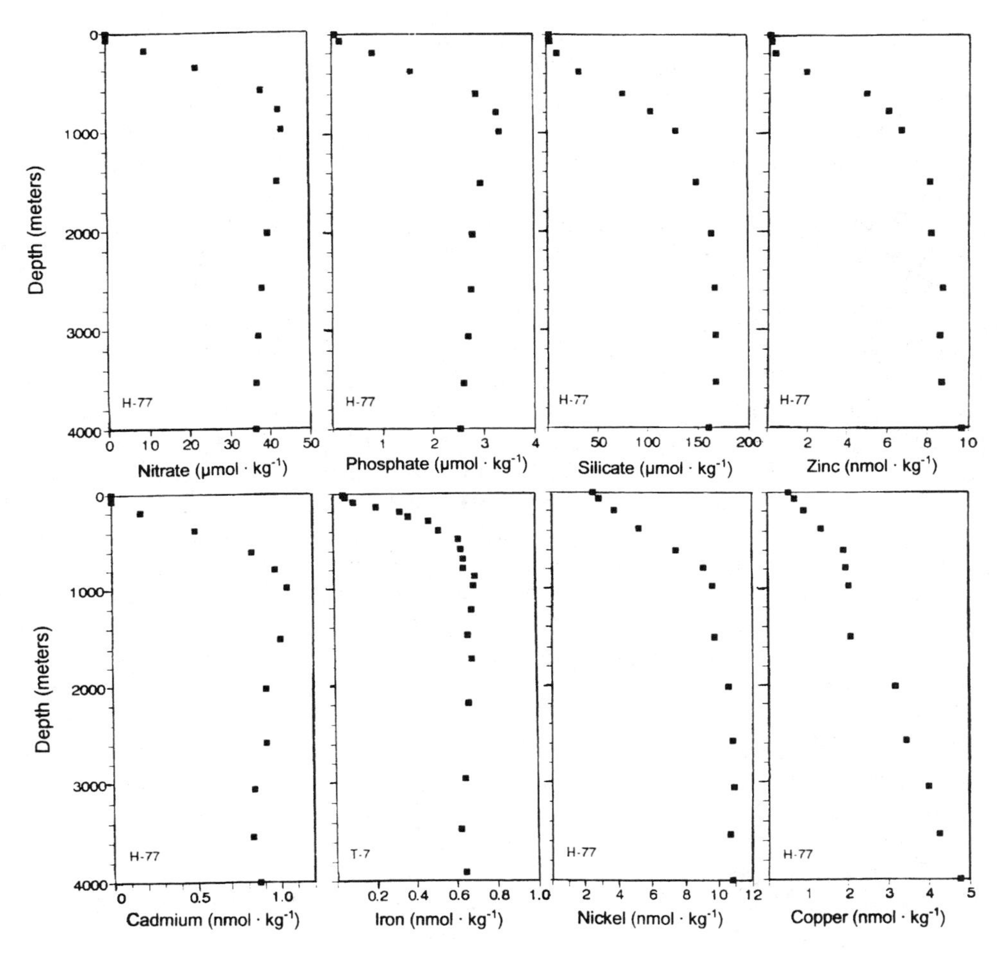

Figure 2. Depth profiles for total dissolved concentrations of major nutrients (nitrate, phosphate, and silicate), trace metal nutrients (zinc, iron, nickel, and copper), and the nutrient analog cadmium in the North Pacific. Data for station H-77 (32.7°N; 145.0°W) are from reference 14, while those for station T-7 (50.0°N; 145.0°W) are from reference 65.

to salt-induced flocculation and subsequent particulate settling (9). The large differences in the concentrations of iron and other trace metals between riverine, neritic, and oceanic environments have had an important influence on phytoplankton evolution, as discussed below.

Metal Speciation

Trace metals exist in a variety of chemical species in natural waters, which strongly influences their availability to phytoplankton. Most occur as cations that are complexed, to various degrees, by inorganic and organic ligands or are adsorbed on or bound within particles. Many biologically active metals (iron, manganese, copper, cobalt, molybdenum, chromium, mercury, and silver) can cycle between different oxidation states. Both complexation and redox cycling affect the behavior of these metals in aquatic systems because of the large differences in the reactivity, kinetic lability, and solubility of different metal oxidation states and coordination species.

The inorganic complexation of trace metals in marine and freshwater systems has been generally well characterized through the use of thermodynamic models. For seawater, these models indicate that Ni, Mn(II), Zn, Co(II), and Fe(II) occur largely as free aquo ions while other metals are heavily complexed (20). Based on these models, Cd, Cu(I), Ag(I), and Hg(II) are heavily complexed by chloride ions, while Cu(II) and Pb(II) are complexed by carbonate and Fe(III) and Al are complexed by hydroxide ions. Inorganic complexation of trace metals varies little over most of the ocean's surface due to the relatively constant pH and major ion composition of seawater. However, in freshwater and estuarine water, large variations in chloride concentration, alkalinity, and pH can lead to substantial variations in inorganic complexation. Of particular relevance is the large chloride gradient in estuaries, which results in sizable variations in the extent of chloride complexation of Cd, Hg, Cu(I), and Ag. Freshwater can vary in pH from <5 to >9, which leads to large differences in hydroxide and carbonate complexation.

Less is known about complexation of trace metals by organic ligands. This situation is rapidly changing, however, with the recent development of a variety of sensitive, chemically selective metal speciation techniques. Copper(II) speciation has been the most widely studied. Research with a number of methods has revealed that $\geq 99\%$ of this metal is complexed to organic ligands in most near-surface aquatic systems (22, 23, 72, 73, 109, 110, 129, 137, 138). The majority of the Cu is bound by unidentified organic ligands which occur at low concentrations and possess extremely high conditional stability constants ($\log K \approx 13$ in seawater [22, 23, 73, 110] and 14 to 15 in some Swiss lakes [138] at pH 8). The distribution of strong Cu-binding ligands generally covaries with productivity. In the North Pacific, the ligands occur at a maximum concentration near the depth of maximum productivity and disappear below the euphotic zone (22, 23). Similarly, in lakes, the levels of Cu-binding ligands are highest in the euphotic zone and covary with productivity both seasonally and among lakes (137).

The covariance of ligand concentrations with productivity suggests a biological source. This hypothesis is supported by results of culture studies with several species of the cyanobacterium *Synechococcus*, which produce extracellular ligands with stability constants similar to those of the strong Cu-binding chelators in seawater (73, 74). *Synechococcus* is widely distributed in seawater, and most species are highly sensitive to copper toxicity (11). The presence of strong Cu-binding ligands benefits the algal community by protecting it from copper toxicity. In the presence of the ligands, the free cupric ion concentration ($[Cu^{2+}]$) in surface seawater is reduced to $10^{-13.5}$ to $10^{-12.5}$ M, levels that are low enough to prevent toxicity but sufficient to meet algal Cu nutritional requirements (11, 111). In the absence of organic complexation, $[Cu^{2+}]$ in seawater would increase by 1,000-fold to $10^{-10.7}$ to $10^{-9.7}$ M at typical Cu concentrations of 0.5 to 5 nM. These levels are toxic to many phytoplankton, including species of *Synechococcus* (11). The ligand is released by *Synechococcus* in response to elevated [Cu^{2+}], and thus its production provides a mechanism by which cyanobacteria can regulate [Cu^{2+}] in their environment within a biologically favorable range (74).

Recent measurements obtained by a ligand competition/cathodic stripping voltammetric method indicate that iron(III), like copper, is >99% complexed in seawater by as yet unidentified organic ligands (40, 98, 135). This chelation benefits the biological community by maintaining dissolved iron in a less reactive state, thereby minimizing its loss from seawater by hydroxide precipitation and adsorption onto particle surfaces and subsequent settling of iron-containing particles (53). Without this chelation, iron concentrations would be substantially reduced, adversely affecting algal growth. Since the organic ligands that solubilize iron are directly or indirectly produced by the biota, their production represents a biological feedback mechanism that enhances productivity. Indeed, in a recent iron enrichment experiment in the equatorial Pacific, the addition of iron resulted in a threefold increase in the concentration of strong Fe-binding chelators, which helped solubilize the iron and prevent its loss from the euphotic zone (99).

Zn and Cd are also appreciably complexed by organic ligands but to a lesser extent than are Cu and Fe. Recent electrochemical studies indicate that 98 to 99% of the Zn in surface seawater (16, 28) and 50 to 99% in estuarine (56, 61, 129) and lacustrine (139) waters is complexed to organic ligands. Likewise, 50 to 70% of Cd in near-surface seawater (17) and 70 to 80% in water from an estuary (56) was found to be organically complexed. In contrast, there is no evidence for organic complexation Mn(II) in marine water or freshwaters (97), in accord with the low reactivity of this metal in complex formation. Inorganic complexation of Mn(II) is also low, and ca 70% of this metal in seawater (20) and >90% in most freshwaters should exist as free Mn^{2+}.

Redox cycling affects the speciation, biological availability, and biogeochemical behavior of at least four essential metals (Fe, Mn, Cu, and Co) which can exist in two or more oxidation states. The thermodynamically stable or metastable forms of Mn in oxic waters, Mn(III) or Mn(IV) oxides, are insoluble and thus not directly available for algal uptake. However, although thermodynamically unstable, Mn(II) can persist in waters below pH $\approx$ 8.5 because of slow oxidation kinetics (106). These kinetics are increased by several orders of magnitude by ubiquitous bacteria that enzymatically catalyze Mn(II) oxidation on their surfaces (25, 33, 75, 112, 113). Due to bacterial catalysis, summertime Mn(II) residence times with respect to oxidation are decreased to hours or days in temperate estuarine waters and weeks or months in the open ocean (75, 112, 113).

In near-surface seawater, there is a photoreductive dissolution of Mn oxides and a photoinhibition of bacterial Mn(II) oxidation, which together markedly decrease Mn oxide concentrations and subsequent removal or Mn via settling of Mn oxide particles (113, 114). These two external light-driven processes contribute to the surface maxima in Mn concentrations observed worldwide in the ocean and thereby increase the supply of Mn needed for photosynthesis (113).

Photosynthesis itself can have the opposite effect and can stimulate Mn(II) oxidation on the surfaces of large algal cells and cell colonies (>20 μm diameter) which possess thick surface diffusion boundary layers (94, 95). Microenvironments with elevated O_2 and pH develop within these boundary layers due to photosynthetic O_2 evolution and CO_2 uptake and to restricted diffusive exchange with the bulk solution. Mn(II) oxidation rates are substantially increased within these microenvironments since these rates are proportional to $[O_2]$ and to the square of $[OH^-]$ (106). In eutrophic lakes with large populations of colonial cyanobacteria, this mechanism can provide the major pathway for Mn(II) oxidation and removal from surface waters (94).

Iron is the most important trace metal nutrient, and its redox and coordination chemistry are perhaps the most complex. Its stable oxidation state, Fe(III), has a low solubility and tends to precipitate as oxyhydroxides. In addition, Fe(III) has a strong tendency to coordinate with surface ligands, such as those found on organic particles. Consequently, iron tends to associate with particles and be scavenged from the water column, accounting for its extremely low concentration in the ocean despite its high crustal abundance.

Reduction of Fe(III) to Fe(II) can greatly increase iron solubility, since Fe(II) hydroxides are much more soluble and Fe(II) forms much weaker and more labile coordination bonds with ligand sites on particle surfaces. There is considerable evidence that iron undergoes a dynamic photoredox cycling that enhances its availability to phytoplankton in both freshwater and saline water (38, 69, 71, 130, 131). Fe(III) associates with organic ligands in natural waters as soluble chelates, complexes with functional groups on organic particles, or organic ligands coordinated to Fe oxyhydroxide surfaces. These Fe(III)-organic associations undergo photoredox reactions to produce Fe(II) and oxidized organics. Much of the Fe(II) dissociates to dissolved inorganic species and is subsequently reoxidized to Fe(III) by reaction with O_2 and H_2O_2 (76). The rate of reoxidation is strongly dependent on $[OH^-]$, allowing appreciable Fe(II) to accumulate at pHs below ~7 but resulting in rapid reversion to Fe(III) at higher pH (71). The photoredox cycling increases the biological availability of iron by increasing the concentrations of labile dissolved inorganic species of both Fe(II) and Fe(III) (4, 38). In addition to photoreduction, Fe(III) chelates can be reduced at algal surfaces by transmembrane reductase enzymes, providing an additional mechanism for enhancing external concentrations of dissolved inorganic Fe(II) and Fe(III) and thereby increasing cellular iron uptake rates (3, 5, 54, 62). Such biologically driven Fe redox cycling may be of particular importance at night or in the lower euphotic zone, where there is insufficient UV light to drive iron photoreduction.

CELLULAR METAL UPTAKE MECHANISMS

The chemical speciation of metals profoundly influences their cellular uptake. The outer membrane of cells is virtually impermeable to charged or polar species, and therefore nutrient metal ions usually enter cells by binding to specialized membrane transport proteins (32) (Fig. 3). Following binding to a receptor site on the protein, the metal ion may either dissociate back into the medium or be transported across the membrane and released into the cytoplasm. The rate of intracellular uptake equals the concentration of metal complexed with the transporter

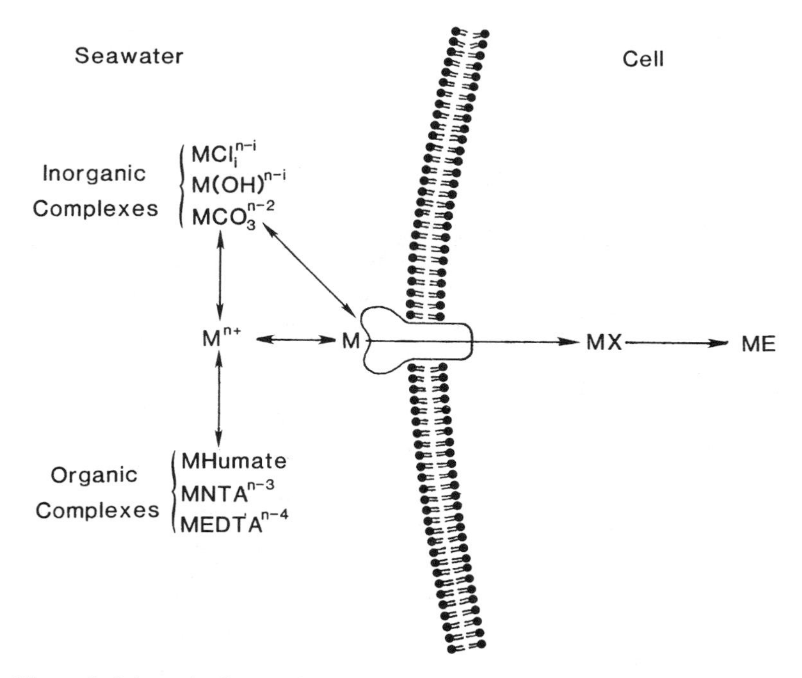

Figure 3. Schematic diagram for the interrelationship between trace metal ion speciation in the external medium and cellular metal uptake by a membrane transport protein. The intracellular uptake rate is controlled by the amount of metal bound to a receptor site on the protein, which in turn is controlled by either the external free metal ion concentration (equilibrium control) or the concentration of kinetically labile free ions plus inorganic complexes (kinetic control). Organic complexation decreases metal uptake by decreasing the concentration of free ions and inorganic complexes. Reprinted from reference 123.

multiplied by the kinetic rate constant for transport across the membrane and release into the cytoplasm. At steady state, the amount of metal bound to the transport site, and hence the metal uptake rate, V, is described by the well-known saturation equation

$$V = \frac{V_{\max} [\mathrm{M}] K_{\mathrm{M}}}{[\mathrm{M}] K_{\mathrm{M}} + 1} \quad (1)$$

where [M] is the concentration of either free metal ions or kinetically labile inorganic species (free ions plus dissolved inorganic complexes), K_M is an affinity constant for the binding of metal M to the transport site, and V_{max} is the maximum rate achieved when all of the transport sites are bound to the metal. Such saturation uptake relationships describe algal uptake of all nutrient metals examined (manganese [115, 116], iron [4, 44, 50], zinc [108, 117], copper [55], and molybdenum [48]) and for competing metals such as cadmium (46, 58). The biology of transport proteins for important micronutrient metals (Fe, Zn, Cu, and Mn) has recently been described at the molecular level in yeast, providing insights into cellular metal transport and transport regulation in other eucaryotes including algae (32).

The affinity constant, K_M, in equation 1 is defined by the equation

$$K_M = k_f/(k_d + k_{in}) \quad (2)$$

where k_f, k_d and k_{in} are rate constants, respectively, for formation of the membrane transport ligand complex, dissociation of metal from the complex back into the medium, and transfer of the bound metal across the membrane and subsequent cytoplasmic release. If $k_d \gg k_{in}$, a pseudo-equilibrium is established between metal species in the medium and the metal bound to the transport ligand; and at equilibrium, the concentration of metal bound to the transporter (and hence, the uptake rate) is related to the external free metal ion concentration (Fig. 3). Such equilibration might be expected for metals such as Cu and Cd, which possess rapid ligand exchange kinetics. However, in recent studies of algal uptake of iron(III), whose ligand exchange rates in seawater are only 1/500 of those for copper (49), k_{in} was found to exceed k_d (50). Consequently, iron uptake was under the kinetic control exerted by the rate of metal binding to the membrane transport site. This rate, in turn, was related to the concentration of labile dissolved inorganic species of Fe(III) and Fe(II), whose exchange kinetics are sufficiently rapid to permit appreciable rates of iron coordination to the transport site. The uptake of other metals may be under similar kinetic control in at least some situations and thus also would be controlled by the concentration of kinetically labile inorganic species. Under kinetic control, organically chelated metals and those bound on or within particles (e.g., metal oxides) would not be directly available for uptake since their dissociation and exchange kinetics (or their diffusion rates in the case of colloids and particulates) are too slow to permit rapid donation of metal ion to membrane transport sites (79).

Thus, in some instances, algal uptake of metals is related to the free metal ion concentration while in others it is related to the concentration of labile inorganic species (free metal ions plus labile inorganic complexes). In constant ionic media such as near-surface seawater equilibrated with the atmosphere (pH $\approx$ 8.2), the inorganic metal speciation is constant and thus concentrations of free ions and inorganic complexes are related to one by constant ratios. Under these conditions, it does not matter in a practical sense whether one defines metal availability in terms of concentrations of free metal ions or of inorganic metal complexes. Furthermore, many bioactive metals [Mn, Zn, Co(II), and Ni] exhibit only minor inorganic complexation in both freshwater and saline water, and thus their inorganic speciation is dominated by the free aquo ion. For these metals, the distinction between thermodynamic or kinetic control may also be of only minor practical importance. However, for metals such as Cd, whose inorganic speciation can vary widely between seawater and freshwater, knowledge of the type of transport control may be critical to predicting changes in metal uptake rates accompanying variations in major-ion composition.

There are some exceptions to the above membrane transporter mechanism for metal uptake. Some neutrally charged, nonpolar complexes, such as $HgCl_2$ and

CH_3HgCl, are soluble in lipids and therefore can diffuse directly across bilayer membranes (43). In experiments with a coastal diatom, the uptake and toxicity of Hg(II) within an Hg/salinity matrix was related to the computed concentration of $HgCl_2$ and not that of Hg^{2+} or of total inorganic Hg species, implying that uptake occurred via diffusion of the neutral complexes through the cell membrane (68). Once inside the cell, the $HgCl_2$ underwent exchange reactions with biological ligands, such as sulfhydryls, providing a sink for the diffusing Hg. Lipophilic metal chelates, such as those with 8-hydroxyquinoline and dithiocarbamate, are taken up via the same diffusion/exchange process (84).

In addition to membrane diffusion and intracellular exchange of lipophilic chelates, a number of metal chelates with biological ligands can be transported into cells by specific membrane transport proteins. Vitamin B_{12}, which contains chelated cobalt within its corrin ring, is taken up by this mechanism (30). Also, under iron-limiting conditions, many microorganisms release strong iron-binding ligands (siderophores) into the surrounding medium to complex and solubilize iron. The siderophore-iron chelates are taken into the cell by specific transport proteins, after which the iron is released for assimilation via Fe reduction or degradation of the siderophore (83). The utilization of such siderophore based Fe acquisition systems is widespread in eubacteria and fungi (83), and also occurs in many but not all cyanobacteria (132, 133). However, there is no definitive evidence for their presence in eucaryotic algae (50; G. Boyer, personal communication).

CELLULAR METAL REGULATION

Cells regulate the uptake rate of nutrient metals to maintain their intracellular concentrations at optimal levels for growth and metabolism. Under steady-state conditions during exponential growth, the external concentration of free metal ions (or labile inorganic species), the cellular metal uptake rate, the cellular metal concentration, and the specific growth rate are all related by a set of interconnecting relationships (80, 116). The cellular metal concentration, [Cell metal], on which the growth rate depends, is equal to the cellular uptake rate, V, divided by the specific growth rate, μ:

$$[\text{Cell metal}] = V/\mu \qquad (3)$$

As outlined above, the uptake of nutrient metals generally occurs via binding to membrane transport proteins and is related to the external concentration of free metal ions (or labile inorganic species) by the saturation kinetics equation (equation 1). For Mn, a micronutrient whose uptake kinetics have been well studied, the metal is taken up by a single high-affinity system (115, 116). The affinity constant (K_{Mn}) is fixed for each algal species, but the maximum uptake rate (V_{max}) increases by up to 30-fold with decreases in available Mn ion concentration ($[Mn^{2+}]$). The increase in V_{max} is under negative feedback control by the cell, allowing internal Mn concentrations to be regulated at optimal levels for growth, despite variations in external $[Mn^{2+}]$ or changes in growth rate. For example, in the oceanic diatom *Thalassiosira oceanica*, there is a 100-fold range in $[Mn^{2+}]$ (10^{-9} to 10^{-7} M) over which the amount of Mn in the cell is roughly constant and optimal for growth (Fig. 4A). The half saturation constant for the Mn uptake system ($10^{-7.95}$ M, the inverse of the affinity constant) falls in the middle of this range. At [Mn^{2+}] above this value, the constant cellular Mn level reflects saturation of the uptake system (i.e., $V \approx V_{max}$)(Fig. 4C). However, at lower [Mn^{2+}], a near-constant cellular Mn level is achieved by compensatory increases in V_{max}, which allow Mn uptake rates to remain constant despite undersaturation of the transport system (Fig. 4B). The ability of the cell to increase V_{max} is finite, and as [Mn^{2+}] is decreased below 10^{-9} M, V_{max} reaches a maximum value. At this point, the Mn uptake rate decreases in proportion to [Mn^{2+}] since the transport system is well undersaturated (Fig. 4B). The growth rate then declines as the cellular Mn level falls below the level required to support photosynthesis.

Cellular iron (44) and zinc (108, 118) levels appear to be under similar feedback regulation. Uptake kinetic studies with marine diatoms, chlorophytes, and coccolithophores indicate the presence of at least two cellular zinc transport systems: a high-affinity system ($K_{Zn} \sim 10^{10}$ M^{-1}) that is under negative feedback regulation and a low-affinity system that is either constitutive or under only minimal cellular control. The presence of both high- and low-affinity systems results in sigmoidal relationships between the zinc uptake rate (and the cellular zinc concentration) and the external [Zn^{2+}] (108, 118) (Fig. 5). Copper in marine algae shows similar sigmoidal relationships with [Cu^{2+}], possibly also due to separate transport systems with differing regulation and affinity constants (111).

Similar high- and low-affinity zinc uptake systems are found in yeast. As in phytoplankton, the V_{max} of the high-affinity system is under substantial negative feedback regulation (140). Molecular studies reveal that the V_{max} is regulated by the binding of zinc to a promoter region of DNA that controls the transcription of zinc transport proteins (140). Zinc binding represses gene transcription, while zinc dissociation promotes the synthesis of additional transport proteins.

The ability of the cell to increase metal uptake rates at low external concentrations ultimately is limited by space on the outer membrane available for transport proteins and by the rate of metal binding to these proteins, as controlled by inherent limits on ligand exchange kinetics (50). Uptake also can be limited by the rate of diffusion of kinetically labile inorganic metal species to the cell surface. Such limits can clearly be seen for iron and zinc uptake in marine phytoplankton (49, 50, 108, 119). Limits on membrane transport kinetics appear to control iron uptake for cells of $\lesssim 30$ μm diameter; and, as predicted from such limitation, measured uptake rates per unit cell volume are directly related to the surface to biovolume ratio and therefore are inversely proportional to the cell diameter (119). However, cell models predict that diffusion begins to dominate as the limiting factor for larger cells (49). This size-related switch in limiting processes occurs because diffusion-limited uptake rates (normalized to cell volume) are inversely proportional to the inverse square of the cell diameter and thus decrease disproportionately for larger cells. Zinc has ligand exchange kinetics that are 40 times those for Fe(III) (49). Thus, for this metal, limits on membrane transport kinetics are less important, increasing the relative importance of diffusion limitation. As a result, diffusion limits zinc uptake by algae at low [Zn^{2+}] for cells as small as 3 μm in diameter (107) (Fig.

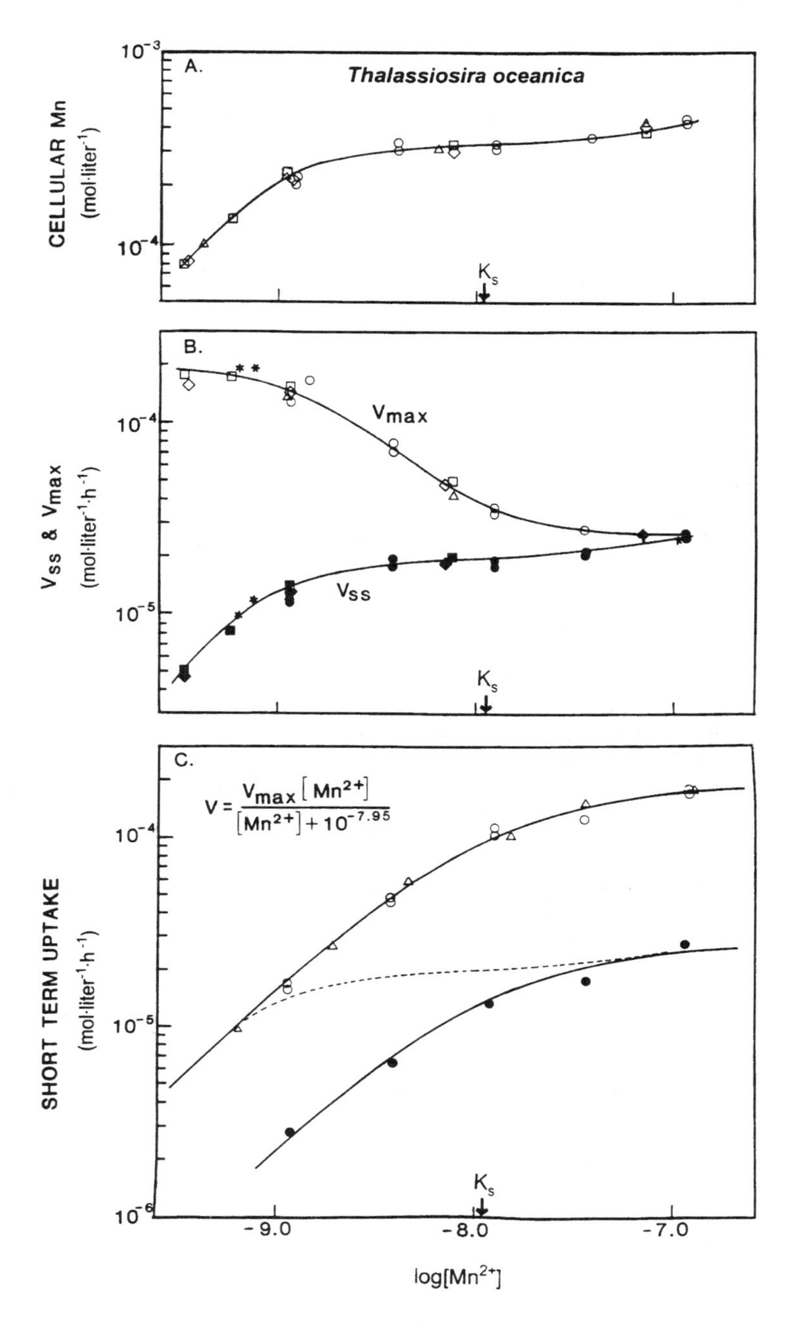

Figure 4. (A) Regulation of cellular Mn in the oceanic diatom *Thalassiosira oceanica*. (B and C) The uptake of Mn follows Michaelis-Menten kinetics, and V_{max}, but not K_s, increases with decreasing $[Mn^{2+}]$ (B). Increases in V_{max} allow the cells to maintain relatively constant steady-state Mn uptake rates (V_{ss}) and therefore constant cellular Mn concentrations at a constant growth rate, down to $[Mn^{2+}]$ of 10^{-9} M. Panel C gives short-term uptake rates for cells preconditioned at $[Mn^{2+}]$ of $10^{-9.2}$ M (open symbols) and $10^{-7.0}$ M (closed symbols). The dotted line in this panel is the steady-state uptake rate for acclimated cells (see panel B). Cellular Mn is given in units of moles per liter of cell volume. Adapted from reference 123 and based on data from reference 116.

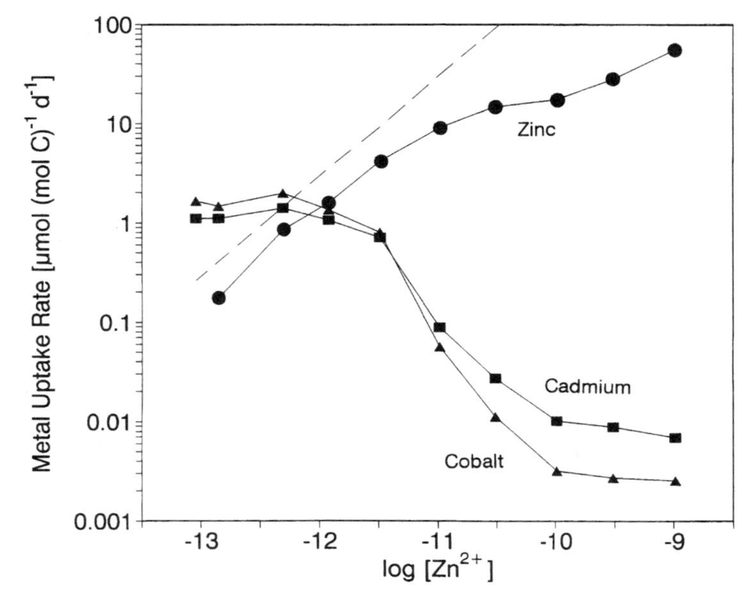

Figure 5. Variations in steady-state uptake rates for zinc, cadmium, and cobalt in the oceanic diatom *Thalassiosira oceanica* as functions of log [Zn^{2+}] in the external seawater medium. At [Zn^{2+}] $< 10^{-11}$ M, the zinc uptake approaches limiting rates for diffusion of labile inorganic zinc species to the cell surface (dashed line). At [Zn^{2+}] below 10^{-10} M, Co and Cd uptake rates increase dramatically due to induction of a transport system (possibly the high affinity Zn system) that is under negative feedback control by cellular Zn. [Co^{2+}] and [Cd^{2+}] were held constant at 10^{-12} and $10^{-13.1}$ M, respectively. Metal uptake was determined with radiotracers, and free ion concentrations were controlled with EDTA-trace metal ion buffer systems at 20°C. Zn and Co data are from reference 107, while Cd data are from Sunda and Huntsman (unpublished).

5). Due to the sensitivity of zinc transport to diffusion limitation, uptake of this metal is adversely affected to a high degree by increasing cell size.

The above size-related physical and chemical limits on Fe and Zn uptake rates have been important forces in phytoplankton evolution, as discussed below. By favoring the growth of small cells, they also have been important factors in regulating the structure of oceanic planktonic food webs (50, 79).

LIMITATION OF GROWTH RATE

Biochemical Roles of Metals

A number of trace metals (Fe, Zn, Mn, Co, Cu, Mo, and Ni) are essential micronutrients and thus exert important controls on algal growth and metabolism. These metals function as catalytic centers in metalloproteins and metal activated enzymes involved in virtually all aspects of cell metabolism (27). Many micronutrient metals (Fe, Cu, Mn, and Mo) can exist in different oxidation states within metalloproteins and function primarily in electron transport or in metabolic redox catalysis. The most important of these is iron, which is present in the active centers of cytochromes and Fe/S proteins involved in numerous metabolic pathways, including photosynthetic and respiratory electron transport, nitrate and nitrite reduction, nitrogen fixation, and sulfate and sulfite reduction. Consequently, iron is an important controller of cellular biosynthetic rates including those for photosynthesis and nitrogen assimilation. Iron is also present in peroxidases, catalase, and some superoxide dismutases responsible for the removal of toxic oxygen species (hydrogen peroxide and superoxide radicals) formed as by-products of metabolism. Manganese, which is quantitatively less important than iron, is present in the water-oxidizing centers in photosystem II and in Mn-superoxide dismutase. Thus, it is essential for photosynthesis and is also important in protection of the cell from free radical oxidative damage. Molybdenum (along with iron) functions as a redox catalyst in nitrogenase (needed for N_2 fixation) and nitrate reductase, while copper is present in some electron transport proteins (cytochrome *c* oxidase and plastocyanin) and in several oxidases (e.g., ascorbate oxidase).

Zinc has an importance second only to iron. It exists in cells in only one oxidation state (+2) and thus does not directly participate in electron transfer reactions. Instead, it acts as a high-affinity Lewis acid in hydrolytic reactions, including the dehydration of bicarbonate ions (in carbonic anhydrase), the hydrolysis of proteins and phosphate esters, and DNA and RNA polymerase reactions (27). It also plays a structural role in numerous proteins through the formation of intramolecular chelate bonds. It appears to play a central role in the regulation of DNA replication and transcription, as exemplified by the euucaryotic Zn finger proteins involved in reading of DNA during transcription (7, 27, 128). It also plays a central role in regulation of the mitotic cycle; consequently, zinc limitation results in inhibition of cellular division (35).

Cobalt and cadmium can metabolically replace zinc in at least some metalloenzymes (notably carbonic anhydrase) and thus can help alleviate zinc limitation in low-zinc oceanic environments (58, 81, 85, 107). In addition, cobalt has the more traditionally recognized function as the catalytic center of vitamin B_{12}, which is needed for methyl transfer reactions, reduction of ribose to deoxyribose, and some organic rearrangement reactions (27).

Nickel plays a limited role as a micronutrient. Its only known requirement in eucaryotes is in urease, which catalyzes the hydrolysis of urea, an important organic nitrogen source. Consequently, in culture experiments, nickel was required for diatom growth only when urea was the limiting nitrogen source (86).

Cellular Metal Requirements and Their Relation to Other Limiting Resources

Due to their catalytic role in metabolism and biosynthesis, trace metals play an important role in controlling the growth rate of phytoplankton. However, compared to major nutrients (N and P) (41), they are generally less important in controlling cell yield. Relationships between growth rate and cellular Fe/C ratios in two coastal diatoms (*Thalassiosira pseudonana* and *T. weissflogii*) and dinoflagellates (*Proro-*

centrum minimum and *P. micans*) provide a good case in point (119). Under Fe limitation, the specific growth rate (μ) was linearly related to the cellular Fe/C ratio for all four species (Fig. 6). Linear regression of the data for cells grown on nitrate at saturating light under a 14-h/10-h light/dark cycle gave the relationship

$$\mu = B([\mathrm{Fe/C}] - m)\ (R^2 = 0.91) \tag{4}$$

where m (3 μmol mol^{-1}) is the average Fe/C ratio needed for cell maintenance at zero growth rate. The slope of this expression (B = 49,360 mol of C [mol of Fe]$^{-1}$ day^{-1}) defines the daily iron use efficiency (IUE), i.e., the moles of carbon fixed per mole of cellular iron per day. The experimental value was determined for a daily photoperiod of 14 h, which translates to an hourly IUE of 3,526 (49,360/14), ignoring nighttime respiration. The latter value compares favorably to the IUE [3,000 mol of C [mol of Fe]$^{-1}$ h^{-1}) computed from cell models based on the Fe needed to support photosynthesis, respiration, and nitrate assimilation under saturating light (89). At high Fe/C ratios, measured growth rates reached maximum values that were inversely related to cell size (Fig. 6).

At a 10-fold-lower light intensity, Fe-limited growth rates of *T. pseudonana* and *P. minimum* also fit equation 4 but the slope of the μ versus Fe/C relationship (9,700 mol of C [mol of Fe]$^{-1}$ day^{-1}) was reduced five-fold (Fig. 6). Maximum growth rates were also reduced but by smaller amounts (1.9- to 3.8-fold). The higher Fe/C ratio needed to support the algal growth rate at low light intensities results from reduced light capture and electron flow per photosynthetic unit (PSU) and a resultant decrease in photosynthetic rate. Cells acclimate to low light inten-

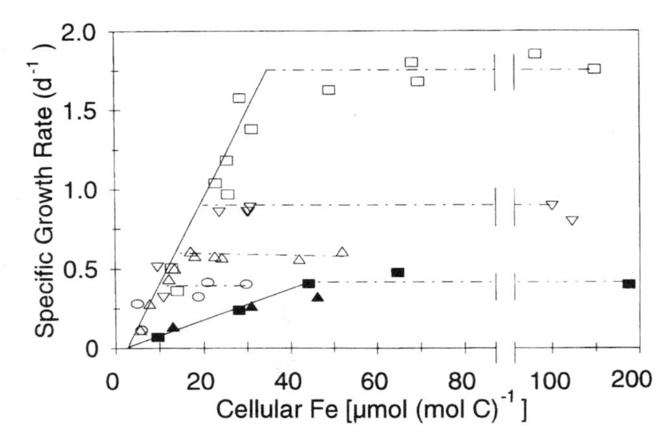

Figure 6. Specific growth rate as a function of intracellular Fe/C for coastal diatoms (□ and ■, *Thalassiosira pseudonana*; ▽, *T. weissflogii*) and dinoflagellates (△ and ▲, *Prorocentrum minimum*; ○, *P. micans*) representing a range of cell diameters (3.5, 11 to 12, 12 to 13, and 28 to 32 μm, respectively). Measurements were made at high and low light levels (500 and 50 microeinsteins m^{-2} s^{-1} [open and solid symbols, respectively]). Cells were grown in EDTA-metal ion buffered seawater media at 20°C under a 14-h/10-h light/dark cycle. Adapted from reference 119.

sities largely by increasing the number of photosynthetic units (37, 90). However, since these units contain much of the cell's iron as cytochromes and Fe/S proteins (89, 90), a low-light adaptive increase in PSUs involves a large increase in cellular iron demand. As a result, cells may often be colimited by iron and light in the environment (119). Iron-limited cells have greater difficulty in acclimating to low light intensity, while low-light-acclimated cells require higher iron levels and therefore are more likely to be iron limited.

Iron is also needed in large amounts for nitrate assimilation and for nitrogen fixation; thus, a similar colimitation can occur for iron and nitrogen. The cellular iron use efficiency models of Raven (89) predict that cells growing on nitrate need ~60% more metabolic iron than do those growing on ammonia due the additional requirements for iron in nitrate and nitrite reductase enzymes and in the photosynthetic production of the reducing equivalents and ATP needed for nitrate reduction. Culture experiments confirm these predictions: coastal and oceanic diatoms grown on nitrate required 70% more cellular iron, on average, to support a given growth rate than did cells grown on ammonia (63). These findings predict that cells utilizing nitrate will be more likely to be limited by iron and that iron-limited cells will preferentially use ammonia or other reduced N sources such as urea. This prediction was verified in bottle incubation experiments conducted in the equatorial Pacific, an iron-limited region of the ocean (87, 88). Phytoplankton in this region preferentially use ammonium for growth despite an abundance of nitrate. The addition of iron doubled the biomass specific rates of nitrate utilization without affecting ammonia utilization, resulting in a shift from ammonia to nitrate as the primary nitrogen source (87, 88).

At the bottom of the euphotic zone in stratified regions of the ocean, light is limiting, iron concentrations are typically low (18), and the primary nitrogen source is nitrate diffusing upwards from aphotic depths. Here phytoplankton are likely to be simultaneously limited by light, iron, and nitrogen. Furthermore, the light-limited cells growing on nitrate will contain unusually high levels of metabolic iron relative to carbon and major nutrients, and the settling of these cells out of the lower euphotic zone will deplete the available iron, driving the system toward further iron limitation (119). Such a process probably accounts for the low concentrations of iron observed within the deep chlorophyll maxima in stratified regions of the ocean (119). Iron and light colimitation may also largely explain why deep chlorophyll maxima in the ocean are populated by very small algal cells (104, 126), which have more favorable iron uptake kinetics than do larger cells.

Nitrogen fixation by cyanobacteria has an exceptionally high iron requirement and thus is particularly prone to iron limitation. Iron use efficiency calculations by Raven (89) indicate that cells utilizing N_2 as a nitrogen source require substantially higher cellular Fe/C ratios to support growth than do cells utilizing ammonia. Because of the low concentrations of iron in the ocean, N_2 fixation may be limited by iron in most marine environments (36, 100). The resulting low levels of N_2 fixation may explain why nitrogen is the primary major nutrient limiting algal growth in the ocean (102), in contrast to lakes, where phosphate limitation predominates (103). Iron concentrations in lakes are much higher than those in the ocean, allowing higher rates of N_2 fixation and enhanced supplies of fixed nitrogen for algal growth. The linkage between iron availability and N_2 fixation has been demonstrated in Clear Lake, Calif., a low-iron eutrophic lake in which algal production was generally limited by nitrogen rather than phosphorus (136). Addition of iron to large-volume, in- situ enclosures of lake water increased N_2 fixation by up to sixfold and carbon fixation by lesser amounts, indicating that the supply of fixed nitrogen to the lake was fundamentally limited by iron availability.

Recent evidence indicates that linkages also exist between iron limitation and silica utilization in diatoms, the algal group responsible for most biological production in the ocean. Diatoms are more heavily silicified and contain higher Si/N and Si/C ratios under iron limiting conditions in the ocean, resulting in high rates of silicon depletion relative to depletion of major nutrients (N and P) (51, 127). Consequently, iron-limited marine systems, such as the equatorial Pacific, Southern Ocean, and some coastal upwelling regimes in the eastern margins of the Pacific, are driven more rapidly toward silica limitation than are comparable iron-replete systems. The rapid depletion of silica promotes the emergence of non-diatom species, such as coccolithophores and flagellates, in iron-limited ocean waters.

Cellular growth requirements for zinc, manganese, cobalt, and copper are substantially lower than those for iron (Table 1). Like iron, the primary requirement for Mn is in photosynthetic electron transport (in the water oxidizing centers of photosystem II), and, like iron, molecular use efficiency models (90) and culture experiments (120) indicate that the amount of Mn needed to support algal growth rate increases with decreasing light. Furthermore, increases in photosynthetic capacity during low-light acclimation substantially increase the cellular demand for manganese, thereby increasing cellular Mn concentrations (120). Due to this increased demand at low light, phytoplankton are more likely to be Mn limited near the bottom of the euphotic zone. Manganese concentrations are usually low in this region relative to those at the surface, further aggravating Mn limitation. Much of the low manganese concentration at the base of the euphotic zone results from increased formation of Mn oxides at low light levels (113), as discussed above. However, at least some of the decrease in Mn concentration within this region may be due to heightened removal of Mn by low-light acclimated algal cells (120), as argued above for iron.

A major requirement for zinc is in carbonic anhydrase, needed for transport and delivery of inorganic carbon (CO_2 and HCO_3^-) to RUBISCO during carbon fixation (81, 91) Consequently, high cellular zinc concentrations are needed to support high rates of photosynthesis and growth (107). Cells need especially high levels of carbonic anhydrase to support high rates of carbon fixation at low CO_2 concentrations, and cells may become colimited by CO_2 and zinc under these conditions (81). Such colimitation may occur in eutrophic systems, where both CO_2 and zinc become depleted by high levels of algal growth. High algal biomass in eutrophic systems also generally leads to light limitation and enhanced iron depletion; and therefore, algal growth rates in such systems may be simultaneously limited by CO_2, zinc, light, and iron. The interactions among these factors, however, are likely to be complex, since the low photosynthetic rates under iron and light limitation also decrease the cellular demand for CO_2 and thus for carbonic anhydrase.

Table 1. Nutrient metal/carbon molar ratios needed to support specific growth rates (μ) of 0.5 and 1.2 day^{-1} in coastal and oceanic phytoplankton[a]

Algal group	Species	μ_{max}	Metal	Metal/C ($\mu mol\ mol^{-1}$) at:		Reference(s)
				$\mu = 0.5$	$\mu = 1.2$	
Coastal diatom	*Thalassiosira pseudonana*	1.7–1.8	Fe	15 ± 2	28	5, 41
		1.6–1.8	Zn[b]	2–3	4 ± 1	107; Sunda and Huntsman, unpublished
		1.9–2.1	Mn	1	4	120, 122
		1.87	Cu		<0.5	111
Coastal diatom	*Thalassiosira weissflogii*	0.89	Fe	13		5
Coastal dinoflagellate	*Prorocentrum minimum*	0.6	Fe	14		5, 41
Coastal cyanobacterium	*Synechococcus bacillaris*	0.62	Fe	100		Sunda and Huntsman, unpublished
		0.53	Co[c]	0.12		107
Oceanic diatom	*Thalassiosira oceanica*	1.55	Fe	~2	4	5
		1.31	Zn[b]		~1	107
		1.48	Mn		~1	116
Oceanic coccolithophorid	*Emiliania huxleyi*	1.1	Fe	~2		5
		1.17	Co[c]	1.4		107

[a]Cells were grown in trace metal ion-buffered seawater medium at 20°C under 14 h of saturating light per day.
[b]Determined in the absence of added Co.
[c]Determined in the absence of added Zn.

METAL-METAL INTERACTIONS AND THEIR ROLE IN METAL INHIBITION

Cellular metal-binding sites are never entirely specific for a single metal, and so, biological ligands designed to bind an "intended" nutrient metal can also bind competing metals with similar ionic radii and coordination geometry. Such competitive binding can occur for cellular metal uptake sites, active sites on metalloproteins, or intracellular regulatory sites such as those controlling the synthesis of metal transport proteins. Competitive binding to cellular metal uptake sites and to intracellular regulatory sites can lead to inhibition of nutrient metal uptake as observed in copper, zinc, and cadmium interactions with the nutrient metal manganese in several diatoms and chlorophytes (46, 121, 122). In experiments examining cadmium and zinc interactions with manganese, the cellular manganese uptake rate was described by a competitive saturation equation:

$$V_{Mn} = \frac{V_{maxMn}\,[Mn^{2+}]K_{Mn}}{[Mn^{2+}]K_{Mn} + [Cd^{2+}]K_{Cd} + [Zn^{2+}]K_{Zn} + 1} \qquad (5)$$

where K_{Mn}, K_{Zn}, and K_{Cd} are affinity constants for the binding of the subscripted metals to the transport site and V_{maxMn} is the saturation uptake rate. For the coastal diatom *Thalassiosira pseudonana*, values for K_{Mn}, K_{Zn}, and K_{Cd} were $10^{7.1}$, $10^{7.5}$, and $10^{8.1}$ M^{-1}, respectively (122), while similar K_{Mn} and K_{Zn} values ($10^{7.0}$ and $10^{7.7}$ M^{-1}) were found for the chlorophyte *Chlamydomonas* sp. (118). According to the above equation, substantial competitive inhibition of manganese uptake occurs only when $[Cd^{2+}]K_{Cd} + [Zn^{2+}]K_{Zn}$ exceeds $[Mn^{2+}]K_{Mn} + 1$. For *T. pseudonana* at low manganese ion concentrations, this would occur only at $[Cd^{2+}] > 1/K_{Cd}$ (i.e., $>10^{-8.1}$ M) or for $[Zn^{2+}] > 1/K_{Zn}$ (i.e., $>10^{-7.5}$ M). However for this species, most of the inhibition of manganese uptake by these metals at low $[Mn^{2+}]$ results from decreases in V_{maxMn}, which initially occurs at much lower threshold concentrations (i.e., at $[Cd^{2+}] \geq 10^{-9}$ M [W. G. Sunda and S. A. Huntsman, unpublished data]). The decreases in V_{max} apparently are caused by the intracellular binding of the competing metals to the internal negative feedback control site that regulates the capacity of the Mn uptake system (122). As a result, the cell loses its capability for feedback regulation of cellular manganese.

An examination of the measured affinity constants reveals an apparent paradox: the Mn transport system of the cells has a higher affinity for Cd and Zn than it does for Mn. The reason for this appears to lie in the low reactivity of Mn, compared to Zn and Cd, toward binding to organic ligands; this is the so-called Irving-Williams order of affinity (27). Because of these inherent differences in reactivity, it may be impossible for an Mn transporter to have a higher affinity for Mn^{2+} than it does for more reactive, similarly sized, divalent metal ions. However, in most nonpolluted waters, the much higher concentrations of Mn^{2+} relative to those of competing metal ions (Zn^{2+}, Cd^{2+}, and Cu^{2+}) should more than compensate for the lower affinity of the transport site for Mn.

Cadmium not only binds to and competitively inhibits the Mn uptake system but also is transported into the cell by this system. In *T. pseudonana*, Cd uptake at high $[Zn^{2+}]$ ($\geq 10^{-10}$ M) is described by a competitive saturation equation similar to that (equation 5) describing Mn uptake by the same system (122):

$$V_{Cd} = \frac{V_{maxCd}\,[Cd^{2+}]K_{Cd}}{[Mn^{2+}]K_{Mn} + [Cd^{2+}]K_{Cd} + [Zn^{2+}]K_{Zn} + 1} \quad (6)$$

where K values are metal affinity constants, as defined above, and V_{maxCd} is the saturation uptake rate, equal to the concentration of uptake sites times the kinetic rate constant (k_{in}) for transport of bound metal across the membrane and subsequent intracellular release (equation 2). In *T. pseudonana*, V_{maxCd} equaled V_{maxMn}, indicating equivalent k_{in} values for both metals (122). This suggests that k_{in} and thus V_{max} are controlled by a common rate-limiting reaction step, such as physical movement of bound metal across the membrane. Decreases in Mn ion concentrations substantially increase cellular uptake of cadmium due initially to decreased competition for binding to membrane uptake sites and then increasingly to negative feedback increases in V_{max} of the system.

Metals are often taken up by more than one transport system, each important under different sets of conditions (Fig. 7). Thus, while Cd uptake at high $[Zn^{2+}]$ is controlled by the Mn system, uptake at low $[Zn^{2+}]$ is dominated by a high affinity Cd system, which is under negative feedback control by cellular Zn (Fig. 5). In *T. pseudonana*, Cd uptake by this system increases by over 1,000-fold as $[Zn^{2+}]$ is decreased from 10^{-10} to 10^{-12} M (117). Cobalt also appears to be taken up by this system, and thus its uptake also increases substantially with decreasing $[Zn^{2+}]$ (107, 117) (Fig. 5). Although the identity of the low-Zn inducible system is unknown, it may be the high-affinity Zn transport system mentioned above, which is known to be under negative feedback control by cellular Zn. Because of its uptake by Mn- and Zn-controlled systems, Cd uptake by cells can be as strongly influenced by external $[Mn^{2+}]$ and $[Zn^{2+}]$ as it is by external $[Cd^{2+}]$ (117).

Inside the cell, the competing metal can bind to nutrient metal coordination sites such as those on metalloproteins. The bound competing metal often will lack the proper coordination geometry, redox behavior, Lewis acidity, or ligand exchange kinetics to confer activity to the site, resulting in a loss or alteration of metabolic function. The competitive interactions between nutrient and competing metals can thus be doubly damaging, since the intracellular uptake of the nutrient metal is inhibited while that of competing metals is enhanced, resulting in high intracellular ratios of inhibitory competing metals to nutrient metal. To counteract this situation, cells have evolved inducible detoxification mechanisms, such as the intracellular production of thiol-containing, heavy metal binding ligands (e.g., phytochelatins [1, 2]) or the export of toxic metals from the cell by efflux systems (59, 122). However, even when these mechanisms are completely successful, the cell is often still left with a deficiency of the nutrient metal as observed in the Cd, Zn, and Cu interactions with Mn in *T. pseudonana* discussed above. The toxic metals inhibited the growth rate of the diatom only at low Mn ion concentrations, and despite cellular accumulation of the competing metals, the growth inhibition was solely accounted for by an induced Mn deficiency (121, 122).

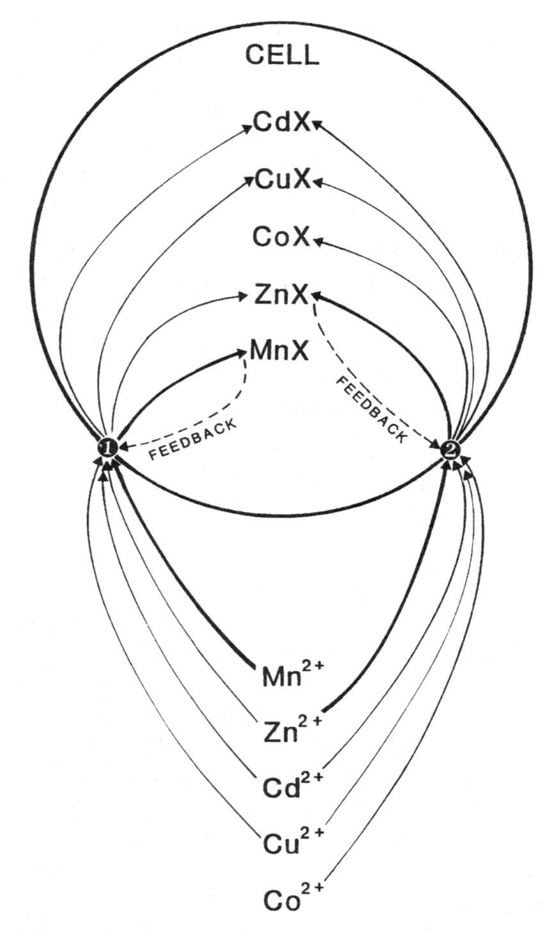

Figure 7. Hypothetical model showing competition among metals for uptake by the Mn transport system (site 1) and the zinc transport system (site 2) in phytoplankton. Both systems are under negative feedback control by either cellular Mn or Zn. CuX, CdX, ZnX, etc. refer to various bound forms of intracellular metals.

Interactions between nutrient and competing metals are quite common and appear to be major factors in metal inhibition of growth and metabolism. Such interactions have been identified in phytoplankton between Mn and Cu, Mn and Zn, Mn and Cd (see above), Zn and Cu (101, 118), Zn and Cd (58; Sunda and Huntsman, unpublished), Co and Zn (107), Fe and Cd (45), and Fe and Cu (82). In addition, algal uptake of molybdate (48) and chromate (96) (the thermodynamically stable forms of Mo and Cr in oxic waters) is inhibited by sulfate, a steriochemically similar oxyanion. Chromate appears to be taken up by the cell's sulfate transport system, explaining the inhibitory effect of sulfate on chromate uptake rates (96). As a consequence of such competitive interactions, variations in sulfate concentrations, such as those accompanying changes in salinity, should influence both Mo nutrition (48) and Cr toxicity (96). Metal-metal interactions probably play an important role in controlling metal toxicity in rivers, lakes, and estuaries contaminated with heavy metals. Competition between toxic metals and Mn may also influence the growth and species composition of phytoplankton in marine upwelling systems, where subsurface waters with high free ion levels of Cu, Zn, and Cd and low $[Mn^{2+}]$ are advected to the surface (123).

However, not all competition between nutrient metals and their chemical analogs is deleterious, since, in some cases, the competing metal will have the right chemical attributes to confer activity to the nutritional site. A notable example is competition among Cd, Co, and Zn in binding to carbonic anhydrase, an important zinc metalloenzyme (58, 81, 85). Co and Cd can substitute for Zn in this and other enzymes, and as a result, both metals can at least partially alleviate Zn deficiency in many algae (60, 107). The situation with Cd is complex, however, since in some situations it can metabolically substitute for Zn while in others it antagonizes Zn nutrition (58), apparently due to its interaction with different Zn enzymes that have different capacities for Cd activation.

METAL CONTROLS ON PHYTOPLANKTON COMMUNITIES

Results of both culture and field studies indicate that trace metals can profoundly affect the productivity and species composition of algal communities. Culture studies with trace metal ion-buffered media show a wide range in the sensitivity of marine algal species to metal toxicity (11, 39) and nutritional limitation (10, 12, 107), indicating that trace metals could have a major impact on community species composition. In metal limitation experiments, algal species isolated from oceanic environments, where trace metal concentrations are very low, were able to grow at much lower concentrations of ionic Fe, Zn, and Mn than were species from coastal environments, where metal concentrations are much higher (10). This matching of requirements of species with metal concentrations in their habitats provides strong evidence that these metals have been important selective factors in phytoplankton evolution. It also suggests that the availability of trace metal nutrients plays an important role in determining the species composition of neritic and oceanic phytoplankton communities and in preventing neritic species from proliferating in oceanic regions.

The ability of oceanic species to grow faster than coastal ones at low iron and zinc concentrations was not due to an increased capability for metal uptake, because similar-sized coastal and oceanic species had similar uptake rates (108, 124). The observed rates approached the limits on transport imposed by diffusion and ligand exchange kinetics, as discussed above. Therefore, oceanic species have been unable to increase their iron and zinc uptake kinetics because the transport systems of neritic species already had evolved to near their chemical and physical limits. To adapt to the low concentrations of iron and zinc in their environments, oceanic species have had essentially three choices. (i) They could become smaller, which increases their surface to volume ratios and decreases the thickness of their surface diffusion boundary layers. (ii) They could grow slower, which increases cellular concentrations at a given uptake rate (see equation 3) and decreases the metabolic demand for metalloenzyme catalysts. (iii) They could decrease their cellular metabolic requirements for limiting metals by altering metabolic pathways or by changing the metalloenzyme content of key pathways. All of these have occurred: oceanic species tend to be smaller and to have lower maximum growth rates than coastal species, and recent experiments indicate that they have substantially reduced cellular growth requirements for iron (12, 63, 124) and zinc (107, 108) (Table 1).

The reduction in cellular growth requirements for limiting metals is not surprising, since any reduction in growth rate places a species at a disadvantage in competition with its neighbors and since decreases in cell size can make it more easily grazed by zooplankton. There is also a limit on how small a cell can be and still perform all of its essential metabolic functions. Therefore, there has been a substantial evolutionary pressure for oceanic species to economize, but the exact mechanisms by which they have done so are obscure. It is known that under iron limiting conditions, many phytoplankton species are able to replace ferredoxin, an iron-sulfur protein, with flavodoxin, a non-metalloprotein (29, 34, 57), and that some species can replace soluble cytochrome *c*-553, which contains iron, with the copper protein plastocyanin (134). However, these examples come largely from studies with coastal or freshwater species, and the trace metal biochemistry of oceanic phytoplankton remains virtually unstudied.

Recent field iron addition experiments provide direct evidence for iron controls on the growth and species composition of oceanic phytoplankton. Historically, nitrogen and phosphorus have been considered the primary limiting nutrients in the ocean, and thus it has long been puzzling why major oceanic upwelling systems (the subarctic Pacific, the equatorial Pacific, and the Southern Ocean) had unexpectedly low productivity despite high surface nitrate and phosphate concentrations (26, 47). However, in recent bottle incubation experiments with water from these regions, additions of 1 to 10 nM iron enhanced photosynthetic efficiency (42) and stimulated phytoplankton growth (19, 24, 65–67), providing strong evidence that these systems were iron limited. Growth stimulation by added iron was also observed in coastal upwelling systems off California, although the results were quite variable depending on the local conditions (52). The iron enrichments caused shifts in species dominance from picoplankton to larger cells (88), especially diatoms (52, 65, 67), and shifts in nitrogen utilization from ammonia to nitrate (87, 88). Similar effects of added iron also have been observed in in situ mesoscale experiments conducted in 50 to 100-km^2 patches of the equatorial Pacific, providing definitive proof for iron limitation (6, 21, 64). In the second of these, the addition of three successive 1 to 2 nM iron enrichments resulted in over a 20-fold increase in chlorophyll *a* levels, mostly from the growth of large diatoms (21).

The effect of added Fe in preferentially stimulating the growth of larger cells in the at-sea experiments is exactly what would be expected, since as discussed above, larger cells are at a disadvantage with respect to iron uptake and therefore are more likely to be iron limited. By favoring smaller cells, iron limitation has an important influence on trophic interactions because small cells are efficiently grazed by microzooplankton (protozoans), capable of specific growth rates exceeding those of many phytoplankton organisms (70). The resulting continuous efficient grazing helps prevent algal blooms from developing. Also, because of the dependence of sinking rates on particle diameter (Stokes' law), the small sizes of algae and grazing zooplankton retard the loss of nutrients from the euphotic zone via settling of intact cells or cellular debris (e.g., zooplankton fecal material). Thus, by favoring small cells, iron limitation favors retention of iron and other nutrients within the euphotic zone, with obvious benefit to the local ecosystem (70, 88). This effect represents an inherent negative feedback that promotes ecosystem stability.

Compared to iron, additions of other essential trace metals (Zn and Mn) to seawater caused only a minor stimulation of phytoplankton growth in bottle incubation experiments (19, 24). However, as with iron, algal species can exhibit large differences in their requirements for other metals (Mn, Zn, and Co), suggesting that variations in these metals could have important influences on the composition of phytoplankton communities (10, 107). One notable example is the requirement of marine algal species for Zn and Co, two metals that can functionally replace one another in many enzymes. Diatoms were observed to have a primary requirement for Zn that could be partially replaced by Co, while *Emiliania huxleyi*, the most abundant oceanic coccolithophore, had a primary requirement for Co that was only partially met by Zn (107). The cyanobacterium *Synechococcus bacillaris* showed a different pattern and had an absolute requirement for Co but no demonstrable need for Zn. Thus, variations in Co and Zn concentrations and Co/Zn ratios could have important impacts on the composition of marine algal communities. Of particular note are the potential effects on the abundance of coccolithophores, the organisms primarily responsible for $CaCO_3$ formation in the ocean. By controlling $CaCO_3$ formation, these algae have a major influence on the alkalinity of surface seawater and therefore on air-sea exchange of CO_2 (31).

Differences among algal species in their sensitivity to trace metal toxicity can also influence algal species composition. For example, culture studies show that species of the cyanobacterium *Synechococcus* are particularly sensitive to copper toxicity, suggesting that copper pollution could have a disproportionate impact on this genus (11). Field studies confirm this prediction: *Synechococcus* populations were severely reduced in Cu-contaminated embayments where $[Cu^{2+}]$ exceeded 10^{-10} M, levels shown to be highly toxic in culture experiments (77). Because of the susceptibility of cyanobacteria to copper toxicity, addition of copper salts has been routinely used to mitigate nuisance blue-green algal blooms in eutrophic lakes.

CONCLUSIONS

It is clear that trace metal nutrients (notably iron) and toxicants (e.g., copper) exert important controls on the productivity and species composition of phytoplankton communities. Moreover, since trace metal nutrients are integrally involved in the metabolic pathways for major algal resources (light, inorganic carbon, and nitrogen), they exert important controls on growth limitation by these resources. Limitation by iron or other micronutrients (e.g., Zn) can impact food web structure and nutrient cycling by favoring the growth of small-celled algal species which are efficiently grazed by protozoans and which do not readily settle from the water column. However, not only do trace metals influence phytoplankton communities, but also these communities in turn exert important controls on metal chemistry, providing feedbacks between the two (Fig. 1). Many of these controls are beneficial, such as the detoxification of Cu by the release of Cu-binding organic ligands or the solubilization of Fe by chelators produced either directly or indirectly from algal activities. To fully understand the effects of trace metals on phytoplankton communities, it is necessary to take into account numerous complex interactions, ranging from those at the molecular level to those involving whole ecosystems.

Acknowledgments. I am indebted to Susan Huntsman for her sound advice and editorial assistance in the preparation of this chapter.

REFERENCES

1. **Ahner, B. A., S. Kong, and F. M. Morel.** 1995. Phytochelatin production in marine algae. I. An intraspecies comparison. *Limnol. Oceanogr.* **40:**649–657.
2. **Ahner, B. A., and F. M. Morel.** 1995. Phytochelatin production in marine algae. II. Induction by various metals. *Limnol. Oceanogr.* **40:**658–665.
3. **Allnutt, F. C. T., and W. D. Bonner, Jr.** 1987. Evaluation of reductive release as a mechanism for iron uptake from ferrioxamine B by *Chlorella vulgaris. Plant Physiol.* **85:**751–756.
4. **Anderson, M. A., and F. M. M. Morel.** 1982. The influence of aqueous iron chemistry on the uptake of iron by the coastal diatom *Thalassiosira weissflogii. Limnol. Oceanogr.* **27:**789–813.
5. **Anderson, M. A., and F. M. M. Morel.** 1980. Uptake of Fe(II) by a diatom in oxic culture medium. *Mar. Biol. Lett.* **1:**263–268.
6. **Behrenfeld, M. J., A. J. Bale, Z. S. Kolber, J. Aiken, and P. G. Falkowski.** 1996. Confirmation of iron limitation of phytoplankton photosynthesis in the equatorial Pacific. *Nature* **383:**508–511.
7. **Berg, J. M., and Y. Shi.** 1996. The galvanization of biology: a growing appreciation of the roles of zinc. *Science* **271:**1081–1085.
8. **Boyle, E. A., F. Sclater, and J. M. Edmond.** 1976. On the marine geochemistry of cadmium. *Nature* **263:**42–44.
9. **Boyle, E. A., J. M. Edmond, and E. R. Sholkovitz.** 1977. The mechanism of iron removal in estuaries. *Geochim. Cosmochim. Acta* **41:**1313–1324.
10. **Brand, L. E., W. G. Sunda, and R. R. L. Guillard.** 1983. Limitation of marine phytoplankton reproductive rates by zinc, manganese and iron. *Limnol. Oceanogr.* **28**:1182–1198.
11. **Brand, L. E., W. G. Sunda, and R. R. L. Guillard.** 1986. Reduction of marine phytoplankton reproduction rates by copper and cadmium. *J. Exp. Mar. Biol. Ecol.* **96:**225–250.
12. **Brand, L. E.** 1991. Minimum iron requirements of marine phytoplankton and the implications for the biogeochemical control of new production. *Limnol. Oceanogr.* **36:**1756–1772.
13. **Bruland, K. W., G. A. Knauer, and J. H. Martin.** 1978. Zinc in north-east Pacific water. *Nature* **271:**741–743.
14. **Bruland, K. W.** 1980. Oceanographic distributions of cadmium, zinc, nickel and copper in the North Pacific. *Earth Planet. Sci. Lett.* **47:**176–198.
15. **Bruland, K. W., and R. P. Franks.** 1983. Mn, Ni, Cu, Zn, and Cd in the western North Atlantic, p. 395–414. *In* C. S. Wong, E. Boyle, K. W. Bruland, J. D. Burton, and E. D. Goldberg (ed.), *Trace Metals in Sea Water.* Plenum Press, New York, N.Y.
16. **Bruland, K. W.** 1989. Complexation of zinc by natural organic ligands in the central North Pacific. *Limnol. Oceanogr.* **34:**269–285.
17. **Bruland, K. W.** 1992. Complexation of cadmium by natural organic ligands in the central North Pacific. *Limnol. Oceanogr.* **37:**1008–1017.
18. **Bruland, K. W., K. J. Orians, and J. P. Cowen.** 1994. Reactive trace metals in the stratified central North Pacific. *Geochim. Cosmochim. Acta* **58:**3171–3182.
19. **Buma, A. G. J., H. J. W. de Baar, R. F. Nolting, and A. J. van Bennekom.** 1991. Metal enrichment experiments in the Weddell-Scotia Seas: effects of iron and manganese on various plankton communities. *Limnol. Oceanogr.* **36:**1865–1878.
20. **Byrne R. H., L. R. Kump, and K. J. Cantrell.** 1988. The influence of temperature and pH on trace metal speciation in seawater. *Mar. Chem.* **25:**163–181.

21. **Coale, K. H., et al.** 1996. A massive phytoplankton bloom induced by an ecosystem-scale iron fertilization experiment in the equatorial Pacific Ocean. *Nature* **383:**495–501.
22. **Coale, K. H., and K. W. Bruland.** 1988. Copper complexation in the northeast Pacific. *Limnol. Oceanogr.* **33:**1084–1101.
23. **Coale, K. H., and K. W. Bruland.** 1990. Spatial and temporal variability in copper complexation in the North Pacific. *Deep-Sea Res.* **34:**317–336.
24. **Coale, K. H.** 1991. Effects of iron, manganese, copper, and zinc enrichments on productivity and biomass in the subarctic Pacific. *Limnol. Oceanogr.* **36:**1851–1864.
25. **Cowen, J. P, and K. W. Bruland.** 1985. Metal deposits associated with bacteria: implications for Fe and Mn marine biogeochemistry. *Deep-Sea Res.* **32:**253–272.
26. **Cullen, J. T.** 1991. Hypotheses to explain high nutrient conditions in the open sea. *Limnol. Oceanogr.* **36:**1579–1599.
27. **da Silva, J. J. R. F., and R. J. P. Williams.** 1991. *The Biological Chemistry of the Elements.* Clarendon Press, Oxford, United Kingdom.
28. **Donat, J. R., and K. W. Bruland.** 1990. A comparison of two voltammetric techniques for determining zinc speciation in northeast Pacific Ocean waters. *Mar. Chem.* **28:**301–323.
29. **Doucette, E. J., D. L. Erdner, M. L. Peleato, J. J. Hartman, and D. M. Anderson.** 1996. Quantitative analysis of iron-stress related proteins in *Thalassiosira weissflogii*: measurement of flavodoxin and ferredoxin using HPLC. *Mar. Ecol. Prog. Ser.* **130:**269–276.
30. **Droop, M. R.** 1968. Vitamin B_{12} and marine ecology. IV. Kinetics of uptake, growth and inhibition in *Monochrysis lutheri*. *J. Mar. Biol. Assoc. U. K.* **48:**689–733.
31. **Dymond, J., and M. Lyle.** 1985. Flux comparisons between sediments and sediment traps in the eastern tropical Pacific: implications for atmospheric CO_2 variations during the Pleistocene. *Limnol. Oceanogr.* **30:**699–712.
32. **Eide, D., and M. L. Guerinot.** 1997. Metal ion uptake in eucaryotes. *ASM News* **63:**199–205.
33. **Emerson, S., S. Kalhorn, L. Jacobs, B. M. Tebo, K. H. Nealson, and R. A. Rosson.** 1982. Environmental oxidation rate of manganese (II): bacterial catalysis. *Geochim. Cosmochim. Acta* **46:** 1073–1079.
34. **Entsch, B., R. G. Sim, and B. G. Hatcher.** 1983. Indications from photosynthetic components that iron is a limiting nutrient in primary producers on coral reefs. *Mar. Biol.* **73:**17–30.
35. **Falchuk, K. H., A. Krishan, and B. L. Vallee.** 1975. DNA distribution in the cell cycle of *Euglena gracilis*. Cytofluorometry of zinc deficient cells. *Biochemistry* **14:**3439–3444.
36. **Falkowski, P. G.** 1997. Evolution of the nitrogen cycle and its influence on the biological sequestration of CO_2 in the ocean. *Nature* **387:**272–274.
37. **Falkowski, P. G., T. G. Owens, A. C. Ley, and D. C. Mauzerall.** 1981. Effects of growth irradiance levels on the ratio of reaction centers in two species of marine phytoplankton. *Plant Physiol.* **68:** 969–973.
38. **Finden, D. A. S., E. Tipping, G. H. M. Jaworski, and C. S. Reynolds.** 1984. Light-induced reduction of natural iron(III) oxide and its relevance to phytoplankton. *Nature* **309:**783–784.
39. **Gavis, J., R. R. L. Guillard, and B. L. Woodward.** 1981. Cupric ion activity and the growth of phytoplankton clones isolated from different marine environments. *J. Mar. Res.* **39:**315–333.
40. **Gledhill, M., and C. M. G. van den Berg.** 1994. Determination of complexation of iron (III) with natural organic complexing ligands in seawater using cathodic stripping voltammetry. *Mar. Chem.* **47:**41–54.
41. **Goldman, J. C., J. J. McCarthy, and D. G. Peavey.** 1979. Growth rate influence on the chemical composition of phytoplankton in oceanic waters. *Nature* **279:**210–215.
42. **Green, R. M., Z. S. Kolber, D. G. Swift, N. W. Tindale, and P. G. Falkowski.** 1994. Physiological limitation of phytoplankton photosynthesis in the eastern equatorial Pacific determined from variability in the quantum yield of flourescence. *Limnol. Oceanogr.* **39:**1061–1074.
43. **Gutknecht, J.** 1981. Inorganic mercury (Hg^{2+}) transport through lipid bilayer membranes. *J. Membr. Biol.* **61:**61–66.
44. **Harrison, G. I., and F. M. M. Morel.** 1986. Response of the marine diatom *Thalassiosira weissflogii* to iron stress. *Limnol. Oceanogr.* **31:**989–997.
45. **Harrison, G. I., and F. M. M. Morel.** 1983. Antagonism between cadmium and iron in the marine diatom *Thalassiosira weissflogii*. *J. Phycol.* **19:**495–507.

46. **Hart, B. A., P. E. Bertram, and B. D. Scaife.** 1979. Cadmium transport by *Chlorella pyrenoidosa*. *Environ. Res.* **18:**327–335.
47. **Hart, T. J.** 1934. On the phytoplankton in the southwest Atlantic and Bellinghausen Sea, 1929–31. *Discovery Rep.* **8:**1–268.
48. **Howarth, R. W., R. Marino, and J. J. Cole.** 1988. Nitrogen fixation in fresh water, estuarine, and marine ecosystems. 2. Biogeochemical controls. *Limnol. Oceanogr.* **33:**688–701.
49. **Hudson, R. J. M. and F. M. M. Morel.** 1993. Trace metal transport by marine microorganisms: implications of metal coordination kinetics. *Deep-Sea Res.* **40:**129–151.
50. **Hudson, R. J. M., and F. M. M. Morel.** 1990. Iron transport in marine phytoplankton: kinetics of cellular and medium coordination reactions. *Limnol. Oceanogr.* **35:**1002–1020.
51. **Hutchins, D. A., and K. W. Bruland.** 1998. Iron-limited diatom growth and Si:N uptake ratios in a coastal upwelling regime. *Nature* **393:**561–564.
52. **Hutchins, D. A., G. R. Ditullio, Y. Zhang, and K. W. Bruland.** 1998. An iron limitation mosaic in the California upwelling regime. *Limnol. Oceanogr.* **43:**1037–1054.
53. **Johnson, K. S., R. M. Gordon, and K. H. Coale.** 1997. What controls dissolved iron concentrations in the world ocean? *Mar. Chem.* **57:**137–161.
54. **Jones, G. J., B. P. Palenik, and F. M. M. Morel.** 1987. Trace metal reduction by phytoplankton: the role of plasmalemma redox enzymes. *J. Phycol.* **23:**237–244.
55. **Knauer, K., R. Behra, and L. Sigg.** 1997. Adsorption and uptake of copper by the green alga *Scenedesmus subspicatus* (Chlorophyta). *J. Phycol.* **33:**596–601.
56. **Kozelka, P. B., and K. W. Bruland.** 1998. Chemical speciation of dissolved Cu, Zn, Cd, Pb, in Narragansett Bay, Rhode Island. *Mar. Chem.* **60:**267–282.
57. **LaRoche, J., P. W. Boyd, R. M. L. McKay, and R. J. Geider.** 1996. Flavodoxin as in situ marker for iron stress in phytoplankton. *Nature* **382:**802–805.
58. **Lee, J. G., S. B. Roberts, and F. M. M. Morel.** 1995. Cadmium: a nutrient for the marine diatom *Thalassiosira weissflogii*. *Limnol. Oceanogr.* **40:**1056–1063.
59. **Lee, J. G., B. A. Ahner, and F. M. M. Morel.** 1996. Export of cadmium and phytochelatin by the marine diatom *Thalassiosira weissflogii*. *Environ. Sci. Technol.* **30:**1814–1821.
60. **Lee, J. G., and F. M. M. Morel.** 1995. Replacement of zinc by cadmium in marine phytoplankton. *Mar. Ecol. Prog. Ser.* **127:**305–309.
61. **Lewis, B. L., G. W. Luther, and T. M. Church.** 1995. Determination of metal-organic complexation in natural waters by SWASV with pseudopolarograms. *Electroanalysis* **7:**166–177.
62. **Maldonado, M. T., and N. M. Price.** Utilization of iron bound to strong organic ligands by plankton communities in the subarctic Pacific Ocean. *Deep-Sea Res.*, in press.
63. **Maldonado, M. T., and N. M. Price.** 1996. Influence of N substrate on Fe requirements of marine centric diatoms. *Mar. Ecol. Prog. Ser.* **141:**161–172.
64. **Martin, J. H., et al.** 1994. Testing the iron hypothesis in ecosystems of the Equatorial Pacific Ocean. *Nature* **371:**123–129.
65. **Martin, J. H., R. M. Gordon, S. E. Fitzwater, and W. W. Broenkow.** 1989. VERTEX: phytoplankton/iron studies in the Gulf of Alaska. *Deep-Sea Res.* **36:**649–680.
66. **Martin, J. H., S. E. Fitzwater, and R. M. Gordon.** 1990. Iron deficiency limits phytoplankton growth in Antarctic waters. *Global Biogeochem. Cycles* **4:**5–12.
67. **Martin, J. H., R. M. Gordon, and S. E. Fitzwater.** 1991. The case for iron. *Limnol. Oceanogr.* **36:**1793–1802.
68. **Mason, R. P., J. R. Reinfelder, and F. M. M. Morel.** 1996. Uptake, toxicity, and trophic transfer of mercury in a coastal diatom. *Environ. Sci. Technol.* **30:**1835–1845.
69. **Miles, C. J., and P. L. Brezonik.** 1981. Oxygen consumption in humic colored waters by a photochemical ferrous-ferric catalytic cycle. *Environ. Sci. Technol.* **15:**1089–1095.
70. **Miller, C. B., B. W. Frost, P. A. Wheeler, M. R. Landry, N. Welschmeyer and T. M. Powell.** 1991. Ecological dynamics in the subarctic Pacific, a possibly iron-limited ecosystem. *Limnol. Oceanogr.* **36:**1600–1615.
71. **Miller, W. L., D. W. King, J. Lin, and D. R. Kester.** 1995. Photochemical redox cycling of iron in coastal seawater. *Mar. Chem.* **50:**63–77.
72. **Moffett, J. W.** 1995. Temporal and spatial variability of copper complexation by strong chelators in the Sargasso Sea. *Deep-Sea Res.* **42:**1273–1295.

73. **Moffett, J. W., L. E. Brand, and R. G. Zika.** 1990. Distribution and potential sources and sinks of copper chelators in the Sargasso Sea. *Deep-Sea Res.* **37:**27–36.
74. **Moffett, J. W., and L. E. Brand.** 1996. Production of strong, extracellular Cu chelators by marine cyanobacteria in response to Cu stress. *Limnol. Oceanogr.* **41:**388–395.
75. **Moffett, J. W.** 1994. A radiotracer study of cerium and manganese uptake onto suspended particles in Chesapeake Bay. *Geochim. Cosmochim. Acta* **58:**695–703.
76. **Moffett, J. W., and R. G. Zika.** 1987. Reaction kinetics of hydrogen peroxide with copper and iron in seawater. *Environ. Sci. Technol.* **21:**804–810.
77. **Moffett, J. W., L. E. Brand, P. L. Croot, and K. A. Barbeau.** 1997. Cu speciation and cyanobacterial distribution in harbors subject to anthropogenic Cu inputs. *Limnol. Oceanogr.* **42:**789–799.
78. **Morel, F. M. M., and R. J. M. Hudson.** 1984. The geobiological cycle of trace elements in aquatic systems: Redfield revisited, p. 251–281. *In* W. Stumm (ed.), *Chemical Processes in Lakes.* John Wiley & Sons, Inc., New York, NY.
79. **Morel, F. M. M., R. J. M. Hudson, and N. M. Price.** 1991. Limitation of productivity by trace metals in the sea. *Limnol. Oceanogr.* **36:**1742–1755.
80. **Morel, F. M. M.** 1987. Kinetics of nutrient uptake and growth in phytoplankton. *J. Phycol.* **23:**137–150.
81. **Morel, F. M. M., J. R. Reinfelder, S. B. Roberts, C. P. Chamberlain, J. G. Lee, and D. Yee.** 1994. Zinc and carbon co-limitation of marine phytoplankton. *Nature* **369:**740–742.
82. **Murphy, L. S., R. R. L. Guillard, and J. F. Brown.** 1984. The effects of iron and manganese on copper sensitivity in diatoms: differences in the responses of closely related neritic and oceanic species. *Biol. Oceanogr.* **3:**187–202.
83. **Nielands, J. B.** 1981. Iron absorption and transport in microorganisms. *Annu. Rev. Nutr.* **1:**27–46.
84. **Phinney, J. T. and K. W. Bruland.** 1994. Uptake of lipophilic organic Cu, Cd, and Pb complexes in the coastal diatom *Thalassiosira weissflogii. Environ. Sci. Technol.* **28:**1781–1790.
85. **Price, N. M., and F. M. M. Morel.** 1990. Cadmium and cobalt substitution for zinc in a marine diatom. *Nature* **344:**658–660.
86. **Price, N. M., and F. M. M. Morel.** 1991. Co-limitation of phytoplankton growth by nickel and nitrogen. *Limnol. Oceanogr.* **36:**1071–1077.
87. **Price, N. M., L. F. Andersen, and F. M. M. Morel.** 1991. Iron and nitrogen nutrition of equatorial Pacific plankton. *Deep-Sea Res.* **38:**1361–1378.
88. **Price, N. M., B. A. Ahner, and F. M. M. Morel.** 1994. The equatorial Pacific Ocean: grazer controlled phytoplankton populations in an iron-limited ecosystem. *Limnol. Oceanogr.* **39:**520–534.
89. **Raven, J. A.** 1988. The iron and molybdenum use efficiencies of plant growth with different energy, carbon, and nitrogen sources. *New Phytol.* **109:**279–287.
90. **Raven, J. A.** 1990. Predictions of Mn and Fe use efficiencies of phototrophic growth as a function of light availability for growth and C assimilation pathway. *New Phytol.* **116:**1–18.
91. **Raven, J. A., and A. M. Johnson.** 1991. Mechanisms of inorganic carbon acquisition in marine phytoplankton and their implications for the use of other resources. *Limnol. Oceanogr.* **36:**1701–1714.
92. **Redfield, A. C.** 1934. On the proportions of organic derivatives in seawater and their relation to the composition of plankton. p. 177–192. *In* R. J. Daniel (ed.), *James Johnson Memorial Volume*, Liverpool University Press, Liverpool, United Kingdom.
93. **Redfield, A. C., B. H. Ketchum, and F. A. Richards.** 1963. The influence of organisms on the composition of seawater, p. 26–77. *In* M. N. Hill (ed.), *The Sea*, vol. 2. John Wiley & Sons, Inc., New York, N.Y.
94. **Richardson, L. L., C. Aguilar, and K. H. Nealson.** 1988. Manganese oxidation in pH and O_2 microenvironments produced by phytoplankton. *Limnol. Oceanogr.* **33:**352–363.
95. **Richardson, L. L., and K. D. Stolzenbach.** 1995. Phytoplankton cell size and the development of microenvironments. *FEMS Microb. Ecol.* **16:**185–192.
96. **Riedel, G. F.** 1985. The relationship between chromium(VI) uptake, sulfate uptake, and chromium toxicity in the estuarine diatom *Thalassiosira pseudonana. Aquat. Toxicol.* **7:**191–204.
97. **Roitz, J. S., and K. W. Bruland.** 1997. Determination of dissolved manganese(II) in coastal and estuarine waters by differential pulse cathodic stripping voltammetry. *Anal. Chim. Acta* **344:**175–180.

98. **Rue, E. L., and K. W. Bruland.** 1995. Complexation of iron(III) by natural organic ligands in the Central North Pacific as determined by a new competitive ligand equilibration/adsorptive cathodic stripping voltammetric method. *Mar. Chem.* **50:**117–138.
99. **Rue, E. L., and K. W. Bruland.** 1997. The role of organic complexation on ambient iron chemistry in the equatorial Pacific Ocean and the response of a mesoscale iron addition experiment. *Limnol. Oceanogr.* **42:**901–910.
100. **Rueter, J. G., D. A. Hutchins, R. W. Smith, and N. L. Unsworth.** 1992. Iron nutrition of *Trichodesmium*, p. 289–306. *In* E. J. Carpenter, D. G. Capone, and J. G. Rueter (ed.), *Marine Pelagic Cyanobacteria: Trichodesmium and Other Diazotrophs*. Kluwer Academic Publishers, Boston, Mass.
101. **Rueter, J. G., and F. M. M. Morel.** 1982. The interaction between zinc deficiency and copper toxicity as it affects the silicic acid uptake mechanisms in *Thalassiosira pseudonana. Limnol. Oceanogr.* **26:**67–73.
102. **Ryther, J. H., and W. M. Dunstan.** 1971. Nitrogen, phosphorus, and eutrophication in the coastal marine environment. *Science* **171:**1008–1013.
103. **Schindler, D. W.** 1976. The evolution of phosphorus limitation in lakes. *Science* **195:**260–262.
104. **Shimada, A., T. Hasegawa, I. Umeda, N. Kadoya, and T. Maruyama.** 1993. Spatial mesoscale patterns of West Pacific picoplankton as analyzed by flow cytometry: their contribution to subsurface chlorophyll maxima. *Mar. Biol.* **115:**209–215.
105. **Sigg, L., A. Kuhn, H. Xue, E. Kiefer, and D. Kistler.** 1995. Cycles of trace elements (copper and zinc) in a eutrophic lake, p. 177–194. *In* C. P. Huang, C. R. O'Melia, and J. J. Morgan (ed.), *Aquatic Chemistry: Interfacial and Interspecies Processes.* American Chemical. Society, Washington, D.C.
106. **Stumm, W., and J. J. Morgan.** 1981. *Aquatic Chemistry.* John Wiley & SOns, Inc., New York, N.Y.
107. **Sunda, W. G., and S. A. Huntsman.** 1995. Cobalt and zinc interreplacement in marine phytoplankton: Biological and geochemical implications. *Limnol. Oceanogr.* **40:**1404–1417.
108. **Sunda, W. G., and S. A. Huntsman.** 1992. Feedback interactions between zinc and phytoplankton in seawater. *Limnol. Oceanogr.* **37:**25–40.
109. **Sunda, W. G., and P. J. Hanson.** 1979. Chemical speciation of copper in river water: Effect of total copper, pH, carbonate and dissolved organic matter, p. 147–180. *In* E. A. Jenne (ed.), *Chemical Modeling in Aqueous Systems: Speciation, Sorption, Solubility and Kinetics.* American Chemical Society, Washington, D.C.
110. **Sunda, W. G., and S. A. Huntsman.** 1991. The use of chemiluminescence and ligand competition with EDTA to measure copper concentration and speciation in seawater. *Mar. Chem.* **36:**137–163.
111. **Sunda, W. G., and S. A. Huntsman.** 1995. Regulation of copper concentration in the oceanic nutricline by phytoplankton uptake and regeneration cycles. *Limnol. Oceanogr.* **40:**135–137.
112. **Sunda, W. G., and S. A. Huntsman.** 1987. Microbial oxidation of manganese in a North Carolina estuary. *Limnol. Oceanogr.* **32:**552–564.
113. **Sunda, W. G., and S. A. Huntsman.** 1988. Effect of sunlight on redox cycles of manganese in the southwestern Sargasso Sea. *Deep-Sea Res.* **35:**1297–1317.
114. **Sunda, W. G., and S. A. Huntsman.** 1994. Photoreduction of manganese oxides in seawater. *Mar. Chem.* **46:**133–152.
115. **Sunda, W. G., and S. A. Huntsman.** 1985. Regulation of cellular manganese and manganese transport rates in the unicellular alga *Chlamydomonas. Limnol. Oceanogr.* **30:**71–80.
116. **Sunda, W. G., and S. A. Huntsman.** 1986. Relationships among growth rate, cellular manganese concentrations and manganese transport kinetics in estuarine and oceanic species of the diatom *Thalassiosira. J. Phycol.* **22:**259–270.
117. **Sunda, W. G., and S. A. Huntsman.** 1998. Control of Cd concentrations in a coastal diatom by interactions among free ionic Cd, Zn, and Mn in seawater. *Environ. Sci. Technol.* **32:**2961–2968.
118. **Sunda, W. G., and S. A. Huntsman.** 1998. Interactions among Cu^{2+}, Zn^{2+}, and Mn^{2+} in controlling cellular Mn, Zn, and growth rate in the coastal alga *Chlamydomonas. Limnol Oceanogr.* **43:**1055–1064.
119. **Sunda, W. G., and S. A. Huntsman.** 1997. Interrelated influence of iron, light, and cell size on growth of marine phytoplankton. *Nature* **390:**389–392.

120. **Sunda, W. G., and S. A. Huntsman.** 1998. Interactive effects of Mn, light and toxic metals on cellular manganese and growth in a coastal diatom. *Limnol. Oceanogr.* **43:**1467–1475.
121. **Sunda, W. G., and S. A. Huntsman.** 1983. Effect of competitive interactions between manganese and copper on cellular manganese and growth in estuarine and oceanic species of the diatom *Thalassiosira. Limnol. Oceanogr.* **28:**924–934.
122. **Sunda, W. G., and S. A. Huntsman.** 1996. Antagonisms between cadmium and zinc toxicity and manganese limitation in a coastal diatom. *Limnol. Oceanogr.* **41:**373–387.
123. **Sunda, W. G.** 1988/89. Trace metal interactions with marine phytoplankton. *Biol. Oceanogr.* **6:** 411–442.
124. **Sunda, W. G., and S. A. Huntsman.** 1995. Iron uptake and growth limitation in oceanic and coastal phytoplankton. *Mar. Chem.* **50:**189–206.
125. **Sunda, W. G.** 1994. Trace metal/phytoplankton interactions in the sea, p. 213–247. *In* G. Bidoglio and W. Stumm (ed.), *Chemistry of Aquatic Systems: Local and Global Perspectives*. Kluwer Academic, Boston, Mass.
126. **Takahashi, M., and T. Hori.** 1984. The abundance of picoplankton in the subsurface chlorophyll maximum layer in subtropical and tropical waters. *Mar. Biol.* **79:**177–186.
127. **Takeda, S.** 1998. Influence of iron availability on the nutrient consumption ratio of diatoms in oceanic waters. *Nature* **393:**774–777.
128. **Vallee, B. L., and K. H. Falchuk.** 1981. Zinc and gene expression. *Philos. Trans. R. Soc. London* **294:**185–197.
129. **Van den Berg, C. M. G., A. G. A. Merks, and E. K. Duursma.** 1987. Organic complexation and its control of the dissolved concentration of copper and zinc in the Scheldt estuary. *Estuarine Coastal Shelf Sci.* **24:**785–797.
130. **Waite, T. D., and F. M. M. Morel.** 1984. Photoreductive dissolution of colloidal iron oxides in natural waters. *Environ. Sci. Technol.* **18:**860–868.
131. **Wells, M. L., and L. M. Mayer.** 1991. The photoconversion of colloidal iron oxyhydroxides in seawater. *Deep-Sea Res.* **38:**1379–1395.
132. **Wilhelm, S. W., and C. G. Trick.** 1994. Iron-limited growth of cyanobacteria: multiple siderophore production is a common response. *Limnol. Oceanogr.* **39:**1979–1984.
133. **Wilhelm, S. W.** 1995. Ecology of iron-limited cyanobacteria: a review of physiological responses and implications for aquatic systems. *Aquat. Microb. Ecol.* **9:**265–303.
134. **Wood, P. M.** 1978. Interchangeable copper and iron proteins in algal photosynthesis. Studies on plastocyanin and cytochrome c-552 in *Chlamydomonas*. *Eur. J. Biochem.* **87:**9–19.
135. **Wu, J., and G. W. Luther.** 1995. Complexation of Fe(III) by natural organic ligands in the Northwest Atlantic Ocean (as determined) by a competitive equilibration method and kinetic approach. *Mar. Chem.* **50:**159–177.
136. **Wurtsbaugh, W. A., and A. J. Horne.** 1983. Iron in eutrophic Clear Lake, California: its importance for algal nitrogen fixation and growth. *Can. J. Fish. Aquat. Sci.* **40:**1419–1429.
137. **Xue, H., A. Oestreich, D. Kistler, and L. Sigg.** 1996. Free cupric ion concentrations and Cu complexation in selected Swiss Lakes. *Aquat. Sci.* **58:**69–87.
138. **Xue, H., and W. G. Sunda.** 1997. Comparison of [Cu^{2+}] measurements in lake water determined by ligand exchange and cathodic stripping voltammetry and by ion-selective electrode. *Environ. Sci. Technol.* **31:**1902–1909.
139. **Xue, H., D. Kistler, and L. Sigg.** 1996. Competition of copper and zinc for strong ligands in a euthrophic lake. *Limnol. Oceanogr.* **40:**1142–1152.
140. **Zhao, H., and D. Eide.** 1996. The yeast ZRT1 gene encodes the zinc transporter protein of a high-affinity uptake system induced by zinc limitation. *Proc. Natl. Acad. Sci. USA* **93:**2454–2458.

Environmental Microbe-Metal Interactions
Edited by Derek R. Lovley

Chapter 5

Biologically Controlled Mineralization of Magnetic Iron Minerals by Magnetotactic Bacteria

Dennis A. Bazylinski and Richard B. Frankel

The biomineralization of several magnetic iron minerals, including the iron oxide magnetite (Fe_3O_4; $Fe^{2+}Fe_2{}^{3+}O_4$) and the iron sulfides greigite (Fe_3S_4) and pyrrhotite (Fe_7S_8), can be mediated by bacteria as well as by other organisms. These minerals are known to be synthesized by procaryotes in one of two fundamentally different modes of biomineralization. The first is uncontrolled mineral formation, called biologically induced mineralization (BIM) (92), while the second is directed-mineral formation, termed biologically controlled mineralization (BCM), also referred to as organic matrix-mediated mineralization (92) or boundary-organized biomineralization (96). Whereas in BIM the organism does not appear to control the biomineralization process, in BCM the organism exerts a high degree of control over the mineralization process.

Different physiological types of bacteria are responsible for BIM and BCM of magnetic minerals. Dissimilatory iron-reducing and dissimilatory sulfate-reducing bacteria cause the formation of magnetite and greigite, respectively, as well as many other nonmagnetic iron minerals, by BIM processes. Among the procaryotes, only the magnetotactic bacteria biomineralize magnetite and greigite by BCM. There is no apparent function to the magnetite and greigite particles produced via BIM by microorganisms. However, there is ecological, evolutionary, and perhaps physiological significance to the BCM particles produced by the magnetotactic bacteria. Additionally, the biomineralization of magnetic minerals by bacteria can be an important source of fine-grained (<1 μm) magnetic material in sediments and soils, which contributes to the paleomagnetic record of ancient geomagnetic field behavior and to the mineral magnetic record of paleoclimate changes (37, 38, 94, 137). It is important to be able to discern the differences between particle types so as to be able to recognize biogenic magnetic mineral particles and also to determine

Dennis A. Bazylinski • Department of Microbiology, Iowa State University, Ames, IA 50011. ***Richard B. Frankel*** • Department of Physics, California Polytechnic State University, San Luis Obispo, CA 93407.

which types of microorganisms are responsible for the deposition of the magnetic mineral particles. In addition, recent mineral discoveries in the Martian meteorite ALH84001 (117), which have been interpreted as evidence for ancient life on Mars, make this need even greater. The purpose of this chapter is to examine and review features of BIM- and BCM-type magnetic particles, to describe the microorganisms that produce magnetic minerals (focusing mainly on the magnetotactic bacteria), and to describe the biomineralization processes involved in the synthesis of magnetic minerals. We also review the physics and function of magnetotaxis in light of recent findings.

BIOLOGICALLY INDUCED MINERALIZATION

In BIM, biomineralization occurs extracellularly as a result of metabolic processes of the organism and subsequent chemical reactions involving metabolic byproducts. In most cases, the organisms secrete a metabolic product(s) that reacts with a specific ion or compound in the surrounding environment, resulting in the production of extracellular mineral particles. Thus, mineralization is an unintended consequence of metabolic activities that are characterized by being poorly crystallized, having a broad particle size distribution, and lacking specific crystalline morphologies. In addition, the lack of control over biomineralization often results in decreased mineral specificity and/or the inclusion of impurities in the particles. Because BIM processes are not controlled by the organisms and because most mineral particles produced by BIM can also be produced chemically with the same crystallochemical features without bacterial catalysts (103), BIM is equivalent to nonbiogenic mineralization under the same environmental conditions. Particles formed by BIM are therefore likely to be indistinguishable from particles produced nonbiogenically in inorganic chemical reactions under similar conditions, such as during pedogenesis (93). The implication is that in BIM the minerals nucleate in solution or form from poorly crystallized mineral species already present. However, bacterial surfaces can also act as important sites for the adsorption of ions and mineral nucleation (25, 78). The importance of bacterial surfaces in most cases of BIM has not been elucidated.

Magnetite

Magnetite can be formed by the dissimilatory iron-reducing bacteria and probably by other microorganisms that reduce iron in nondissimilatory, non-energy-yielding reactions as well. Dissimilatory iron-reducing microorganisms respire with oxidized iron, Fe(III), in the form of amorphous Fe(III) oxyhydroxide (87, 88) under anaerobic conditions and secrete reduced iron, Fe(II), into the surrounding environment, where it subsequently reacts with excess Fe(III) oxyhydroxide to form magnetite. Magnetite particles, formed extracellularly by these microorganisms, are (i) irregular in shape with a relatively broad size distribution, with most particles being in the superparamagnetic size range (<35 nm), and (ii) poorly crystallized (124, 166).

Many different species and physiological types of bacteria reduce Fe(III) (see reference 86 for a complete list and chapter 2 of this volume for a more detailed discussion of these microorganisms). However, it seems that only a handful of these organisms conserve energy and grow from the reduction of this environmentally abundant terminal electron acceptor (125) and produce magnetite. *Geobacter metallireducens* and *Shewanella* (formerly *Alteromonas*) *putrefaciens* are the most extensively studied members of this group and are phylogenetically associated with the δ and γ subdivisions, respectively, of the *Proteobacteria* (84, 89, 125), a diverse, vast assemblage of gram-negative procaryotes in the domain *Bacteria* (196). *Shewanella* and *Geobacter* species appear to be very common in aquatic and sedimentary environments (43), and several new Fe(III) species of these genera have been isolated in recent years (36, 152), suggesting that members of these genera may be the most environmentally significant microbes involved in Fe(III) reduction and extracellular magnetite precipitation. Although magnetite formation by BIM has been demonstrated only in cultures of the mesophiles *Shewanella*, *Geobacter* (87, 91), and *Geothrix fermentans* (J. D. Coates, personal communication), several thermophiles, including the Fe(III)-reducing bacterium strain TOR-39 (198), the archaeon *Pyrobaculum islandicum* (185), and the bacterium *Thermotoga maritima* (185), and mixed cultures or consortia containing Fe(III) reducers (15, 81, 197), it is likely that magnetite can be produced by a pure culture of any Fe(III)-reducing bacterium, regardless of whether the organism can grow with Fe(III), under suitable environmental conditions when the Fe(III) source is insoluble amorphous Fe(III) oxides. It should also be noted that black, unidentified magnetic precipitates commonly observed forming in enrichment cultures or pure cultures of Fe(III) reducing bacteria containing insoluble amorphous Fe(III) oxide as the Fe(III) source (see, e.g., references 64, 163, and 164) probably consist of primarily magnetite. The halotolerant, facultatively anaerobic, iron-reducing bacterium described by Rossello-Mora et al. (152) most probably produces nonstoichiometric particles of magnetite with a composition intermediate between those of magnetite and maghemite (γ-Fe_2O_3) (68). *S. putrefaciens* is able to reduce and grow on the Fe(III) in magnetite which contains both Fe(II) and Fe(III) (79), while it appears that *G. metallireducens* is unable to do so (90). Thus, *Shewanella* species may be also be important in the dissolution as well as the formation of magnetite in sediments.

Cells of one magnetotactic species, *Magnetospirillum magnetotacticum*, reduce iron in growing cultures, and there is some evidence that iron reduction may be linked to energy conservation and growth in this bacterium (66). Extracellular BIM-type magnetite has never been observed in cultures of this organism, although, like all magnetite-producing magnetotactic species, cells of *M. magnetotacticum* synthesize intracellular particles of magnetite (55) through BCM.

Greigite and Pyrrhotite

Some anaerobic sulfate-reducing bacteria produce particles of the magnetic iron sulfide greigite by using BIM processes. In this case, the sulfate-reducing bacteria respire with sulfate anaerobically, releasing hydrogen sulfide. The sulfide ions react with excess iron present in the external environment, forming magnetic particles of greigite and pyrrhotite as well as a number of other nonmagnetic iron sulfides including mackinawite (approximately FeS), pyrite (cubic FeS_2), and marcasite (orthorhombic FeS_2) (58, 148, 149). Mineral species formed in these bacterially catalyzed reactions appear to be dependent on the pH of the growth medium, the incubation temperature, the E_h (redox potential), the presence of specific oxidizing and reducing agents, and the type of iron source in the growth medium. For example, cells of *Desulfovibrio desulfuricans* produce greigite when grown in the presence of ferrous salts but not when the iron source is goethite, FeO(OH) (148). Microorganisms can clearly modify pH, E_h, etc., during growth. Berner (16–19) reported the chemical synthesis of a number of iron sulfide minerals including tetragonal FeS, marcasite, a magnetic cubic iron sulfide of the spinel type (probably greigite), pyrrhotite, amorphous FeS, and even framboidal pyrite, a globular form of pyrite that was once thought to represent fossilized bacteria (47, 85). Not surprisingly, Rickard (148, 149) concluded that extracellular biogenic and abiogenic particles of iron sulfides could not be distinguished from one another. However, none of the magnetic iron sulfides produced by the sulfate-reducing bacteria appear to have been examined in any detail using modern electron microscopy techniques.

The ubiquitous, anaerobic sulfate-reducing bacteria are a physiological group of microorganisms that are phylogenetically and morphologically very diverse and include species in the domains *Bacteria* (δ subdivision of *Proteobacteria* and gram-positive group) and *Archaea.* Most of the studies on the extracellular production of iron sulfides by sulfate-reducing bacteria, including the studies described above, involved the motile gram-negative bacterium *Desulfovibrio desulfuricans*, a member of the δ subgroup of the *Proteobacteria.* However, because all sulfate-reducing bacteria respire with sulfate and release sulfide ions, it is likely that all species, regardless of phylogeny or classification, produce iron sulfide minerals through BIM under appropriate environmental conditions when excess iron is available. The sulfate-reducing magnetotactic bacterium strain RS-1 can produce extracellular magnetic iron sulfides, presumably through BIM, as well as intracellular particles of magnetite through BCM (153). The iron sulfides produced by this microorganism, however, were not identified.

Other Nonmagnetic Iron Minerals

Other nonmagnetic iron minerals formed by BIM have been observed in bacterial cultures. In these cases, Fe(II), excreted by iron-reducing bacteria during the reduction of Fe(III), reacts with anions present in the growth medium. Two examples are carbonate, which reacts with Fe(II) to form the mineral siderite ($FeCO_3$), and phosphate, which reacts with Fe(II) to form vivianite [$Fe_3(PO_4)_2 \cdot 8H_2O$]. Siderite was observed in cultures of *Geobacter metallireducens* along with magnetite when the organism was grown in a bicarbonate buffering system (90, 166), while only vivianite was produced, with no production of magnetite or siderite, by the same organism when Fe(III) citrate was provided as the terminal electron acceptor (87, 90). Growing cells of the magnetotactic species *Magnetospirillum magnetotacticum* produced significant amounts of extracellular, needle-like crystals of vivianite while actively reducing Fe(III) in the form of Fe(III) oxyhydroxides (28). Although it is likely that many other iron-reducing bacteria form siderite and vivianite, most microbiologists are not looking for such compounds during routine culturing.

BIOLOGICALLY CONTROLLED MINERALIZATION

In BCM, the organisms exert a great degree of crystallochemical control over the nucleation and growth of the mineral particles. For the most part, the minerals are directly synthesized at a specific location in or on the cell and only under specific conditions. The mineral particles produced by bacteria in BCM are characterized by being well-ordered crystals (not amorphous) and having narrow size distributions and very specific, consistent particle morphologies. Because of these features, BCM processes are likely to be under specific metabolic and genetic control. In the microbial world, the most widely recognized example of BCM is magnetosome production by the magnetotactic bacteria.

Magnetotactic Bacteria

Classification, Phylogeny, and General Features

The magnetotactic bacteria represent a heterogeneous group of procaryotes that align and swim along the Earth's geomagnetic field lines (27, 29). They have a myriad of cellular morphologies including coccoid, rod shaped, vibrioid, spirilloid (helical), and even multicellular (3, 27, 29, 49, 150). Thus, the term "magnetotactic bacteria" has no taxonomic significance and should be interpreted as a term for a collection of diverse bacteria that possess the apparently widely distributed trait of magnetotaxis (3). While all that have been examined to date are gram-negative members of the domain *Bacteria*, the possibility of the existence of a magnetotactic member of the *Archaea* cannot be excluded.

Despite the great diversity of the magnetotactic bacteria, they all share several features: (i) they are all motile (although this does not preclude the possibility of the existence of nonmotile bacteria that synthesize magnetosomes, which, by definition, would be magnetic but not magnetotactic); (ii) they all exhibit a negative tactic and/or growth response to atmospheric concentrations of oxygen; and (iii) they all possess a number of intracellular structures called magnetosomes. Magnetosomes are the hallmark feature of the magnetotactic bacteria and are responsible for their behavior in magnetic fields. The bacterial magnetosome consists of a single-magnetic-domain crystal of a magnetic iron mineral enveloped by a membrane (2). Magnetotactic bacteria generally synthesize particles of either magnetite or greigite. There is one known exception: a large, rod-shaped magnetotactic bacterium collected from the Pettaquamscutt Estuary was found to contain both magnetite and greigite (9, 13).

Phylogenetic analysis, based on the sequences of 16S rRNA genes of many cultured and uncultured magnetotactic bacteria, initially showed that the magnetite-producing strains were associated with the α subdivision of the *Proteobacteria* (34, 40, 46, 155, 169), while an uncultured greigite-producing bacterium was found to be associated with the sulfate-reducing bacteria in the δ subdivision of the *Proteobacteria* (40). Since the different subdivisions of the *Proteobacteria* are considered to be coherent, distinct evolutionary lines of descent (194, 196), DeLong et al. (40) proposed that the evolutionary origin of magnetotaxis was polyphyletic and that magnetotaxis based on iron oxide (magnetite) production evolved separately from that based on iron sulfide (greigite) production. However, more recent studies (158, 170–172) have shown that many but not all magnetite-producing magnetotactic bacteria are associated with the α subgroup of the *Proteobacteria.* Kawaguchi et al. (74) showed that a cultured, magnetite-producing sulfate-reducing magnetotactic bacterium, strain RS-1 (153), is a member of the δ subgroup of the *Proteobacteria*, while Spring et al. (168) described an uncultured magnetite-producing magnetotactic bacterium called *Magnetobacterium bavaricum*, which is in the domain *Bacteria* but apparently is not phylogenetically associated with the *Proteobacteria.* These results suggest that magnetotaxis as a trait may have evolved several times and, moreover, may indicate that there is more than one biochemical-chemical pathway for the BCM of magnetic minerals by magnetotactic bacteria.

Ecology

Magnetotactic bacteria are ubiquitous in aquatic habitats that are close to neutrality in pH and that are not thermal, strongly polluted, or well oxygenated (27, 121). They are cosmopolitan in distribution (27), and because magnetotactic bacterial cells are easy to observe and separate from mud and water by exploiting their magnetic behavior using common laboratory magnets (121), there are frequent continued reports of their occurrence in various, sometimes exotic, freshwater and marine locations (see, e.g., references 1, 72, 107, and 134). Magnetotactic bacteria are generally found in the largest numbers at the oxic-anoxic transition zone (OATZ), also referred to as the microaerobic zone or the redoxocline (3). In many freshwater habitats, the OATZ is located at or just below the sediment/water interface. However, in some brackish-to-marine systems, the OATZ is found or is seasonally located in the water column, as shown in Fig. 1. The Pettaquamscutt Estuary (Narragansett Bay, R.I.) (44) and Salt Pond (Woods Hole, Mass.) (186, 187) are good examples of the latter situation. Hydrogen sulfide, produced by sulfate-reducing bacteria in the anaerobic zone and sediment, diffuses upward and oxygen diffuses downward from the surface, resulting in a double, vertical chemical concentration gradient (Fig. 1) with a coexisting redox gradient. In addition, strong pycnoclines and other physical factors, probably including the microorganisms themselves, stabilize the vertical chemical gradients and the resulting OATZ.

Many types of magnetotactic bacteria are found at both the Pettaquamscutt Estuary (9) and Salt Pond (D. A. Bazylinski, unpublished results). Generally, the magnetite-producing magnetotactic bacteria prefer the OATZ proper (Fig. 1) and behave as microaerophiles. Information obtained from culture experiments supports this observation. Two strains of magnetotactic bacteria have been isolated from the Pettaquamscutt Estuary. One is a vibrio designated strain MV-2 (40, 118), and the other is a coccus designated strain MC-1 (40, 52, 119). Both strains grow as microaerophiles, although strain MV-2 can also grow anaerobically with nitrous oxide (N_2O) as a terminal electron acceptor. Other cultured magnetotactic bacterial strains, including spirilla (106, 155, 158) and rods (153), are microaerophiles or anaerobes or both. The greigite producers are probably anaerobes preferring the

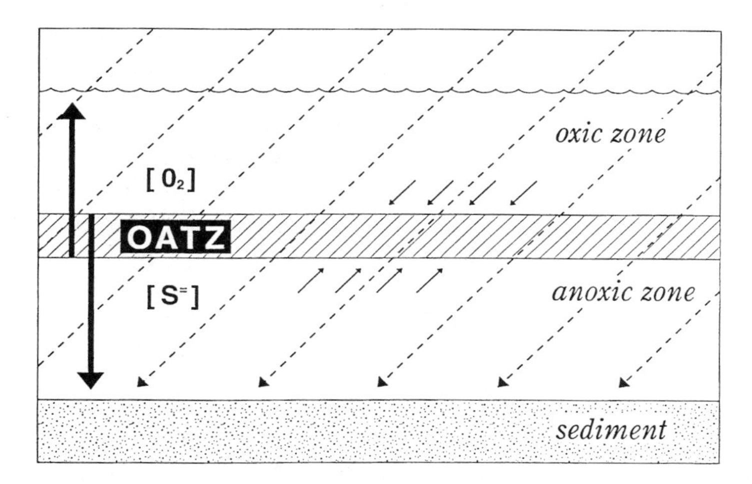

Figure 1. Depiction of the OATZ in the water column as typified by Salt Pond (Woods Hole, Mass.). Note the inverse double concentration gradients of oxygen ([O_2]) diffusing from the surface and sulfide ([S^-]) generated by sulfate-reducing bacteria in the anaerobic zone (vertical arrows). Magnetite-producing magnetotactic bacteria exist in their greatest numbers at the OATZ, where microaerobic conditions predominate, and greigite-producers are found just below the OATZ, where S^{2-} becomes detectable. When polar-magneto-aerotactic, magnetite-producing coccoid cells are above the OATZ in vertical concentration gradients of O_2 and S^{--} (higher [O_2] than optimal), they swim downward (small arrows above OATZ) along the inclined geomagnetic field lines (dashed lines). When they are below the OATZ (lower [O_2] than optimal), they reverse direction (by reversing the direction of their flagellar motor) and swim upward (small arrows below the OATZ) along the inclined geomagnetic field lines. The direction of flagellar rotation is coupled to a aerotactic sensory system that acts as a switch when cells are at a suboptimal position in the gradient as defined in the text. The magnetotactic spirilla (and other axial magneto-aerotactic microorganisms) align along the geomagnetic field lines and swim up and down, relying on a temporal sensory mechanism of aerotaxis to find and maintain position at their optimal oxygen concentration at the OATZ.

more sulfidic waters just below the OATZ (Fig. 1), where oxygen is barely or not detectable (3, 7–9). No greigite-producing magnetotactic bacterium has been grown in pure culture.

Function and Physics of Magnetotaxis

There is important physical significance to the size of the magnetosome mineral phase, which is reflected in its physical and magnetic properties. The magnetosome mineral crystals occur in a very narrow size range, from about 35 to 120 nm (12, 14, 51, 57, 122). Magnetite and greigite particles in this range are stable single magnetic domains (35, 42). Smaller particles would be superparamagnetic at ambient temperature and would not have stable, remanent magnetizations. Larger particles would tend to form multiple domains, reducing the remanent magnetization. Thus, by producing single-magnetic-domain particles, cells have maximized the remanent magnetization of the magnetosome mineral phase.

In most magnetotactic bacteria, the magnetosomes are arranged in one or more chains (3, 14). Magnetic interactions between the magnetosome particles in a chain cause their magnetic dipole moments to orient parallel to each other along the chain length. In this chain motif, the total magnetic dipole moment of the cell is simply the sum of the permanent dipole moments of the individual, single-magnetic-domain magnetosome particles. Magnetic measurements (142), magnetic-force microscopy (147, 176), and electron holography (45) have confirmed this conclusion and show that the chain of magnetosomes in a magnetotactic bacterium functions like a single magnetic dipole and causes the cell to behave similarly. Therefore, the cell has maximized its magnetic dipole moment by arranging the magnetosomes in chains. The magnetic dipole moment of the cell is generally large enough that its interaction with the Earth's geomagnetic field overcomes the thermal forces tending to randomize the cell's orientation in its aqueous surroundings (50). Magnetotaxis results from the passive alignment of the cell along geomagnetic field lines while it swims. It is important to realize that cells are neither attracted nor pulled toward either geomagnetic pole. Dead cells, like living cells, align along geomagnetic field lines but do not move. In essence, living cells behave like very small, self-propelled magnetic compass needles.

Originally, magnetotactic bacteria were thought to have one of two magnetic polarities, north or south seeking, depending on the magnetic orientation of the magnetic dipoles of the cells with respect to their direction of motion. The vertical component of the inclined geomagnetic field appeared to select for a predominant polarity in each hemisphere by favoring cells whose polarity causes them to migrate down toward the microaerobic sediments and away from potentially toxic concentrations of oxygen in surface waters. This scenario appears to be at least partially true: north-seeking magnetotactic bacteria predominate in the Northern hemisphere, while south-seeking cells predominate in the Southern hemisphere (30, 75). At the Equator, where the vertical component of the geomagnetic field is zero, equal numbers of both polarities coexist (54). However, the discovery of stable populations of magnetotactic bacteria existing at specific depths in the water columns of chemically stratified aquatic systems at higher latitudes (9) and the observation that virtually all magnetotactic bacteria in pure culture form microaerophilic bands of cells some distance from the meniscus of the growth medium (52) are not consistent with this original simple model of magnetotaxis. For example, according to this model, persistent North-seeking magnetotactic bacteria in the Northern hemisphere should always be found in the sediments or at the bottom of culture tubes.

Like most other free-swimming bacteria, magnetotactic bacteria propel themselves forward in their aqueous surroundings by rotating their helical flagella (162). Unlike cells of *Escherichia coli* and other chemotactic bacteria, magnetotactic bacteria do not exhibit the characteristic "run and tumble" motility (3). Because of their magnetosomes, magnetotactic bacteria are oriented and migrate along the local magnetic field **B**. Some magnetotactic spirilla, such as *Magnetospirillum magnetotacticum*, swim parallel or antiparallel to **B** and form aerotactic bands (31, 167) at a preferred oxygen concentration [O_2]. In a homogeneous medium, roughly equal numbers of cells swim in either direction along **B** (31, 167). This behavior has been termed axial magneto-aerotaxis (52), because the cells use the magnetic field as an axis for migration, with aerotaxis determining the direction of migration along the axis. Most microaerophilic bacteria form aerotactic bands at a preferred or optimal [O_2] where the proton motive force is maximal (199), using a temporal sensory mechanism (159) that samples the local environment as they swim and compares the present [O_2] with that in the recent past (180).The change in [O_2] with time determines the sense of flagellar rotation (105). The behavior of individual cells of *M. magnetotacticum* in aerotactic bands in thin capillaries is consistent with the temporal sensory mechanism (52). Thus, cells in the band which are moving away from the optimal [O_2], to either higher or lower [O_2], eventually reverse their swimming direction and return to the band.

In contrast, ubiquitous freshwater and marine bilophotrichously flagellated (having two flagellar bundles on one hemisphere of the cell) magnetotactic cocci (26, 52, 120, 121) and some other magnetotactic strains swim persistently in a preferred direction relative to **B** when examined microscopically in wet mounts (26, 52). In fact, this persistent swimming of cocci in a preferred direction led to the discovery of magnetotaxis in bacteria (26). However, it has recently been shown that magnetotactic cocci in oxygen gradients, like cells of *M. magnetotacticum*, can swim in both directions along **B** without turning around (52). The cocci also form microaerophilic, aerotactic bands, like cells of *M. magnetotacticum*, and seek a preferred [O_2] along the concentration gradient (52). However, while the aerotactic behavior of *M. magnetotacticum* is consistent with the temporal sensory mechanism, the aerotactic behavior of the cocci is not. Instead, their behavior is consistent with a two-state aerotactic sensory model in which the [O_2] determines the sense of the flagellar rotation and hence the swimming direction relative to **B**. Cells at [O_2] higher than optimum swim persistently in one direction relative to **B** until they reach a low [O_2] threshold, where they reverse flagellar rotation and hence swimming direction relative to **B**. They continue until they reach a high [O_2] threshold, where they reverse again. In wet mounts, the [O_2] is above optimal and the cells swim persistently in one direction relative to **B**. This model, termed polar magneto-aerotaxis (52), accounts for the ability of the magnetotactic cocci to migrate to and maintain position at the preferred [O_2] at the OATZ in chemically stratified, semianaerobic basins. An assay using chemical gradients in thin capillaries has been developed that distinguishes between axial and polar magnetoaerotaxis (53).

For both aerotactic mechanisms, migration along magnetic field lines reduces a three-dimensional search problem to a one-dimensional search problem. Thus, magnetotaxis is presumably advantageous to motile microorganisms in vertical concentration gradients because it increases the efficiency of finding and maintaining an optimal position in such concentration gradients, in this case, vertical oxygen gradients (Fig. 1). It is likely that there are other forms of magnetically assisted chemotaxis to molecules or ions other than oxygen, such as sulfide, or magnetically assisted redox- or phototaxis in bacteria that inhabit the anaerobic zone (e.g., greigite producers) in chemically stratified waters and sediments.

It is important to realize that the function of cellular magnetotaxis described above appears to be a consequence of the cell possessing magnetosomes. Bacteria can only react to a stimulus and therefore do not make the magnetosomes for magnetotaxis (a teleological argument). In addition, there is some conflicting evidence about the role of magnetosomes in magnetotaxis. For example, many obligately microaerophilic bacteria find and maintain an optimal position at the OATZ without the help of magnetosomes, cultured magnetotactic bacteria form microaerophilic bands of cells in the absence of a magnetic field, and some greigite-producing magnetotactic bacteria collected from the environment produce gas vacuoles, presumably for buoyancy (4). Thus, it appears likely that the enormous amount of iron uptake and magnetosome production is somehow also linked to the physiology of the cell and to other, as yet unknown, cellular functions.

Composition and Morphology of the Magnetosome Mineral Phase

As stated above, there are two compositional types of magnetosomes in magnetotactic bacteria: iron oxides and iron sulfides. The mineral composition of the magnetosome is probably under genetic control in that cells of several cultured magnetite-producing magnetotactic bacteria still synthesize an iron oxide (magnetite) and not an iron sulfide (grigite) even when hydrogen sulfide is present in the growth medium (118, 119). The iron oxide-type magnetosomes consist solely of magnetite. The particle morphology of magnetite varies but is consistent within cells of a single bacterial species or strain (12). Three general morphologies of magnetite particles have been observed in magnetotactic bacteria using transmission electron microscopy (TEM) (12, 29, 102, 173). These are include (i) roughly cuboidal (2, 98), (ii) parallelepipedal (rectangular in the horizontal plane of projection) (10, 120, 121, 183), and (iii) tooth, bullet, or arrowhead shaped (anisotropic) (100, 101, 181). Examples are shown in Fig. 2.

High-resolution TEM and selected area electron diffraction studies have revealed that the magnetite particles within magnetotactic bacteria are of high structural perfection and have been used to determine their idealized morphologies (98–101, 108, 118, 119). These morphologies are all derived from combinations of {111}, {110}, and {100} forms with suitable distortions (41). The roughly cuboidal particles are cubo-octahedra ({100} + {111}), and the parallelepipedal particles are either truncated pseudo-hexahedral or pseudo-octahedral prisms. Examples are shown in Figs. 3a to d. The cubo-octahedral crystal morphology preserves the symmetry of the face-centered cubic spinel structure; i.e., all equivalent crystal faces develop equally. The pseudo-hexahedral and pseudo-octahedral prismatic particles represent anisotropic growth in which equivalent faces develop unequally (41, 97).

The synthesis of the tooth-, bullet-, and arrowhead-shaped magnetite particles (Fig. 2c) appears to be more complex than that of the other forms. They have been examined by high-resolution TEM in one uncultured organism (100, 101), and their idealized morphology suggests that the growth of these particles occurs in two stages. The nascent crystals are cubo-octahedra which subsequently elongate along the [111] axis parallel to the chain direction.

Whereas the cubo-octahedral form of magnetite can occur in inorganically formed magnetites (138), the prevalence of elongated pseudo-hexahedral or pseudo-octahedral habits in magnetosome crystals imply anisotropic growth conditions, e.g., a temperature gradient, a concentration gradient, or anisotropic ion flux (97).

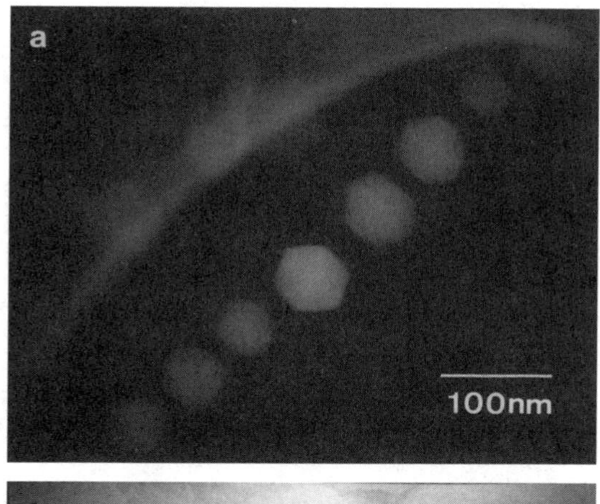

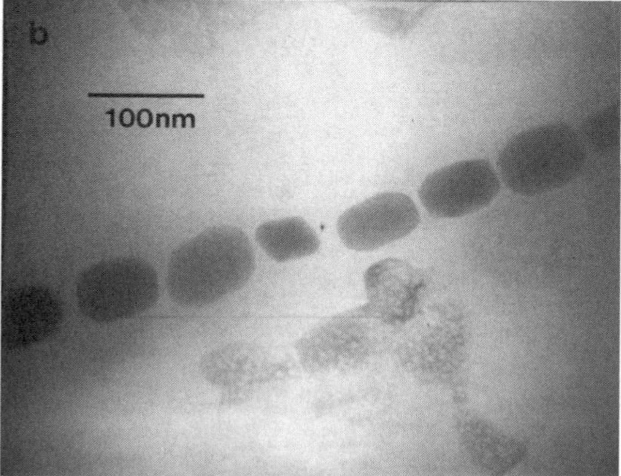

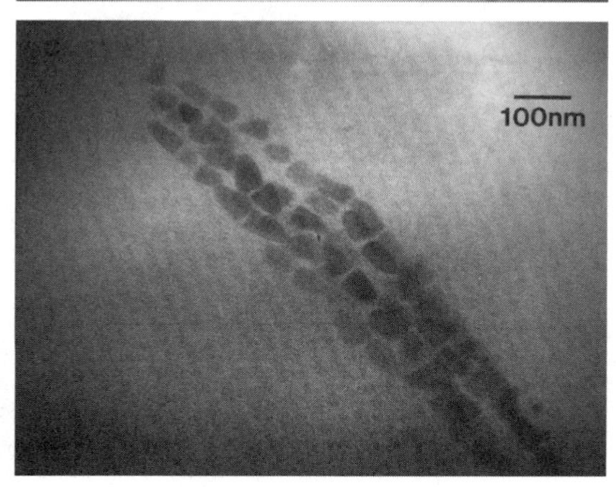

Figure 2. Morphologies of intracellular magnetite (Fe_3O_4) particles produced by magnetotactic bacteria collected from the OATZ of the Pettaquamscutt Estuary. (a) Dark-field scanning-transmission electron micrograph (STEM) of a chain of cubo-octahedra in cells of an unidentified rod-shaped bacterium, viewed along a [111] zone axis for which the particle projections appear hexagonal. (b) Bright-field STEM of a chain of crystals within a cell of an unidentified marine vibrio, with parallelepipedal projections. (c) Bright-field STEM of tooth-shaped (anisotropic) magnetosomes from an unidentified marine rod-shaped bacterium.

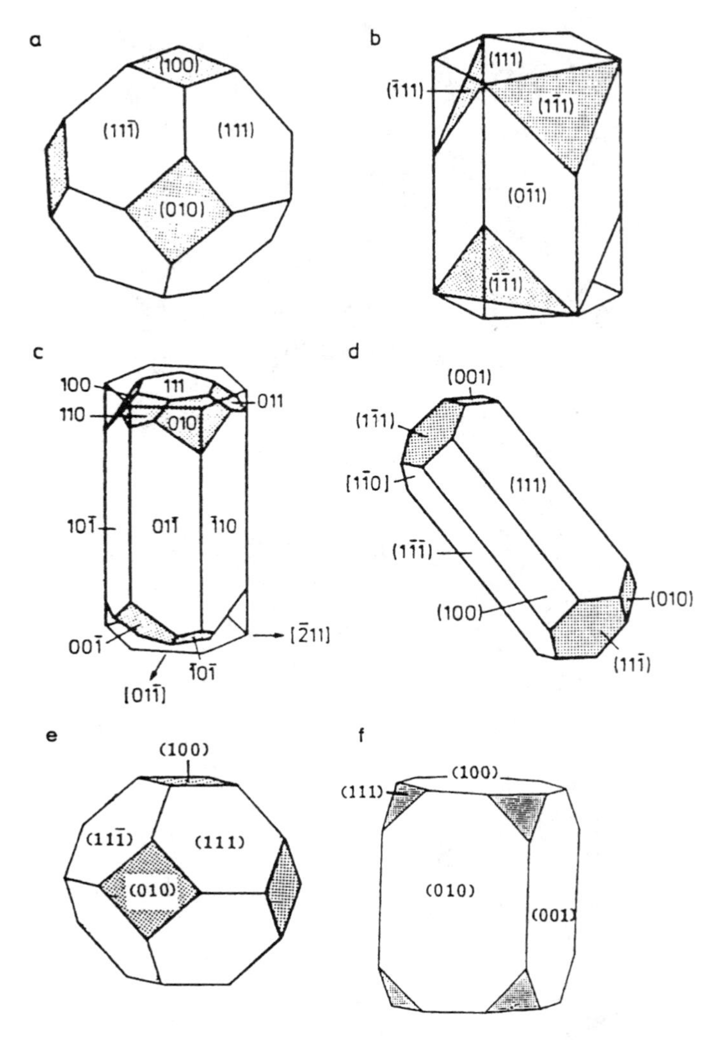

Figure 3. Idealized magnetite (a to d) and greigite (e and f) crystal morphologies derived from high-resolution TEM studies of magnetosomes from magnetotactic bacteria. (a and e) Cubo-octahedrons. (b, c, and f) Variations of pseudo-hexagonal prisms. (d) Elongated cubo-octahedron. Adapted from references 70 and 97.

This aspect of magnetosome particle morphology has been used to distinguish magnetosome magnetite from detrital or BIM-type magnetite using electron microscopy of magnetic extracts from sediments (37, 38, 141, 143, 173–175).

Although all freshwater magnetotactic bacteria synthesize magnetite as the mineral phase of their magnetsomes, many marine, estuarine, and salt marsh species produce an iron sulfide-type magnetosome which consists primarily of the magnetic iron sulfide greigite (7, 8, 11, 69, 70, 104, 145, 146). Reports of iron pyrite (104) and pyrrhotite (48) have not been confirmed and may represent misidentifications of additional iron sulfide species occasionally observed with greigite in cells (145, 146) (discussed below). Currently recognized greigite-producing magnetotactic bacteria include a many-celled magnetotactic procaryote (MMP) that consists of an aggregation of about 20 to 30 cells arranged in a spherical manner, which is motile as an entire unit and not as separate cells (40, 49, 150, 151), and a variety of relatively large, rod-shaped bacteria (7, 8, 11, 69, 70).

The iron sulfide-type magnetosomes contain either particles of greigite (69, 70) or a mixture of greigite and transient nonmagnetic iron sulfide phases that probably represent mineral precursors to greigite (145, 146). These phases include mackinawite and probably a sphalerite-type cubic FeS (145, 146). Based on TEM observations, electron diffraction, and known iron sulfide chemistry (18, 20, 21), the reaction scheme for greigite formation in the magnetotactic bacteria appears to be cubic FeS → mackinawite (tetragonal FeS) → greigite (Fe_3S_4) (145, 146). The de novo synthesis of nonmagnetic crystalline iron sulfide precursors to greigite aligned along the magnetosome chain indicates that chain formation and the orientation of the magnetosomes in the chain do not necessarily involve magnetic interactions. Interestingly, under the strongly reducing, sulfidic conditions at neutral pH in which the greigite-producing magnetotactic bacteria are found (7, 8), greigite particles would be expected to transform into pyrite (18, 20). It is not known if and how cells prevent this transformation.

As with magnetite, three particle morphologies of greigite have been observed in magnetotactic bacteria: (i) cubo-octahedral (the equilibrium form of face-centered cubic greigite) (69, 70), (ii) pseudo-rectangular prismatic (Fig. 3e to f and 4) (69, 70), and (iii) tooth shaped (145, 146). Like that of their magnetite counterparts, the morphology of the greigite particles also appears to be species- and/or strain specific, although confirmation of this observation will require controlled studies of pure cultures of greigite-producing magnetotactic bacteria, none of which

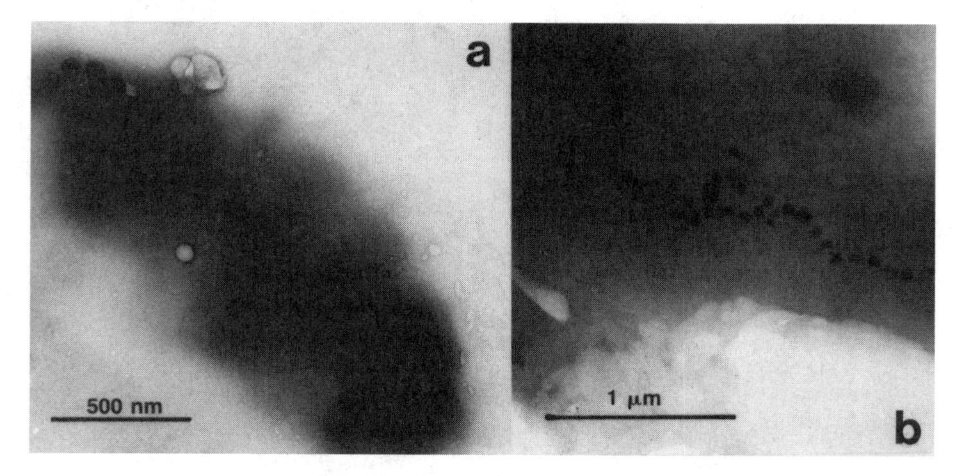

Figure 4. Morphologies of greigite (Fe_3S_4) particles within cells of unidentified rod-shaped bacteria collected from the sulfidic waters of a salt marsh pool. (a) TEM of cubo-octahedra. (b) TEM of rectangular prisms.

is currently available. One clear exception to this rule is the MMP (see above) (7, 8, 11, 49, 104, 150, 151). This unusual microorganism, found in salt marsh pools all over the world and in some deep-sea sediments, contains pleomorphic, pseudo rectangular prismatic, tooth-shaped, and cubo-octahedral greigite particles. Some of the these particle morphologies are shown in Fig. 5. Therefore, the biomineralization process(es) appears to be more complicated in this organism than in the rods with greigite-containing magnetosomes or in magnetite-producing magnetotactic bacteria.

One slow-swimming, rod-shaped bacterium, collected from the OATZ from the Pettaquamscutt Estuary, was found to contain arrowhead-shaped crystals of magnetite and rectangular prismatic crystals of greigite co-organized within the same chains of magnetosomes (this organism usually contains two parallel chains of magnetosomes), as shown in Fig. 6 (9, 13). In cells of this uncultured organism, the magnetite and greigite crystals occur with different, mineral-specific morphologies and sizes and are positioned with their long axes oriented along the chain direction. Both particle morphologies have been found in organisms with single-mineral-component chains (69, 70, 100, 101), which suggests that the magnetosome membranes surrounding the magnetite and greigite particles contain different nucleation templates and that there are differences in magnetosome vesicle biosynthesis. Thus, it is likely that two separate sets of genes control the biomineralization of magnetite and greigite in this organism.

In sum, the consistent narrow size range (41) and morphologies of the magnetosome particles represent features typical of BCM and are clear indications that

Figure 5. High-magnification bright-field STEM of pleomorphic greigite-mackinawite particles within MMP cells.

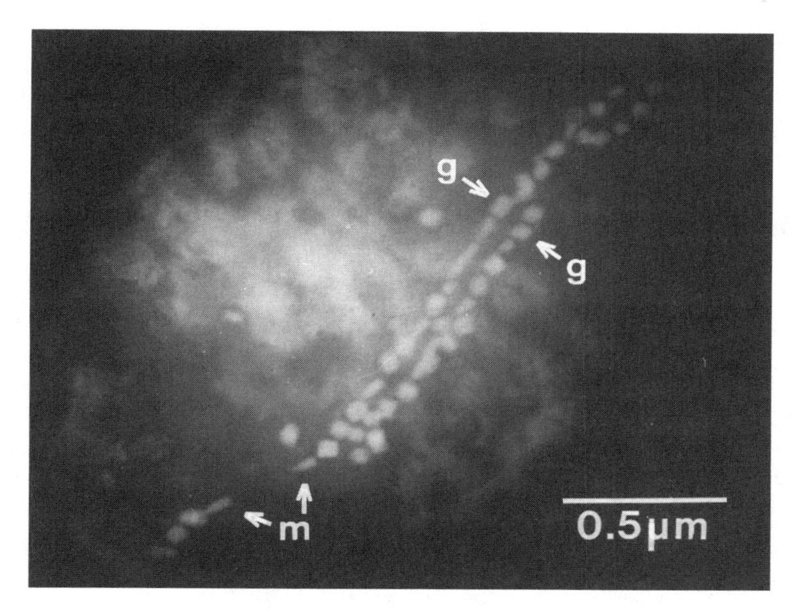

Figure 6. Dark-field STEM of rectangular prismatic greigite (g) and tooth-shaped magnetite particles (m) co-organized within the same chains of magnetosomes in an unusual rod-shaped magnetotactic bacterium collected from the Pettaquamscutt Estuary (see the text).

the magnetotactic bacteria exert a high degree of control over the biomineralization processes involved in magnetosome synthesis.

Effect of Environmental Conditions on Biomineralization

Can local environmental conditions affect the biomineralization of the mineral phase of the magnetosome? This might occur in the form of variations in the stoichiometry of the metal and/or the nonmetal components of the mineral phase or through the replacement of the either the metal or nonmetal component of the magnetosome mineral phase. Here we discuss stoichiometry changes and specifically whether iron can be replaced with other transition metal ions and whether sulfur and oxygen can replace each other as the nonmetal component in the magnetosome mineral phase.

The replacement of iron in magnetosomes has not been studied in great detail in pure cultures. Gorby (62) showed that iron could not be replaced by other transition metal ions, including titanium, chromium, cobalt, copper, nickel, mercury, and lead, in the magnetite crystals of *Magnetospirillum magnetotacticum* when cells were grown in the presence of these ions. However, Towe and Moench (183) reported very small amounts of titanium in the magnetite particles of an uncultured freshwater magnetotactic coccus collected from a wastewater treatment pond. Significant amounts of copper have been found in the greigite particles of the MMP (11) described above. The amount of copper was extremely variable and ranged from about 0.1 to 10 atomic% relative to iron. The copper appeared to be concentrated mostly on the surface of the particles and was present only in organisms collected from a salt marsh in Morro Bay, Calif., and not in those collected from other sites. Interestingly, magnetosomes in rod-shaped magnetotactic bacteria that also produce greigite collected from the same site in Morro Bay did not contain copper. The presence of copper did not appear to affect the function of the magnetosomes, since the organisms were still magnetotactic. More recently, copper has also been found in the greigite particles of rod-shaped magnetotactic bacteria from sites other than Morro Bay (146). These observations may indicate that the mineral phase of the magnetosomes in these organisms is more susceptible to chemical and redox conditions in the external environment and/or that the magnetosomes could function in transition metal detoxification.

The conversion of cubic FeS and mackinawite to greigite (145, 146) in greigite-producing magnetotactic bacteria clearly represents an example of changes in the stoichiometry of the metal and nonmetal components of the magnetosome particle. Based on thermodynamic considerations, cubic FeS and mackinawite transform to greigite under strongly reducing sulfidic conditions at neutral pH (21, 145, 146), conditions that exist where these organisms are found (7, 8). There is also the possibility that these transformations are catalyzed by the cell.

Cells of the slow-moving magnetotactic rod from the Pettaquamscutt Estuary that biomineralize both magnetite and greigite were found to extend well below the OATZ proper at this chemically stratified site (9). Cells collected from the more oxidized regions of the OATZ contained more or exclusively arrowhead-shaped magnetite particles, while cells collected from below the OATZ in the anaerobic sulfidic zone contained more or exclusively greigite particles. This finding suggests that environmental parameters such as local molecular oxygen and/or hydrogen sulfide concentrations and/or redox conditions somehow regulate the relative biomineralization of iron oxides and iron sulfides in this bacterium (9).

Chemistry and Biochemistry of Magnetosome Formation

The first step in the synthesis of magnetite (and probably greigite as well) in magnetotactic bacteria is the uptake of iron. Free reduced Fe(II) is very soluble (up to 100 mM at neutral pH [132]) and is easily taken up by bacteria, usually by nonspecific means. However, because free oxidized Fe(III) is so insoluble, most microbes have to rely on iron chelators which bind and solubilize Fe(III) for uptake. Microbially produced Fe(III) chelators are called siderophores and are defined as low-molecular-mass (<1 kDa), virtually specific ligands that facilitate the solubilization and transport of Fe(III) (67). Siderophores are generally produced under iron-limited conditions, and their synthesis is repressed under high-iron conditions. Several studies, all involving magnetotactic spirilla, have focused on iron uptake in the magnetotactic bacteria.

A hydroxamate siderophore was found to be produced by cells of *M. magnetotacticum* grown under high-iron but not under low-iron conditions (139). Thus, the siderophore production pattern here is the reverse of what is normally observed. Later studies have not confirmed this finding. Earlier, Frankel et al. (56) assumed that iron uptake by this organism probably occurred via a nonspecific transport system. Although iron is supplied as Fe(III) chelated to quinic acid, the growth medium also contains chemical reducing agents (e.g., thioglycolate or ascorbic acid) potent enough to reduce Fe(III) to Fe(II). Thus, both forms of iron are present in the growth medium, and it is not known which form is taken up by the cell.

Nakamura et al. (128) did not detect siderophore production by *Magnetospirillum* strain AMB-1 and concluded that iron was taken up as Fe(III) by cells and that Fe(III) uptake was mediated by a periplasmic binding protein-dependent iron transport system. Schüler and Baeuerlein (156) found that spent medium stimulated iron uptake to a high degree in cells of *M. gryphiswaldense* but found no evidence for the presence of a siderophore. They also showed that the major portion of iron for magnetite synthesis was taken up as Fe(III) and that Fe(III) uptake appears to be an energy-dependent process. Fe(II) was also taken up by cells but by a slow, diffusion-like process, while Fe(III) uptake followed Michaelis-Menten kinetics with a K_m of 3 M and a V_{max} of 0.86 nmol min^{-1} mg of dry cell weight^{-1}, suggesting that Fe(III) uptake by cells of *M. gryphiswaldense* is a low-affinity but high-velocity transport system (156). In a later study (157) using the same organism, Schüler and Baeuerlein showed that magnetite formation was induced in nonmagnetotactic cells by a low threshold oxygen concentration of about 2 to 7 μM (at 30°C) and was tightly linked to Fe(III) uptake.

Few studies have addressed what occurs in the cell after iron is taken up by magnetotactic bacteria. Frankel et al. (56) examined the nature and distribution of major iron compounds in *M. magnetotacticum* by using ^{57}Fe Mössbauer spectroscopy. They proposed a model in which Fe(III) is taken up by the cell (by nonspecific means as described above) and reduced to Fe(II) as it enters the cell. It is then thought to be reoxidized to form a low-density hydrous Fe(III) oxide, which is then dehydrated to form a high-density Fe(III) oxide (ferrihydrite), which was directly observed in cells. In the last step, one-third of the Fe(III) ions in ferrihydrite are reduced and, with further dehydration, magnetite is produced. However, Schüler and Baeuerlein (157) showed that in cells of *M. gryphiswaldense*, Fe(III) is taken up and rapidly converted to magnetite without any apparent delay, suggesting that there is no significant accumulation of a precursor to magnetite inside the cell, at least under the conditions of the experiment, which appeared to be optimal for magnetite production by that organism.

The mechanism by which specific shapes of magnetite are formed is unknown at present, but it is thought that the final two steps occur in the magnetosome membrane vesicle, which apparently nucleates and also constrains crystal growth. It is clear that many more studies involving several different organisms are required before we fully understand the precipitation of magnetite and greigite in magnetotactic bacteria.

Physiology

Many features common to the magnetotactic bacteria were described above. There are also some physiological features common to all that have been studied. However, since only a handful of strains are in pure culture and most of these are microaerophilic freshwater spirilla that produce cubo-octahedral particles of magnetite, the amount of data is limited. Although magnetite synthesis has not yet been linked to the physiology of a magnetotactic bacterium, it is important to understand the physiology of these bacteria and the conditions under which they synthesize magnetosomes in order to find this link. One point is clear, though; magnetite is formed by physiologically diverse magnetotactic bacteria only under microaerobic and/or anaerobic conditions, depending on the species.

The first magnetotactic bacterium to be isolated and grown in pure culture was *Magnetospirillum* (formerly *Aquaspirillum* [155]) *magnetotacticum* strain MS-1 (31, 106). This obligately microaerophilic species represents the most extensively studied magnetotactic bacterium. It was isolated from a freshwater swamp and synthesizes cubo-octahedral crystals of magnetite (97, 98). Its cells are helical and possess an unsheathed polar flagellum at each end. This organism is obligately respiratory and cannot ferment and is nutritionally a chemoorganoheterotroph that uses organic acids as a source of energy and carbon (31). The cells do not produce catalase, which might explain their microaerophilic nature, but they produce several periplasmic superoxide dismutases of the iron and manganese types (161). Cells fix atmospheric dinitrogen, as evidenced by their ability to reduce acetylene to ethylene under nitrogen-limited conditions (5, 6a). This organism uses oxygen or nitrate as a terminal electron acceptor and is a denitrifier that produces nitrous oxide and dinitrogen from nitrate (6). Unlike most denitrifying bacteria, cells require a small amount of molecular dioxygen for nitrate-dependent growth (6). Interestingly, cells produce more magnetite when grown with nitrate than with oxygen as a terminal electron acceptor. However, molecular oxygen must still be present for magnetite synthesis when nitrate is present in the growth medium, with the optimal concentration for maximum magnetite yields being 1% oxygen in the headspace of cultures and concentrations greater than 5% being inhibitory (32). Recent isotope experiments clearly show that molecular O_2 is not incorporated into magnetite, however, and that the oxygen in magnetite is derived from water (95). The role that molecular O_2 plays in magnetite synthesis is thus unknown, but it has been clearly shown to affect the synthesis of specific proteins in *M. magnetotacticum*. For example, Sakaguchi et al. (154) showed that the presence of oxygen in nitrate-grown cultures repressed the synthesis of a 140-kDa membrane protein in *M. magnetotacticum* whose function is unknown and Short and Blakemore (161) showed that increasing the oxygen tension from 1 to 10% of saturation in cultures of *M. magnetotacticum* caused cells to express increased activity of the manganese-type superoxide dismutase relative to that of the iron type.

Guerin and Blakemore (66) reported anaerobic, iron-dependent growth of *M. magnetotacticum* in the absence of nitrate. Although cells grown with "amorphous" Fe(III) were extremely magnetic and produced nearly twice as many magnetosomes as did nitrate-grown cells with optimal [O_2] (32), they grew very slowly under these conditions and the growth yields were poor compared to those on nitrate and/or oxygen. These authors further showed that iron oxidation may also be linked to aerobic respiratory processes, energy conservation, and magnetite synthesis in *M. magnetotacticum*. However, given that cells produce so much magnetite during growth with Fe(III) under anaerobic conditions, it would appear that aerobic iron oxidation is not necessary for magnetite synthesis.

In an effort to understand the relationship between nitrate and oxygen utilization and magnetite synthesis, Fukumori and coworkers examined electron transport and cytochromes in *M. magnetotacticum*. Tamegai et al. (179) reported a novel "cytochrome a_1-like" hemoprotein that was present in greater amounts in magnetic cells than nonmagnetic cells. They did not find any true cytochrome a_1 or any *o*-type cytochromes, which were once considered to be terminal oxidases in *M. magnetotacticum* by others (135). A new *ccb*-type cytochrome *c* oxidase (178) and a cytochrome cd_1-type nitrite reductase (195) were isolated and purified from *M. magnetotacticum*. The latter protein was of particular interest since it showed Fe(II): nitrite oxidoreductase activity, which may be linked to the oxidation of Fe(II) in the cell and thus to magnetite synthesis (195).

Cells of *M. magnetotacticum* actively reduce Fe(III) (28) and translocate protons when Fe(III) is introduced into them anaerobically (160), suggesting that cells conserve energy during the reduction of Fe(III). Growth yields on Fe(III) suggest that iron reduction is also linked to growth, as in the dissimilatory iron-reducing bacteria (66). Fe(III) reductase activity has also been demonstrated in cell extracts of *M. magnetotacticum* (140), and recently, Noguchi et al. (133) purified a Fe(III) reductase from this microorganism. The enzyme appears to be loosely bound on the cytoplasmic face of the cytoplasmic membrane, has an apparent molecular mass of 36 kDa, and requires reduced NADH and flavin mononucleotide as an electron donor and cofactor, respectively. Enzyme activity was inhibited by zinc, which also reduced the number of magnetite-containing magnetosomes when included in the growth medium as $ZnSO_4$.

Other microaerophilic magnetotactic spirilla, physically identical to *M. magnetotacticum*, have been isolated in recent years. Cells of *Magnetospirillum* strain AMB-1, isolated by Matsunaga et al. (113), are apparently much more oxygen tolerant than are those of other magnetotactic species and, unlike other magnetotactic bacteria, can form colonies on the surface of agar plates under a fully aerobic atmosphere. This species seems very similar to *M. magnetotacticum* in that it appears to synthesize cubo-octahedral crystals of magnetite, is obligately respiratory, and has a chemoorganoheterotrophic mode of nutrition using organic acids as sources of energy and carbon. Its cells, like those of *M. magnetotacticum*, form more magnetosomes when grown with nitrate, but unlike cells of *M. magnetotacticum*, they grow anaerobically with nitrate and synthesize magnetite without molecular oxygen (113). Growth and inhibitor studies (113, 115) show that *Magnetospirillum* strain AMB-1 uses nitrate as a terminal electron acceptor, although the products of nitrate reduction were not reported. Cells of another microaerophilic, freshwater magnetotactic spirillum, described by Schleifer et al. (155), *M. gryphiswaldense*, are very similar in morphology, ultrastructure, and physiology to the other *Magnetospirillum* species described above and, like cells of all the freshwater magnetotactic spirilla, produce cubo-octahedral crystals of magnetite. As in *M. magnetotacticum*, magnetite production in *M. gryphiswaldense* is induced under microaerobic conditions (156, 157). Whether *M. gryphiswaldense* uses nitrate as a terminal electron acceptor and grows anaerobically with it or any other terminal electron acceptor or whether nitrate affects magnetite synthesis in this bacterium, as it does in *M. magnetotacticum*, was not reported. Like cells of *M. magnetotacticum*, those of both *M. gryphiswaldense* and *Magnetospirillum* strain AMB-1 appear to fix atmospheric dinitrogen (6a).

A marine magnetotactic vibrioid bacterium, strain MV-1, was isolated by Bazylinski et al. (10). Cells of strain MV-1 possess a single, unsheathed polar flagellum and grow and synthesize pseudo-hexahedral prismatic crystals of magnetite in their magnetosomes microaerobically and anaerobically with nitrous oxide as the terminal electron acceptor. The cells appear to produce more magnetite under anaerobic than under microaerobic conditions (10). This species is nutritionally versatile, being able to grow chemoorganoheterotrophically with organic and amino acids as carbon and energy sources and chemolithoautotrophically with thiosulfate or sulfide as energy sources (oxidizing them to sulfate) and carbon dioxide as the sole carbon source (L. K. Kimble and D. A. Bazylinski, Abstr. 96th Gen. Meet. Am. Soc. Microbiol. 1996, abstr. K-174, 1996). Cells produce intracellular sulfur deposits when grown with sulfide (Kimble and Bazylinski, Abstr. 96th Gen. Meet. Am. Soc. Microbiol. 1996). As do virtually all aerobic chemolithoautotrophic bacteria, strain MV-1 uses the Calvin-Benson cycle for autotrophic carbon dioxide fixation (116). Cell extracts from thiosulfate-grown cells of strain MV-1 show ribulose bisphosphate carboxylase/oxygenase (Rubisco) activity (Kimble and Bazylinski, Abstr. 96th Gen. Meet. Am. Soc. Microbiol. 1996), and recently Dean and Bazylinski (A. J. Dean and D. A. Bazylinski, Abstr. 96th Gen. Meet. Am. Soc. Microbiol. 1996, abstr. H-207, p. 369, 1996) cloned and sequenced a *cbbM* gene (which encodes form II Rubisco enzymes) from strain MV-1. They found no evidence for a *cbbL* gene (which encodes form I Rubisco enzymes) in Southern analyses despite using *cbbL* gene probes from several different organisms. Because many uncultured magnetotactic bacteria collected from natural habitats thrive in oxygen-sulfide inverse gradients and contain internal sulfur deposits (9, 51, 71, 120, 168), it seems likely that many species are chemolithoautotrophs that derive electrons from the oxidation of sulfide.

By using pulsed-field gel electrophoresis, the genome of strain MV-1 was found to consist of a single, circular chromosome of approximately 3.7 Mb (39). There was no evidence of linear chromosomes or extrachromosomal DNA such as plasmids.

A virtually identical strain to strain MV-1, designated MV-2, was isolated from the Pettaquamscutt Estuary (40, 118). Cells of this strain produce the same morphological type of magnetite crystals as do cells of strain MV-1 (118) and display many of the same phenotypic traits as do cells of strain MV-1, such as anaerobic growth with nitrous oxide as a terminal electron acceptor, heterotrophic growth with organic and amino acids, and chemolithoautotrophic growth on reduced sulfur compounds. However, strain MV-2 shows slightly different restriction fragment patterns in pulsed-field gels from those of strain MV-1 with the same restriction enzymes (39). As with strain MV-1, the genome of strain MV-2 consists of a single circular chromosome of a similar size, about 3.6 Mb (39).

Strain RS-1 is a gram-negative, sulfate-reducing, rod-shaped bacterium that grows and produces bullet-shaped particles of magnetite only under anaerobic conditions (153). Its cells are helicoid to rod shaped and possess a single polar flagellum. Little is known about the physiology of this strain. Cells grow chemoorganoheterotrophically using certain organic acids and alcohols as carbon and energy sources and cannot use nitrate as a terminal electron acceptor.

Several other pure cultures of magnetotactic bacteria exist, but they appear to be obligate microaerophiles and grow poorly (Bazylinski, unpublished). Hence, very little is known about them. Strain MC-1, a marine coccus, produces pseudo hexahedral prisms of magnetite and grows chemolithoautotrophically with thiosulfate or sulfide as an electron and energy source (40, 52, 119). It has a genome size of approximately 4.5 Mb as determined by pulsed-field gel electrophoresis (39). Strain MV-4, a small marine spirillum, produces elongated octahedrons of magnetite and can grow chemolithoautotrophically with thiosulfate or chemoorganoheterotrophically with succinate (118).

Magnetosome Membrane and BCM of Magnetite

In all the magnetite-producing magnetotactic bacteria examined to date, the magnetosome crystals appear to be encased in a coating or membrane. In *M. magnetotacticum*, this structure, the so-called magnetosome membrane, consists of a lipid bilayer containing phospholipids and numerous proteins, some of which appear to be unique to this membrane and are not found in the outer or cell membrane (63). The magnetosome membrane does not appear to be contiguous with the cell membrane and is presumably the locus of control over the size and morphology of the inorganic particle as well as the structural entity that anchors the magnetosome at a particular location within the cell. However, it is not known whether the magnetosome membrane is premade as an empty membrane vesicle prior to the biomineralization of the mineral phase. Empty and partially filled vesicles have been observed in iron-starved cells of *M. magnetotacticum* (63) but have not been commonly observed in other magnetotactic strains. The unlikely alternative would be that nucleation of the mineral phase occurs before it is surrounded by the membrane. In any case, most biochemical and molecular biological studies directed toward the understanding of the biomineralization processes involved in magnetosome formation are focused on aspects of the magnetosome membrane, particularly on the functions of specific proteins present on this membrane.

The magnetosome membrane appears to be a universal feature of at least the magnetite-producing species of magnetotactic bacteria, since it has been found in virtually all cultured and some noncultured strains (Fig. 7). Although it is not known whether greigite-producing magnetotactic bacteria actually have a magnetosome membrane like their magnetite-producing counterparts, it seems likely that they do, based on the consistent particle morphologies observed in these bacteria.

Molecular Biology of Magnetosome Formation and Genetics of Magnetotactic Bacteria

It is not known how many genes are required for magnetosome synthesis, how these genes are regulated, etc. Establishing a genetic system with the magnetotactic bacteria is an absolute necessity before these questions can be answered and although several laboratories (including ours) have persisted in trying to achieve this goal, they have been hampered by many problems including the lack of a significant number of magnetotactic bacterial strains, the fastidiousness of the organisms in culture and the elaborate techniques required for the growth of these organisms, and the inability of almost all these strains to grow on the surface of agar plates in experiments to screen for mutants etc. While the researchers involved have taken

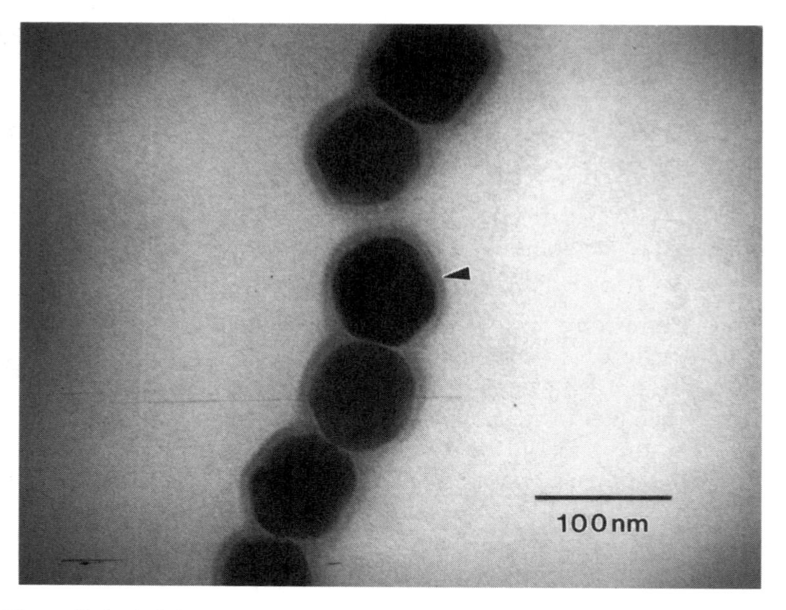

Figure 7. Dark-field STEM of pseudo-hexagonal prismatic magnetite particles from the marine magnetotactic coccus strain MC-1, surrounded by the magnetosome membrane (arrowhead).

different approaches, most of the genetic and molecular studies have involved one or more of the magnetotactic spirilla.

Waleh and coworkers (23, 190) initiated the first studies in the establishment of a genetic system with magnetotactic bacteria in order to understand the molecular biology of magnetosome synthesis in these microorganisms. Working with *M. magnetotacticum* strain MS-1, they showed that at least some of the genes of this organism can be functionally expressed in *E. coli* and that the transcriptional and translational elements of the two microorganisms are compatible (necessary features for a good genetic system). They were able to clone, characterize, and sequence the *recA* gene from *M. magnetotacticum* (23, 24). They later examined iron uptake by *M. magnetotacticum*. They cloned and characterized a 2-kb DNA fragment from *M. magnetotacticum* that complemented the *aroD* (biosynthetic dehydroquinase) gene function in *E. coli* and *Salmonella enterica* serovar Typhimurium. *aroD* mutants of these strains cannot take up iron from the growth medium. In other words, when the 2-kb DNA fragment from *M. magnetotacticum* was introduced into these mutants, the ability of the mutants to take up iron from the growth medium was restored (22), suggesting that the 2-kb DNA fragment may be important in iron uptake (and therefore possibly in magnetite synthesis) in *M. magnetotacticum*. Although the cloned fragment restored iron uptake deficiencies in siderophoreless, iron uptake-deficient mutants of *E. coli*, it did not mediate siderophore biosynthesis (22).

As mentioned above, cells of *Magnetospirillum* strain AMB-1 (113) are apparently much more oxygen tolerant than are those of other magnetotactic species and form colonies on the surface of agar plates under air. Colonies on plates incubated under fully aerobic conditions (21% oxygen) were white and contained cells that were nonmagnetic. However, when the oxygen concentration of the incubation atmosphere was decreased to 2%, the cells formed black-brown colonies that were made up of magnetic cells. This feature facilitated the selection of nonmagnetic mutants of *Magnetospirillum* strain AMB-1, obtained by the introduction of transposon Tn*5* into the genome of *Magnetospirillum* AMB-1 by the conjugal transfer of plasmid pSUP1021 which contained the transposon (112). This plasmid and its transposon was also introduced into *M. magnetotacticum*, but this strain did not form colonies. Using Tn*5*-derived nonmagnetic mutants of *Magnetospirillum* strain AMB-1, Nakamura et al. (126) found that at least three regions of the *Magnetospirillum* strain AMB-1 chromosome were required for the successful synthesis of magnetosomes. One of these regions, 2,975 bp in length, contained two putative open reading frames. One of these, designated *magA*, encoded a protein that is homologous to two cation efflux proteins, the *E. coli* potassium ion-translocating protein, KefC, and the putative sodium ion/proton antiporter, NapA, from *Enterococcus hirae*. *magA* was expressed in *E. coli*, and membrane vesicles prepared from these cells that contained the *magA* gene product took up iron only when ATP was supplied, indicating that energy was required for the uptake of iron. The same group, using a *magA-luc* fusion protein, showed that the *MagA* protein is a membrane protein that is localized in the cell membrane and perhaps the magnetosome membrane as well (127). Interestingly, the *magA* gene appears to be expressed to a much greater degree when wild-type *Magnetospirillum* strain AMB-1 cells are grown under iron-limited conditions rather than iron-sufficient conditions, in which they would produce more magnetosomes. Moreover, the nonmagnetotactic Tn*5* mutant overexpressed the *magA* gene under iron-limited conditions, although it did not make magnetosomes. Thus, although some evidence suggests that the MagA protein is involved in iron transport, the role of the *magA* gene in magnetosome synthesis, if there is one, is unclear.

Okuda et al. (136) took a different "reverse genetics" approach to the magnetosome problem. They found three proteins, with apparent molecular masses of 12, 22, and 28 kDa, that appear to be unique to the magnetosome membranes of *M. magnetotacticum* and are not present in the cellular membrane fraction. They were able to determine enough of the N-terminal amino acid sequence of the 22-kDa protein to construct a 17-bp oligonucleotide probe for the genomic cloning of the gene encoding that protein. They also found that the protein exhibited significant homology to a number of proteins that belong to the tetratricopeptide repeat protein family, which include mitochondrial protein import receptors and peroxisomal protein import receptors. Thus, although the role of the 22-kDa magnetosome membrane protein in magnetosome synthesis has not been elucidated, it may function as a receptor interacting with associated cytoplasmic proteins (136).

Cells of the marine vibrio strain MV-1, like those of *M. magnetotacticum*, also produce a number of magnetosome membrane proteins that are not found in any other cellular fractions (B. L. Dubbels and D. A. Bazylinski, Abstr. 98th Gen. Meet.

Am. Soc. Microbiol. 1998, abstr. H-82, p. 290, 1998) (Fig. 8). N-terminal sequences have been reported (Dubbels and Bazylinski, Abstr. 98th Gen. Meet. Am. Soc. Microbiol. 1998) for at least three of these proteins, but the use of degenerate primers based on these sequences to clone the genes responsible for encoding these proteins has so far proved unsuccessful.

A stable, spontaneous, nonmagnetotactic mutant of strain MV-1 that does not synthesize magnetosomes has been isolated and partially characterized (Dubbels and Bazylinski, Abstr. 98th Gen. Meet. Am. Soc. Microbiol. 1998). Protein profiles of the different cell fractions of wild-type MV-1 and the mutant were compared. Cells of the mutant strain do not produce the magnetosome membrane proteins or

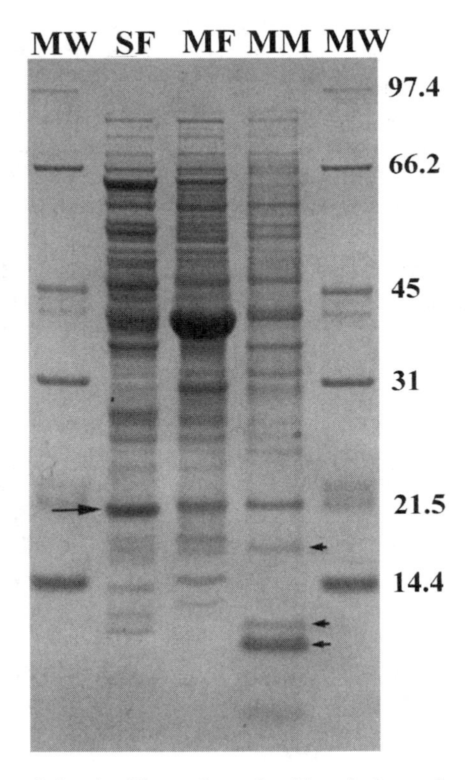

Figure 8. Sodium dodecyl sulfate-polyacrylamide gel electrophoresis of wild-type strain MV-1 cell fractions. Cell fractions were separated by differential ultracentrifugation. Lanes: SF, soluble fraction; MF, membrane fraction (excluding magnetosome membranes); MM, magnetosome membranes; MW, molecular mass standards, with masses in kilodaltons shown on the right. Magnetosomes were purified as previously described (12), and magnetosome membranes were extracted from magnetosomes with 1% sodium dodecyl sulfate in 20 mM HEPES (pH 7.2) (Dubbels and Bazylinski, Abstr. 98th Gen. Meet. Am. Soc. Microbiol. 1998). The large arrow denotes a soluble 19-kDa protein that is not produced by a nonmagnetotactic mutant of strain MV-1 (see the text), and the small arrows denote some proteins that appear to be associated with the magnetosome membrane but not with the other cell fractions.

a periplasmic 19-kDa protein produced in abundance by wild-type cells (Fig. 8). Degenerate primers, constructed from N-terminal sequence of this protein and conserved sequences from homologous proteins, were used to generate a 255-bp PCR product that was used to probe a genomic cosmid library of strain MV-1. The encoding the 19-kDa protein was cloned and sequenced (B. L. Dubbels and D. A. Bazylinski, Abstr. 99th Gen. Meet. Am. Soc. Microbiol. 1999, abstr. H-219, p. 371, 1999). A similar 19-kDa periplasmic protein was purified from *Campylobacter jejuni* (73). The protein is acidic, with an isoelectric point of 4.8, and is iron regulated through the Fur (ferric uptake regulator) protein, its synthesis is repressed by high iron concentrations in the growth medium (73, 184). Interestingly, synthesis of the 19-kDa protein from strain MV-1 also appears to be regulated by iron (Dubbels and Bazylinski, Abstr. 99th Gen. Meet. Am. Soc. Microbiol. 1999). The function of this protein in either microorganism is unknown, but similar proteins and genes encoding similar proteins have been found in several gram-negative pathogens including *Treponema pallidum* (177).

Applications of Magnetotactic Bacteria, Magnetosomes, and Magnetic Particles

Magnetotactic bacteria and their magnetic inclusions have novel magnetic properties. Some of these are discussed above, and others are described in more detail in various references (see, e.g., references 14, 123, and 124). Because of these properties, magnetotactic bacterial cells and their magnetic crystals have been exploited in several scientific and commercial applications (109). For example, north-seeking magnetotactic bacteria have been used to determine south magnetic poles in meteorites and rocks containing fine-grained (<1,000 nm) magnetic minerals (59, 60). They have also been used in magnetic cell separations after introducing magnetotactic bacteria into granulocytes and monocytes by phagocytosis (110).

Magnetotactic bacterial magnetite particles have been used in studies of magnetic domain analysis (61). The crystals have been used in many commercial applications including the immobilization of enzymes including glucose oxidase and uricase (111), the formation of magnetic antibodies in various fluoroimmunoassays (114) involving the detection of allergens (131) and squamous cell carcinoma cells (109) and the quantification of immunoglobulin G (130), the detection and removal of bacterial cells (*E. coli*) with a fluorescein isothiocyanate-conjugated monoclonal antibody immobilized on magnetotactic bacterial magnetite particles (129), and the transfer of genes into cells, a technology in which magnetosomes are coated with DNA and injected or "shot" (using a particle gun) into cells that are difficult to transform using more standard methods (109).

Considering that the bacterial magnetosome and its magnetic properties have been refined and optimized in the course of evolution by the cells that synthesize them by controlling the chemical composition, size, and morphology of the magnetic particles as well as their position within the cell, the elucidation of how magnetosomes form within the cell will yield clues to the production of ordered arrays of inorganic electronic, optical, and magnetic materials at the nanometer scale that could prove useful in many additional scientific and commercial applications.

Magnetosomes as Evidence of Ancient Life on Mars

As noted in the introduction, a number of mineralogical and other features from the Martian meteorite ALH84001 were suggested to result from biological processes on ancient Mars (117). Included in these features was the presence of ultrafine-grained magnetite, pyrrhotite, and possibly greigite. The magnetite crystals ranged from about 10 to 100 nm, were in the superparamagnetic (<35 nm) and single-magnetic-domain size ranges, and were cuboid, teardrop, and irregular in shape. The iron sulfide particles varied in size and shape, with the pyrrhotite particles ranging up to 100 nm. Both mineral particles were embedded in a fine-grained carbonate matrix on the rim of carbonate inclusions within the meteorite.

The magnetite particles, in particular, were invoked as evidence of life because of the similarities in size and morphology between these particles and those produced by the magnetotactic bacteria on Earth. In addition, the magnetite particles in ALH84001 differ greatly from those found in other meteorites. Other nanocrystalline forms of magnetite, including platelets and whiskers, are also present in ALH84001 (33). Some of these contain specific crystalline defects such as axial screw dislocations indicating that they were formed nonbiologically at high temperature by vapor phase growth (33). Although one study shows that statistical analysis of the sizes and shapes of of fine-grained magnetite crystals might prove to be a robust criterion for distinguishing between biogenic and nonbiogenic magnetite (41), the findings described above illustrate the need for additional criteria by which to distinguish between biogenic and nonbiogenic particles of iron minerals. One of these criteria might be whether magnetotactic bacteria isotopically fractionate iron and/or oxygen in the magnetite crystals they produce. In a recent study, Mandernack et al. (95) showed an oxygen isotopic fractionation that was temperature dependent and closely matched that for magnetite produced by a bacterial consortium containing thermophilic iron-reducing bacteria (197). Because these results contrast with those for magnetite thought to be formed inorganically, bacteria may fractionate oxygen in magnetite differently from what occurs in magnetite formed abiotically, which, in turn, might reflect different redox pathways of magnetite formation. No detectable fractionation in iron isotopes was observed in magnetite produced by *M. magnetotacticum* and strain MV-1. However, Beard et al. (14a) measured low ^{56}Fe values in the soluble ferrous iron produced by dissimilatory iron-reducing bacteria growing with the iron oxide mineral ferrihydrite, meaning that cells prefer to reduce the lighter, rarer ^{54}Fe (rather than ^{56}Fe). How this isotopic fractionation is reflected in magnetite formed by iron-reducing bacteria remains to be seen. There is no doubt that discussion and debate on the interpretation of these fascinating findings will continue. It is nonetheless intriguing to think that microbes may be or have been responsible for the geochemical cycling of iron on other planets as well as on Earth.

CONCLUSIONS

Magnetic effects have been reported in a large number of organisms other than bacteria, including invertebrates such as protists (182), honeybees (191), and mollusks (83) and vertebrates such as fish (192), amphibians (144), reptiles (82), birds (188, 189), and mammals (193) including humans (76, 77). Magnetite crystals virtually morphologically indistinguishable from those produced by the magnetotactic bacteria have been found in a euglenoid protist (182), in the ethmoid tissues of tuna and salmon and other animals (165), and in the human brain (76, 77). In addition, other animals such as honeybees (65, 80) and pigeons (188) appear to contain particles of magnetite or hydrous iron oxides that could be precursors to magnetite. The fact that many higher creatures biomineralize single-magnetic-domain magnetite crystals of similar morphologies suggests the intriguing idea that all these organisms have the same or a similar set of genes responsible for magnetite biomineralization that would probably have originated in the magnetotactic bacteria. Thus, studying how magnetotactic bacteria biomineralize magnetite might have a scientific impact far beyond the studies of microbiology and geology.

Acknowledgments. This chapter is dedicated to the memory of a most distinguished microbiologist, Holger W. Jannasch, who had a significant influence on the study of the magnetotactic bacteria.

We thank our many collaborators over the years for their participation in much of the work described herein. We particularly appreciate the efforts of graduate students A. J. Dean and B. L. Dubbels; our most recent associations with T. J. Beveridge, P. R. Buseck, J. D. Coates, R. Guerrero, B. L. Hoyle, K. Mandernack, D. S. McKay, M. Pósfai, and D. Schüler, K. L. Thomas-Keprta; and our continued collaborations with D. R. Lovley and B. M. Moskowitz.

We acknowledge support from the U.S. Office of Naval Research (grant ONR N00014-91-J-1290), the U.S. National Science Foundation (grant CHE-9714101), and the NASA Johnson Space Center (grant NAG 9-1115).

REFERENCES

1. **Akai, J., S. Takaharu, and S. Okusa.** 1991. TEM study on biogenic magnetite in deep-sea sediments from the Japan Sea and the western Pacific Ocean. *J. Electron Microsc.* **40:**110–117.
2. **Balkwill, D. L., D. Maratea, and R. P. Blakemore.** 1980. Ultrastructure of a magnetic spirillum. *J. Bacteriol.* **141:**1399–1408.
3. **Bazylinski, D. A.** 1995. Structure and function of the bacterial magnetosome. *ASM News* **61:**337–343.
4. **Bazylinski, D. A.** 1999. Synthesis of the bacterial magnetosome: the making of a magnetic personality. *Int. Microbiol.* **2:**71–80.
5. **Bazylinski, D. A., and R. P. Blakemore.** 1983. Nitrogen fixation (acetylene reduction) in *Aquaspirillum magnetotacticum*. Curr. Microbiol. **9:**305–308.
6. **Bazylinski, D. A., and R. P. Blakemore.** 1983. Denitrification and assimilatory nitrate reduction in *Aquaspirillum magnetotacticum*. *Appl. Environ. Microbiol.* **46:**1118–1124.

6a. **Bazylinski, D. A., A. J. Dean, D. Schüler, E. J. P. Phillips, and D. R. Lovley.** N_2-dependent growth and nitrogenase activity in the metal-metabolizing bacteria, *Geobacter* and *Magnetospirillum* species. *Environ. Microbiol.*, in press.

7. **Bazylinski, D. A., and R. B. Frankel.** 1992. Production of iron sulfide minerals by magnetotactic bacteria from sulfidic environments, p. 147–159. *In* H. C. W. Skinner and R. W. Fitzpatrick (ed.), *Biomineralization Processes of Iron and Manganese: Modern and Ancient Environments.* Catena Verlag, Cremlingen-Destedt, Germany.

8. **Bazylinski, D. A., R. B. Frankel, A. J. Garratt-Reed, and S. Mann.** 1990. Biomineralization of iron sulfides in magnetotactic bacteria from sulfidic environments, p. 239–255. *In* R. B. Frankel and R. P. Blakemore (ed.), *Iron Biominerals.* Plenum Press, New York, N.Y.
9. **Bazylinski, D. A., R. B. Frankel, B. R. Heywood, S. Mann, J. W. King, P. L. Donaghay, and A. K. Hanson.** 1995. Controlled biomineralization of magnetite (Fe_3O_4) and greigite (Fe_3S_4) in a magnetotactic bacterium. *Appl. Environ. Microbiol.* **61:**3232–3239.
10. **Bazylinski, D. A., R. B. Frankel, and H. W. Jannasch.** 1988. Anaerobic production of magnetite by a marine magnetotactic bacterium. *Nature* (London) **334:**518–519.
11. **Bazylinski, D. A., A. J. Garratt-Reed, A. Abedi, and R. B. Frankel.** 1993. Copper association with iron sulfide magnetosomes in a magnetotactic bacterium. *Arch. Microbiol.* **160:**35–42.
12. **Bazylinski, D. A., A. J. Garratt-Reed, and R. B. Frankel.** 1994. Electron microscopic studies of magnetosomes in magnetotactic bacteria. *Microsc. Res. Tech.* **27:**389–401.
13. **Bazylinski, D. A., B. R. Heywood, S. Mann, R. B. Frankel.** 1993. Fe_3O_4 and Fe_3S_4 in a bacterium. *Nature* (London) **366:**218.
14. **Bazylinski, D. A., and B. M. Moscowitz.** 1997. Microbial biomineralization of magnetic iron minerals: microbiology, magnetism and environmental significance. *Rev. Mineral.* **35:**181–223.
14a. **Beard, B. L., C. M. Johnson, L. Cox, H. Sun, K. H. Nealson, and C. Aguilar.** 1999. Iron isotope biosignatures. *Science* **285:**1889–1892.
15. **Bell, P. E., A. L. Mills, and J. S. Herman.** 1987. Biogeochemical conditions favoring magnetite formation during anaerobic iron reduction. *Appl. Environ. Microbiol.* **53:**2610–2616.
16. **Berner, R. A.** 1962. Synthesis and description of tetragonal iron sulfide. *Science* **137:**669.
17. **Berner, R. A.** 1964. Iron sulfides formed from aqueous solution at low temperatures and atmospheric pressure. *J. Geol.* **72:**293–306.
18. **Berner, R. A.** 1967. Thermodynamic stability of sedimentary iron sulfides. *Am. J. Sci.* **265:**773–785.
19. **Berner, R. A.** 1969. The synthesis of framboidal pyrite. *Econ. Geol.* **64:**383–393.
20. **Berner, R. A.** 1970. Sedimentary pyrite formation. *Am. J. Sci.* **268:**1–23.
21. **Berner, R. A.** 1974. Iron sulfides in Pleistocene deep Black Sea sediments and their palaeooceanographic significance, p. 524–531. *In* E. T. Degens and D. A. Ross (ed.), *The Black Sea: Geology, Chemistry and Biology. AAPG Memoirs 20.* American Association of Petroleum Geologists, Tulsa, Okla.
22. **Berson, A. E., D. V. Hudson, and N. S. Waleh.** 1991. Cloning of a sequence of *Aquaspirillum magnetotacticum* that complements the *aroD* gene of *Escherichia coli. Mol. Microbiol.* **5:**2261–2264.
23. **Berson, A. E., M. R. Peters, and N. S. Waleh.** 1989. Cloning and characterization of the *recA* gene of *Aquaspirillum magnetotacticum. Arch. Microbiol.* **152:**567–571.
24. **Berson, A. E., M. R. Peters, and N. S. Waleh.** 1990. Nucleotide sequence of *recA* gene of *Aquaspirillum magnetotacticum. Nucleic Acids Res.* **18:**675.
25. **Beveridge, T. J.** 1989. Role of cellular design in bacterial metal accumulation and mineralization. *Annu. Rev. Microbiol.* **43:**147–171.
26. **Blakemore, R. P.** 1975. Magnetotactic bacteria. *Science* **190:**377–379.
27. **Blakemore, R. P.** 1982. Magnetotactic bacteria. *Annu. Rev. Microbiol.* **36:**217–238.
28. **Blakemore, R. P., and N. A. Blakemore.** 1990. Magnetotactic magnetogens, p. 51–67. *In* R. B. Frankel and R. P. Blakemore (ed.), *Iron Biominerals.* Plenum Press, New York, N.Y.
29. **Blakemore, R. P., N. A. Blakemore, D. A. Bazylinski, and T. T. Moench.** 1989. Magnetotactic bacteria, p. 1882–1889. *In* J. T. Staley, M. P. Bryant, N. Pfennig, and J. G. Holt (ed.), *Bergey's Manual of Systematic Bacteriology*, vol. 3. The Williams & Wilkins Co., Baltimore, Md.
30. **Blakemore, R. P., R. B. Frankel, and A. J. Kalmijn.** 1980. South-seeking magnetotactic bacteria in the southern hemisphere. *Nature* (London) **236:**384–385.
31. **Blakemore, R. P., D. Maratea, and R. S. Wolfe.** 1979. Isolation and pure culture of a freshwater magnetic spirillum in chemically defined medium. *J. Bacteriol.* **140:**720–729.
32. **Blakemore, R. P., K. A. Short, D. A. Bazylinski, C. Rosenblatt, and R. B. Frankel.** 1985. Microaerobic conditions are required for magnetite formation within *Aquaspirillum magnetotacticum. Gemicrobiol. J.* **4:**53–71.

33. **Bradley, J. P., R. P. Harvey, and H. Y. McSween, Jr.** 1996. Magnetite whiskers and platelets in the ALH84001 martian meteorite: evidence for vapor phase growth. *Geochim. Cosmochim. Acta* **60:** 5149–5155.
34. **Burgess, J. G., R. Kanaguchi, T. Sakaguchi, R. H. Thornhill, and T. Matsunaga.** 1993. Evolutionary relationships among *Magnetospirillum* strains inferred from phylogenetic analysis of 16S rRNA sequences. *J. Bacteriol.* **175:**6689–6694.
35. **Butler, R. F., and S. K. Banerjee.** 1975. Theoretical single-domain grain size range in magnetite and titanomagnetite. *J. Geophys. Res.* **80:**4049–4058.
36. **Caccavo, F., Jr., D. J. Lonergan, D. R. Lovley, M. Davis, J. F. Stolz, and M. J. McInerny.** 1994. *Geobacter sulfurreducens* sp. nov., a hydrogen- and acetate-oxidizing dissimilatory metal reducing microorganism. *Appl. Environ. Microbiol.* **60:**3752–3759.
37. **Chang, S.-B. R., and J. L. Kirschvink.** 1989. Magnetofossils, the magnetization of sediments, and the evolution of magnetite biomineralization. *Annu. Rev. Earth Planet. Sci.* **17:**169–195.
38. **Chang, S.-B. R., J. F. Stolz, J. L. Kirschvink, and S. M. Awramik.** 1989. Biogenic magnetite in stromatolites. 2. Occurrence in ancient sedimentary environments. *Precambrian Res.* **43:**305–312.
39. **Dean, A. J., and D. A. Bazylinski.** 1999. Genome analysis of several marine, magnetotactic bacterial strains by pulsed-field gel electrophoresis. *Curr. Microbiol.* **39:**219–225.
40. **DeLong, E. F., R. B. Frankel, and D. A. Bazylinski.** 1993. Multiple evolutionary origins of magnetotaxis in bacteria. *Science* **259:**803–806.
41. **Devouard, B., M. Pósfai, X. Hua, D. A. Bazylinski, R. B. Frankel, and P. R. Buseck.** 1998. Magnetite from magnetotactic bacteria: size distributions and twinning. *Am. Mineral.* **83:**1387–1398.
42. **Diaz-Ricci, J. C., and J. L. Kirschvink.** 1992. Magnetic domain state and coercivity predictions for biogenic greigite (Fe_3S_4): a comparison of theory with magnetosome observations. *J. Geophys. Res.* **97:**17309–17315.
43. **DiChristina, T. J., and E. F. DeLong.** 1993. Design and application of rRNA-targeted oligonucleotide probes for the dissimilatory iron- and manganese-reducing bacterium *Shewanella putrefaciens*. *J. Bacteriol.* **59:**4152–4160.
44. **Donaghay, P. L., H. M. Rines, and J. M. Sieburth.** 1992. Simultaneous sampling of fine scale biological, chemical and physical structure in stratified waters. *Arch. Hydrobiol. Beih. Ergebn. Limnol.* **36:**97–108.
45. **Dunin-Borkowski, R. E., M. R. McCartney, R. B. Frankel, D. A. Bazylinski, M. Pósfai, and P. R. Buseck.** 1998. Magnetic microstructure of magnetotactic bacteria by electron holography. *Science* **282:**1868–1870.
46. **Eden, P. A., T. M. Schmidt, R. P. Blakemore, and N. R. Pace.** 1991. Phylogenetic analysis of *Aquaspirillum magnetotacticum* using polymerase chain reaction-amplified 16S rRNA-specific DNA. *Int. J. Syst. Bacteriol.* **41:**324–325.
47. **Fabricus, F.** 1961. Die Strukturen des "Rogenpyrits" (Kossener Schichten, Rat) als Betrag zum Problem der "Vererzten Bakterien." *Geol. Rundschau* **51:**647–657.
48. **Farina, M., D. M. S. Esquivel, and H. G. P. Lins de Barros.** 1990. Magnetic iron-sulphur crystals from a magnetotactic microorganism. *Nature* (London) **343:**256–258.
49. **Farina, M., H. Lins de Barros, D. M. S. Esquivel, and J. Danon**. 1983. Ultrastructure of a magnetotactic bacterium. *Biol. Cell* **48**:85–88.
50. **Frankel, R. B.** 1984. Magnetic guidance of organisms. *Annu. Rev. Biophys. Bioeng.* **13:**85–103.
51. **Frankel, R. B., and D. A. Bazylinski.** 1994. Magnetotaxis and magnetic particles in bacteria. *Hyperfine Interact.* **90:**135–142.
52. **Frankel, R. B., D. A. Bazylinski, M. Johnson, and B. L. Taylor.** 1997. Magneto-aerotaxis in marine, coccoid bacteria. *Biophys. J.* **73:**994–1000.
53. **Frankel, R. B., D. A. Bazylinski, and D. Schüler.** 1998. Biomineralization of magnetic iron minerals in bacteria. *Supramol. Sci.* **5:**383–390.
54. **Frankel, R. B., R. P. Blakemore, F. F. Torres de Araujo, D. M. S. Esquivel, and J. Danon.** 1981. Magnetotactic bacteria at the geomagnetic equator. *Science* **212:**1269–1270.
55. **Frankel, R. B., R. P. Blakemore, and R. S. Wolfe**. 1979. Magnetite in freshwater magnetotactic bacteria. *Science* **203:**1355–1356.
56. **Frankel, R. B., G. C. Papaefthymiou, R. P. Blakemore, and W. O'Brien.** 1983. Fe_3O_4 precipitation in magnetotactic bacteria. *Biochim. Biophys. Acta* **763:**147–159.

57. **Frankel, R. B., J.-P. Zhang, and D. A. Bazylinski.** 1998. Single magnetic domains in magnetotactic bacteria. *J. Geophys. Res.* **103:**30601–30604.
58. **Freke, A. M., and D. Tate.** 1961. The formation of magnetic iron sulphide by bacterial reduction of iron solutions. *J. Biochem. Microbiol. Technol. Eng.* **3:**29–39.
59. **Funaki, M., H. Sakai, and T. Matsunaga.** 1989. Identification of the magnetic poles on strong magnetic grains from meteorites using magnetotactic bacteria. *J. Geomagn. Geoelectr.* **41:**77–87.
60. **Funaki, M., H. Sakai, T. Matsunaga, and S. Hirose.** 1992. The S pole distribution on magnetic grains in pyroxenite determined by magnetotactic bacteria. *Phys. Earth Planet. Interiors* **70:**253–260.
61. **Futschik, K., H. Pfützner, A. Doblander, P. Schönhuber, T. Dobeneck, N. Petersen, and H. Vali.** 1989. Why not use magnetotactic bacteria for domain analyses? *Phys. Scr.* **40:**518–521.
62. **Gorby, Y. A.** 1989. *Regulation of Magnetosome Biogenesis by Oxygen and Nitrogen,* p. 72–88. Ph.D. dissertation. University of New Hampshire, Durham.
63. **Gorby, Y. A., T. J. Beveridge, and R. P. Blakemore.** 1988. Characterization of the bacterial magnetosome membrane. *J. Bacteriol.* **170:**834–841.
64. **Greene, A. C., B. K. C. Patel, and A. J. Sheehy.** 1997. *Deferribacter thermophilus* gen. nov., sp. nov., a novel thermophilic manganese- and iron-reducing bacterium isolated from a petroleum reservoir. *Int. J. Syst. Bacteriol.* **47:**505–509.
65. **Gould, J. L., J. L. Kirschvink, and K. S. Deffeyes.** 1978. Bees have magnetic remanence. *Science* **201:**1026–1028.
66. **Guerin, W. F., and R. P. Blakemore.** 1992. Redox cycling of iron supports growth and magnetite synthesis by *Aquaspirillum magnetotacticum. Appl. Environ. Microbiol.* **58:**1102–1109.
67. **Guerinot, M. L.** 1994. Microbial iron transport. *Annu. Rev. Microbiol.* **48:**743–772.
68. **Hanzlik, M. M., N. Petersen, R. Keller, and E. Schmidbauer.** 1996. Electron microscopy and ^{57}Fe Mössbauer spectra of 10 nm particles, intermediate in composition between Fe_3O_4–γ-Fe_2O_3, produced by bacteria. *Geophys. Res. Lett.* **23:**479–482.
69. **Heywood, B. R., D. A. Bazylinski, A. J. Garratt-Reed, S. Mann, and R. B. Frankel.** 1990. Controlled biosynthesis of greigite (Fe_3S_4) in magnetotactic bacteria. *Naturwissenschaften* **77:**536–538.
70. **Heywood, B. R., S. Mann, and R. B. Frankel.** 1991. Structure, morphology and growth of biogenic greigite (Fe_3S_4), p. 93–108. *In* M. Alpert, P. Calvert, R. B. Frankel, P. Rieke, and D. Tirrell (ed.), *Materials Synthesis Based on Biological Processes.* Materials Research Society, Pittsburgh, Pa.
71. **Iida, A., and J. Akai.** 1996. Crystalline sulfur inclusions in magnetotactic bacteria. *Sci. Rep. Niigata Univ. Ser. E* **11:**35–42.
72. **Iida, A., and J. Akai.** 1996. TEM study on magnetotactic bacteria and contained magnetite grains as biogenic minerals, mainly from Hokuriku-Niigata region, Japan. *Sci. Rep. Niigata Univ. Ser. E* **11:**43–66.
73. **Janvier, B., P. Constantinidou, P. Aucher, Z. V. Marshall, C. W. Penn, and J. L. Fauchère.** 1998. Characterization and gene sequencing of a 19-kDa periplasmic protein of *Campylobacter jejuni/coli. Res. Microbiol.* **149:**95–107.
74. **Kawaguchi, R., J. G. Burgess, T. Sakaguchi, H. Takeyama, R. H. Thornhill, and T. Matsunaga.** 1995. Phylogenetic analysis of a novel sulfate-reducing magnetic bacterium, RS-1, demonstrates its membership of the δ-Proteobacteria. *FEMS Microbiol. Lett.* **126:**277–282.
75. **Kirschvink, J. L.** 1980. South-seeking magnetic bacteria. *J. Exp. Biol.* **86:**345–347.
76. **Kirschvink, J. L., A. Kobayashi-Kirschvink, and B. J. Woodford.** 1992. Magnetite biomineralization in the human brain. *Proc. Natl. Acad. Sci. USA* **89:**7683–7687.
77. **Kobayashi, A., J. L. Kirschvink.** 1995. Magnetoreception and elctromagnetic field effects: sensory perception of the geomagnetic field in animals and humans. *Adv. Chem. Ser.* **250:**367–394.
78. **Konhauser, K. O.** 1998. Diversity of bacterial iron mineralization. *Earth Sci. Rev.* **43:**91–121.
79. **Kostka, J. E., and K. H. Nealson.** 1995. Dissolution and reduction of magnetite by bacteria. *Environ. Sci. Technol.* **29:**2535–2540.
80. **Kuterbach, D. A., B. Walcott, R. J. Reeder, and R. B. Frankel.** 1982. Iron-containing cells in the honeybee (*Apis mellifera*). *Science* **218:**695–697.

81. **Liu, S. V., J. Zhou, C. Zhang, D. R. Cole, M. Gajdarziska-Josifovska, and T. J. Phelps.** 1997. Thermophilic Fe(III)-reducing bacteria from the deep subsurface: the evolutionary implications. *Science* **277:**1106–1109.
82. **Lohmann, K. J.** 1991. Magnetic orientation by hatchling loggerhead sea turtles (*Caretta caretta*). *J. Exp. Biol.* **155:**37–49.
83. **Lohmann, K. J., and A. O. D. Willows.** 1987. Lunar-modulated geomagnetic orientation by a marine mollusk. *Science* **235:**331–334.
84. **Lonergan, D. J., H. L. Jenter, J. D. Coates, E. J. P. Phillips, T. M. Schmidt, and D. R. Lovley.** 1996. Phylogenetic analysis of dissimilatory Fe(III)-reducing bacteria. *J. Bacteriol.* **178:**2402–2408.
85. **Love, L. G., and D. O. Zimmerman.** 1961. Bedded pyrite and microrganisms from the Mount Isa Shale. *Econ. Geol.* **56:**873–896.
86. **Lovley, D. R.** 1987. Organic matter mineralization with the reduction of ferric iron: a review. *Geomicrobiol. J.* **5:**375–399.
87. **Lovley, D. R.** 1990. Magnetite formation during microbial dissimilatory iron reduction, p. 151–166. *In* R. B. Frankel and R. P. Blakemore (ed.), *Iron Biominerals.* Plenum Press, New York, N.Y.
88. **Lovley, D. R.** 1991. Dissimilatory Fe(III) and Mn(IV) reduction. *Microbiol. Rev.* **55:**259–287.
89. **Lovley, D. R., S. J. Giovannoni, D. C. White, J. E. Champine, E. J. P. Phillips, Y. A. Gorby, and S. Goodwin.** 1993. *Geobacter metallireducens* gen. nov. sp. nov., a microorganism capable of coupling the complete oxidation of organic compounds to the reduction of iron and other metals. *Arch. Microbiol.* **159:**336–344.
90. **Lovley, D. R., and E. J. P. Phillips.** 1988. Novel mode of microbial energy metabolism: organic carbon oxidation coupled to dissimilatory reduction of iron or manganese. *Appl. Environ. Microbiol.* **54:**1472–1480.
91. **Lovley D. R., J. F. Stolz, G. L. Nord, Jr., and E. J. P. Phillips.** 1987. Anaerobic production of magnetite by a dissimilatory iron-reducing microorganism. *Nature* (London) **330:**252–254.
92. **Lowenstam, H. A.** 1981. Minerals formed by organisms. *Science* **211:**1126–1131.
93. **Maher, B. A.** 1990. Inorganic formation of ultrafine-grained magnetite, p. 179–192. *In* R. B. Frankel and R. P. Blakemore (ed.), *Iron Biominerals.* Plenum Press, New York, N.Y.
94. **Maher, B. A., and R. M. Taylor.** 1988. Formation of ultra-fine grained magnetite in soils. *Nature* (London) **336:**368–370.
95. **Mandernack, K. W., D. A. Bazylinski, W. C. Shanks, and T. D. Bullen.** 1999. Oxygen and iron isotope studies of magnetite produced by magnetotactic bacteria. *Science* **285:**1892–1896.
96. **Mann, S.** 1986. On the nature of boundary-organised biomineralization. *J. Inorg. Chem.* **28:**363–371.
97. **Mann, S., and R. B. Frankel.** 1989. Magnetite biomineralization in unicellular organisms, p. 389–426. *In* S. Mann, J. Webb, and R. J. P. Williams (ed.), *Biomineralization: Chemical and Biochemical Perspectives.* VCH Publishers, New York, N.Y.
98. **Mann, S., R. B. Frankel, and R. P. Blakemore.** 1984. Structure, morphology and crystal growth of bacterial magnetite. *Nature* (London) **310:**405–407.
99. **Mann, S., T. T. Moench, and R. J. P. Williams.** 1984. A high resolution electron microscopic investigation of bacterial magnetite. Implications for crystal growth. *Proc. R. Soc. Lond. Sec.* B **B 221:**385–393.
100. **Mann, S., N. H. C. Sparks, and R. P. Blakemore.** 1987. Ultrastructure and characterization of anisotropic inclusions in magnetotactic bacteria. *Proc. R. Soc. Lond. Ser.* B **231:**469–476.
101. **Mann, S., N. H. C. Sparks, and R. P. Blakemore.** 1987. Structure, morphology and crystal growth of anisotropic magnetite crystals in magnetotactic bacteria. *Proc. R. Soc. Lond. Ser.* B **231:**477–487.
102. **Mann, S., N. H. C. Sparks, and R. G. Board.** 1990. Magnetotactic bacteria: microbiology, biomineralization, palaeomagnetism, and biotechnology. *Adv. Microb. Physiol.* **31:**125–181.
103. **Mann, S., N. H. C. Sparks, S. B. Couling, M. C. Larcombe, and R. B. Frankel.** 1989. Crystallochemical characterization of magnetic spinels prepared from aqueous solution. *J. Chem. Soc. Faraday Trans.* **85:**3033–3044.
104. **Mann, S., N. H. C. Sparks, R. B. Frankel, D. A. Bazylinski, and H. W. Jannasch.** 1990. Biomineralization of ferrimagnetic greigite (Fe_3S_4) and iron pyrite (FeS_2) in a magnetotactic bacterium. *Nature* (London) **343:**258–260.

105. **Manson, M. D.** 1992. Bacterial chemotaxis and motility. *Adv. Microb. Physiol.* **33:**277–346.
106. **Maratea, D., and R. P. Blakemore.** 1981. *Aquaspirillum magnetotacticum* sp. nov., a magnetic spirillum. *Int. J. Syst. Bacteriol.* **31:**452–455.
107. **Matitashvili, E. A., D. A. Matojan, T. S. Gendler, T. V. Kurzchalia, and R. S. Adamia.** 1992. Magnetotactic bacteria from freshwater lakes in Georgia. *J. Basic Microbiol.* **32:**185–192.
108. **Matsuda, T., J. Endo, N. Osakabe, A. Tonomura, and T. Arii.** 1983. Morphology and structure of biogenic magnetite particles. *Nature* (London) **302:**411–412.
109. **Matsunaga, T.** 1991. Applications of bacterial magnets. *Trends Biotechnol.* **9:**91–95.
110. **Matsunaga, T., K. Hashimoto, N. Nakamura, K. Nakamura, and S. Hashimoto.** 1989. Phagocytosis of bacterial magnetite by leucocytes. *Appl. Microbiol. Biotechnol.* **31:**401–405.
111. **Matsunaga, T., and S. Kamiya.** 1987. Use of magnetic particles isolated from magnetotactic bacteria for enzyme mobilization. *Appl. Microbiol. Biotechnol.* **26:**328–332.
112. **Matsunaga, T., C. Nakamura, J. G. Burgess, and K. Sode.** 1992. Gene transfer in magnetic bacteria: transposon mutagenesis and cloning of genomic DNA fragments required for magnetite synthesis. *J. Bacteriol.* **174:**2748–2753.
113. **Matsunaga, T., T. Sakaguchi, and F. Tadokoro.** 1991. Magnetite formation by a magnetic bacterium capable of growing aerobically. *Appl. Microbiol. Biotechnol.* **35:**651–655.
114. **Matsunaga, T., F. Tadokoro, and N. Nakamura.** 1990. Mass culture of magnetic bacteria and their application to flow type immunoassays. *IEEE Trans. Magn.* **26:**1557–1559.
115. **Matsunaga, T., and N. Tsujimura.** 1993. Respiratory inhibitors of a magnetic bacterium *Magnetospirillum* sp. AMB-1 capable of growing aerobically. *Appl. Microbiol. Biotechnol.* **39:**368–371.
116. **McFadden, B. A., and J. M. Shively.** 1991. Bacterial assimilation of carbon dioxide by the Calvin cycle, p. 25–49. *In* J. M. Shively and L. L. Barton (ed.), *Variations in Autotrophic Life.* Academic Press, Inc., San Diego, Calif.
117. **McKay, D. S., E. K. Gibson Jr., K. L. Thomas-Keprta, H. Vali, C. S. Romanek, S. J. Clemett, X. D. F. Chillier, C. R. Maechling, and R. N. Zare.** 1996. Search for past life on Mars: possible relic biogenic activity in Martian meteorite ALH84001. *Science* **273:**924–930.
118. **Meldrum, F. C., B. R. Heywood, S. Mann, R. B. Frankel, and D. A. Bazylinski.** 1993. Electron microscopy study of magnetosomes in two cultured vibrioid magnetotactic bacteria. *Proc. R. Soc. Lond. Ser.* B **251:**237–242.
119. **Meldrum, F. C., B. R. Heywood, S. Mann, R. B. Frankel, and D. A. Bazylinski.** 1993. Electron microscopy study of magnetosomes in a cultured coccoid magnetotactic bacterium. *Proc. R. Soc. Lond. Ser.* B **251:**231–236.
120. **Moench, T. T.** 1988. *Biliphococcus magnetotacticus* gen. nov. sp. nov., a motile, magnetic coccus. *Antonie Leeuwenhoek* **54:**483–496.
121. **Moench, T. T., and W. A. Konetzka.** 1978. A novel method for the isolation and study of a magnetic bacterium. *Arch. Microbiol.* **119:**203–212.
122. **Moskowitz, B. M.** 1995. Biomineralization of magnetic minerals. *Rev. Geophys. Suppl.* **33:**123–128.
123. **Moskowitz, B. M., R. B. Frankel, D. A. Bazylinski.** 1993. Rock magnetic criteria for the detection of biogenic magnetite. *Earth Planet. Sci. Lett.* **120:**283–300.
124. **Moskowitz, B. M., R. B. Frankel, D. A. Bazylinski, H. W. Jannasch, and D. R. Lovley.** 1989. A comparison of magnetite particles produced anaerobically by magnetotactic and dissimilatory iron-reducing bacteria. *Geophys. Res. Lett.* **16:**665–668.
125. **Myers, C. R., and K. H. Nealson.** 1990. Iron mineralization by bacteria: metabolic coupling of iron reduction to cell metabolism in *Alteromonas putrefaciens* MR-1, p. 131–149. *In* R. B. Frankel and R. P. Blakemore (ed.), *Iron Biominerals.* Plenum Press, New York, N.Y.
126. **Nakamura, C., J. G. Burgess, K. Sode, and T. Matsunaga.** 1995. An iron-regulated gene, *magA,* encoding an iron transport protein of *Magnetospirillum* AMB-1. *J. Biol. Chem.* **270:**28392–28396.
127. **Nakamura, C., T. Kikuchi, J. G. Burgess, and T. Matsunaga.** 1995. Iron-regulated expression and membrane localization of the MagA protein in *Magnetospirillum* sp. strain AMB-1. *J. Biochem.* **118:**23–27.
128. **Nakamura, C., T. Sakaguchi, S. Kudo, J. G. Burgess, K. Sode, and T. Matsunaga.** 1993. Characterization of iron uptake in the magnetic bacterium *Aquaspirillum* sp. AMB-1. *Appl. Biochem. Biotechnol.* **39/40:**169–176.

129. **Nakamura, N., J. G. Burgess, K. Yagiuda, S. Kudo, T. Sakaguchi, and T. Matsunaga.** Detection and removal of *Escherichia coli* using fluorescein isothiocyanate conjugated monoclonal antibody immobilized on bacterial magnetic particles. *Anal. Chem.* **65:**2036–2039.
130. **Nakamura, N., K. Hashimoto, and T. Matsunaga.** 1991. Immunoassay method for the determination of immunoglobulin G using bacterial magnetic particles. *Anal. Chem.* **63:**268–272.
131. **Nakamura, N., and T. Matsunaga.** 1993. Highly sensitive detection of allergen using bacterial magnetic particles. *Anal. Chim. Acta* **281:**585–589.
132. **Neilands, J. B.** 1984. A brief history of iron metabolism. *Biol. Metals* **4:**1–6.
133. **Noguchi, Y., T. Fujiwara, K. Yoshimatsu, and Y. Fukumori.** 1999. Iron reductase for magnetite synthesis in the magnetotactic bacterium *Magnetospirillum magnetotacticum. J. Bacteriol.* **181:** 2142–2147.
134. **Oberhack, M., R. Süssmuth, and F. Hermann.** 1987. Magnetotactic bacteria from freshwater. *Z. Naturforsch. Ser.* **42E:**300–306.
135. **O'Brien, W., L. C. Paoletti, and R. P. Blakemore.** 1987. Spectral analysis of cytochromes in *Aquaspirillum magnetotacticum. Curr. Microbiol.* **15:**121–127.
136. **Okuda, Y., K. Denda, and Y. Fukumori.** 1996. Cloning and sequencing of a gene encoding a new member of the tetratricopeptide protein family from magnetosomes of *Magnetospirillum magnetotacticum. Gene* **171:**99–102.
137. **Oldfield, F.** 1992. The source of fine-grained magnetite in sediments. *Holocene* **2:**180–182.
138. **Palache, C., H. Berman, and C. Frondel**. 1944. *Dana's System of Mineralogy.* John Wiley & Sons, Inc., New York, N.Y.
139. **Paoletti, L. C., and R. P. Blakemore.** 1986. Hydroxamate production by *Aquaspirillum magnetotacticum. J. Bacteriol.* **167:**73–76.
140. **Paoletti, L. C., and R. P. Blakemore.** 1988. Iron reduction by *Aquaspirillum magnetotacticum. Curr. Microbiol.* **17:**339–342.
141. **Peck, J. A., and J. W. King.** 1996. Magnetofossils in the sediments of Lake Baikal, Siberia. *Earth Planet. Sci. Lett.* **140:**159–172.
142. **Penninga, I., H. deWaard, B. M. Moskowitz, D. A. Bazylinski, and R. B. Frankel.** 1995. Remanence curves for individual magnetotactic bacteria using a pulsed magnetic field. *J. Magn. Magn. Mater.* **149:**279–286.
143. **Petersen, N., T. von Dobeneck, and H. Vali.** 1986. Fossil bacterial magnetite in deep-sea sediments from the South Atlantic Ocean. *Nature* (London) **320:**611–615.
144. **Phillips, J. B.** 1986. Two magnetoreception pathways in a migratory salamander. *Science* **233:** 765–767.
145. **Pósfai, M., P. R. Buseck, D. A. Bazylinski, and R. B. Frankel.** 1998. Reaction sequence of iron sulfide minerals in bacteria and their use as biomarkers. *Science* **280:**880–883.
146. **Pósfai, M., P. R. Buseck, D. A. Bazylinski, and R. B. Frankel.** 1998. Iron sulfides from magnetotactic bacteria: structure, compositions, and phase transitions. *Am. Mineral.* **83:**1469–1481.
147. **Proksch, R. B., B. M. Moskowitz, E. D. Dahlberg, T. Schaeffer, D. A. Bazylinski, and R. B. Frankel.** 1995. Magnetic force microscopy of the submicron magnetic assembly in a magnetotactic bacterium. *Appl. Phys. Lett.* **66:**2582–2584.
148. **Rickard, D. T.** 1969. The microbiological formation of iron sulfides. *Stockh. Contrib. Geol.* **20:** 50–66.
149. **Rickard, D. T.** 1969. The chemistry of iron sulfide formation at low temperatures. *Stockh. Contrib. Geol.* **20:**67–95.
150. **Rogers, F. G., R. P. Blakemore, N. A. Blakemore, R. B. Frankel, D. A. Bazylinski, D. Maratea, and C. Rogers.** 1990. Intercellular structure in a many-celled magnetotactic procaryote. *Arch. Microbiol.* **154:**18–22.
151. **Rogers, F. G., R. P. Blakemore, N. A. Blakemore, R. B. Frankel, D. A. Bazylinski, D. Maratea, and C. Rogers.** 1990. Intercellular junctions, motility and magnetosome structure in a multicellular magnetotactic procaryote, p. 239–255. *In* R. B. Frankel and R. P. Blakemore (ed,) *Iron Biominerals.* Plenum Press, New York, N.Y.
152. **Rossello-Mora, R. A., F. Caccavo Jr., K. Osterlehner, N. Springer, S. Spring, D. Schüler, W. Ludwig, R. Amann, M. Vannacanneyt, and K.-H. Schleifer.** 1994. Isolation and taxonomic

characterization of a halotolerant, facultative anaerobic iron-reducing bacterium. *Syst. Appl. Microbiol.* **17:**569–573.

153. **Sakaguchi, T., J. G. Burgess, and T. Matsunaga.** 1993. Magnetite formation by a sulphate-reducing bacterium. *Nature* (London) **365:**47–49.
154. **Sakaguchi, H., H. Hagiwara, Y. Fukumori, Y. Tamaura, M. Funaki, and S. Hirose.** 1993. Oxygen concentration-dependent induction of a 140-kDa protein in magnetic bacterium *Magnetospirillum magnetotacticum* MS-1. *FEMS Microbiol. Lett.* **107:**169–174.
155. **Schleifer, K.-H., D. Schüler, S. Spring, M. Weizenegger, R. Amann, W. Ludwig, and M. Kohler.** 1991. The genus *Magnetospirillum* gen. nov., description of *Magnetospirillum gryphiswaldense* sp. nov., and transfer of *Aquaspirillum magnetotacticum* to *Magnetospirillum magnetotacticum* comb. nov. *Syst. Appl. Microbiol.* **14:**379–385.
156. **Schüler, D., and E. Baeuerlein.** 1996. Iron-limited growth and kinetics of iron uptake in *Magnetospirillum gryphiswaldense*. *Arch. Microbiol.* **166:**301–307.
157. **Schüler, D., and E. Baeuerlein.** 1998. Dynamics of iron uptake and Fe_3O_4 mineralization during aerobic and microaerobic growth of *Magnetospirillum gryphiswaldense*. *J. Bacteriol.* **180:**159–162.
158. **Schüler, D., S. Spring, and D. A. Bazylinski.** 1999. Improved technique for the isolation of magnetotactic spirilla from a freshwater sediment and their phylogenetic characterization. *Syst. Appl. Microbiol* **22:**466–471.
159. **Segall, J. E., S. M. Block, and H. C. Berg.** 1986. Temporal comparisons in bacterial chemotaxis. *Proc. Natl. Acad. Sci. USA* **83:**8987–8991.
160. **Short, K. A., and R. P. Blakemore.** 1986. Iron respiration-driven proton translocation in aerobic bacteria. *J. Bacteriol.* **167:**729–731.
161. **Short, K. A., and R. P. Blakemore.** 1989. Periplasmic superoxide dismutases in *Aquaspirillum magnetotacticum*. *Arch. Microbiol.* **152:**342–346.
162. **Silverman, M., and M. Simon.** 1974. Flagellar rotation and the mechanism of bacterial motility. *Nature* (London) **249:**73–74.
163. **Slobodkin, A. I., C. Jeanthon, S. L'Haridon, T. Nazina, M. Miroschnichenko, and E. Bonch-Osmolovskaya.** 1999. Dissimilatory reduction of Fe(III) by thermophilic bacteria and archaea in deep subsurface petroleum reservoirs of western Siberia. *Curr. Microbiol.* **39:**99–102.
164. **Slobodkin, A. I., A.-L. Reysenbach, N. Strutz, M. Dreier, and J. Wiegel.** 1997. *Thermoterrabacterium ferrireducens* gen. nov., sp. nov., a thermophilic anaerobic dissimilatory Fe(III)-reducing bacterium from a continental hot spring. *Int. J. Syst. Bacteriol.* **47:**541–547.
165. **Sparks, N. H. C.** 1990. Structural and morphological characterization of biogenic magnetite crystals, p. 167–177. *In* R. B. Frankel and R. P. Blakemore (ed,) *Iron Biominerals.* Plenum Press, New York, N.Y.
166. **Sparks, N. H. C., S. Mann, D. A. Bazylinski, D. R. Lovley, H. W. Jannasch, and R. B. Frankel.** 1990. Structure and morphology of magnetite anaerobically-produced by a marine magnetotactic bacterium and a dissimilatory iron-reducing bacterium. *Earth Planet. Sci. Lett.* **98:**14–22.
167. **Spormann, A. M., and R. S. Wolfe.** 1984. Chemotactic, magnetotactic and tactile behavior in a magnetic spirillum. *FEMS Microbiol. Lett.* **22:**171–177.
168. **Spring, S., R. Amann, W. Ludwig, K.-H. Schleifer, H. van Gemerden, and N. Petersen.** 1993. Dominating role of an unusual magnetotactic bacterium in the microaerobic zone of a freshwater sediment. *Appl. Environ. Microbiol.* **59:**2397–2403.
169. **Spring, S., R. Amann, W. Ludwig, K.-H. Schleifer, and N. Petersen.** 1992. Phylogenetic diversity and identification of non-culturable magnetotactic bacteria. *Syst. Appl. Microbiol.* **15:**116–122.
170. **Spring, S., R. Amann, W. Ludwig, K.-H. Schleifer, D. Schüler, K. Poralla, and N. Petersen.** 1994. Phylogenetic analysis of uncultured magnetotactic bacteria from the alpha-subclass of Proteobacteria. *Syst. Appl. Microbiol.* **17:**501–508.
171. **Spring, S., U. Lins, R. Amann, K.-H. Schleifer, L. C. S. Ferreira, D. M. S. Esquivel, and M. Farina.** 1998. Phylogenetic affiliation and ultrastructure of uncultured magnetic bacteria with unusually large magnetosomes. *Arch. Microbiol.* **169:**136–147.
172. **Spring, S., and K.-H. Schleifer.** 1995. Diversity of magnetotactic bacteria. *Syst. Appl. Microbiol.* **18:**147–153.
173. **Stolz, J. F.** 1993. Magnetosomes. *J. Gen Microbiol.* **139:**1663–1670.

174. **Stolz, J. F., S.-B. R. Chang, and J. L. Kirschvink.** 1986. Magnetotactic bacteria and single domain magnetite in hemipelagic sediments. *Nature* (London) **321:**849–850.
175. **Stolz, J. F., D. R. Lovley, and S. E. Haggerty.** 1990. Biogenic magnetite and the magnetization of sediments. *J. Geophys. Res.* **95:**4355–4361.
176. **Suzuki, H., T. Tanaka, T. Sasaki, N. Nakamura, T. Matsunaga, and S. Mashiko.** 1998. High resolution magnetic force microscope images of a magnetic particle chain extracted from magnetic bacteria AMB-1. *Jpn. J. Appl. Phys.* **37:**L1343–L1345.
177. **Swancutt, M. A., B. S. Riley, J. D. Radolf, and M. V. Norgard.** 1989. Molecular characterization on the pathogen-specific, 34–kilodalton membrane immunogen of *Treponema pallidum. Infect. Immun.* **57:**3314–3323.
178. **Tamegai, H., and Y. Fukumori.** 1994. Purification, and some molecular and enzymatic features of a novel *ccb*-type cytochrome *c* oxidase from a microaerobic denitrifier, *Magnetospirillim magnetotacticum. FEBS Lett.* **347:**22–26.
179. **Tamegai, H., T. Yamanaka, and Y. Fukumori.** 1993. Purification and properties of a "cytochrome a_1"-like hemoprotein from a magnetotactic bacterium, *Aquaspirillum magnetotacticum. Biochim. Biophys. Acta* **1158:**237–243.
180. **Taylor, B. L.** 1983. How do bacteria find the optimal concentration of oxygen? *Trends Biochem. Sci.* **8:**438–441.
181. **Thornhill, R. H., J. G. Burgess, T. Sakaguchi, and T. Matsunaga.** 1994. A morphological classification of bacteria containing bullet-shaped magnetic particles. *FEMS Microbiol. Lett.* **115:** 169–176.
182. **Torres de Araujo, F. F., M. A. Pires, R. B. Frankel, and C. E. M. Bicudo**. 1986. Magnetite and magnetotaxis in algae. *Biophys. J.* **50**:385–378.
183. **Towe, K. M., and T. T. Moench.** 1981. Electron-optical characterization of bacterial magnetite. *Earth Planet. Sci. Lett.* **52:**213–220.
184. **van Vliet, A. H., K. G. Woolridge, and J. M. Ketley.** 1998. Iron-responsive gene regulation in a *Campylobacter jejuni fur* mutant. *J. Bacteriol.* **180:**5291–5298.
185. **Vargas, M., K. Kashefi, E. L. Blunt-Harris, and D. R. Lovley.** 1998. Microbiological evidence for Fe(III) reduction on early Earth. *Nature* (London) 395:65–67.
186. **Wakeham, S. G., B. L. Howes, and J. W. H. Dacey.** 1984. Dimethyl sulphide in a stratified coastal salt pond. *Nature (London)* **310:**770–772.
187. **Wakeham, S. G., B. L. Howes, J. W. H. Dacey, R. P. Schwarzenbach, and J. Zeyer.** 1987. Biogeochemistry of dimethylsulfide in a seasonally stratified coastal salt pond. *Geochim. Cosmochim. Acta* **51:**1675–1684.
188. **Walcott, C., J. L. Gould, and J. L. Kirschvink.** 1979. Pigeons have magnets. *Science* **205:**1027–1029.
189. **Walcott, C., and R. P. Green.** 1974. Orientation of homing pigeons altered by a change in the direction of an applied magnetic field. *Science* **184:**108–182.
190. **Waleh, N. S.** 1988. Functional expression of *Aquaspirillum magnetotacticum* genes in *Escherichia coli* K12. *Mol. Gen. Genet.* **214:**592–594.
191. **Walker, M. M., and M. E. Bitterman.** 1989. Honeybees can be trained to respond to very small changes in geomagnetic field intensity. *J. Exp. Biol.* **145:**489–494.
192. **Walker, M. M., C. E. Diebel, C. V. Haugh, P. M. Pankhurst, and J. C. Montgomery.** 1997. Structure and function of the vertebrate magnetic sense. *Nature* **390:**371–376.
193. **Walker, M. M., J. L. Kirschvink, A. E. Dizon, and G. Ahmed.** 1992. Evidence that fin whales respond to the geomagnetic field during migration. *J. Exp. Biol.* **171:**67–78.
194. **Woese, C. R.** 1987. Bacterial evolution. *Microbiol. Rev.* **51:**221–271.
195. **Yamazaki, T., H. Oyanagi, T. Fujiwara, and Y. Fukumori.** 1995. Nitrite reductase from the magnetotactic bacterium *Magnetospirillum magnetotacticum*: a novel cytochrome cd_1 with Fe(II): nitrite oxidoreductase activity. *Eur. J. Biochem.* **233:**665–671.
196. **Zavarzin, G. A., E. Stackebrandt, and R. G. E. Murray.** 1991. A correlation of phylogenetic diversity in the Proteobacteria with the influences of ecological forces. *Can. J. Microbiol.* **37:**1–6.
197. **Zhang, C., S. Liu, T. J. Phelps, D. R. Cole, J. Horita, and S. M. Fortier.** 1997. Physiochemical, mineralogical, and isotopic characterization of magnetite-rich iron oxides formed by thermophilic iron-reducing bacteria. *Geochim. Cosmochim. Acta* **61:**4621–4632.

198. **Zhang, C., H. Vali, C. S. Romanek, T. J. Phelp, and S. V. Lu.** 1998. Formation of single domain magnetite by a thermophilic bacterium. *Am. Mineral.* **83:**1409–1418.
199. **Zhulin, I. B., V. A. Bespelov, M. S. Johnson, and B. L. Taylor.** 1996. Oxygen taxis and proton motive force in *Azospirillum brasiliense*. *J. Bacteriol.* **178:**5199–5204.

Environmental Microbe-Metal Interactions
Edited by Derek R. Lovley

Chapter 6

The Role of Siderophores in Iron Oxide Dissolution

Larry E. Hersman

Iron(Fe) is an essential element for all but a few living organisms and is important for numerous physiological functions. Used in all heme enzymes, including cytochromes and hydroperoxides, a constituent of ribonucleotide reductase, and important for the activity of nitrogenases, Fe is required by all microorganisms except lactobacilli lacking heme and using a cobalt type of ribonucleotide reductase (5, 10). The biological importance of iron is the result of its electronic structure, which is capable of reversible changes in oxidation state over a wide range of oxidation-reduction potentials: +300 mV in *a*-type cytochromes to −490 mV in some iron-sulfur proteins (23, 48). Owing to this redox versatility, Fe plays a unique and vital role in biological systems; however, too much Fe can be problematic. Both ferrous [Fe(II)] and ferric [Fe(III)] iron can generate lethal oxidizing hydroxyl radicals (23). While it is possible to store nontoxic forms of Fe intracellularly, most cells have evolved systems that carefully regulate the uptake of this potentially dangerous metal.

Perhaps an even more daunting challenge posed by Fe to living cells is its acquisition because Fe is very insoluble at physiological pH and in oxic environments. Although abundant at the Earth's surface, Fe oxides (the most common form of Fe) have solubility products (ranging from 10^{-39} to 10^{-44}) that limit the aqueous equilibrium concentration of Fe to approximately 10^{-17} M (51). Most microorganisms require micromolar concentrations of Fe to support growth (43). In fact, many species will suffer Fe deprivation when the culture medium contains less than 0.1 μM available Fe. Thus, microorganisms are faced with overcoming approximately 10 orders of magnitude discrepancy between the Fe that is available in aerobic environments and their metabolic requirement. This is a problem faced not only by microorganisms but also by all living organisms, including plants and animals.

Larry E. Hersman • Environmental Molecular Biology Group, Los Alamos National Laboratory, Los Alamos, NM 87545.

For over 50 years microbiologists have been aware of a class of compounds, produced by microorganisms, called siderophores. In 1981, Neilands (42) stated that "siderophores are viewed as the evolutionary response to the appearance of O_2 in the atmosphere, the concomitant oxidation of Fe(II) to Fe(III), and the precipitation of the latter as ferric hydroxide, $K_s = <10^{-38}$." Siderophores are low-molecular-mass, ferric iron-specific ligands that are induced at low iron concentrations for the purpose of biological assimilation of Fe(III). Over 200 siderophores have been isolated, of which several have been fully characterized, with most being classified chemically as either catechols or hydroxamates. As an effective ligand of Fe(III) chelation, siderophores have molecular masses ranging between 0.5 and 1.5 kDa (40); however, such a molecular size exceeds the free diffusion limit of the outer membrane of microorganisms. To compensate, microorganisms have developed specialized surface proteins (receptors) and associated internal enzymes and proteins that transport the siderophore-Fe(III) complex to the sites of physiological utilization.

The most common siderophores contain the hydroxamic acid functional group,—R—CO—N(OH)—R′, which forms a five-member ring with Fe^{3+} between the two oxygen atoms, with the hydroxyl proton being displaced. Frequently, three hydroxamate groups are found on a single siderophore molecule and thus supply six oxygen ligands to satisfy the preferred hexacoordinate octahedral geometry of Fe(III), each having partial double-bond characteristics. Once complexation occurs, a "more or less distorted coordination octahedron is formed at the metal site by (these) three 5-membered chelate rings" (39). Other bidentate ligand moieties [e.g., catechol, α-hydroxy acid, 2-(2-hydroxyphenyl)oxazoline, and fluorescent chromophore] also are found in siderophores (9). Like the hydroxamates, catechols can occur in triplicate as the only unit in a siderophore molecule (e.g., the siderophore enterobactin). It is also common to find siderophores composed of mixtures of functional moieties, such as pseudobactin, which contains a hydroxamate, an α-hydroxy acid, and a fluorescent chromophore. While there are significant measurable differences in the Fe coordination structural features of these various siderophores (56), the common characteristic of all siderophores is that they are virtually ferric ion specific, with some reported to have 1:1 complex formation constants with Fe(III) that exceed 10^{50}. "Iron (III) has a higher charge and smaller radius than Fe (II) and hence ranks as a harder acid which will prefer to bind to an atom classified as a harder base, such as O^-. . . What this means is that siderophores (through their various functional moieties) . . . will offer mainly oxygen atoms as electron donors to the metal ion. . . Unloading the iron from the ferric-siderophore is most efficiently accomplished by reduction since the relatively soft Fe(II) has little affinity for oxygen" (44).

Indeed, much is known about the physiological functions of siderophores. Also, the genetic controls that regulate siderophore production and regulate the production of the membrane-bound siderophore receptor and transport proteins are well characterized. While there are many variations, the most widely accepted paradigm is as follows: (i) in the absence of Fe, siderophore production is derepressed and the newly produced siderophore binds to Fe(III), (ii) once bound to a siderophore, the Fe(III)-siderophore complex binds to its cognate receptor on the cell's outer membrane and is transported into the cytoplasm, where (iii) Fe(III) is released via reduction to Fe(II) or the siderophore is disassembled (e.g., esterase cleavage of ester linkages in the siderophore enterochelin). For a more detailed description of these processes, the reader is referred to reviews by Weiberg (59), Briat (10), and Guerinot (22) and books by Winkelmann (60), Winkelmann et al. (61), and Barton and Hemming (7).

In soils, microorganisms rarely exist in pure culture, alone with their own siderophores; rather, potentially hundreds of species are present and competing for Fe. Hence, the extracellular complexation of Fe with a siderophore has profound effects not only on the microbial community but also on the greater soil ecosystem. So important is the acquisition of Fe in soil systems that competition for its complexation is a matter of life and death for the organisms involved. Thus, within the general domain of the ecological significance of siderophores, no subject commands greater research interest than does that of plant-microbe interactions, as exemplified by the impressive number of publications on "plant growth-promoting" and "plant-deleterious" activities by soil microorganisms (see, e.g., references 6, 11–15, 17–19, 24, 31–38, 46, 49, 50, 62, and 63). And rightfully so, because the role that siderophores play in the movement of iron (both microbe-to-plant and microbe-to-microbe movement) in the rhizosphere is clearly significant for plant growth and nutrition. While a discussion of Fe availability in the rhizosphere is beyond the purpose of this chapter, two points should be made. First, even though there are scores of publications on this subject, there is still much to be learned. As Marschner and Romheld (38) pointed out, "our understanding is poor concerning the processes taking place. . . (such as) the role of low-molecular-weight root exudates and siderophores on Fe acquisition of plants growing in soils of differing Fe availability." Which reflects an earlier frustration expressed by Crowley et al. (17), that ". . .microorganisms can be an unpredictable confounding factor in experiments examining mechanisms for utilization of microbial siderophores or phytosiderophores under nonsterile conditions."

Second, Loper and Lindow (35) and Marschner and Crowley (37) have made an interesting observation. Using the ice nucleation reporter gene as an indicator of siderophore production, both of these investigations suggested that Fe is not limiting in the root zone, if one assumes that low levels of siderophore concentration are inversely proportional to high levels of Fe availability. Marschner and Crowley (37) concluded that the ". . . nutritional competition for iron in the rhizosphere may not be a major factor influencing root colonization." The implication is that at this time the role of siderophores in the root zone is not clearly understood, which is not surprising when one considers the difficulty and complexity in working with soil, microorganisms, and plants under nonsterile conditions. Even so, in reviewing the literature on plant growth-promoting and plant-deleterious microorganisms, one comes away with a sense of the enormous importance of siderophores as extracellular Fe-sequestering and shuttling molecules, which are important for Fe transfer and competition.

In contrast to the discussions common to previous reviews, this chapter examines the type of Fe to which siderophores bind and the location where they do so. The assumption in the literature has been that "the purpose of siderophores is to supply

Fe to the cell—so often stated that this quote is rarely referenced. Of course, there exists no controversy here, siderophores do supply Fe to the cell. However, if one looks for quantitative investigations of the source of that Fe, until recently one would have come up empty-handed. To supply "what" Fe to the cell? What was its source? If mineral, what are the dissolution mechanism(s) and rate-limiting step(s)? These are the ecological questions of interest to biogeochemists and geomicrobiologists and therefore are the subjects of this chapter. However, not covered in this chapter is the research is the dissolution of Fe when used as a terminal electron acceptor by facultative anaerobes (e.g., *Shewanella putrefaciens*), because this type of Fe oxide dissolution occurs most commonly in microaerophilic and anoxic environments, where reductive processes dominate Fe equilibrium.

SIDEROPHORE CONCENTRATION IN SOILS

Bossier et al. (9) discussed the concentration of fungal and bacterial siderophores in soil. Fungal hydroxamate siderophore concentrations ranged between 0 and 250 μg kg of soil^{-1}, or 0.04 to 0.32 mg g of biomass C^{-1}—the latter derived by comparison to an axenic culture of *Boletus edulis* (55). In nearly perfect agreement, Ackers (1–3) estimated the fungal siderophore concentration to be between 30 and 240 μg kg of soil^{-1} (0.9 to 7.5 $\times$ 10^{-10} mol g of soil^{-1}). Later, using bioassays, Nelson et al. (45) estimated the concentration of bacterial siderophores to be between 0.4 $\times$ 10^{-10} and 3.0 $\times$ 10^{-10} mol g of soil^{-1}. More recently, siderophore concentrations in soils have been measured by using monoclonal antibodies to ferric pseudobactin (a siderophore produced by plant growth-promoting strain B10). Buyer et al. (11) estimated the concentration of pseudobactin to be 3.5 $\times$ 10^{-10} mol g of soil^{-1} (wet weight; approximately 14% moisture content). Finally, using an ice nucleation reporter gene assay, Marschner and Crowley (37) reported pyroverdin concentrations between 2.0 $\times$ 10^{-10} and 5.0 $\times$ 10^{-10} mol g (dry weight) of root material^{-1}. All of these studies compare favorably to the early hydroxamate estimate by Szaniszlo et al. (55), but the concentrations are significantly lower than the siderophore production measured in vitro by Meyer and Abdallah (41) for microorganisms grown under Fe stress (approximately 875 mg g of biomass C^{-1}).

Using this information, it is possible to make a rough approximation of the siderophore content of a common soil. The fungal siderophore concentration is approximately 2.7 $\times$ 10^{-10} mol g of soil^{-1} (42–44, 55), while the bacterial concentration is approximately 2.4 $\times$ 10^{-10} mol g of soil^{-1} (11.45). Together, they equal a total concentration of 5.1 $\times$ 10^{-10} mol g^{-1}. Assuming a typical gravimetric soil water content of 0.4 kg kg of soil^{-1} (28), soils would therefore contain an approximate siderophore aqueous concentration of 1.275 $\times$ 10^{-6} mol liter^{-1}. If one also assumes a 1:1 binding of siderophore to Fe and assumes single use of the siderophore (however, intuitively one would suspect that siderophores can be used repeatedly for multiple chelation events), an Fe concentration of 1.275 $\times$ 10^{-6} mol liter^{-1} would supply an order of magnitude above the minimum Fe concentration (0.1 μmol liter^{-1}) needed for microbial growth.

OCCURRENCE OF IRON IN SOILS

While there may be enough siderophore in the soil to chelate Fe, another factor that must be considered is the occurrence and availability of Fe in soils. The most abundant metal oxides in soils are the iron oxides, which include oxides, oxyhydroxides, and hydrated oxides. Iron is present in most rocks of the Earth's crust. The initial source of this Fe is magmatic rocks, which contain Fe bound in silicates in the divalent (reduced) state. Then, through protolysis and oxidation, Fe is released during weathering. The liberated Fe(II) may be oxidized if O_2 is present or, if O_2 is not present, as in deeper parts of the soil profile, may migrate until it reaches an oxygenated zone. As stated above, the resulting iron(III) oxides have solubility products ranging from 10^{-39} to 10^{-44}. It is improbable that complete redissolution of these oxides would occur, because the pH is seldom low enough to permit the reverse hydrolysis reaction (53).

Iron(III) oxides consist of densely packed O layers with Fe^{3+} (and Fe^{2+}) in the interstices. With the exception of magnetite, O is usually arranged octahedrally around Fe. The Fe oxides differ from one another in the way the $Fe(O, OH)_6$ octahedral are arranged in space through sharing of corners, edges and/or faces of the octahedral (51). The properties of these oxides often are modified (i.e., weakened) by isomorphous replacement of Fe^{3+} in its octahedral position by Al^{3+}.

The most common and stable Fe oxides found in soil are goethite and hematite, whole leidocrocite, ferrihydrite, and maghemite are less common. Fe oxide concentrations in soils vary between <0.1 and >50%. Owing to their high energy of crystallization and inhibitory effects of soil solution chemistry, Fe oxide crystals are usually small, about 5 to 1.50 nm, and crystal disorder is common. Because of their large specific surface area, Fe oxides contribute significantly to the overall surface area of soils.

The conditions under which a mineral will be in the most stable phase usually are determined from thermodynamic data, which suggest that in most aerobic soils goethite (α-FeOOH) should be the most stable phase. These data may also be used to describe the goethite-hematite (δ-Fe_2O_3) relationship: hematite + H_2O = 2 goethite. The standard Gibbs energy ($\Delta G°$) of this reaction lies between +1.7 and −0.4 kJ mol^{-1}. Negative values indicate a higher stability of goethite at a H_2O activity of unity. As the H_2O activity decreases below unity, hematite becomes the more stable phase. Other factors such as particle size, foreign ions (particularly Al), temperature, and pH also affect the distribution of Fe between these two phases (51).

IRON OXIDE DISSOLUTION

Even though the dissolution of Fe oxides has been reviewed extensively in the literature (16, 51, 53, 54), it is worthwhile to briefly review dissolution mechanisms, since they apply to the potential involvement of siderophores. During dissolution of Fe oxides, the coordination environment of Fe(III) changes, where the Fe^{3+} in the crystalline lattice exchanges its O^{2-} or OH^- ligands for water or another ligand or changes its oxidation state to Fe^{2+}. Thus, H^+, OH^-, ligands, and reductants are important reactants participating in the three general categories of dissolution: proton-promoted, ligand-promoted, and reductive dissolution. For each of these, the first step involves the formation of a surface complex by fast adsorption of protons, ligands, or electron donors or a combination of these. This leads to a polarizing and thereby a weakening of the Fe-O bond and then to the detachment of the Fe atom into solution. In dissolving an iron-(III)-(hydr) oxide complex, the coordinative environment of Fe(III) changes. The Fe^{3+} in the crystalline structure replaces its O^{2-} or OH^- by water or another ligand. In reductive dissolution, Fe also changes its oxidation state to Fe^{2+} (54).

During proton-promoted dissolution, protons are adsorbed to the surface and eventually polarize neighboring Fe-OH groups; this is followed by the detachment of the Fe(III) cation. The original surface structure with neutral OH or OH^+ groups is then restored by the adsorption of aqueous protons and the process starts again.

Through surface adsorption, organic ligands may either accelerate or retard dissolution, depending upon whether they weaken the Fe-O bond or block adsorption sites. Generally, ligands, including reducing ligands, that form complexes bring negative charge into the coordination sphere of the metal oxide surface (a Lewis acid center) and also are able to polarize the critical Fe-O bond. This bond is now less stable, enabling the detachment of the central ion into the surrounding solution (54). Bidentate ligands that form mononuclear complexes generally are more effective in enhancing such dissolution than are monodentate ones.

Reductive dissolution of iron(III) oxides proceeds by a mechanism similar to that of ligand-promoted dissolution. Electron transfer occurs following the adsorption of the electron donor species onto the oxide surface. The reduction of Fe(III) will destabilize the coordination sphere and induce the detachment of the Fe as Fe^{2+}.

SIDEROPHORE-PROMOTED DISSOLUTION OF IRON OXIDES

Due to the high formation constants of siderophore-Fe(III) complexes, it has been assumed that the thermodynamic pressure needed to trap ferric Fe is so great that Fe oxides will undergo ligand-promoted dissolution in the presence of a siderophore ligand. One should, however, keep in mind that for siderophores to provide Fe (by the dissolution of metal oxides), siderophores would have to promote chelation at a rate equal to or greater than the metabolic requirements of the total microbial and plant population, assuming that siderophores are used by plants directly or indirectly as a source of Fe.

In a personal communication, G. Sposito wrote:

> Experimental studies on the ligand-promoted dissolution of metal oxides have shown that the organic ligands most effective in enhancing the rate of nonreductive dissolution are those which can form five-membered rings in a chelate with a metal cation at the periphery of the oxide structure (20). Thus, for example, catechol is more effective than salicylate (six-membered ring), which, in turn, is more effective than phthalate (seven-membered ring). This

ordering of catalyzing ability is the same as that of chelate stability (8), suggesting by implication that siderophores also should be effective ligands in promoting iron oxide dissolution.

On the other hand, Fe^{3+} in aqueous solution differs from Fe^{3+} at the periphery of an iron oxide, in that the latter metal cation is not accessible to a complexing ligand from all directions in space. This "reduction in dimensionality" that attends any reaction between a ligand and a metal cation at the oxide-water interface may lead to a reduction in the rate of the reaction and in the stability of its product (58). Thus, the stereochemistry of a siderophore-Fe^{3+} chelate at an iron oxide surface may differ enough from that in aqueous solution to impede the effective promotion of dissolution. In particular, hexadentate chelation may be impossible.

Conversely, stereochemistry may well favor siderophore attachment. For example, it may be possible for two or three of the hydroxamate groups of desferrioxamine to bind multiple surface Fe atoms. Holmen and Casey (29) estimate the distances between the hydroxamate groups in desferrioxamine, compared those distances to the known distances between adjacent Fe atoms in goethite, and concluded that a linear desferrioxamine molecule could form simultaneously bidentate surface complexes with three adjacent Fe atoms.

Therefore, although there is considerable uncertainty, it does seem plausible that Fe oxide dissolution could be affected by siderophores. Unfortunately, microbial acquisition of Fe from Fe oxides has been reported sparingly in the literature. Page and Huyer (47) were among the first to compare microbial growth and siderophore production as a function of the source of Fe. In this study, Fe release and siderophore production by *Azotobacter vinelandii* (a gram-negative, asymbiotic nitrogen-fixing soil bacterium with an absolute requirement for Fe) was investigated by using several Fe oxide minerals as sources of Fe. The minerals tested fell into four different groupings with respect to their effect on siderophore repression: (i) *A. vinelandii* solubilized Fe from certain minerals (FeS and marcasite [FeS_2]) by producing only dihydroxybenzoic acid (DHBA), which was produced constitutively; (ii) solubilization of other minerals, vivianite [$Fe_2(PO_4)_2{\cdot}H_2O$], olivine [$(Mg,Fe)_2SiO_4$], and Fe_3O_2, occurred due to the derepression of the siderophore azotochelin plus the continued production of DHBA; (iii) the minerals hematite (Fe_2O_3), siderite ($FeCO_3$), pyrite (FeS_2), and goethite (FeOOH) caused full derepression of azotochelin, partial derepression of a second siderophore, azotobactin, and continued production of DHBA; and (iv) the minerals ilmentite ($FeTi(O_3)$, micaceous hematite (unknown), and illite [$2K_2O{\cdot}3(Mg,Fe)O{\cdot}8(Al,Fe)2O_3{\cdot}12H_2O$] caused full derepression of azotobactin, hyperproduction of azotochelin, and continued production of DHBA. The sequential production of DHBA, azotochelin, and azotobactin led the authors to suggest that as the availability of Fe decreased (with decreasing solubility of the minerals) the microorganism responded by producing a more powerful siderophore. Even though the authors did not measure directly the dissolution of the minerals by the siderophores, the fact that the different minerals elicited different responses from *A. vinelandii* was indirect evidence for the involvement of siderophores in mineral dissolution.

Later, Adams et al. (4) assayed for the presence of siderophores on the surfaces of desert rocks. Their analyses confirmed that both hydroxamates and catechols were present on most of the surfaces tested, although no quantitative information was presented. Again, although their evidence was circumstantial, the authors suggest that siderophores are responsible for the liberation of Fe from desert rock surfaces.

Hersman et al. (26) have measured the use of hematite by an aerobic *Pseudomonas* sp. A simple way to demonstrate that Fe is being obtained from a mineral is to monitor microbial growth. If, under Fe limitation, iron is supplied in the form of a mineral and one observes increased microbial growth, one can assume that the microorganism is obtaining iron from the mineral. Synthetic hematite, prepared by the method of Schwertmann and Cornell (52), consisted of hexagonal plates up to 1.5×10^{-6} m in size and had a surface area of approximately 29 m^2 g^{-1}. The hematite was examined using scanning electron microscopy equipped with energy-dispersive spectroscopy. Energy-dispersive spectroscopy analyses confirmed that the mineral was pure, but scanning electron microscopy analysis revealed that some goethite was also present; however, according to Schwertmann (personal communication), his process commonly produces goethite.

Growth on larger surface areas suggested that the bacteria (strict aerobes similar to *P. stutzeri*) were acquiring Fe for growth, whereas the low growth on the lower hematite concentrations was interpreted as iron limitation. Before the experiment was started, the control medium was mixed with hematite for 6 days, the hematite was removed by filtration, and the medium was inoculated. Lack of growth in the control flasks demonstrated the iron was not being removed from the hematite by the growth medium, otherwise iron would have been leached from the hematite in the 6-day preconditioning period and would have been available for microbial growth. These results demonstrated that this aerobic microorganism was able to obtain iron from hematite to support growth. Based on comparisons to growth on Fe-EDTA, the area-base dissolution rate of Fe was approximately 3.0×10^{-9} mol $\cdot$ m^{-2} h^{-1}. Furthermore, growth on hematite was accompanied by the production of pseudobactin-like siderophore. Throughout the course of the experiments, the siderophore concentration (estimated by high-performance liquid chromatography) in the growth medium increased with time, consistent with the earlier observations by Page and Huyer (47). The siderophore concentration was not determined quantitatively; however, one can estimate that the concentration was approximately 4.4×10^{-8} mol$\cdot$liter^{-1} (assuming an average of 440 mg of siderophore of cell biomass^{-1}, approximately 10^8 cells per ml, 1.0×10^{-12} g cell^{-1}, and a gram molecular weight of 1,000 for the siderophore).

Although Page and Huyer (47) and Hersman et al. (26) demonstrated the production of siderophore in the presence of Fe oxide, neither study demonstrated that siderophores were produced as a result of the Fe oxide itself. It is common knowledge that siderophores are produced in environments where there are few mineral surfaces (e.g., laboratory broth cultures and the open ocean). It remains to be investigated whether the Fe oxide alone elicits siderophore production. It is possible that the microorganisms were unaware of the mineral surface as a source of Fe and were simply responding to Fe limitation.

In a separate study, Hersman et al. (25) reported that the dissolution of hematite by the siderophore compared favorably to the dissolution of hematite by oxalic and ascorbic acids and to proton-promoted dissolution. Briefly, siderophore from the *Pseudomonas* sp. was purified as described by Hersman et al. (27). Triplicate batch dissolution experiments were conducted in acid-washed 250-ml Teflon flasks containing 50 ml of H_2O (ionic strength = 10^{-3} M $NaNO_3$). This experiment was performed at pH 3 to allow a comparison to published results of other ligand-promoted dissolution experiments performed at pH 3 (64). The hematite and siderophore concentrations were 4.32×10^{-4} g ml^{-1} (1.3×10^{-2} ml^{-1}) and 240×10^{-6} M, respectively. At 2.5 h, the concentration of soluble iron had increased to greater than 3.0 μmol of Fe $liter^{-1}$, and it then slowly increased to 6 μmol of Fe $liter^{-1}$ by 24 h. Clearly, the siderophore had enhanced Fe dissolution. Typical linear plots (54) gave an excellent fit to a regression equation ($R^2 > 0.992$), and the slope of this equation corresponded to area-based dissolution rates of 10^{-8} mol m^{-2} h^{-1}. This rate was an order of magnitude greater than the rate reported by Watteau and Berthelin (57), whose area-based dissolution rate of goethite by desferrioxamine was approximately 10^{-9} mol$\cdot m^{-2}$ h^{-1}, averaged over a 48-h period.

While the naturally occurring siderophore did compare favorably with oxalate dissolution rates, the concentration of siderophore used (240×10^{-6} mol $liter^{-1}$) was nearly 190 times greater than the previously estimated amount of siderophore expected in soils (1.275×10^{-6} mol $liter^{-1}$). Furthermore, while the dissolution rate (10^{-8} mol m^{-2} h^{-1}) was approximately three times greater than the dissolution rate mediated by the *Pseudomonas* sp. (25) grown in the presence of hematite (3.0×10^{-9} mol $m^{-2} \cdot h^{-1}$), this threefold greater dissolution rate is disappointing given that the siderophore concentration was approximately 30,000 times greater (240×10^{-6} versus the estimated 4.4×10^{-8} mol $liter^{-1}$) that in the *Pseudomonas*-mediated dissolution experiment.

In more recent work, Holmen and Casey (29) discussed in detail the dissolution of goethite (FeOOH) by acetohydroxamic acid, a monomeric hydroxamic acid that is smaller than but similar to the naturally produced siderophore, deferriferrioxamine B. Their results speak to the complexity of dissolution processes, even those mediated by simple organic acids. For example, in their initial experiments, dissolution below pH 4 could be explained by the reduction of Fe(III) to Fe(II) caused by hydroxylamine, formed from surface- or metal-enhanced hydrolysis of the hydroxamate ligand. Later, however, using cylindrical internal reflection Fourier-transformed infrared spectroscopy, they found no evidence to support the formation of a hydroxylamine ligand on the goethite surface (30). Therefore, the question of different dissolution rates below pH 4.0 remains unresolved. Above pH 4, dissolution appeared to proceed via ligand-induced reactions, involving three steps: (i) formation of the monohydroxamate surface complex; (ii) detachment of the Fe-hydroxamate complex, accompanied by movement and dissociation of water molecules to replace the eliminated oxygen sites; and (iii) formation of another monohydroxamate surface complex.

Needless to say, in the pH range of the entire experiment, dissolution as a function of pH was not linear. As the pH was increased from 2.2 to 4.0, the dissolution rate declined sharply from approximately 5×10^{-11} to less than 1×10^{-11} mol m^{-1} h^{-2}. It then jumped dramatically to a maximum of 6×10^{-11} mol m^{-2} h^{-1}, at pH 4.2. Then, as the pH was increased to 9, it again declined to less than 1×10^{-11} mol m^{-2} h^{-1}. This decline was interrupted at both pH 5.4 and 6.4 by a sharp but brief increase followed by an immediate resumption in the decline. The maximum dissolution rate (approximately 6×10^{-11} mol m^{-2} h^{-1}) occurred at about 200 min. This early-stage high Fe flux is common to virtually all dissolution experiments (W. H. Casey, personnel communication); however, the relatively low overall dissolution rate compared to that of oxalate was surprising (29).

CONCLUSIONS

Microorganisms have been evolving for nearly 4 billion years on Earth. Their evolution has led to a diversity and complexity in biochemical, physiological, and molecular biology processes that are sources of amazement, and sometimes bewilderment, to scientists. The acquisition of Fe in aerobic environments for metabolic functions (as opposed to the use of Fe as a terminal electron acceptor in anaerobic respiration) certainly provides a challenge to microorganisms and also to researchers attempting to understand this process.

Undoubtedly, siderophores are used by microorganisms to acquire Fe and are produced by microorganisms in response to a limited availability of Fe. These two observations are supported by hundreds of publications in the open literature. However, the results discussed in this chapter suggest that siderophores may not be entirely responsible for Fe oxide dissolution, that the role that siderophores play in the dissolution of Fe oxides remains unclear, and thus that the microbial dissolution of Fe oxides merits further investigation.

There is recent evidence for a relationship between attachment of the microorganism to the mineral surface and microbially mediated dissolution (21) in that attachment appears to be associated with dissolution. If this is indeed true, by establishing both oxidation-reduction and pH gradients in the microenvironment between themselves and the mineral surface, attachment to the mineral surface would allow microorganisms to use the other two dissolution processes discussed above (proton promoted and reductive), in addition to ligand-promoted dissolution. This process would be much the same as the corrosion processes used by microorganisms attached to metal surfaces. Attachment does not diminish the role of siderophores but, rather, adds to our appreciation of the ability of microorganisms to meet the challenge of limited Fe in an aerobic environment.

REFERENCES

1. **Ackers, H. A.** 1981. The effect of waterlogging on the quantity of microbial iron chelators (siderophores) in soils. *Soil Sci.* **132:**150–152.
2. **Ackers, H. A.** 1983. Multiple hydroxamic acid microbial chelators (siderophores) in soils. *Soil Sci.* **135:**156–160.

3. **Ackers, H. A.** 1983. Isolation of the siderophore schizokinen from soil of rice field. *Appl. Environ. Microbiol.* **45:**1704–1706.
4. **Adams, J. B., F. Palmer, and J. T. Staley.** 1992. Rock weathering in deserts: Mobilization and concentration of ferric iron by microorganisms. *Geomicrobiol. J.* **10:**99–114.
5. **Archibald, F.** 1983. *Micrococcus lysodeikticus*, an organism not requiring iron. *FEMS Microbiol. Lett.* **19:**29–32.
6. **Bar-Ness, E., Y. Hadar, Y. Chen, V. Romheld, and H. Marschner.** 1992. Short-term effects of rhizosphere microorganisms on the Fe uptake from microbial siderophores by maize and oat. *Plant Physiol.* **100:**451–456.
7. **Barton, L. L., and B. C. Hemming.** 1993. *Iron Chelation in Plants and Soil Microorganisms.* Academic Press, Inc., San Diego, Calif.
8. **Bennett, P. C., and W. Casey.** 1994. Chemistry and mechanisms of low-temperature dissolution of silicates by organic acids, p. 162–200. *In* E. D. Pittman and M. D. Lewan (ed.), *Organic Acids in Geochemical Processes.* Springer-Verlag, New York, N.Y.
9. **Bossier, P., M. Hoft, and W. Verstraete.** 1988. Ecological significance of siderophores in soil. *Adv. Microb. Ecol.* **10:**385–414.
10. **Briat, J.-F.** 1992. Iron assimilation and storage in prokaryotes. *J. Gen. Microbiol.* **138:**2475–2483.
11. **Buyer, J. S., M. G. Kratzke, and L. J. Sikora.** 1993. A method for detection of pseudobactin, the siderophore produced by a plant-growth-promoting *Pseudomonas* strain, in the barley rhizosphere. *Appl. Environ. Microbiol.* **59:**677–681.
12. **Buyer, J. S., and J. Leong.** 1986. Iron transport-mediated antagonism between plant growth-promoting and plant-deleterious *Pseudomonas* strains. *J. Biol. Chem.* **261:**791–794.
13. **Buyer, J. S., and L. J. Sikora.** 1990. Rhizosphere interactions and siderophores. *Plant Soil* **129:** 101–107.
14. **Buysens, S., K. Heungens, K. Poppe, and M. Hofte.** 1996. Involvement of pyrochelin and pyroverdin in suppression of *Pythium*-induced damping-off of tomato by *Pseudomonas aeruginosa* 7NSK2. *Appl. Environ. Microbiol.* **62:**865–871.
15. **Carson, K. C., A. R. Glen, and M. J. Dilworth.** 1994. Specificity of siderophore-mediated transport of iron in rhizobia. *Arch. Microbiol.* **161:**333–339.
16. **Casey, W. H., and C. Ludwig.** 1996. The mechanism of dissolution of oxide minerals. *Nature* **381:** 506–509.
17. **Crowley, D. E., V. Romheld, H. Marschner, and P. J. Szanislo.** 1992. Root-microbial effects on plant iron uptake from siderophores and phytosiderophores. *Plant Soil* **142:**1–7.
18. **De Weger, L. A., J. J. C. M. van Arendonk, K. Recourt, J. A. J. M. van der Hofstad, P. J. Weisbeek, and B. Lugtenberg.** 1988. Siderophore-mediated uptake of Fe^{3+} by plant growth-stimulating *Pseudomonas putida* strain WCS358 and by other rhizosphere microorganisms. *J. Bacteriol.* **170:**4693–4698.
19. **Duijff, B. J., W. J. de Kogel, P. A. H. M. Bakker, and B. Schippers.** 1994. Influence of pseudobactin 358 on the iron nutrition of barley. *Soil Biol. Biochem.* **26:**1681–1688.
20. **Furrer, G., and W. Stumm.** 1986. The coordination chemistry of weathering. I. Dissolution of delta-Al_2O_3 and BeO. *Geochim. Cosmochim. Acta* **50:**1847–1860.
21. **Grantham, M. C., and P. M. Dove.** 1996. Investigation of bacterial-mineral interactions using fluid tapping mode atomic force microscopy. *Geochim. Comochim. Acta* **60:**2473–2480
22. **Guerinot, M. L.** 1994. Microbial iron transport. *Annu. Rev. Microbiol.* **48:**743–772.
23. **Guerinot, M. L., and Y. Yi.** 1994. Iron: nutritious, noxious and not readily available. *Plant Physiol.* **104:**815–820.
24. **Handelsman, J., and E. V. Stabb.** 1996. Biocontrol of soilborne plant pathogens. *Plant Cell* **8:** 1855–1869.
25. **Hersman, L., T. Lloyd, and G. Sposito.** 1995. Siderophore-promoted dissolution of hematite. *Geochim. Cosmochim. Acta* **59:**3327–3330.
26. **Hersman, L., P. Maurice, and G. Sposito.** 1996. Iron acquisition from hydrous Fe(III) oxides by an aerobic *Pseudomonas* sp. *Chem. Geol.* **132:**25–31.
27. **Hersman, L. E., P. D. Palmer, and D. E. Hobart.** 1993. The role of siderophores in the transport of radionuclides. *Mater. Res. Soc. Proc.* **294:**765–770.
28. **Hillel, D.** 1980. *Fundamentals of Soil Physics*, p. 12. John Wiley & Sons, Inc., New York, N.Y.

29. **Holmen, B. A., and W. H. Casey.** 1996. Hydroxymate ligands, surface chemistry, and the mechanism of ligand-promoted dissolution of goethite [α-FeOOH(s)]. *Geochim. Cosmochim. Acta* **60:** 4403–4416.
30. **Holmen, B. A., M. I. Tejedor-Tejedor, and W. H. Casey.** 1997. Hydroxymate complexes in solution and at the goethite-water interface: a cylindrical internal reflection Fourier transform infrared spectroscopy study. *Langmuir* **13:**2197–2206.
31. **Jurkevitch, E., Y. Hadar, and Y. Chen.** 1992. Differential siderophore utilization and iron uptake by soil and rhizosphere bacteria. *Appl. Environ. Microbiol.* **58:**119–124.
32. **Loper, J. E.** 1988. Role of fluorescent siderophore production in biological control of *Pythium ultimum* by a *Pseudomonas fluorescens* strain. *Phytopathology* **78:**166–172.
33. **Loper, J. E., and J. S. Buyer.** 1991. Siderophores in microbial interactions on plant species. *Mol. Plant-Microbe Interac.* **4:**5–13.
34. **Loper, J. E., and M. D. Henkels** 1997. Availability of iron to *Pseudomonas fluorescens* in rhizosphere and bulk soil evaluated with an ice nucleation reporter gene. *Appl. Environ. Microbiol.* **63:** 99–105.
35. **Loper, J. E., and S. E. Lindow.** 1994. A biological sensor for iron available to bacteria in the habitats on plant surfaces. *Appl. Environ. Microbiol.* **60:**1934–1941.
36. **Manthey, J. A., D. E. Crowley, and D. G. Luster.** 1994. *Biochemistry of Metal Nutrients on the Rhizosphere.* Lewis Publishers, Boca Raton, Fla.
37. **Marschner, H., and D. S. Crowley.** 1997. Iron stress and pyroverdin production by a fluorescent pseudomonad in the rhizosphere of white lupin (*Lupinus albus L.*) and barley (*Hordeum vulgare* L.). *Appl. Environ. Microbiol.* **63:**277–281.
38. **Marschner, H. and V. Romheld.** 1994. Strategies of plants for acquisition of iron. *Plant Soil* **165:** 261–274.
39. **Matzanke, B. F.** 1991. Structures, coordination chemistry and functions of microbial iron chelates, p. 15–60. *In* G. Winkelmann (ed.), *Handbook of Microbial Iron Chelates.* CRC Press, Inc., Boca Raton, Fla.
40. **Matzanke, B. F., G. Muller-Matzanke, and K. N. Raymond.** 1989. Siderophore mediated iron transport, p. 1–121. *In* T. M. Loehr (ed.), *Iron Carriers and Proteins.* VCH Publishers, New York, N.Y.
41. **Meyer, J. M., and M. A. Abdallah.** 1978. The fluorescent pigment of *Pseudomonas fluorescens*: biosynthesis, purification and physico-chemical properties. *J. Gen. Microbiol.* **107:**321–331.
42. **Neilands, J. B.** 1981. Microbial iron compounds. *Annu. Rev. Biochem.* **50:**715–732.
43. **Neilands, J. B.** 1995. Siderophores: structure and function of microbial iron transport compounds. *J. Biol. Chem.* **270:**26723–26726.
44. **Neilands, J. B., and K. Nakamura.** 1991. Detection, determination, isolation, characterization and regulation of microbial chelates, p. 1–14. *In* G. Winkelmann (ed.), *Handbook of Microbial Iron Chelates.* CRC Press, Inc., Boca Raton, Fla.
45. **Nelson, M., C. R. Cooper, D. E. Crowley, C. P. P. Reid, and P. J. Szaniszlo.** 1988. An *Escherichia coli* bioassay of individual siderophores in soil. *J. Plant Nutr.* **11:**915–924.
46. **O'Sullivan, D. J., and F. O'Gara.** 1992. Traits of fluorescent *Pseudomonas* spp. involved in suppression of plant root pathogens. *Microbiol. Rev.* **56:**662–676.
47. **Page, W. J., and M. Huyer,** 1984. Derepression of the *Azotobacter vinelandii* siderophore system, using iron-containing minerals to limit repletion. *J. Bacteriol.* **158:**496–502.
48. **Payne, S. M.** 1988. Iron and virulence in the family Enterobacteriaceae. *Crit. Rev. Microbiol.* **16:** 81–111.
49. **Raaijmakers, J. M., I. Van der Sluis, M. Koster, P. A. H. M. Bakker, P. J. Weisbeek, and B. Schippers.** 1995. Utilization of heterologous siderophore and rhizosphere competence of fluorescent *Pseudomonas* spp. *Can. J. Microbiol.* **41:**126–135.
50. **Raaijmakers, J. M., D. M. Weller, and L. S. Thomashow.** 1997. Frequency of antibiotic-producing *Pseudomonas* spp. in natural environments. *Appl. Environ. Microbiol.* **63:**881–887.
51. **Schwertmann, U.** 1991. Stability and dissolution of iron oxides. *Plant Soil* **129:**1–25.
52. **Schwertmann, U., and R. M. Cornell.** 1991. *Iron Oxides in the Laboratory: Preparation and Characterization.* VCH Publishers, Weinheim, Germany.

53. **Schwertmann, U., and R. M. Taylor.** 1989. *Minerals in Soil Environments*, 2nd ed. Soil Science Society of America, Madison, Wis.
54. **Stumm W., and B. Sulzberger.** 1992. The cycling of iron in natural environments: considerations based on laboratory studies of heterogeneous redox processes. *Geochem. Cosmochim. Acta* **56:**3233.
55. **Szaniszlo, P. J., P. E. Powell, C. P. P. Reid, and G. R. Cline.** 1981. Production of hydroxamate siderophore iron chelators by ectomycorrhizal fungi. *Mycologia* **73:**1158–1175.
56. **van der Helm, D., M. A. F. Jahal, and M. B. Hossain.** 1987. The crystal structure, conformations, and configurations of siderophores, p. 135–165. *In* G. Winkelmann, D. van der Helm, and J. B. Neilands (ed.), *Iron Transport in Microbes, Plants and Animals.* VCH Publishers, Weinheim, Germany.
57. **Watteau, F., and J. Berthelin.** 1994. Microbial dissolution of iron and aluminum from soil minerals: efficiency and specificity of hydroxamate siderophores compared to aliphatic acids. *Eur. J. Soil Biol.* **30:**1–9.
58. **Wehrli, B.** 1990. Redox reactions of metal ions at mineral surfaces, p. 311–336. *In* W. Stumm (ed.), *Aquatic Chemical Kinetics.* John Wiley & Sons, Inc., New York, N.Y.
59. **Weinberg, E. D.** Cellular regulation of iron assimilation. *Q. Rev. Biol.* **63:**261–290.
60. **Winkelmann, G.** 1991. *CRC Handbook of Microbial Iron Chelates.* CRC Press, Inc., Boca Raton, Fla.
61. **Winkelmann, G., D. van der Helm, and J. B. Neilands (ed.).** 1987. *Iron Transport in Microbes, Plants and Animals.* VCH Publishers, Weinheim, Germany.
62. **Yehuda, Z., M. Shenker, V. Romheld, H. Marschner, Y. Hadar, and Y. Chen.** 1996. The role of ligand exchange in the uptake of iron from microbial siderophores by gramineous plants. *Plant Physiol.* **112:**1273–1280.
63. **Yeoman, K. H., M.-J. Delgado, M. Wexler, J. A. Downie, and W. B. Johnston.** 1996. High affinity iron acquisition in *Rhizobium leguminosarum* requires the *cycHJKL* peron and the *feuPQ* gene products, which belong to the family of two-component transcriptional regulators. *Microbiology* **143:** 127–134.
64. **Zinder, B., G. Furrer, and W. Stumm.** 1986. The coordination chemistry of weathering. II. Dissolution of Fe(III) oxides. *Geochim. Cosmochim. Acta* **50:**1861–1869

Environmental Microbe-Metal Interactions
Edited by Derek R. Lovley

Chapter 7

Microbially Influenced Corrosion of Steel

Ralf Cord-Ruwisch

This chapter, on microbially influenced corrosion (MIC), presents traditional and new concepts on this topic from the microbiological viewpoint. For the description of the various electrochemical processes involved in corrosion and on the electrochemical monitoring methods, please refer to the general corrosion literature. Many review articles on microbially influenced corrosion take the viewpoint of metal destruction. This chapter also investigates corroding iron as a food source for hydrogen-consuming bacteria. Some calculations are presented to visualize quantitative aspects of this economically important process.

CORROSION IN THE PRESENCE OF OXYGEN

On the surface of the Earth, iron does not occur in significant amounts in its metallic form (oxidation state zero) but is found either as ferrous (oxidation state +II) or as ferric (oxidation state +III) iron. To produce the highly reduced metallic iron, energy has to be invested to transfer two or three electrons to ferrous or ferric iron, respectively. This energy is part of the driving force for the naturally occurring corrosion processes when metallic iron in aqueous environments is exposed to suitable electron acceptors such as oxygen. The process of corrosion is best known as rust formation of steel when in contact with oxygen and water. In this process, oxygen is the oxidizing agent responsible for accepting the electrons from metallic iron.

Many studies have investigated the effect of aerobic bacterial activity on the corrosion of iron. However, because the process also occurs at relatively high rates in the absence of bacteria, the bacterial effect is not easy to monitor or predict. In fact reports show that bacteria may stimulate the aerobic corrosion process (29, 33) or inhibit it (14, 19). In general, due to their oxygen uptake activity, bacterial biofilms on the metal surface can create localized environments of differential aeration, which are likely to produce cathodic areas (where the reducing power of iron

Ralf Cord-Ruwisch • Biotechnology, Murdoch University, Perth, WA 6150, Australia.

reduces suitable electron acceptors) separate from anodic areas (where the oxidized iron dissolves), resulting in a corrosion current and the dissolution of the metal.

CORROSION IN ANAEROBIC ENVIRONMENTS: MICROBIALLY INFLUENCED CORROSION

In the absence of oxygen, iron is usually much more stable. However, in oxygen-free environments that are rich in anaerobic microbial activity, particularly with active bacterial sulfate reduction, iron is repeatedly found to deteriorate at high rates, sometimes higher than those due to oxygen alone (24). This is largely due to the formation of localized pitting corrosion, rendering pipelines, tanks, etc. useless within a fraction of their designed lifetime. This increased material deterioration has been related mainly to the activity of sulfate-reducing bacteria (SRB). This microbially influenced corrosion is estimated to account for damages costing many billions of dollars in the United Kingdom and the United States (18, 20). One of the industries most heavily affected is the petroleum industry (8, 21), with new types of SRB being found regularly from oil pipelines and oilfields (9, 44, 45). Some of the newly isolated SRB are capable of degrading petroleum hydrocarbons (34, 36) and thus potentially generating organic acids and CO_2. With these new strains of SRB now becoming available, the role of SRB in MIC will need to be reinvestigated using appropriate strains and organic substrates.

ROLE OF HYDROGEN IN MICROBIALLY INFLUENCED CORROSION

Polarization

In the absence of oxygen, metallic iron can be corroded only if alternative oxidizing agents are available to accept the electrons from the metal oxidation. Its negative redox potential enables iron to reduce even protons from the dissociation of water to molecular hydrogen. Hence, in principle, protons represent constant potential electron acceptors for anaerobic iron corrosion. Thus, it may seem almost surprising that iron does not corrode completely in buffered, oxygen free water. The reason for the relative stability of iron in the absence of oxygen is explained by its polarization. According to classical corrosion theory, when iron is immersed in an oxygen-free aqueous solution, traces of the metal will dissolve as Fe^{2+}, leaving negative charges on the metal surface (18, 20). These negative charges represent a strong reducing agent, with oxygen being the electron acceptor in rust formation. In the absence of oxygen, the negative charges reduce protons from the dissociation of water to form a protective film of adsorbed hydrogen. The formation of this hydrogen layer is believed to prevent the metal from further dissolving. The iron is polarized. Any process that can interfere with this protective hydrogen film is therefore likely to influence the corrosion process.

Cathodic Depolarization Theory

After it had been discovered that the strictly anaerobic SRB of the genus *Desulfovibrio* are potent hydrogen consumers, Von Wolzogen Kuehr and van der Vlugt

(43) developed the theory of cathodic depolarization as the main principle of bacterial stimulation of anaerobic corrosion of steel. In this theory, SRB in close proximity to a hydrogen-coated polarized steel surface oxidize the cathodic hydrogen, which, according to this theory, stimulates the cathodic reaction

$$Fe^0 + 2H^+ \rightarrow Fe^{2+} + H_2 \quad (1)$$

A continued cathodic reaction maintains an electron flow between the anodic and cathodic areas, which in turn will favor the anodic reaction, the actual dissolution of positively charged ferrous iron. Because of the involvement of an electric current, corrosion is defined as an electrochemical process with the current flow (corrosion current) reflecting the corrosion rate. Many attempts to quantify microbially influenced corrosion by measuring the corrosion current using electrochemical methods have not succeeded in replacing weight loss measurements and solution chemistry methods for corrosion monitoring. In part, the reason lies in the difficulty of obtaining reproducible electrochemical data in an undefined complex environment such as a bacterial culture. The production of ferric iron does usually not occur, and if it is observed, it is likely to be due to laboratory artifacts (e.g., entry of oxygen).

For many years, large amounts of anecdotal and scientific evidence have been gathered supporting the view that the destruction of steel structures in anaerobic environments is linked to the presence of SRB (18, 20, 24). However, conclusive evidence that the cathodic depolarization is the principal mechanism of microbially influenced corrosion has not been produced. This is in part because SRB may also influence corrosion in other ways (see below).

Can Bacteria Feed on Metallic Iron?

When a substrate is defined as the bacterial feed, it is typically oxidized by the bacterial activity, with the electrons obtained being used for the generation of ATP, (e.g., via an electron transport phosphorylation). In the cathodic depolarization theory, SRB clearly trigger the oxidation of the metal by removing the protective hydrogen layer and link the electron flow from the metal to the respiratory reduction of sulfate to sulfide with hydrogen as an electron carrier. In combination with sulfate, the metal is the energy source for bacterial growth. Laboratory experiments using pure cultures of SRB have demonstrated that bacterial reduction of sulfate and thus conservation of energy for growth can be obtained from the corrosion process via the consumption of cathodic hydrogen (7, 32). Hydrogen-consuming SRB were able to reduce sulfate with metallic iron as the only electron donor, while hydrogenase negative SRB were not (7). In summary, hydrogen-consuming bacteria such as the SRB can clearly feed on metallic iron by utilizing the electrons released in the corrosion process for their own respiration.

Cathodic Hydrogen as an Energy Source for Biofilm Buildup

The extent to which the supply of energy via the electron transfer from the corroding iron to the bacterial cell can contribute significantly to the formation of bacterial biomass such as a biofilm will depend largely on the electron supply rate (corrosion rate). For example, in environments that do not contain significant concentrations of easily degradable material for bacterial growth, the electron supply from a fast corroding metal surface may well be the major energy source for the biofilm ecosystem.

For example, for each square decimeter of uniform anaerobic iron corrosion to a depth of 0.1 mm, approximately 0.14 mol (3.14 liters) of hydrogen gas can be produced. Assuming a net ATP gain of 0.5 ATP per H_2 molecule and a net biomass formation of 10 g of dry biomass/mol of ATP, approximately 0.7 g of dry biomass could be produced for the formation of a biofilm. With a biofilm dry weight of 5%, this could result in about 39 g of biofilm, which (at a density of 1 g/cm^3) could be up to 3.9 mm thick. Although this calculation does not account for biomass loss due to bacterial death, it indicates that the energy supplied from corroding iron can be sufficient for the buildup of a biofilm with a maximum value of 40 μm of biofilm per μm of iron corroded. At a high uniform corrosion rate of 100 mg/dm^2/day (0.46 mm/year), this would mean a typical biofilm 40 to 50 μm thick could build up about once per day.

It may be argued that a suitable carbon source must be available. However, CO_2 and bicarbonate from the water or from the iron could serve as the carbon source, for autotrophic metabolism of SRB or homoacetogenic bacteria (using H_2 for bicarbonate reduction to acetate), which are likely to be part of a biofilm. In cases where cathodic protection (the use of current or a sacrificial anode to encourage the cathodic reaction on the steel while suppressing the anodic dissolution reaction) is applied, it is even more likely that the protective current serves as the significant energy source for a bacterial biofilm. While the cathodic protection will protect the metal as long as the current is switched on, one could expect a well-nourished biofilm to be aggressively corrosive when the cathodic protection is interrupted.

Comparison of Cathodic Depolarization and Interspecies Hydrogen Transfer

In anaerobic environments, organic material is broken down by different groups of bacteria, with molecular hydrogen being one of the most important fermentation end products. Hydrogen plays a particularly delicate role in the degradation of volatile fatty acids such as propionate or butyrate, alcohols, or aromatic acids such as benzoate (38). The anaerobic degradation of these compounds is mediated by two groups of bacteria, the syntrophic hydrogen producers and the hydrogen-consuming bacteria. The relationship between these two different types of bacteria is termed syntrophy, since each group depends on the presence and activity of the other. The degradation of the organic substrate under hydrogen (and acetate) production is thermodynamically favorable only when the dissolved hydrogen concentration is kept at an extremely low level. Under standard conditions (100 kPa of H_2 partial pressure), the hydrogen production is endergonic. As a consequence, the hydrogen-producing reaction is directly controlled by the hydrogen uptake activity of the hydrogen-consuming bacteria. Without the hydrogen uptake by hydrogenophilic bacteria such as methanogenic bacteria or SRB, the hydrogen-producing reaction can not proceed. Thus, in their natural environment, hydrogen-consuming anaerobic bacteria can be described as ''pulling'' the hydrogen-producing reactions such as the conversion of propionate to acetate. This natural phenomenon of interspecies hydrogen transfer or interspecies electron transfer is a common feature observed in many anaerobic environments such as anaerobic digestors and anaerobic sediments.

An obvious analogy can be observed between the cathodic depolarization theory and the interspecies hydrogen transfer by anaerobic bacteria (Fig. 1). It may seem that anaerobic microbial communities are designed to utilize cathodic hydrogen and hence potentially stimulate the corrosion reaction by effectively oxidizing trace concentrations of molecular hydrogen. However, from a thermodynamic point of view, there is a substantial difference between the two hydrogen transfer reactions: the syntrophic degradation of organic compounds is thermodynamically difficult and therefore dependent on a constantly low concentration of hydrogen, while the production of hydrogen from metallic iron is thermodynamically favorable under standard conditions (pH 0):

$$Fe + 2H^+ \rightarrow Fe^{2+} + H_2, \qquad \Delta G^\circ = -78.9 \text{ kJ/mol} \tag{2}$$

Although the reaction has virtually no driving force at pH 7 ($\Delta G^{\circ\prime} = 0.9$ kJ/mol), it is quite favorable ($\Delta G = -39.0$ kJ/mol) under conditions more similar to the field situation (pH 7, H_2 partial pressure of 10 Pa, and dissolved Fe^{2+} concentration of 1 mM). Theoretically, several megapascals of hydrogen could be produced under these conditions before the reaction becomes thermodynamically limited. In other words, the bacterial removal of dissolved hydrogen cannot have a large effect on

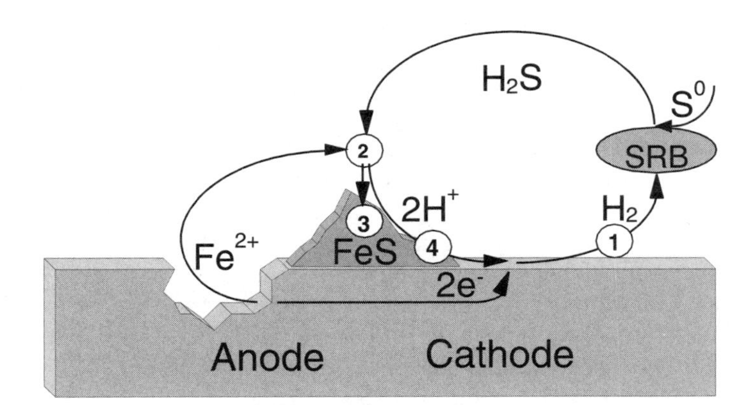

Figure 1. Role of hydrogen-consuming SRB in the anaerobic corrosion of steel. This diagram uses elemental sulfur rather than sulfate as the electron acceptor for SRB to demonstrate a possible mechanism for proton recycling after precipitation of hydrogen sulfide with ferrous iron as an iron sulfide deposit. In principle, four different ways of stimulating the corrosion process can be visualized: 1, consumption of cathodic hydrogen (cathodic depolarization); 2, anodic depolarization by Fe^{2+} removal; 3, stimulation by the formation of an iron sulfide layer, which may be cathodic; and 4, supply of protons to the cathode.

the energetics of cathodic production of hydrogen, while it is absolutely essential for the syntrophic degradation of organic material. From a purely thermodynamic point of view, it is unlikely that the removal of molecular hydrogen that diffuses from the metal surface will substantially increase the corrosion rate.

Does Bacterial H_2 Uptake Stimulate the Corrosion Process?

Biofilms containing SRB were investigated in terms of the severity of corrosion obtained. The corrosion was not necessarily related to the number of SRB (23, 28) but was related to the hydrogenase activity within the biofilm (6). The authors used this finding as evidence for the cathodic depolarization theory of microbially induced corrosion. However, the presence of high bacterial hydrogenase activity may also be the consequence rather than the cause of the hydrogen-producing corrosion reaction.

To investigate the pure effect of hydrogen consumption separate from the effect of SRB metabolites such as HS^- or FeS, hydrogen-consuming nonsulfidogenic bacteria or sulfidogenic non-hydrogen-consuming bacteria are good tools. However, no clear picture has been established in the literature. On the one hand, hydrogenase-negative SRB cause similar levels of corrosion to those caused by *Desulfovibrio* species (13), and on the other hand, only hydrogenase-positive species increased the corrosion rate whereas the hydrogenase-negative *Desulfobacter postgatei* did not cause significant corrosion (17). In fact, non-sulfide-producing bacteria have also been found to utilize cathodic hydrogen as an electron donor for their energy metabolism. Methanogenic bacteria were described to reduce CO_2 to CH_4 (3, 10, 12), SRB could reduce nitrate to ammonia (W. A. Johnston, R. Cord-Ruwisch, and I. M. Ritchie, First NACE Asian Conf. 1992, paper 1138) and other anaerobic bacteria could couple the reduction of chloroform (46) with the cathodic hydrogen oxidation. If bacterial hydrogen consumption can stimulate the cathodic reaction, then the above-described non-sulfide-producing bacteria should also be able to cause corrosion.

Studies using hydrogen-consuming methanogenic bacteria (12, 13) have indicated that the bacterial cathodic hydrogen removal does not necessarily stimulate the corrosion reaction. It could be argued here that the methanogenic bacteria are less capable of effectively oxidizing hydrogen than are the SRB (8, 27). However, even experiments with extremely powerful hydrogen-consuming non-sulfide-producing bacterial cultures failed to produce high corrosion rates in the absence of sulfide (Johnston et al., First NACE Asian Conf. 1992). Here, the same strain of SRB was grown under sulfate-reducing or nitrate-reducing (to ammonia) conditions. The results showed that the increased efficiency of hydrogen uptake under nitrate-reducing conditions did not enable the strain to oxidize cathodic hydrogen faster. In combination, the above results imply that the corrosion reaction is not stimulated merely by bacterial hydrogen uptake activity but that other mechanisms, such as the excretion of metabolites by the SRB, are the main factors for MIC. Furthermore, the dramatic pitting corrosion observed in the field has not been reproduced in the laboratory using the typical metabolites of SRB such as H_2S, FeS, and bicarbonate.

AVAILABILITY OF THE ELECTRON ACCEPTOR, THE TRUE CORRODING AGENT

Possible Rate-Controlling Step

The suggestion that microbial removal of hydrogen does not necessarily stimulate the hydrogen production rate from the cathode was also supported by headspace hydrogen measurements (5; R. Cord-Ruwisch, G. E. Strong, and I. M. Ritchie, First NACE Asian Conf. 1992, paper 1137). When exposed to oxygen-free electrolytes, metallic iron in powder, filings, or coupon form produced dissolved hydrogen that did not remain on the metal surface as a polarizing hydrogen film but instead diffused into the bulk liquid, where it could be measured. This hydrogen release rate was closely related to the corrosion rate measured by dissolved-iron monitoring (Cord-Ruwisch et al., First NACE Asian Conf. 1992). The fact that the slow but measurable corrosion of 4.4 $mg/dm^2/day$ continued at a constant rate in pH- controlled experiments (pH 6.8) indicated that neither the hydrogen nor the ferrous iron buildup in the liquid significantly slowed the corrosion reaction. According to these findings, bacterial stimulation of corrosion is unlikely to be due to hydrogen or Fe^{2+} removal (such as by precipitation). Stimulation of corrosion by the removal of these two end products has also been excluded by an analytical review of MIC (47). The fact that in the above experiments the corrosion rate slowed relatively quickly in the absence of pH control or buffer compounds (resulting in a pH increase) indicated that the supply rate of oxidizing agents (here protons) to the metal surface was rate limiting. Accordingly, the effect of microbial activity in increasing the overall corrosion rate is likely to be linked to the supply of suitable electron acceptors such as protons to the metal surface.

Kinetics of Proton Diffusion to the Metal

If the corrosion process under anaerobic conditions is controlled by the rate of proton transport to the metal surface, as described above, the theoretical corrosion rate resulting from the diffusion transport of protons can be calculated. Presuming a constant supply of these oxidizing agents (e.g., oxygen or protons) in the bulk water surrounding the metal, the transport of these oxidizing agents to the metal surface will be by diffusion through the typical boundary layer (the layer close to surfaces where convection is absent and all mass transfer is by diffusion). The slower the flux of the oxidizing agents through this diffusion layer, the more likely it is that the corrosion process is diffusion controlled. With no oxidizing agents, diffusing to the metal surface the iron will remain stable, since the electrons at the cathode are not removed, resulting in the iron being polarized.

According to Fick's first law of diffusion, the flux of a diffusing species across a given area is the product of the diffusion coefficient (9.31×10^{-9} m^2/s = 33.5 mm^2/h for protons) (1) and the diffusion gradient (e.g., in $nmol/mm^4$). Assuming a diffusion layer thickness of 10 μm, the proton diffusion gradient at pH 7 would be at most 10 $pmol/mm^4$. The corresponding proton flux of 80 $\mu mol/dm^2/day$ would account for a uniform corrosion rate of 2.25 $mg/dm^2/day$, which is comparable to rates of 2 to 4 $mg/dm^2/day$ determined under sterile anaerobic conditions at pH 6.8 to 7.2 (3, 4; Cord-Ruwisch et al., First NACE Asian Conf. 1992). Corrosion rates of this order (4 mg/dm^2/day) are generally considered to be acceptable for steel construction, since they would result in a uniform wall thickness loss of about 18.4 μm/year. In contrast, a diffusion model with oxygen as the electron acceptor would predict rates that are several orders of magnitude higher merely based on the higher concentration and thus higher flux of the oxidizing agent O_2 (about 0.25 mM at saturation, i.e., 2,500 times higher than that of protons).

In the field situation, there may be chemical species such as buffers (e.g., HCO_3^-) or weak acids (e.g., H_2S) that may act as proton transporters. In laboratory experiments, phosphate and bicarbonate buffers increased the corrosion rate significantly over that for a distilled-water control (5; Cord-Ruwisch et al., First NACE Asian Conf. 1992). Because of the usually much higher concentrations of buffer species than protons (proton concentration, 0.1 μM at pH 7), such species are expected to accelerate proton transport through the boundary manyfold. The ability of bacteria to supply protons to the metal surface is discussed below.

CORROSION MODELS NOT BASED ON CATHODIC DEPOLARIZATION

From the previous section, it can be concluded that mechanisms other than bacterial hydrogen consumption must be responsible for the steel destruction in the presence of anaerobic bacteria. Furthermore, the fact that a buildup of dissolved ferrous iron in the medium does not substantially inhibit the corrosion process (see above) indicates that bacterial Fe^{2+} removal from the metal surface by precipitation as FeS is also not the principal mechanism of microbially influenced corrosion. According to the corrosion literature, many other possible factors are suspected to contribute to MIC. However, none of the suggested mechanisms has been accepted as a generally applicable corrosion mechanism by electrochemists, corrosion engineers, and microbiologists.

Bacterial Sulfide Production

Virtually all authors agree that sulfide, the principal metabolic end product of SRB, plays a significant role in the corrosion process (18, 20, 24). From laboratory experiments and other observations, the following conclusions have been drawn.

1. Dissolved sulfide stimulates the anodic reaction by precipitating Fe^{2+} as FeS (Fig. 1, step 2).
2. Dissolved sulfide stimulates the cathodic reaction by oxidizing metallic iron (equation 6) (24, 40).
3. Precipitated iron sulfides stimulate the corrosion by creating a galvanic cell with iron sulfide as the cathode and iron as the (dissolving) anode (Fig. 1, step 3) (22, 24).
4. Iron sulfide layers, in particular pyrite (FeS_2), can form a protective layer inhibiting the corrosion process (Fig. 1, step 3) (35).

5. The formation of localized anodes occurs after destruction of protective iron sulfide layers.

However, agreement about the main principle of sulfide-enhanced corrosion is not established. It seems certain that sulfide can have stimulating as well as protective effects, depending on the conditions and the presence of other chemical species such as Fe^{2+}. Also, in the presence of oxygen, sulfide gives rise to a number of other corrosive sulfur compounds, such as elemental sulfur, sulfuric acid, and polysulfides.

Bacterial Enzymes

The exact mechanism of proton reduction at the cathodic site of corroding iron is not understood. According to traditional corrosion theory, protons are reduced to form adsorbed atomic hydrogen (20). The fusion of atomic to molecular hydrogen is suspected to be the corrosion rate-limiting step and to be influenced by the bacterial hydrogenase activity. Although tests with bacterial hydrogenases showed that cathodic hydrogen was converted (5), it is not established that this enzyme does stimulate the corrosion in the field. With evidence accumulating against the cathodic depolarization theory, the role of hydrogenase might be of minor importance in future investigations. However, the monitoring of hydrogenase activity within biofilms may still be a useful indicator of MIC.

Bacterial Polysaccharides

With the focus on bacterial biofilms being responsible for MIC, bacterial polysaccharides have been suspected to play a direct role and are being investigated in more detail (2, 31, 48). However, the exact role of such polysaccharides in MIC has not been elucidated. It will be interesting to follow research efforts in this area.

POSSIBLE MECHANISMS OF BACTERIAL PROTON SUPPLY TO THE CATHODE

Acid-Producing Fermentative Bacteria

The production of acids is known to cause iron corrosion. The typical anaerobic acid-forming bacteria such as *Clostridium* need fermentable organic compounds such as carbohydrates for acid production. While this may contribute to corrosion in wastewater pipes and other environments, it is not likely to be a sustainable process within a biofilm or other environments low in carbohydrates such as petroleum and freshwater pipes.

Production of Buffer Species and Weak Acids by SRB

Observation (Cord-Ruwisch et al., First NACE Asian Conf. 1992) and calculation (see above) support the view that anaerobic steel corrosion is limited by proton transport to the metal surface. Evidence against this mechanism has not been produced. Although proton diffusion at neutral pH is slow, protons can also be transported by the diffusion of weak acids or buffer species that dissociate when reaching the proton-depleted cathode at the iron surface. Because of the significantly higher concentrations in the bulk liquid, a much higher proton flux can be established via the buffer species in spite of up to 10-fold-lower diffusion coefficients. This would explain the increased iron dissolution under hydrogen production in the presence of phosphate or bicarbonate buffers as reported previously (5; Cord-Ruwisch et al., First NACE Asian Conf. 1992). A more direct reaction of bicarbonate with the metal cathode ($2HCO_3^- + 2e^- \rightarrow H_2 + CO_3^{2-}$) has also been suggested (30).

The anaerobic conversion of organic compounds under sulfate-reducing conditions leads to HCO_3^- and H_2S as the two major end products. Both species could help transfer protons through a diffusion layer to the metal. It should be noted that for proper consideration of these reactions, in particular within biofilms, the possible consumption of protons during the overall reaction has to be taken into account. It has been observed that the corrosion rate in pure cultures of SRB was significantly increased when large amounts of HCO_3^- and H_2S were produced due to increased organic substrate concentration (7).

Proton Excretion by SRB

SRB can be involved in the supply (or removal) of protons. The observation that, when growing on lactate, SRB can adjust the pH of the medium to values close to neutral (11) suggested that SRB is involved in the stimulation of the corrosion reaction due to proton production. While the production of protons and bicarbonate ions can be responsible for acidification by SRB growing on organic substrates such as lactate, the release of protons from the metabolism of cathodic hydrogen, as would occur in a biofilm, is unlikely. In fact the cathodic hydrogen consumption utilizes protons:

$$4H_2 + SO_4^{2-} + 2H^+ \rightarrow H_2S + 4H_2O \quad (3)$$

The disproportionation of elemental sulfur (42) into sulfate and sulfuric acid

$$4S^0 + 4H_2O \rightarrow H_2SO_4 + 3H_2S \quad (4)$$

by SRB is an acid-producing reaction that is likely to occur in biofilms and may strongly influence anaerobic corrosion rates. Traditionally, the production of sulfuric acid has been attributed to oxygen-dependent reactions and thus has been related to aerobic corrosion. The above disproportionation reaction could theoretically occur in the oil industry, where elemental sulfur is present in the produced oil. In the presence of SRB biofilms in the pipeline, local production of sulfuric acid could be expected. No research has looked at this possibility yet.

H_2S as an Electron Acceptor for the Cathodic Reaction

Some literature sources conclude that bacterial hydrogen sulfide can be the oxidizing agent for the electrons from the cathode (18, 24, 40, 47).

$$Fe + H_2S \rightarrow FeS + H_2 \quad (5)$$

Strictly speaking, it is not the sulfur atom that accepts the electrons from the corrosion process but the protons that are carried by the hydrogen sulfide molecule. H_2S can be seen as a weak acid and thus as a suitable proton carrier, as indicated above. It should also be considered that at neutral pH only about half of the dissolved sulfide produced will be present as H_2S, with the other half being present as HS^-. At pH 8, which more closely reflects the situation of many bicarbonate-buffered systems such as marine environments, less than 10% of the total dissolved sulfide is present as H_2S. Nonetheless, the buffer system $H_2S/HS^-/S^{2-}$ may represent a possible proton shuttle through boundary layers and biofilms similar to other buffer systems.

Recycling of Cathodic H_2

In principle, any reaction that allows the recycling of cathodic hydrogen back to protons would be likely to stimulate anaerobic corrosion. A possible way for SRB to recycle the protons consumed by the cathodic reaction is not with sulfate (equation 3) but with elemental sulfur as the electron acceptor:

$$H_2 + S \rightarrow H_2 \quad (6)$$

Most SRB have the capacity for sulfur reduction. If this reaction is close to the anodic site, the protons could be recycled (Fig. 1, step 4):

$$H_2S + Fe^{2+} \rightarrow FeS + 2H^+ \quad (7)$$

Hence, the complete corrosion reaction would be independent of external proton supply:

$$Fe + S \rightarrow FeS \quad (8)$$

and would therefore be sustainable in a closed system (Fig. 1) or a biofilm until the elemental sulfur is depleted. In laboratory experiments, a substantial increase in corrosion and the formation of visible pits was observed after replacing sulfate by elemental sulfur in experiments with SRB and mild steel coupons (R. Cord-Ruwisch, Proc. Corros, Prev. 95, 1995, paper 7). However, the aggressiveness of elemental sulfur toward metal in aqueous systems has also been observed in the absence of bacteria (37, 39).

The generation of elemental sulfur and polysulfides in biofilms is typically due the reaction of oxygen from the bulk liquid with bacterial sulfides:

$$2HS^- + O_2 \rightarrow 2S^0 + 2OH^- \quad (9)$$

and is likely to occur in biofilms (24) where the bacterial sulfide meets the oxygen.

Proton Translocation

A potential mechanism of exposing protons to the metal surface that has not been discussed in the literature is the proton translocation that is a general feature in aerobic and anaerobic respiring bacteria such as the SRB (15). The cells excrete protons during the electron transport in the anaerobic sulfate respiration with the aim of using use the proton gradient for ATP synthesis. In close proximity to the metal surface (e.g., in a biofilm), this measurable local excretion of protons might stimulate cathodic hydrogen production, in particular if the reaction is diffusion controlled, as indicated above.

BIOFILMS

Over the last several years of study of MIC, more attention has been given to the effect of biofilms on the corrosion process. Some authors believe that most cases of MIC are due to the formation of biofilms containing SRB (24). Independent of the different hypotheses of biofilm-related corrosion, it should be noted that from a purely kinetic point of view, the biofilm will increase the diffusion layer substantially and thus slow the flux of electron-accepting agents from the bulk liquid to the metal surface. For example, in an environment with a diffusion layer thickness of 50 μm, the production of a biofilm an additional 50 μm thick would slow all diffusion processes by a factor of 2, thus inhibiting diffusion-controlled corrosion accordingly. In fact, an agar layer on the metal surface was shown to be protecting (41).

There must be other processes of chemical and biological origin inside the biofilm that stimulate corrosion. Laboratory experiments that simulated very closely the field situation (aerobic bulk liquid, flowthrough system with biofilm buildup) showed that biofilms developed aerobic and sulfate-reducing activity, with the main corrosion process being attributed to inconsistencies in the biofilm development as a result of FeS production (25) or elemental sulfur formation as the product of sulfide reacting with oxygen (26). The reason why pitting corrosion was again associated with the occurrence of elemental sulfur (26) is not entirely clear. Reactions resulting in sulfuric acid formation (equations 4 and 10) or in the recycling of protons (Fig. 1) may be part of the explanation. These experiments indicate that even in fully aerobic systems (e.g., rusting of marine structures), SRB may play a role in influencing the corrosion process and possibly stimulating pitting corrosion. The possibility that SRB will metabolize oxygen and even oxidize sulfide in those semi-aerobic microenvironments (16) certainly warrants further research.

Typical corroding biofilms consist of a number of different bacterial species. These may include SRB and fermenting and iron-reducing bacteria. Some evidence has been presented that the coexistence of several species, rather than SRB alone, is responsible for the aggressive nature of a biofilm.

LOCALIZED VERSUS UNIFORM CORROSION

Up to now, this chapter has been concerned with the ability of microbes to stimulate the overall corrosion rate. However, increased loss of steel structures also occurs if bacteria do not markedly stimulate the overall corrosion but merely initiate the formation of a localized anode (Fig. 1). For example, a corrosion rate that may be acceptable to users of a steel structure (e.g., 20-year designed lifetime) would result in unacceptable failure if localized or pitting corrosion with a large cathode/anode ratio is established instead of uniform corrosion. Leaving the overall corrosion rate the same, a 20-fold smaller anodic area could result in a 1-year lifetime of the steel structure. Localized corrosion is typical for MIC. Although the exact bacterial involvement in the formation of pitting corrosion is not known, it is likely to be related to the formation of uneven biofilms and sulfide layers. Also, it cannot be excluded that the formation of localized corrosion may be the only mechanism by which SRB significantly stimulate corrosion. The actual increase in the corrosion rate in laboratory experiments is usually less than a factor of 2 over that of the sterile control. By contrast, the damage caused, for example, in oil production equipment occurs on the order of 10 to 50 times faster than in sterile laboratory tests (unpublished discussions with local corrosion engineers).

A number of reports have suggested that the formation of pitting corrosion is related to the occurrence of other sulfur species such as elemental sulfur or polysulfides. The presence of such compounds usually indicates that a strong oxidizing agent such as oxygen was available. Thus corrosion due to such compounds may often be indirectly linked to the presence of oxygen, even if this occurs only intermittently. This view has been expressed in other reviews (18, 47) and has been supported in experiments that closely reflect the field situation (25, 26).

The presence of oxygen and sulfide has long been known to cause severe corrosion and may even include the formation of sulfuric acid by aerobic bacteria of the genus *Thiobacillus*:

$$H_2S + 2O_2 \rightarrow H_2SO_4 \qquad (10)$$

Since *Thiobacillus* species are usually found only in small numbers in anaerobic environments of neutral pH, it is possible that the newly discovered abilities of *Desulfobulbus propionicus* to utilize oxygen and nitrate for the oxidation of sulfide (16) will be be involved in MIC in environments that are believed to be anaerobic but may be exposed to oxygen on occasions. In future research on MIC in the petroleum industry, these new capacities and the oil degradation capability of SRB (34, 36) will have to be considered.

REFERENCES

1. **Atkins, P. W.** 1994. *Physical Chemistry*, 5th ed. Oxford University Press, Oxford, United Kingdom.
2. **Beech, I. B., and C. Gaylarde.** 1991. Microbial polysaccharides and corrosion. *Int. Biodeterior.* **27:** 95–107.
3. **Boopathy, R., and L. Daniels.** 1991. Effect of pH on anaerobic mild steel corrosion by methanogenic bacteria. *Appl. Environ. Microbiol.* **57:**2104–2108.
4. **Booth, G. H., P. M. Cooper, and D. S. Wakerley.** 1966. Corrosion of mild steel by actively growing cultures of sulfate-reducing bacteria: the influence of ferrous iron. *Br. Corros. J.* **1:**345–349.
5. **Bryant, R. D., and E. J. Laishley.** 1990. The role of hydrogenase in anaerobic biocorrosion. *Can. J. Microbiol.* **36:**259–264.

6. **Bryant, R. D., W. Jansen, J. Boivin, E. J. Laishley, and J. W. Costerton.** 1991. Effect of hydrogenase and mixed sulfate-reducing bacterial populations on the corrosion of steel. *Appl. Environ. Microbiol.* **57:**2804–2809.
7. **Cord-Ruwisch, R., and F. Widdel.** 1986. Corroding iron as hydrogen source for sulfate reduction in growing cultures of sulfate-reducing bacteria. *Appl. Microbiol. Biotechnol.* **25:**169–174.
8. **Cord-Ruwisch, R., H.-J. Seitz, and R. Conrad.** 1988. The capacity of hydrogenotrophic bacteria to compete for traces of hydrogen depends on the redox potential of the terminal electron acceptor. *Arch. Microbiol.* **149:**350–357.
9. **Cord-Ruwisch, R., W. Kleinitz, and F. Widdel.** 1987. Sulfate-reducing bacteria and their activities in oil production. *J. Petrol. Technol.* **39:**97–102.
10. **Daniels, L., N. Belay, and B. S. Rajogopal.** 1987. Bacterial methanogenesis and growth from CO_2 with elemental iron as the sole source of electrons. *Science* **237:**509–511.
11. **Daumas, S., M. Magot, and J. L. Crolet.** 1993. Measurement of the net production of acidity by a sulfate-reducing bacterium—experimental checking of theoretical models of microbially influenced corrosion. *Res. Microbiol.* **144:**327–332.
12. **Deckena, S., and K.-H. Blotevogel.** 1990. Growth of methanogenic and sulfate-reducing bacteria with cathodic hydrogen. *Biotechnol. Lett.* **12:**615–620.
13. **Deckena, S., and K.-H. Blotevogel.** 1992. Fe^0-oxidation in the presence of methanogenic and sulfate-reducing bacteria and its possible role in anaerobic corrosion. *Biofouling* **5:**287–293.
14. **Eashwar, M., S. Maruthamuthu, S. Sathiyanarayanan, and K. Balakrishnan.** 1995. The enoblement of stainless alloys by marine biofilms—the neutral pH and passivity enhancement model. *Corros. Sci.* **37:**1169–1176.
15. **Fitz, R. M., and H. Cypionka.** 1989. A study on electron transport-driven proton translocation in *Desulfovibrio desulfuricans. Arch. Microbiol.* **152:**369–376.
16. **Fuseler, K., and H. Cypionka.** 1995. Elemental sulfur as an intermediate of sulfide oxidation with oxygen by *Desulfobulbus propionicus. Arch. Microbiol.* **164:**104–109.
17. **Gaylarde, C. C.** 1992. Sulfate-reducing bacteria which do not induce accelerated corrosion. *Int. Biodeterior. Biodegrad.* **30:**331-338.
18. **Hamilton, W. A.** 1985. Sulfate-Reducing bacteria and anaerobic corrosion. *Annu. Rev. Microbiol.* **39:**195–217.
19. **Hernandez, G., V. Kucera, D. Thierry, A. Pedersen, and M. Hermansson.** 1994 Corrosion inhibition of steel by bacteria. *Corrosion* **50:**603–608.
20. **Iverson, W. P.** 1987. Microbial corrosion of metals. *Adv. Appl. Microbiol.* **32:**1–36.
21. **Iverson, W. P., and G. J. Olson.** 1984. Problems related to sulfate-reducing bacteria in the petroleum industry, p. 619–641. *In* R. M. Atlas (ed.), *Petroleum Microbiology*. Macmillan, New York, N.Y.
22. **King, R. A., J. D. A. Miller, and D. S. Wakerley.** 1973. Corrosion of mild steel in cultures of sulfate-reducing bacteria, effect of changing the soluble iron concentration during growth. *Br. Corros. J.* **8:**89–93.
23. **Lee, W., and W. G. Characklis.** 1993. Corrosion of mild steel under anaerobic biofilm. *Corrosion* **49:**186–198.
24. **Lee, W., Z. L. Andowski, P. H. Nielsen, and W. A. Hamilton.** 1995. Role of sulfate-reducing bacteria in corrosion of mild steel: a review. *Biofouling* **8:**165–194.
25. **Lee, W., Z. Lewandowski, M. Morrison, W. G. Characklis, R. Avci, and P. H. Nielsen.** 1993. Corrosion of mild steel underneath aerobic biofilms containing sulfate-reducing bacteria. Part II: At high dissolved oxygen concentration. *Biofouling* **7:**217–239.
26. **Lee, W., Z. Lewandowski, S. Okabe, W. G. Characklis, and R. Avci.** 1993. Corrosion of mild steel underneath aerobic biofilms containing sulfate-reducing bacteria. I. At low dissolved oxygen concentration. *Biofouling* **7:**197–216.
27. **Lovley, D. R.** 1985. Minimum threshold for hydrogen metabolism in methanogenic bacteria. *Appl. Environ. Microbiol.* **49:**1530–1531.
28. **Moosavi, A. N., R. S. Pirrie, and W. A. Hamilton.** 1990. Effect of sulfate reducing bacteria activity on performance of sacrified anodes, p. 3–13. *In* N. Dowling, M. M. Mittelman, and J. C. Dank (ed.), *Proceedings of the International Symposium on Microbially Influenced Corrosion.*

29. **Morales, J., P. Esparza, S. Gonzalez, R. Salvarezza, and M. P. Arevalo.** 1993. The role of *Pseudomonas aeruginosa* on the localized corrosion of 304-stainless steel. *Corros. Sci.* **34:**1531–1540.
30. **Ogundele, G. I., and W. E. White.** 1986. Some observations on corrosion of carbon steel in aqueous environments containing carbon dioxide. *Corros. NACE* **42:**71–88.
31. **Okabe, S., P. H. Nielsen, W. L. Jones, and W. G. Characklis.** 1994. Estimation of cellular and extracellular carbon contents in sulfate-reducing bacteria biofilms by lipopolysaccharide assay and epifluorescence microscopic technique. *Water Res.* **28:**2263–2266.
32. **Pankhania, I. P., A. N. Moosavi, and W. A. Hamilton.** 1986. Utilization of cathodic hydrogen by *Desulfovibrio vulgaris* (Hildenborough). *J. Gen. Microbiol.* **132:**3357–3365.
33. **Parra, A., J. Carpio, and L. Martinez.** 1996. Microbial corrosion of metals exposed to air in tropical marine environments. *Mater. Performance* **35:**44–49.
34. **Rabus, R., M. Fukui, H. Wilkes, and F. Widdel.** 1996. Degradative capacities and 16S rRNA-targeted whole-cell hybridization of sulfate-reducing bacteria in an anaerobic enrichment culture utilizing alkylbenzenes from crude oil. *Appl. Environ. Microbiol.* **62:**3605–3613.
35. **Ramanarayanan, T. A., and S. N. Smith.** 1990. Corrosion of iron in gaseous environments and in gas-saturated aqueous environments. *Corrosion* **46:**66–74.
36. **Rueter, P., R. Rabus, H. Wilkes, F. Aeckersberg, F. A. Rainey, H. W. Jannasch, and F. Widdel.** 1994. Anaerobic oxidation of hydrocarbons in crude oil by new types of sulfate-reducing bacteria. *Nature* **372:**455–458.
37. **Schaschl, E.** 1980. Elemental sulfur as a corrodent in deaerated neutral aqueous solutions. *Mater. Performance* **19:**9–12.
38. **Schink, B.** 1997. Energetics of syntrophic cooperations in methanogenic degradation. *Microbiol. Mol. Biol. Rev.* **61:**262– 280.
39. **Schmitt, G.** 1991. Effect of elemental sulfur on corrosion in sour gas systems. *Corrosion* **47:**285–308.
40. **Shoesmith, D. W., P. Taylor, M. G. Baily, and D. G. Owen.** 1980. The formation of ferrous monosulfide polymorphs during the corrosion of iron by aqueous hydrogen sulfide at 21°C. *J. Electrochem. Soc.* **127:**1007–1015.
41. **Smith, C. A., K. G. Compton, and F. H. Coley.** 1973. Aerobic marine bacteria and the corrosion of carbon steel in seawater. *Corros. Sci.* **13:**677–685.
42. **Thamdrup, B., K. Finster, J. Wuergler-Hansen, and F. Bak.** 1993. Bacterial disproportionation of elemental sulfur coupled to chemical reduction of iron or manganese. *Appl. Environ. Microbiol.* **59:**101–108.
43. **Von Wolzogen Kuehr, C. A. H., and I. S. van der Vlugt.** 1934. The graphitization of cast iron as an electrobiochemical process in anaerobic soils. *Water* **18:**147–165.
44. **Voordouw, G., J. K. Voordouw, T. R. Jack, J. Foght, P. M. Fedorak, and D. W. S. Westlake.** 1992. Identification of distinct communities of sulfate-reducing bacteria in oil fields by reverse sample genome probing. *Appl. Environ. Microbiol.* **58:**3542–3552.
45. **Voordouw, G., Y. Shen, C. S. Harrington, A. J. Telang, T. R. Jack, and D. W. S. Westlake.** 1993. Quantitative reverse sample genome probing of microbial communities and its application to oil field production waters. *Appl. Environ. Microbiol.* **59:**4101–4114.
46. **Weathers, L. J., G. F. Parkin, and P. J. Alvarez.** 1997. Utilization of cathodic hydrogen as an electron donor for chloroform cometabolism by a mixed methanogenic culture. *Environ. Sci. Technol.* **31:**880–885.
47. **Widdel, F.** 1988. Microbial corrosion, p. 277–318. *In* P. Praeve, M. Schlingmann, W. Crueger, K. Esser, R. Thauer, and F. Wagner (ed.), *Jahrbuch Biotechnologie*, vol. 3. Carl Hanser Verlag, Munich, Germany.
48. **Zinkevich, V., I. Bogdarina, H. Kang, M. A. W. Hill, R. Tapper, and I. B. Beech.** 1996. Characterization of exo-polymers produced by different isolates of marine sulfate-reducing bacteria. *Int. Biodeterior. Biodegrad.* **37:**163–172.

II. MICROBIAL INTERACTIONS WITH TOXIC METALS: BIOMINERALIZATION AND BIOREMEDIATION

Environmental Microbe-Metal Interactions
Edited by Derek R. Lovley

Chapter 8

Microbial Mercury Reduction

Jon L. Hobman, Jon R. Wilson, and Nigel L. Brown

Over the past 30 years considerable research effort has been expended in the study of the biochemistry, genetics, and environmental aspects of microbial mercury reduction. As a result, mercury resistance has become one of the best understood of the microbial metal resistances. The literature on mercuric ion resistance and reduction is considerable, and a number of reviews cover various aspects of mercuric ion resistance (Hg^r) in detail (4, 37, 51, 72, 92, 103, 117, 119, 122–124). We do not attempt to summarize all of the data that have accumulated on Hg^r but would refer the reader to these reviews, which reflect the breadth of the literature and the evolution of our understanding of mercury resistance. Instead, we concentrate on selected aspects of microbial mercury reduction and other mechanisms of mercuric ion resistance which are relevant to the interaction of metals and microorganisms in the environment.

MERCURY

Mercury, hydrargyrum (Hg), or quicksilver as it was known, is a group IIb element with an atomic weight of 200.59, and an atomic number of 80, and its ions have a valency of 1 or 2. It is the only common metal that is liquid at physiological temperatures. It has a very high vapor pressure (2×10^{-3}mm Hg at 25°C) and a boiling point of 356.58°C. It rarely occurs as the pure metal in nature but is found most commonly as the ore cinnabar (HgS) (133). Mercury is simple to refine—the ores are roasted in a current of air, and metallic mercury is condensed from the vapor. The simplicity of refining and the unusual properties of this metal (e.g., mercury metal readily forms amalgams with other metals such as gold [Au] and silver [Ag]) probably account for the long and fascinating human relationship with mercury. Earliest records indicate that mercury and cinnabar were used in alchemy in China as early as the second century B.C. (45), and references to

Jon L. Hobman, Jon R. Wilson, and Nigel L. Brown • School of Biological Sciences, University of Birmingham, Edgbaston, Birmingham B15 2TT, United Kingdom.

cinnabar mines and the medicinal use of mercury were made by Pliny the Elder in the first century A.D. (45).

On average, mercury is present at a concentration of ~0.5 ppm in the earth's crust (133), but large deposits of Hg, or its ores, are often found in areas of volcanic activity or tectonic plate movement. Hg has been released into the lithosphere, atmosphere, and hydrosphere over millennia by geochemical processes, and it is therefore an important toxic element in the biosphere.

Toxicity

Mercury can exist in three states: Hg(0) (metallic), Hg_2^{2+} (mercurous) and Hg^{2+} (mercuric). Metallic mercury is relatively nontoxic compared to oxidized forms of Hg, in part due to its low solubility, but it can be converted by catalase and peroxidase enzymes to highly toxic oxidized forms in vivo. Ionic or partially covalent inorganic salts and many of the organic alkyl and aryl derivatives of mercury are highly toxic (136). The human health effects of breathing mercury vapor during refining have been known at least since Dioscorides reported them in the first century A.D. (45), and mass human poisoning incidents with metallic mercury vapor have occurred since then (33). The toxicology of human exposure to mercury is well documented (40, 67) and has manifested in diseases such as acrodynia ("pink disease"), erethism, nephrotic syndrome, and both respiratory and acute renal failure (67), as well as other diverse toxicological effects. The absorption, toxicology, and tissue targets of different forms of mercury are a function of the chemical characteristics and biological interactions of the metal or its organic or inorganic derivatives.

The extreme toxicity of mercury is due to (i) the ability of both organomercurial compounds (14) and inorganic forms of mercury (47) to cross biological membranes and (ii) their ability to bind with a high affinity to thiol and imino nitrogen groups in essential enzymes (62, 79). The consequences of these biological interactions are the ready uptake of mercurials by cells, with consequent damage to membranes as they pass through them, and inactivation of periplasmic and cytoplasmic enzymes. Some organomercurial compounds are lipid soluble, and all mercury compounds bind to lipids and nucleotides to a greater or lesser degree (43, 46). In addition, mercury compounds are genotoxic (31).

The organomercurial monomethylmercury (CH_3Hg^+; MMHg), is particularly problematic. It is the most toxic species of mercury, with a high affinity for central nervous system tissue, high lipid solubility, a high rate of uptake across biological membranes, and a longer residence time in biological tissue than that of inorganic Hg (4, 69). These characteristics of MMHg contribute significantly to bioaccumulation and biomagnification as factors in mercury toxicity. Bioaccumulation can be defined as the increase in the total amount of Hg in an organism over time (69), while biomagnification of mercury is an increase in mercury concentrations in tissue through trophic transfer, from primary producers up to carnivorous terminal consumers, via food webs (13). MMHg is evenly distributed in body tissues. Inorganic mercury is unevenly distributed, because it is less efficient than MMHg at crossing biological membranes. This has two implications. (i) Inorganic mercury tends to be distributed in visceral tissue rather than evenly throughout the body like MMHg, and so there is a potentially lower body burden of inorganic Hg compared to MMHg. (ii) Because of the lower body burden of inorganic Hg, there is a consequent lower potential for its transfer along food chains (69).

Mercury uptake, distribution, and toxicity are therefore dependent on the speciation and transformations of the mercury and the rate of elimination in vivo.

Human Uses of Mercury and Mercury Compounds

Mercury and its inorganic compounds, particularly HgCl (calomel) and $HgCl_2$ (corrosive sublimate), have been used widely in both medicine and industry for many centuries. Before the industrial age, the only large-scale use of Hg was in gold and silver refining. Subsequently, large-scale uses have included the production of sodium hydroxide and chlorine by the chloralkali process, application in electrical apparatus, battery, and scientific equipment manufacture, and use in agricultural chemicals. Organic mercury compounds have only relatively recently been chemically synthesized [with the exception of MMHg and dimethylmercury, $(CH3)_2Hg$ (DMHg), which are naturally occurring] but have enjoyed widespread usage as antimicrobials and agricultural chemicals. Some are still used in antifouling paints, as agents to control fungal infections in plant materials (40), and as preservatives in health care products. Many of the uses of organomercurials have been superceded by other, safer chemical compounds, while other uses, such as alkylmercury seed dressings, have been discouraged (40) and ultimately banned.

Medical uses of mercury have principally been in the treatment of skin disease and syphilis, although mercury or its compounds have been prescribed for many human ailments at one time or another (45, 63). In the past two centuries, mercurials have also been used as antiseptics, antimicrobials (61), diuretics, laxatives, teething powders, and contraceptives, and mercury metal is still used in amalgam tooth fillings (45). Some of the uses detailed above are now declining or have been terminated.

Human Impacts on the Biogeochemical Mercury Cycle

Although the total amount of mercury in the biogeochemical cycle remains constant, the liberation of mercury from the lithosphere to the atmosphere and hydrosphere has been enhanced by human activity. Anthropogenic sources of mercury release have been responsible both for severe localized pollution incidents and for contributing to the elevation and potential destabilization of the overall mercury flux. It has been estimated that global mercury emissions to the atmosphere are in the order of 6.1×10^6 kg $year^{-1}$ (82) and that anthropogenic releases of mercury to the environment now exceed those from natural sources. Coal burning and municipal waste incineration are now perhaps the two biggest single anthropogenic contributors to the worldwide gaseous mercury background (83), although point (or single-source) releases of Hg have had an immediate toxic impact on ecosystems. For example, minor levels of Hg contamination may occur downwind of crematoria due to volatilization of amalgam fillings from cadavers (71). However, very significant losses of mercury, to both the atmosphere and aquatic systems, have occurred during the chloralkali and other chemical processes and in Hg, Au, and Ag refining. These losses have had direct impacts on the local environment and ultimately, in some cases, on the local human population. A tragic example of this was the discharge of methylmercury in effluent from acetaldehyde and vinyl chloride manufacture at Minamata Bay, Japan. The discharged MMHg was responsible for human methylmercury poisonings on a large scale through the consumption of MMHg contaminated fish and shellfish (49, 132).

Speciation and Bioavailability

The mercury cycle has multiple biotic and abiotic components, which have been the subject of recent reviews (36, 57, 69). A critical factor in Hg cycling is its speciation, since this affects its chemical and biological reactivity, its transport, and the residence time of Hg species in the environment.

The absolute levels of mercury, determined using physical methods of measurement, and the bioavailable mercury in environmental samples may differ greatly. Bioavailability is the more relevant to our understanding of how the biological component of the mercury cycle changes chemical speciation of Hg and also to the potential toxicity of Hg contamination to the ecosystem under a given set of physical conditions. The bioavailability of both organic and inorganic mercury is determined by the proportion of Hg species that are in a chemical form that can be taken up by organisms (as a fraction of the total Hg). It is also determined by the abundance of other ligands that might complex with these Hg species to form insoluble or biologically nonreactive species (69). Mercury species have been classified by Lindquist et al. (64) in chemical terms as

volatile species — Hg^0 and $(CH_3)_2Hg$
reactive species — Hg^{2+}, HgX^+, HgX_2, HgX_3^- and HgX_4^{2-} (where X = OH^- Cl^- or Br^-), HgO on aerosol particles, and Hg^{2+} complexes with organic acids (water soluble or particle borne), and
nonreactive species — CH_3Hg^+, CH_3HgCl, CH_3HgOH, and other organomercurials and $Hg(CN)_2$, HgS, and Hg^{2+} bound to S in humus.

However, some mercury species classified above as chemically nonreactive are bioavailable, since they both bioaccumulate and biomagnify. Other chemically nonreactive species may be converted to reactive or volatile species by biotic processes. MMHg is a particular example of a chemically nonreactive but bioavailable form of mercury. MMHg concentrations in vertebrates are determined by the amount of bioavailable MMHg at the base of the food web and the trophic level of the vertebrate (69). The majority of MMHg in fish is taken up via food ingestion (68), and major enrichment of MMHg occurs in aquatic environments between the water and the phytoplankton (48).

Impact of Microbial Processes on the Mercury Cycle

Seven microbial mercury transformations have been demonstrated experimentally:

1. Methylation of inorganic Hg(II) (4, 25–27, 42, 93, 126)
2. Oxidation of Hg(0) to Hg(II) (120)
3. Reduction of Hg(II) to Hg(0) (60)
4. Cleavage of the C-Hg bond in organomercurials followed by reduction to Hg(0) (41)
5. DMHg synthesis from MMHg (5, 6)
6. Demethylation of MMHg (95)
7. Oxidative demethylation of MMHg (89, 90)

Reactions 5 and 6 may be the same (discussed below).

The bacterial transformations of Hg compounds, either to more toxic or bioavailable species or to less toxic or insoluble Hg species, constitute important parts of the biogeochemical mercury cycle. For example, microbial methylation of inorganic Hg will increase the level of bioavailable MMHg and hence increase its transfer through different trophic levels, whereas microbial production of highly volatile DMHg from MMHg will decrease the amount of bioavailable MMHg. Reduction of Hg(II) to Hg(0), which is also volatile, can remove oxidized mercury species to the atmosphere and thus reduce the amount of soluble Hg in the local environment. Figure 1 summarizes microbial transformations of mercury.

MERCURY TOLERANCE AND RESISTANCE IN MICROORGANISMS

The high affinity of mercury for thiol and imino nitrogen groups in proteins and the diverse cellular targets of mercuric ions preclude some of the main strategies used by microorganisms to avoid, eliminate, or detoxify other toxic metals (106). Ion efflux pumps, e.g., the Cd^{2+}, AsO_2^{1-}, AsO_4^{3-}, and Zn^{2+} efflux ATPases (81, 117–119, 121, 137) or the chemiosmotic Cd^{2+}, Zn^{2+}, Co^{2+} resistance in *Ralstonia metallidurans* (formerly *Alcaligenes eutrophus*) strain CH34 (32, 80, 117, 119), have not been reported in an Hg^r mechanism, nor have mercury-specific prokaryotic sequestration systems been reported.

The ability of the microorganism to grow and metabolize at increased levels of toxic Hg species may be due to either a specific resistance or a tolerance to Hg. Metal resistance gene operons (e.g., *mer* operons [15, 51, 92], arsenic resistance [81, 137], cadmium resistance [117–119, 121], and, copper resistance [16, 29]) have specific regulatory components that sense the presence of the toxic metal in the cellular environment and respond to the presence of that metal by induction of specific resistance proteins. Using the criterion that a specific metal resistance is regulated by a specific discriminatory, regulatory component, some of the mechanisms discussed below may be simply tolerance mechanisms. Such tolerance mechanisms may be the result of adventitious by-products of normal cellular metabolism rather than a mercury-specific response.

Mechanisms of Mercury Resistance and Tolerance

Reduction

Reductive mercury resistance (*mer*) determinants have been classified into two groups, narrow and broad spectrum, on the basis of the range of mercury com-

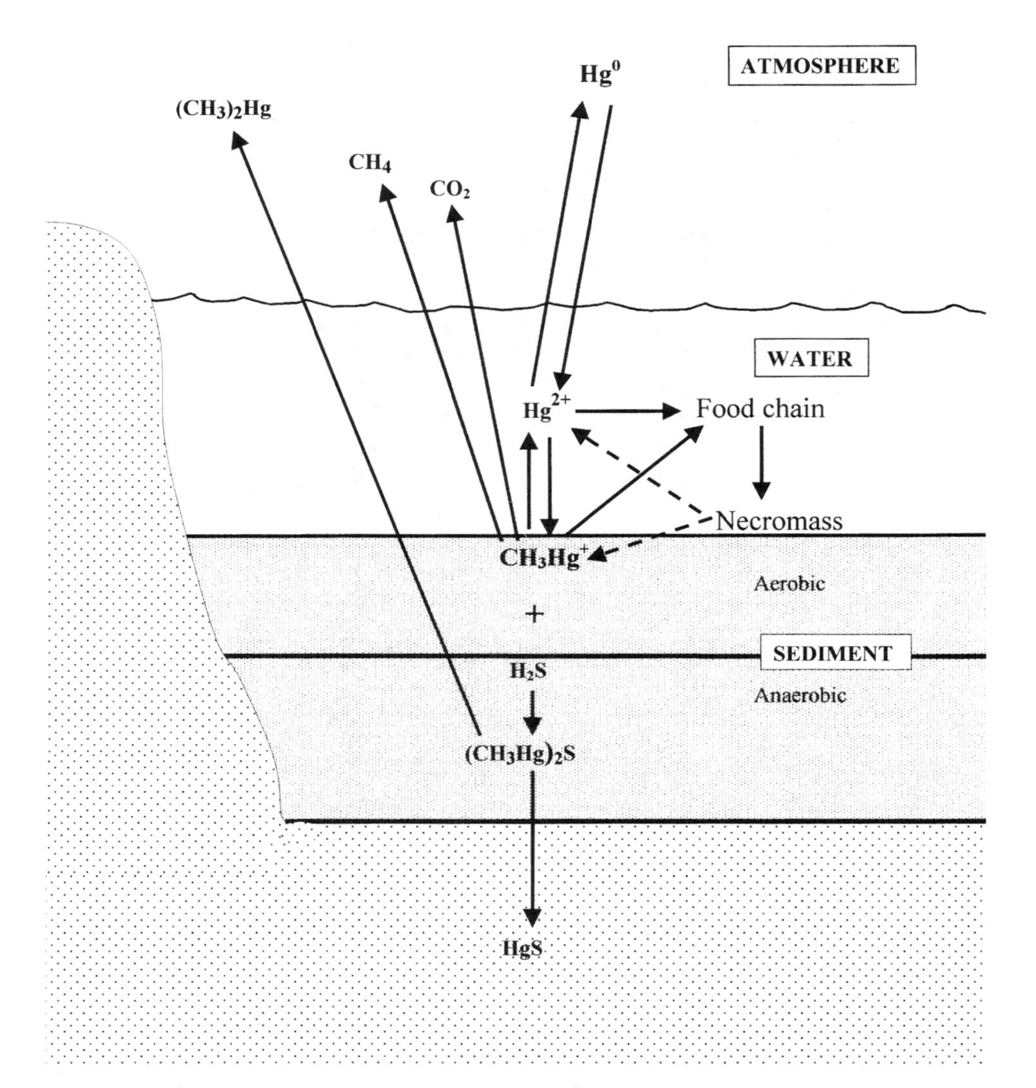

Figure 1. Microbial transformations in the mercury cycle. A schematic summary of microbial mercury transformations is shown. The levels of Hg(II) and MMHg are governed by the balance of reduction and oxidation by aerobic bacteria and by the rate of bacterial methylation and demethylation. Demethylation of MMHg can be reductive (broad-spectrum *mer* operons), generating CH_4 as a metabolite, or oxidative, where CO_2 is produced. Both Hg(II) and MMHg can pass into the food chain or be adsorbed by organic matter (particulate or dissolved). Hg(II) and MMHg will be released back into the mercury cycle when the biomass decays. Under anoxic conditions, the reaction of MMHg with H_2S (generated by sulfate-reducing bacteria from sulfates) produces dimethylmercuric sulfide. This is unstable and degrades to insoluble HgS and volatile DMHg. DMHg can degrade under mild acid conditions to form CH_4 and Hg(II), which can be transformed to Hg(0). Modified from Baldi, 1997 (4).

pounds that they will detoxify (24, 134). All of the characterized reductive Hg^r determinants show evidence of mercuric ion-responsive regulation. Narrow-spectrum Hg^r determinants confer resistance to inorganic cations by reduction to Hg(0). In some bacteria, narrow-spectrum mercury resistance determinants confer resistance to a very limited number of organomercurials, by an unknown mechanism that does not involve reduction to Hg(0) (24, 134).

Broad-spectrum Hg^r determinants confer resistance both to inorganic Hg(II) and to a wide range of organomercurial compounds. Detoxification of organomercurials (including MMHg in some instances [77]) involves enzymatic cleavage of the C-Hg bond of the organomercurial compound, followed by reduction of Hg(II) to Hg(0). For MMHg, this process has been referred to as reductive demethylation.

Methylation

Bacteria from a variety of habitats (including the human gut, water, and soil) can methylate inorganic mercury (126). The biological methylation of Hg(II) to MMHg and/or the highly volatile DMHg by bacteria, occurs by transfer of the carbanion (CH_3^-) to Hg(II) from methylcobalamine (4, 126). The predominant methylators of Hg in anaerobic environments are sulfate-reducing bacteria (25, 26). Mixed cultures of methanogens and sulfate-reducing bacteria in sulfate-limited media methylate Hg, which suggests that interspecies hydrogen and acetate transfer may play an important role in this process (93). While the addition of SO_4^{2-} to freshwater lake sediments stimulated Hg methylation, it was paradoxically inhibited at high SO_4^{2-} concentrations (27).

Metallothioneins

The production of metallothionein, which sequesters heavy metals, is a common resistance mechanism in higher organisms, but metallothioneins appear to be rare in prokaryotes. A notable exception is the Zn^{2+}-binding SmtA protein in *Synechococcus* sp. strain PCC6301 (104). Although experimental evidence shows that a gluthathione *S*-transferase–SmtA fusion protein will bind mercuric ions (116), the regulator of *smtA* expression, *smtB*, was less responsive to Hg than to Zn, Cu, Cd, Co, and Ni (53). No prokaryotic mercuric ion-specific metallothioneins have been reported.

Other Mechanisms

Reduced uptake. Two plasmid-encoded membrane proteins have been implicated in the reduction of cellular permeability to Hg(II) in a strain of *Enterobacter aerogenes* (96). A plasmidless variant of this strain of *E. aerogenes* was Hg(II) sensitive and lacked these two proteins.

Degradation of MMHg. *Desulfovibrio desulfuricans* API produces hydrogen sulfide (H_2S), which reacts with MMHg to form insoluble dimethylmercury sulfide [$(CH_3Hg)_2S$)]. This then degrades to HgS, DMHg, methane, and traces of ionic mercury by abiotic processes (6).

It has been suggested that two functions encoded on the same plasmid in *Clostridium cochlearium* T-2 (94) are responsible for the degradation of MMHg and the formation of insoluble mercuric sulfide (HgS). It was postulated that this transformation occurs due to the action of H_2S (produced during anaerobic respiration of *C. cochlearium* [95]) on inorganic Hg species produced by the demethylation of MMHg in a separate reaction. However, no volatile Hg was detected from cultures of *C. cochlearium*. It has been argued that the process of conversion of MMHg to HgS in *C. cochlearium* T-2 and *D. desulfuricans* API may be the same (4) (Fig. 1).

Oxidative demethylation of MMHg has been reported to take place in anoxic sediments (89, 90). This process can be differentiated from reductive demethylation (broad-spectrum Hg^r) by the use of [^{14}C]MMHg in tracer experiments. In these experiments reductive, demethylation produces $^{14}CH_4$ from [^{14}C]MMHg whereas oxidative demethylation produces $^{14}CO_2$ from the substrate.

The three transformations of MMHg described separately above may be part of the same complex series of reactions that seem to characterize methylation and demethylation of Hg compounds by anaerobic bacteria.

Further mechanisms. Several other mechanisms of mercury tolerance have been reported, though relatively little is known about them. These include intracellular metal sequestration, extracellular binding of Hg, and nonenzymatic volatilization of Hg(0) (reviewed in reference 4).

BIOCHEMISTRY AND GENETICS OF MICROBIAL MERCURY REDUCTION

Common Principles

There is remarkable similarity in the mechanisms of mercuric ion resistance in many genera of both gram-negative and gram-positive bacteria. The fundamental mechanism of Hg(II) detoxification common to narrow spectrum *mer* operons in gram-negative bacteria (Fig. 2) is transport of mercuric ions into the cytoplasm of the cell via specific transport proteins, followed by reduction to metallic mercury [Hg(0)] by the enzyme mercuric reductase (MR) in the cytoplasm. Metallic mercury is eliminated by passive diffusion from the cell (as mercury vapor) under normal physiological conditions. Mercury resistance in gram-positive bacteria is characterized less well, but the general mechanism is equivalent to that in gram-negative bacteria. As would be expected from the differences in their cell envelopes, the detail of Hg transport in gram-positive bacteria is different from that in gram-negative genera, but the reductases appear very similar.

The narrow-spectrum *mer* operons carried by transposons Tn*501* and Tn*21* (Fig. 3) are the most intensively studied of all reductive mercury resistances and have become a paradigm for all other resistances. Mercuric ion resistance in gram-negative baceria can be thought of as three distinct but integrated components: mercuric ion-responsive regulation of *mer* gene expression, transport of mercuric ions into the cell, and reduction of Hg(II) to Hg(0). Each of these components of the resistance mechanism is the function of one or more proteins.

Regulation

The dimeric MerR protein regulates expression of the *mer* structural genes (transport and reduction) as well as of itself. It binds as a complex with RNA

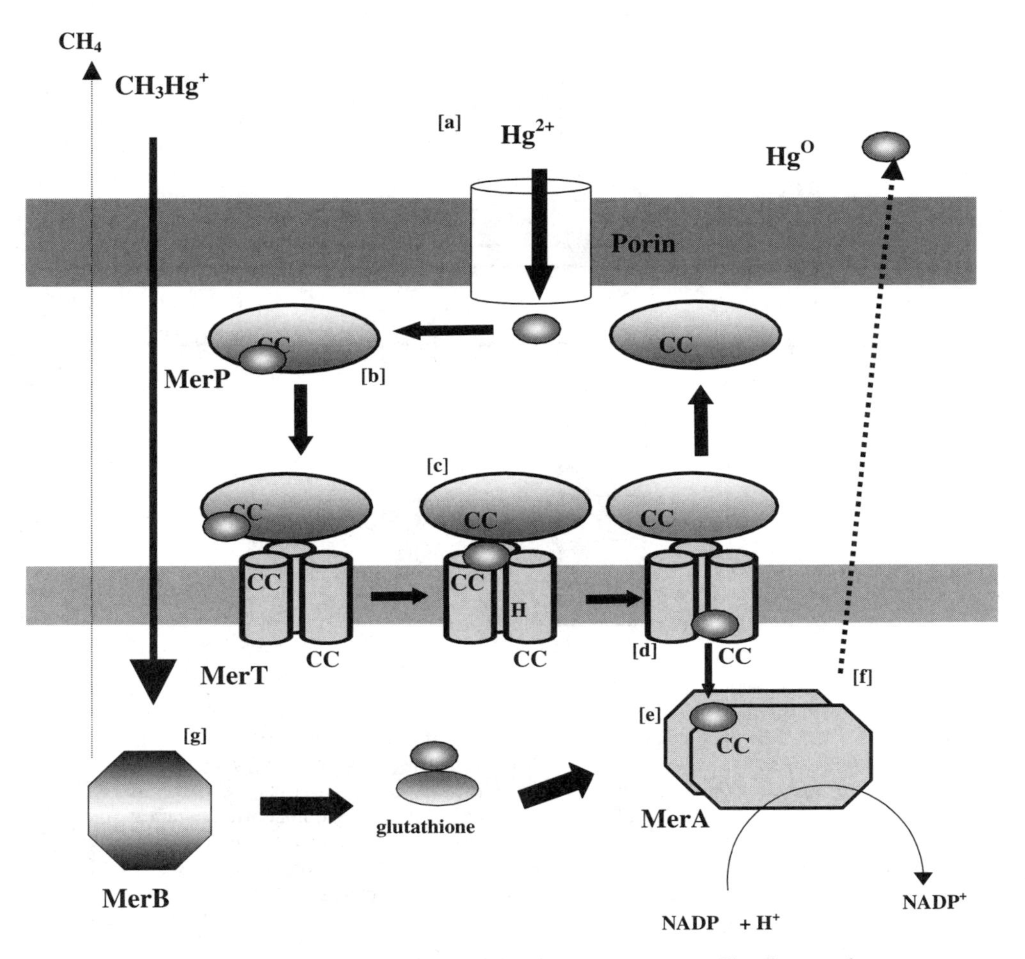

Figure 2. Generalized model of bacterial resistance to mercury. The diagram shows the model for mercuric ion resistance in gram-negative bacteria. Mercuric ions in the environment of a bacterial cell [a] pass through the porins (OmpC and OmpF) in the outer membrane, where they are [b] scavenged by the periplasmic protein, MerP and bind to the cysteine residues in each subunit of the protein. [c] The mercuric ion is then passed from the cysteines in MerP, to those in the transmembrane region of the inner membrane protein, MerT. As part of the transport mechanism, the Hg(II) ion is transferred to the cysteines on the cytoplasmic face of MerT, whence [d] they are passed to the heavy-metal associated motif in the amino-terminal MerP-like domain of mercuric reductase [e]. The mercuric ion is then bound at the active site and reduced to elemental mercury, Hg(0) [f]. The volatile product is released from the enzyme and diffuses through the bacterial membranes to the environment. MMHg can diffuse in through the cell membrane, and with broad-spectrum determinants, is cleaved by organomercurial lyase [g]. The Hg(II) so produced is proposed to bind to glutathione in the cytoplasm and be reduced by MR. Resistance in gram-positive bacteria operates by a similar mechanism, but the detailed structures of the transport proteins are different. Cysteine residues are also present in other mercury transport proteins from gram-negative sources (e.g., MerC and MerF) or from gram-positive sources and are predicted to lie in the transmembrane region. The MR from gram-negative and gram-positive sources are similar but differ in the number (0, 1, or 2) of MerP-like N-terminal domains and in their detailed amino acid sequences.

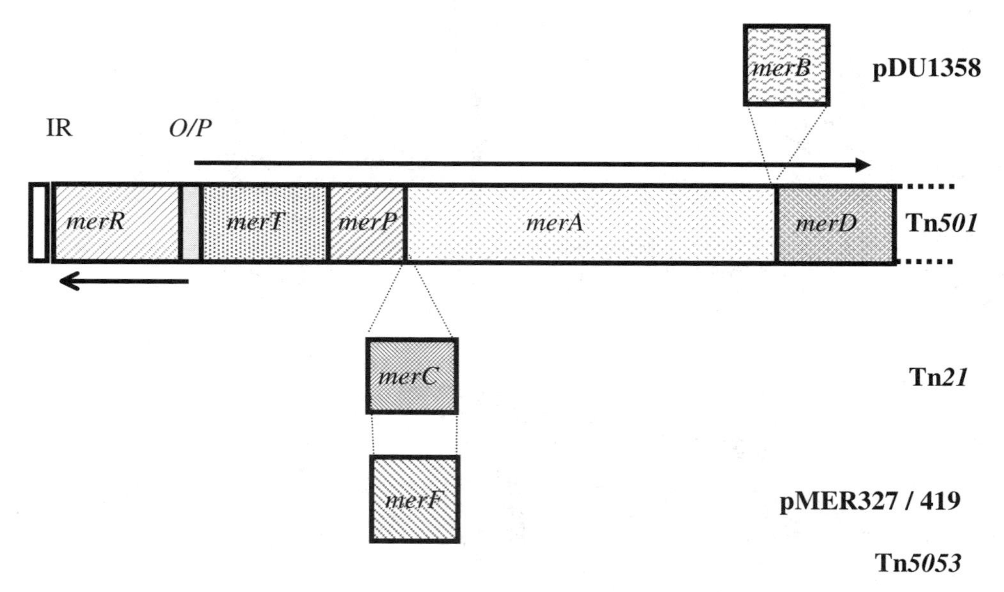

Figure 3. General organization of the Tn*501* and related *mer* operons in gram-negative bacteria. A generalized schematic diagram of the *mer* genes of transposon Tn*501* is shown. Additional genes from closely related *mer* resistances (referred to in the text), and their positions in the operon relative to *merR, mer-T, mer-P, mer-A*, and *mer-D* are marked. The *merR* and structural gene transcripts are shown as arrows from the *mer* operator/promoter (*merO/P*) site. The terminal inverted repeat (IR) of the transposon (Tn*501*) is marked. There are a number of other gram-negative bacterium *mer* operons whose organizations do not conform to this generalized structure yet whose genes are clearly closely related to Tn*501* and Tn*21* (51, 92, 119, 122).

polymerase to a bidirectional operator/promoter, P_{mer}. MerR acts both as a transcriptional activator in the presence of mercuric ions and as a repressor of gene expression in their absence (15, 87, 92, 123). Broad-spectrum *mer* operons are regulated by MerR proteins that respond both to inorganic and organomercury compounds (84). Such broad-spectrum MerR proteins differ from those encoded by narrow-spectrum determinants in having different C-terminal domains. The replacement of the C-terminal sequence of a narrow-spectrum MerR with that from a broad-spectrum protein allows MerR to respond to organomercurials (84).

MerR acts as a hypersensitive biological switch in the presence of Hg(II), with structural gene expression increasing from 5 to 95% of full induction over a two- to four-fold range of Hg(II) concentration (107). Upon binding Hg(II), the protein undergoes a conformational change, which in turn causes a distortion of the *mer* promoter DNA, such that the promoter is correctly recognized by the σ^{70} subunit of RNA polymerase and a transcriptionally active complex is formed (1, 2). The polycistronic mRNA synthesized from the promoter encodes the transport and reductase proteins.

Transport

In Tn*501*, the mercury transport component of the resistance is composed of two proteins, MerP and MerT. MerP is a small (72-amino-acid) periplasmic protein, which binds Hg(II). There is evidence that MerP interacts with the 116-amino-acid cytoplasmic membrane associated protein, MerT and transfers mercuric ions to it (66, 74, 76). MerT facilitates the passage of divalent mercuric ions through the cytoplasmic membrane and into the cytoplasm, where they are reduced to Hg(0) by mercuric reductase. MerT-dependent transport is absolutely required for inorganic mercuric ion resistance, and deletion of *merT* abolishes mercuric ion resistance. Deletion of *merP* reduces resistance by approximately 20%, as measured by the MIC assay (74). MerP and MerT are not involved in the transport of methylmercury in *Pseudomonas* strain K-62, which exhibits broad spectrum Hg^r, but are involved in the transport of phenylmercury (58, 128). Methylmercury appears to enter the cell via diffusion, probably because of its chemical properties, as discussed above.

The transport proteins of Tn*21* include an additional gene product, MerC, encoded between the *merP* and *merA* genes (73). MerC is a transport protein (109) closely related to the mercury transport protein of *Thiobacillus* (54). It is unclear why the Tn*21* system encodes two transmembrane transport proteins, MerT and MerC, whereas Tn*501* encodes only the MerT protein. However, analysis of *mer* determinants found in river sediments indicates that Tn*21*-like determinants are found at higher mercuric ion concentrations than are Tn*501*-like determinants (88). Two transport systems may offer a selective advantage at high mercury concentrations.

Reduction

Reduction of Hg(II) to Hg(0) is catalyzed by the cytoplasmic NADPH-dependent flavoenzyme mercuric reductase. When expression of Tn*501* MR is fully induced on a plasmid-borne operon in *Pseudomonas aeruginosa*, the enzyme constitutes 6% of the soluble protein (38). The enzyme is soluble but may transiently be associated with the cytoplasmic membrane. Transport is rate limiting for resistance, and the capacity for mercuric ion reduction exceeds the rate of transport of Hg(II) into the cell in vivo (98). This ensures that the intracellular Hg(II) concentration remains low and that any Hg(II) diffusing across the cytoplasmic membrane independently of the transport system is reduced to Hg(0). The substrate for mercuric reductase is probably the dimercaptan, since Hg(II) in biological systems associates with thiol groups, and the reaction catalyzed (70, 114) would be

$$\text{RS-Hg}^{\text{II}}\text{-SR}' + \text{NADPH} + \text{H}^+ \rightarrow \text{Hg}^{(0)} + \text{RSH} + \text{R}'\text{SH} + \text{NADP}^+$$

The kinetics of the enzyme have been extensively studied (30, 39, 101, 111), and a crystal structure has been determined (113). The enzyme is structurally and functionally related to glutathione reductase and lipoamide dehydrogenase but differs in detailed mechanism from the dithiol oxidoreductases in the catalytic cycling of the enzyme and in the details of electron transfer. MR contains an additional N-terminal domain related to the MerP-like heavy metal-associated domains found in the transporters of other heavy metals in bacteria and eukaryotes (19). MR is specific for Hg(II) since the substrate and other metals are not reduced. It has been expressed in yeast (100) and plants (108) in which it reduces mercuric ions.

Other Genes

merD. The *merD* gene is distal to *merA* in Tn*501* and Tn*21* (Fig. 3) and has been proposed to be a repressor of the *mer* operon promoter. It may be involved in switching off expression of the *mer* genes when Hg(II) concentrations fall following reduction to Hg(0) and subsequent volatilization. The MerD protein has N-terminal homology to MerR, and partially purified MerD binds to the *mer*O/P region (with lower affinity than MerR does), but it does not activate transcription in the presence of mercuric ions (75, 85).

merB. The *merB* gene is found in broad-spectrum Hg^r determinants and encodes organomercurial lyase, which is responsible for the cleavage of C-Hg bonds in organomercurials (10). The Hg(II) product is then reduced by mercuric reductase. Organomercurial lyase is important in the recycling of organomercurials in the biosphere, and the reaction catalyzed is most unusual. The enzyme has been purified (10) but has been little studied because of its instability.

***merF*.** The *mer* operon of plasmid pMER327/419, isolated from a gram-negative bacteria in the river Mersey, United Kingdom, contains an open reading frame between the *merP* and *merA* genes (50) (Fig. 3). This gene, now designated *merF*, encodes another transport protein (our unpublished data). As with Tn*21 merC*, this is in addition to a *merT* gene in the same *mer* determinant and may offer a selective advantage at high mercuric ion concentrations, although there are no experimental data to support or contradict this.

A generalized scheme for broad- and narrow-spectrum mercuric ion resistance in gram-negative bacteria is summarized in Fig. 2.

Mercury Resistance (*mer*) Operons in Gram-Negative and Gram-Positive Bacteria

The full or partial DNA sequences of at least 10 broad- and narrow-spectrum *mer* operons are now known (51, 92), as well as a number of single *mer* gene sequences (91). These DNA sequences have been derived from different gram-negative bacteria isolated from geographically distant and diverse habitats, yet there is surprising similarity between the *mer* genes of these operons. Many of the *mer* determinants in gram-negative bacteria appear to be pandemic (51, 92) and many of the *mer* operons are located on plasmids or different types of transposable element.

Fewer *mer* operons in gram-positive bacteria have been characterized by DNA sequence analysis than have *mer* operons in gram-negative baceria (51, 92). However, the recent determination of a number of sequences from *Bacillus* spp. isolated from the environment (11) has shown that they are also widely distributed and that mobile genetic elements are again implicated in this horizontal transfer of *mer* sequences.

MERCURY REDUCTION IN THE ENVIRONMENT

Gene Expression

Although the presence of *mer* genes homologous to characterized *mer* operons has been confirmed by gene probing (7, 8, 44, 105) or PCR (17, 18, 77), in bacteria isolated from both Hg polluted and nonpolluted environmental samples, little was known about the expression of these genes in the environment. There have now been several investigations that have confirmed *mer* gene expression in situ, either by quantification of *mer* gene mRNA (55, 56, 127) or by measurement of NADPH consumption in the MR-catalyzed reduction of mercuric ions (86). In addition, comparative studies have examined *mer* mRNA production and mercury volatilization in environmental water samples (9, 78). These have shown that microbial metabolic activity, as influenced by addition of nutrients, rather than mercuric ion concentration, is the essential factor affecting *merA* mRNA abundance and MR activity.

Biosensors

Biosensors have the potential to accurately, simply, and cheaply identify the concentrations of bioavailable Hg or Hg compounds in an environmental sample. There are potentially two sorts of biosensor for Hg: gene- and protein-based systems. Gene-based biosensors are propagated and expressed in viable cells, and those described to date are based upon the expression of a promoterless gene with an assayable phenotype (e.g. *lux*, and *lacZ*) from the *mer* promoter (12, 28, 99, 125, 129). Many of these biosensors also express mercury transport genes as well as the reporter gene, either from the *mer* promoter or in *trans* on another plasmid, thus enhancing the uptake of Hg and increasing the sensitivity of the system (59, 115). Protein-based biosensors have been created which indirectly measure MR activity by coupling light production by luciferase to the rereduction of $NADP^+$ oxidised in the MR reduction of Hg(II) (35). Others utilize the toxic nature of mercuric ions to inhibit other enzymes (e.g., urease) that have a measurable activity (131). Recently biosensors have been constructed which rely on Hg binding to immobilized MerR (E. Csöregi, B. Matthiason, J. R. Lloyd, J. L. Hobman, J. R. Wilson, N. L. Brown, P. Corbisier, and D. Van Der Lelie, 15 September 1997, Swedish Patent Office). Binding of the metal to the protein causes a change in the capacitance of the protein, which is transduced to a measurable electronic signal.

Potential for Bioremediation

Unlike organic compounds, toxic metal pollutants are immutable. Even though a biological transformation may alter the valence state of the metal, it is still present in some form. The extreme toxicity of Hg or its compounds and their chemical reactivity means that there are only a limited number of strategies to remove them from a contaminated environment. The method of choice depends not only on the substrate for remediation (e.g., soil or water) but also on the method of recovery of Hg once removed from the polluted sample. The chemical form of the mercury also dictates the choice of bioremediation strategy. If the Hg compound is an organomercurial (e.g., phenyl mercuric acetate), the aromatic moiety and the Hg produced by degradation will have to be metabolized and removed, respectively (52).

Many strategies for the removal of toxic metals from industrial effluents rely upon sorption by microbial biomass (34, 65, 130) or by ion-exchange resins (102). The disadvantages of both of these methods are that they lack sorption specificity in the presence of other competing metal ions and can be sensitive to ionic strength, pH, and the presence of metal ion-chelating agents (22). There have been several refinements to the general principle of biosorption for Hg(II) removal. These have relied on the principle that expression of mercuric ion transport proteins increases the ability of biomass to take up Hg(II). One approach has used the expression of MerT, MerP, and a plant metallothionein in *Escherichia coli* to enhance Hg(II) accumulation (22, 23). Another method used inactivated *Pseudomonas aeruginosa* PU21 biomass that had expressed Hg^r proteins from plasmid RIP64 (20). In both cases, enhanced ability to bind Hg(II) in the presence of competing ions, compared to the control bacterial strain, was demonstrated. Compared to volatilization, biosorption of Hg(II) compounds offers the advantage of immobilization of the metal to a readily recoverable biomass.

Experimental and pilot scale microbial mercury reduction and volatilization systems have been developed for removal of Hg(II) (21, 112, 135). Essentially, the bacteria are used to reduce soluble mercuric ions to volatile Hg(0). In an open system, Hg(0) is lost to the atmosphere, thus diluting the soluble pool of Hg, whereas in a closed system, Hg(0) can be captured using a mercury capture device. This method has advantages for in situ remediation of contaminated aquatic systems, since it will diminish the pool of Hg that can be methylated. It may only require the addition of suitable nutrients to the contaminated water body to stimulate the indigenous Hg^r bacteria to increase their rate of volatilization.

The transport of mercuric ions into the bacterial cell is rate limiting for volatilization by mercuric reductase (98). One approach to enhancing the volatilization of Hg(0) for bioremediation purposes is to increase the rate of delivery of Hg(II) to MR. There have been several different attempts to do this, including increasing the copy number of the plasmid containing the *mer* genes (97), using permeabilized *E. coli* cells expressing MR (98), and immobilizing MR on a solid support (3).

Phytoremediation of contaminated soils is potentially a useful method of removing and immobilizing toxic metal contaminants in plant biomass (110). This idea has been applied by re-engineering the *merA* gene from transposon Tn*21* to express in *Arabidopsis thaliana*, creating a mercury volatilizing plant (108).

Whether these bioremediation systems will be of commercial use depends on whether they offer improvements over chemical methods, in both cost and efficiency, and whether they are susceptible to "poisoning" by other contaminants. Most contaminated sites have more than one metal present, and many are contaminated with multiple toxic metals and organic compounds. Remediation of contaminated sites may require a combination of approaches, from civil engineering solutions through to biotechnological applications.

CONCLUSIONS AND PERSPECTIVES

Microbial mercuric ion reduction is a complex and sophisticated mechanism for the detoxification of divalent mercuric ions or organomercurials. Unlike many bacterial metal ion resistances, reductive Hg(II) resistance alters the valence state of the metal ion as the resistance strategy, and it requires reducing equivalents to do so. Transport of mercuric ions into the cytoplasm via a specific transport system delivers these ions to a cellular environment that produces the required reducing equivalents for the enzymatic transformation of Hg(II) to Hg(0).

Initial interest in Hg^r arose from the isolation of bacteria resistant to mercurials in hospitals, where mercurial antiseptics were used. Since then, it has become clear that *mer* sequences are widely distributed in the environment and that reductive mercuric ion resistance is an important factor in the cycling of Hg in the biosphere. Reductive mercury resistance is the single most widespread of all antimicrobial determinants, and our current state of knowledge suggests that *mer* operons are pandemic gene sequences, found in gram-negative and gram-positive bacteria. Clearly, reductive mercury resistance gene operons are highly successful. The same *mer* gene sequences are global in their distribution and are associated with mobile DNA elements, either transposons or plasmids, which are recombinogenic. Hg^r is an underexploited model system for the study of the evolution of an environmental trait across different prokaryotic genera.

Great progress has been made in our understanding of the genetics and biochemistry of reductive Hg^r. However, as in the study of other systems, it should be noted that laboratory experiments do not replicate the nutrient-limited, low temperature conditions found in most environments and are therefore not accurate representations of what occurs in nature. Nor do they reproduce the low gene copy number of most *mer* operons found in environmental bacteria. In situ studies of *mer* gene expression and reduction of mercuric ions by Hg^r bacteria have begun to give us insights into the functioning of the systems under the conditions in which they have evolved to operate.

Acknowledgments. Work from our laboratory was supported by the European Union, the Biotechnology and Biological Sciences Research Council, the Medical Research Council, and the Royal Society.

We thank Elena Bogdanova and Anne Summers for communicating data prior to publication.

REEFERENCES

1. **Ansari, A. Z., J. E. Bradner, and T. V. O'Halloran.** 1995. DNA-bend modulation in a repressor-to-activator switching mechanism. *Nature* **374:**371–375.
2. **Ansari, A. Z., M. L. Chael, and T. V. O'Halloran.** 1995. Allosteric underwinding of DNA is a critical step in positive control of transcription by Hg-MerR. *Nature* **355:**87–89.
3. **Anspach, F. B., M. Hukel, M. Brunke, H. Schutte, and W. D. Deckwer.** 1994. Immobilization of mercuric reductase from a *Pseudomonas putida* strain on different activated carriers. *Appl. Biochem. Biotechnol.* **44:**135–150.
4. **Baldi, F.** 1997. Microbial transformation of mercury species and their importance in the biogeochemical cycle of mercury. *Metal Ions Biol. Syst.* **34:**213–257.
5. **Baldi, F., F. Parati, and M. Filippelli.** 1995. Dimethylmercury and dimethylmercury-sulfide of microbial origin in the biogeochemical cycle of Hg. *Water Air Soil Pollut.* **80:**805–815.

6. **Baldi, F., M. Pepi, and M. Filippelli.** 1993. Methylmercury resistance in *Desulfovibrio desulfuricans* strains in relation to methylmercury degradation. *Appl. Environ. Microbiol.* **59:**2479–2485.
7. **Barkay, T., D. L. Fouts, and B. H. Olson.** 1985. Preparation of a DNA gene probe for detection of mercury resistance genes in gram-negative bacterial communities. *Appl. Environ. Microbiol.* **49:** 686–692.
8. **Barkay, T., C. Liebert, and M. Gillman.** 1989. Hybridization of DNA probes with whole-community genome for detection of genes that encode microbial responses to pollutants—*mer* genes and Hg^{2+} resistance. *Appl. Environ. Microbiol.* **55:**1574–1577.
9. **Barkay, T., R. R. Turner, A. Van den Brook, and C. Liebert.** 1991. The relationship of Hg(II) volatilization from a fresh-water pond to the abundance of *mer* genes in the gene pool of the indigenous microbial community. *Microb. Ecol.* **21:**151–161.
10. **Begley, T. P., A. E. Walts, and C. T. Walsh.** 1986. Bacterial organomercurial lyase: overproduction, isolation and characterization. *Biochemistry* **25:**7186–7192.
11. **Bogdanova, E. S., I. A. Bass, L. S. Minakhin, M. A. Petrova, S. Z. Mindlin, A. A. Volodin, E. S. Kalyaeva, J. M. Tiedje, J. L. Hobman, N. L. Brown, and V. G. Nikiforov.** 1998. Horizontal spread of *mer* operons among Gram-positive bacteria in natural environments. *Microbiology* **144:** 609–620.
12. **Bohlander, F. A., A. O. Summers, and R. B. Meagher.** 1981. Cloning a promoter that puts the expression of tetracycline resistance under the control of the regulatory elements of the *mer* operon. *Gene* **15:**395–403.
13. **Boudou, A., and F. Ribeyre.** 1997. Mercury in the food web: accumulation and transfer mechanisms. *Metal Ions Biol. Syst.* **34:**289–320.
14. **Bremner, I.** 1974. Heavy metal toxicities. *Q. Rev. Biophys.* **7:**75–124.
15. **Brown, N. L., K. R. Brocklehurst, B. Lawley, and J. L. Hobman.** 1998. Metal regulation of gene expression in bacterial systems. *NATO ASI Ser. H* **103:**159–173.
16. **Brown, N. L., B. T. O. Lee, and S. Silver.** 1994. Bacterial transport of and resistance to copper. *Metal Ions Biol. Syst.* **30:**405–434.
17. **Bruce, K. D., W. D. Hiorns, J. L. Hobman, A. M. Osborn, P. Strike, and D. A. Ritchie.** 1992. Amplification of DNA from native populations of soil bacteria by using the polymerase chain reaction. *Appl. Environ. Microbiol.* **58:**3413–3416.
18. **Bruce, K .D., A. M. Osborn, A. J. Pearson, P. Strike, and D. A. Ritchie.** 1995. Genetic diversity within *mer* genes directly amplified from communities of noncultivated soil and sediment bacteria. *Mol. Ecol.* **4:**605–612.
19. **Bull, P. C., and D. W. Cox.** 1994. Wilson disease and Menkes disease: new handles on heavy-metal transport. *Trends Genet.* **10:**246–252.
20. **Chang, J.-S., and J. Hong.** 1994. Biosorption of mercury by the inactivated cells of *Pseudomonas aeruginosa* PU21 (Rip64). *Biotechnol. Bioeng.* **44:**999–1006.
21. **Chang, J.-S., and W.-S. Law.** 1998. Development of microbial mercury detoxification processes using a mercury hyperresistant strain of *Pseudomonas aeruginosa* PU21. *Biotechnol. Bioeng.* **57:** 462–470.
22. **Chen, S., and D. B. Wilson.** 1997. Genetic engineering of bacteria and their potential for Hg^{2+} bioremediation. *Biodegradation* **8:**97–103.
23. **Chen, S., and D. B. Wilson.** 1997. Construction and characterization of *Escherichia coli* genetically engineered for bioremediation of Hg^{2+}-contaminated environments. *Appl. Environ. Microbiol.* **63:** 2442–2445.
24. **Clark, D. L., A. A. Weiss, and S. Silver.** 1977. Mercury and organomercurial resistances determined by plasmids in *Pseudomonas. J. Bacteriol.* **132:**186–196.
25. **Compeau, G. C., and R. Bartha.** 1984. Methylation and demethylation of mercury under controlled redox, pH, and salinity conditions. *Appl. Environ. Microbiol.* **48:**1203–1207.
26. **Compeau, G. C., and R. Bartha.** 1985. Sulfate reducing bacteria: principal methylators of mercury in anoxic estuarine sediment. *Appl. Environ. Microbiol.* **50:**498–502.
27. **Compeau, G. C., and R. Bartha.** 1987. Effect of salinity on mercury-methylating activity of sulfate-reducing bacteria in estuarine sediments. *Appl. Environ. Microbiol.* **53:**261–265.

28. **Condee, C. W., and A. O. Summers.** 1992. A *mer-lux* transcriptional fusion for real-time examination of in vivo gene-expression kinetics and promoter response to altered superhelicity. *J. Bacteriol.* **174:**8094–8101
29. **Cooksey, D. A.** 1993. Copper uptake and resistance in bacteria. *Mol. Microbiol.* **7:**1–5.
30. **Cummings, R. T., and C. T. Walsh.** 1992. Interaction of Tn*501* mercuric reductase and dihydroflavin adenine anion with metal ions—implications for the mechanism of mercuric reductase mediated Hg(II) reduction. *Biochemistry* **31:**1020–1030.
31. **De Flora, S., C. Benicelli, and M. Bagnasco.** 1994. Genotoxicity of mercury compounds. A review. *Mutat. Res.* **317:**57–79.
32. **Diels, L., Q. H. Dong, D. Van der Lelie, W. Baeyens, and M. Mergeay.** 1995. The *czc* operon of *Alcaligenes eutrophus* CH34—from resistance mechanism to the removal of heavy metals. *J. Ind. Microbiol.* **14:**142–153.
33. **Earles, M. P.** 1964. A case of mass mercury poisoning with mercury vapour on board *H.M.S. Triumph* at Cadiz, 1810. *Med. Hist.* **8:**281–286.
34. **Eccles, H.** 1995. Removal of heavy metals from effluent streams—why select a biological process? *Int. Biodeterior. Biodegrad.* **35:**5–16.
35. **Eccles, H., G. W. Garnham, C. R. Lowe, and N. C. Bruce.** March 1996. Biosensors for detecting metal ions capable of being reduced by reductase enzymes. U.S. patent 5500351.
36. **Fitzgerald, W. F., and R. P. Mason.** 1997. Biogeochemical cycling of mercury in the marine environment. *Metal Ions Biol. Syst.* **34:**53–111.
37. **Foster, T. J.** 1987. The genetics and biochemistry of mercury resistance. *Crit. Rev. Microbiol* **15:** 117–140.
38. **Fox, B., and C. T. Walsh.** 1982. Mercuric reductase. Purification and characterization of a transposon-encoded flavoprotein containing an oxidation-reduction-active disulfide. *J. Biol. Chem.* **257:** 2498–2503.
39. **Fox, B. S., and C. T. Walsh.** 1983. Mercuric reductase—homology to glutathione reductase and lipoamide dehydrogenase-iodoacetamide alkylation and sequence of the active-site peptide. *Biochemistry* **22:**4082–4088.
40. **Friberg, L. (ed.)** 1991. *Environmental Health Criteria no. 118. Inorganic Mercury.* World Health Organization, Geneva, Switzerland.
41. **Furukawa, K., T. Suzuki, and S. Tonomura.** 1969. Decomposition of organic mercurial compounds by mercury resistant bacteria. *Agric. Biol. Chem.* **33:** 128–130.
42. **Gadd, G. M.** 1993. Microbial formation and transformation of organometallic and organometalloid compounds. *FEMS Microbiol. Rev.* **11:**297–316.
43. **Ganser, A. L., and D. A. Kirschner.** 1985. The interaction of mercurials with myelin—comparison of *in vitro* and *in vivo* effects. *Neurotoxicology* **6:**63–77.
44. **Gilbert, M. P., and A. O. Summers.** 1988. The distribution and divergence of DNA sequences related to the Tn*21* and Tn*501 mer* operons. *Plasmid* **20:**127–136.
45. **Goldwater, L. J.** 1972. *Mercury: a History of Quicksilver.* York Press, Baltimore, Md.
46. **Gruenwendel, D. W., and N. Davidson.** 1966. Complexing and denaturation of DNA by methylmercuric hydroxide. I. Spectrophotometric studies. *J. Mol. Biol.,* **21:**129–144.
47. **Gutknecht, J.** 1981. Inorganic mercury (Hg^{2+}) transport through lipid bilayer membranes. *J. Membr. Biol.* **61:**61–66.
48. **Hall, B. D., R. A. Bodaly, R. J. P. Fudge, J. W. M. Rudd, and D. M. Rosenburg.** 1997. Food as the dominant pathway of methylmercury uptake by fish. *Water Air Soil Pollut.* **100:**3–24.
49. **Harada, M.** 1995. Minamata disease- methylmercury poisoning in Japan caused by environmental pollution. *Crit. Rev. Toxicol.* **25:**1–24.
50. **Hobman, J., G. Kholodii, V. Nikiforov, D. A. Ritchie, P. Strike, and O. Yurieva.** 1994. The nucleotide sequence of the *mer* operon of pMER327/419 and transposon ends of pMEr327/419, 330 and 05. *Gene* **146:**73–78.
51. **Hobman, J. L., and N. L. Brown.** 1997. Bacterial mercury-resistance genes. *Metal Ions Biol. Syst.* **34:**527–568.
52. **Horn, J. M., M. Brunke, W. D. Deckwer, and K. N. Timmis.** 1994. *Pseudomonas putida* strains which constitutively overexpress mercury resistance for biodetoxification of organomercurial pollutants. *Appl. Environ. Microbiol.* **60:**357–362.

53. **Huckle, J. W., A. P. Morby, J. S. Turner, and N. J. Robinson.** 1993. Isolation of a prokaryotic metallothionein locus and analysis of transcriptional control by trace metal ions. *Mol. Microbiol.* **7:** 177–187.
54. **Inoue, C., K. Sugawara, and T. Kusano.** 1990. *Thiobacillus ferrooxidans mer* operon: sequence analysis of the promoter and adjacent genes. *Gene* **96:**115–120.
55. **Jeffrey, W. H., S. Nazaret, and R. Vonhaven**. 1994. Improved method for recovery of messenger RNA from aquatic samples and its application to detection of *mer* expression. *Appl. Environ. Microbiol.* **60:**1814–1821.
56. **Jeffrey, W. H., S. Nazaret, and T. Barkay.** 1996. Detection of the *merA* gene and its expression in the environment. *Microb. Ecol.* **32:**293–303.
57. **Kim, K.-H., P. J. Hanson, M. O. Barnett, and S. E. Lindberg.** 1997. Biogeochemistry of mercury in the air-soil-plant system. *Metal Ions Biol. Syst.* **34:**185–212.
58. **Kiyono, M., T. Omura, H. Fujimori, and H. Pan-Hou.** 1995. Lack of involvement of *merT* and *merP* in methylmercury transport in mercury resistant *Pseudomonas* K-62. *FEMS Microbiol. Lett.* **128:**301–306.
59. **Klein, J., J. Altenbuchner, and R. Mattes.** 1997. Genetically modified *Escherichia coli* for colorimetric detection of inorganic and organic Hg compounds, p. 133–151. *In* F.W. Scheller, F. Schubert, and J. Fedrowitz (ed.), *Frontiers in Biosensorics. 1. Fundamental Aspects.* Birkhäuser Verlag, Basel, Switzerland.
60. **Komura, I., and K. Izaki.** 1971. Mechanism of mercuric chloride resistance in microorganisms. I. Vaporization of a mercury compound from mercuric chloride by multiple drug resistance strain of *Escherichia coli. J. Biochem.* **70:**885–893.
61. **Krönig, B., and T. Paul.** 1897. Die chemischen Grundlagen der Lehre von der Giftwirkung und Desinfection, p. 163–176. *In* T. D. Brock (ed.), *Milestones in Microbiology*, 1961. Prentice-Hall, Inc., Englewood Cliffs, N.J.
62. **Leach, S. J.** 1960. The reaction of thiol and disulphide groups with mercuric chloride and mercuric iodide. *J. Aust. Chem. Soc.* **13:**520.
63. **Lenihan, J.** 1988. *The Crumbs of Creation*, p. 76. Adam Hilger, Bristol, United Kingdom.
64. **Lindquist, O., Å. Jernelöv, K. Johansson, and H. Rohde.** 1984. *Mercury in the Swedish Environment: Global and Local Sources.* Swedish Environmental Protection Board report no. 1816, p. 105. Swedish Environmental Protection Board, Stockholm, Sweden.
65. **Lovley, D. R., and J. D. Coates.** 1997. Bioremediation of metal contamination. *Curr. Opin. Biotechnol.* **8:**285–289.
66. **Lund, P. A., and, N. L. Brown. 1987.** Role of *merT* and *merP* gene products of transposon Tn*501* in the induction and expression of resistance to mercuric ions. *Gene* **52:**207–214.
67. **Magos, L.** 1997. Physiology and toxicology of mercury. *Metal Ions Biol. Syst.* **34:**321–370.
68. **Mason, R. P., J. R. Reinfelder, and F. M. M. Morel.** 1995. Bioaccumulation of mercury and methylmercury. *Water Air Soil Pollut.* **80:**915–921.
69. **Meili, M.** 1997. Mercury in lakes and rivers. *Metal Ions Biol. Syst.* **34:**21–52.
70. **Miller, S. M., D. P. Ballou, V. Massey, C. H. Williams, and C. T. Walsh.** 1986. 2-electron reduced mercuric reductase binds Hg(II) to the active-site dithiol but does not catalyze Hg(II) reduction. *J. Biol. Chem.* **261:**8081–8084.
71. **Mills, A.** 1990. Mercury from crematorium chimneys. *Nature.* **346:**615.
72. **Misra, T. K.** 1992. Bacterial resistance to inorganic mercury salts and organomercurials. *Plasmid* **27:**4–16.
73. **Misra, T. K., N. L. Brown, D. C. Fritzinger, R. D. Pridmore, W. M. Barnes, L. Haberstroh, and S. Silver.** 1984. Mercuric ion resistance operons of plasmid R100 and transposon Tn*501*—the beginning of the operon including the regulatory region and the first two structural genes. *Proc. Natl. Acad. Sci. USA* **81:**5975–5979.
74. **Morby, A. P., J. L. Hobman, and N. L. Brown.** 1995. The role of cysteine residues in the transport of mercuric ions by the Tn*501* MerT and MerP mercury-resistance proteins. *Mol. Microbiol.* **17:** 1153–1162.
75. **Mukhopadhyay, D., H. Yu, G. Nucifora, and T. K. Misra.** 1991. Purification and functional characterization of MerD: a coregulator of the mercury resistance operon in Gram-negative bacteria. *J. Biol. Chem.* **266:**18538–18542.

76. **Nakahara, H., S. Silver, T. Miki, and R. H. Rownd.** 1979. Hypersensitivity to Hg^{2+} and hyperbinding activity associated with cloned fragments of the mercurial resistance operon of plasmid NR1. *J. Bacteriol.* **140:**161–166.
77. **Nakamura, K., and S. Silver.** 1994. Molecular analysis of mercury-resistant *Bacillus* isolates from sediment of Minamata Bay, Japan. *Appl. Environ. Microbiol.* **60:**4596–4599.
78. **Nazaret, S., W. H. Jeffrey, E. Saouter, R. Vonhaven, and T. Barkay.** 1994. *merA* gene expression in aquatic environments measured by messenger RNA production and Hg(II) volatilization. *Appl. Environ. Microbiol.* **60:**4059–4065.
79. **Niebor, E., and D. H. S. Richardson.** 1980. The replacement of the nondescript term "heavy metals" by a biologically and chemically significant classification of metal ions. *Environ. Pollut. Ser. B* **1:**3–26.
80. **Nies, D. H.** 1995. The cobalt, zinc, and cadmium efflux system *czcABC* from *Alcaligenes eutrophus* functions as a cation-proton antiporter in *Escherichia coli. J. Bacteriol.* **177:**2707–2712.
81. **Nies, D. H., and S. Silver.** 1995. Ion efflux systems involved in bacterial metal resistances. *J. Ind. Microbiol.* **14:**186–199.
82. **Nriagu, J. O.** 1989. A global assessment of natural sources of atmospheric trace metals. *Nature* **338:**47–49.
83. **Nriagu, J. O., and J. M. Pacyna.** 1988. Quantitative assessment of worldwide contamination of air, water and soils by trace metals. *Nature* **333:**134–139.
84. **Nucifora, G., L. Chu, S. Silver, and T. K. Misra.** 1989. Mercury operon regulation by the *merR* gene of the organomercurial resistance system of plasmid pDU1358. *J. Bacteriol.* **171:**4241–4247.
85. **Nucifora, G., S. Silver, and T. K. Misra.** 1990. Down regulation of the mercury resistance operon by the most promoter-distal gene *merD. Mol. Gen. Genet.* **220:**69–72.
86. **Ogunseitan, O.** 1998. Protein method for investigating mercuric reductase gene expression in aquatic environments. *Appl. Environ. Microbiol.* **64:**695–702.
87. **O'Halloran, T. V.** 1993. Transition metals in control of gene-expression. *Science* **261:**715–725.
88. **Olson, B. H., J. N. Lester, S. M. Cayless, and S. Ford.** 1988. Distribution of mercury resistance determinants in bacterial communities of river sediments. *Water Res.* **23:**1209–1217.
89. **Oremland, R. S., C. W. Culbertson, and M. R. Winfrey.** 1991. Methylmercury decomposition in sediments and bacterial cultures: involvement of methanogens and sulfate reducers in oxidative demethylation. *Appl. Environ. Microbiol.* **57:**130–137.
90. **Oremland, R. S., L. G. Miller, P. Dowdle, T. Connell, and T. Barkay.** 1995. Methylmercury oxidative degradation potentials in contaminated and pristine sediments of the Carson River, Nevada. *Appl. Environ. Microbiol.* **61:**2745–2753.
91. **Osborn, A. M., K. D. Bruce, P. Strike, and D. A. Ritchie.** 1995. Sequence conservation between regulatory mercury resistance genes in bacteria from mercury polluted and pristine environments. *Syst. Appl. Microbiol.* **18:**1–6.
92. **Osborn, A. M., K. D. Bruce, P. Strike, and D. A. Ritchie.** 1997. Distribution, diversity and evolution of the bacterial mercury resistance (*mer*) operon. *FEMS Microbiol. Rev.* **19:**239–262.
93. **Pak, K.-R., and R. Bartha.** 1998. Mercury methylation by interspecies hydrogen and acetate transfer between sulfidogens and methanogens. *Appl. Environ. Microbiol.* **64:**1987–1990.
94. **Pan-Hou, H. S. K., M. Hosono, and N. Imura.** 1980. Plasmid controlled mercury biotransformation by *Clostridium cochlearium* T-2. *Appl. Environ. Microbiol.* **40:**1007–1011.
95. **Pan-Hou, H. S. K., M. Hosono, and N. Imura.** 1981. Role of hydrogen sulphide in mercury resistance determined by plasmid of *Clostridium cochlearium* T-2. *Arch. Microbiol.* **129:**49–52.
96. **Pan-Hou, H. S. K., M. Nishimoto, and N. Imura.** 1981. Possible role of membrane proteins in mercury resistance of *Enterobacter aerogenes. Arch. Microbiol.* **130:**93–95.
97. **Philippidis, G. P., L. H. Malmberg, W. S. Hu, and J. L. Schottel.** 1991. Effect of gene amplification on mercuric ion reduction activity of *Escherichia coli. Appl. Environ. Microbiol.* **57:**3558–3564.
98. **Philippidis, G. P., J. L. Schottel, and W. S. Hu.** 1990. Kinetics of mercuric reduction in intact and permeabilized *Escherichia coli* cells. *Enzyme Microb. Technol.* **12:**854–859.
99. **Rasmussen, L. D., R. R. Turner, and T. Barkay.** 1997. Cell-density-dependent sensitivity of a *mer-lux* bioassay. *Appl. Environ. Microbiol.* **63:**3291–3293.

100. **Rensing, C., U. Kues, U. Stahl, D. H. Nies, and B. Friedrich.** 1992. Expression of bacterial mercuric ion reductase in *Saccharomyces cerevisiae*. *J. Bacteriol.* **174:**1288–1292.
101. **Rinderle, S. J., J. E. Booth, and J. W. Williams.** 1983. Mercuric reductase from R-plasmid NR1—characterization and mechanistic study. *Biochemistry* **22:**869–876.
102. **Ritter, J. A., and J. P. Bibler.** 1992. Removal of mercury from waste-water: large scale performance of an ion exchange process. *Water Sci. Technol.* **25:**165–172.
103. **Robinson, J. B., and O. H. Tuovinen.** 1984. Mechanisms of microbial resistance and detoxification of mercury and organomercury compounds: physiological, biochemical and genetic analysis. *Microbiol. Rev.* **48:**95–124.
104. **Robinson, N. J., A. Gupta, A. P. Fordham-Skelton, R. R. D. Croy, B. A. Whitton, and J. W. Huckle.** 1990. Prokaryotic metallothionein gene characterization and expression: chromosome crawling by ligation-mediated PCR. *Proc. Ro. Soc. London B Ser.* **242:**241–247.
105. **Rochelle, P. A., M. K. Wetherbee, and B. H. Olson.** 1991. Distribution of DNA sequences encoding narrow-spectrum and broad-spectrum mercury resistance. *Appl. Environ. Microbiol.* **57:** 1581–1589.
106. **Rouch, D. A., B. T. O. Lee, and A. P. Morby.** 1995. Understanding cellular responses to toxic agents: a mechanism-choice in bacterial metal resistances. *J. Ind. Microbiol.* **14:**132–141.
107. **Rouch, D. A., J. Parkhill, and N. L. Brown.** 1995. Induction of bacterial mercury-responsive and copper-responsive promoters–functional differences between inducible systems and implications for their use in gene-fusions for *in-vivo* metal biosensors. *J. Ind. Microbiol.* **14:**349–353.
108. **Rugh, C. L., H. D. Wilde, N. M. Stack, D. M. Thompson, A. O. Summers, and R. B. Meagher.** 1996. Mercuric ion reduction and resistance in transgenic *Arabidopsis thaliana* plants expressing a modified bacterial *merA* gene. *Proc. Natl. Acad. Sci. USA* **93:**3182–3187.
109. **Sahlman, L., W. Wong, and J. Powlowski.** 1997. A mercuric ion uptake role for the integral inner membrane protein, MerC, involved in bacterial mercuric ion resistance. *J. Biol. Chem.* **272:**29518–29526.
110. **Salt, D. E., M. Blaylock, N. P. B. A. Kumar, V. Dushenkov, B. D. Ensley, L. Chet, and I. Raskin.** 1995. Phytoremediation: a novel strategy for the removal of toxic metals from the environment using plants. *Bio/Technology* **13:**468–474.
111. **Sandstrom, A., and S. Lindskog.** 1987. Activation of mercuric reductase by the substrate NADPH. *Eur. J. Biochem.* **173:**411–415.
112. **Saouter, E., R. Turner, and T. Barkay.** 1994. Microbial reduction of ionic mercury for the removal of mercury from contaminated environments. *Ann. N.Y. Acad. Sci.* **721:**423–427.
113. **Schiering, N., W. Kabsch, M. J. Moore, M. D. Distefano, C. T. Walsh, and E. F. Pai.** 1991. Structure of the detoxification catalyst mercuric ion reductase from *Bacillus* sp. strain RC607. *Nature* **352:**168–172.
114. **Schultz, P. G., K. G. Au, and C. T. Walsh.** 1985. Directed mutagenesis of the redox active disulfide in the flavoenzyme mercuric ion reductase. *Biochemistry* **24:**6840–6848.
115. **Selifonova, O., R. Burlage, and T. Barkay.** 1993. Bioluminescent sensors for detection of bioavailable Hg(II) in the environment. *Appl. Environ. Microbiol.* **59:**3083–3090.
116. **Shi, J., W. P. Lindsay, J. W. Huckle, A. P. Morby, and N. J. Robinson.** 1992. Cyanobacterial metallothionein gene expressed in *Escherichia coli*. Metal-binding properties of the expressed protein. *FEBS Lett.* **303:**159–163.
117. **Silver, S.** 1996. Bacterial metal resistance—a review. *Gene* **179:**9–19.
118. **Silver, S., G. Nucifora, and L. T. Phung.** 1993. Human Menkes X-chromosome disease and the staphylococcal cadmium-resistance ATPase—a remarkable similarity in protein sequences. *Mol. Microbiol.* **10:**7–12.
119. **Silver, S., and L. T. Phung.** 1996. Bacterial heavy metal resistance: new surprises. *Annu. Rev. Microbiol.* **50:**753–789.
120. **Smith, T., K. Pitts, J. A. McGarvey, and A. O. Summers.** 1998. Bacterial oxidation of mercury metal vapor, Hg(0). *Appl. Environ. Microbiol.* **64:**1328–1332.
121. **Solioz, M., and C. Vulpe.** 1996. CPX-type ATPases—a class of P-type ATPases that pump heavy metals. *Trends Biochem. Sci.* **21:**237–241.
122. **Summers, A. O.** 1986. Organisation, expression and evolution of genes for mercury resistance. *Annu. Rev. Microbiol.* **40:**607–634.

123. **Summers, A. O.** 1992. Untwist and shout: a heavy metal-responsive transcriptional regulator. *J. Bacteriol.* **174:**3097–3101.
124. **Summers, A. O., and T. Barkay.** 1989. Metal resistance genes in the environment, p. 287–309. *In* S. B. Levy and R. V. Miller (ed.), *Gene Transfer in the Environment.* McGraw-Hill Publishing Co., New York, N.Y.
125. **Tescione, L., and G. Belfort.** 1993. Construction and evaluation of a metal-ion biosensor. *Biotechnol. Bioeng.* **42:**945–952.
126. **Trevors, J. T.** 1986. Mercury methylation by bacteria. *J. Basic Microbiol.* **26:**499–504.
127. **Tsai, Y.-L., M. J. Park, and B. H. Olson.** 1991. Rapid method for direct extraction of mRNA from seeded soils. *Appl. Environ. Microbiol.* **57:**765–768.
128. **Uno, Y., M. Kiyono, T. Tezuka, and H. Pan-Hou.** 1997. Phenylmercury transport mediated by *merT-merP* genes of *Pseudomonas* K-62 plasmid pMR26. *Biol. Pharm. Bull.* **20:**107–109.
129. **Virta, M., J. Lampinen, and M. Karp.** 1995. A luminescence-based mercury biosensor. *Anal. Chem.* **67:**667–669.
130. **Volesky, B., and Z. R. Holan.** 1995. Biosorption of heavy metals. *Biotechnol. Prog.* **11:**235–250.
131. **Volotovsky, V., Y. J. Nam, and N. Kim.** 1997. Urease-based biosensor for mercuric ion determination. *Sensors Actuators Ser. B.* **42:**233–237.
132. **Watanabe, C., and H. Sato.** 1996. Evolution of our understanding of methylmercury as a health threat. *Environ. Health Perspect.* **104**(Suppl. 2)**:**367–379.
133. **Weast, R. C. (ed.).** 1984. *CRC Handbook of Chemistry and Physics*, 65th ed., p. B-24. CRC Press, Inc., Boca Raton, Fla.
134. **Weiss, A. A., J. L. Schottel, D. L., Clark, R. G. Beller, and S. Silver.** 1978. Mercury and organomercurial resistance with enteric, staphylococcal and pseudomonad plasmids, p. 121–124. *In* D. Schlessiger (ed.), *Microbiology–1978.* American Society for Microbiology, Washington, D.C.
135. **Williams, J. W., and S. Silver.** 1984. Bacterial resistance and detoxification of heavy-metals. *Enzyme Microb. Technol.* **6:**530–537.
136. **Windholz, M. (ed.).** 1983. *The Merck Index*, 10th ed. Merck & Co., Rahway, N.J.
137. **Xu, C., T. Q. Zhou, M. Kuroda, and B. P. Rosen.** 1998. Metalloid resistance mechanisms in prokaryotes. *J. Biochem.* **123:**16–23.

Environmental Microbe-Metal Interactions
Edited by Derek R. Lovley

Chapter 9

Dissimilatory Reduction of Selenate and Arsenate in Nature

Ronald S. Oremland and John Stolz

Selenium and arsenic are two elements which are notable in that they are toxic and teratogenic to many metazoans. In oxic environments they occur primarily as the soluble oxyanions selenate [Se(VI)], and arsenate [As(V)]. The presence of both substances in natural waters, sediments, and soils arises from geologic and hydrologic processes which include the weathering of exposed parent rocks and the leaching of subsurface minerals through the action of geothermal fluids. In addition, these elements constitute environmental pollutants which stem from a diversity of sources, including drainage from mines and tailing wastes, the combustion of fossil fuels, the irrigated cultivation of seleniferous soils, and a number of industrial and agricultural applications such as pesticides, medicinals, catalysts, and semiconductors (5, 21). In this chapter, we discuss the biogeochemical reduction of Se(VI) and As(V) when they enter anoxic environments and are used as electron acceptors for the oxidation of organic matter. These reductions are of a dissimilative nature and support the anaerobic growth of selected bacteria which conserve energy from this process. The bacteria which carry out these dissimilatory reductions have only recently been described, and we summarize what is known about their taxonomy, physiology, and biochemistry. In addition, these reactions have important environmental implications. Reduction to the solid, relatively unreactive Se(0) represents a mechanism for the removal of toxic Se(VI) and Se(IV) from natural waters. Just the opposite occurs for arsenic, where its reduction to As(III) represents the formation of a more toxic and mobile species. The environmental ramifications of these issues are also discussed. The reader is referred to the reviews by Oremland (51), Cullen and Reimer (8), Losi and Frankenberger (32), Newman et al. (47), and Stolz and Oremland (69a) for further details on the biogeochemistry of selenium and arsenic.

Ronald S. Oremland • U.S. Geological Survey, ms 480, 345 Middlefield Road, Menlo Park, CA 94025. ***John Stolz*** • Department of Biological Sciences, Duquesne University, Pittsburgh, PA 15282.

BIOGEOCHEMISTRY OF Se(VI) AND As(V) IN ANOXIC ENVIRONMENTS

Within anoxic environments, a transition from oxidized to reduced chemical speciation of these two elements is commonly observed, with Se(VI) and selenite [Se(IV)] being replaced by elemental selenium [Se(0)] and with As(V) being replaced by arsenite [As(III)]. This can be illustrated for selenium by examining a profile of the interstitial porewaters of a core taken from an evaporative agricultural brine in the San Joaquin Valley, Calif. (Fig. 1). Se(VI) and Se(IV) disappear from solution near the surface of the core by a process which causes their reduction to solid Se(0) in the presence of high levels of sulfate (54, 55). A number of studies have demonstrated that Se(0) is the dominant species of selenium in anoxic sediments (71, 74, 77, 78). Similarly, the behavior of arsenic species can be illustrated by the conditions in Mono Lake, Calif., a water body with unusually high levels of dissolved arsenic derived from the combined effects of hydrothermal inputs and high evaporation (Fig. 2). A near stoichiometric balance was achieved between As(V) removed and As(III) produced, with a transition through the oxycline of this stratified lake (40). Comparable results have been reported for other stratified lakes and anoxic soils (42, 56, 61). However, the biogeochemistry of arsenic in sediments is complicated by the strong binding of As(V) to minerals like FeOOH. Hence, bacterial reduction of Fe(III) to Fe(II) should release any bound As(V) and make it available for further chemical or biological reduction (33). The question then arises whether these changes in speciation were caused by chemical reactions or direct biological reductions.

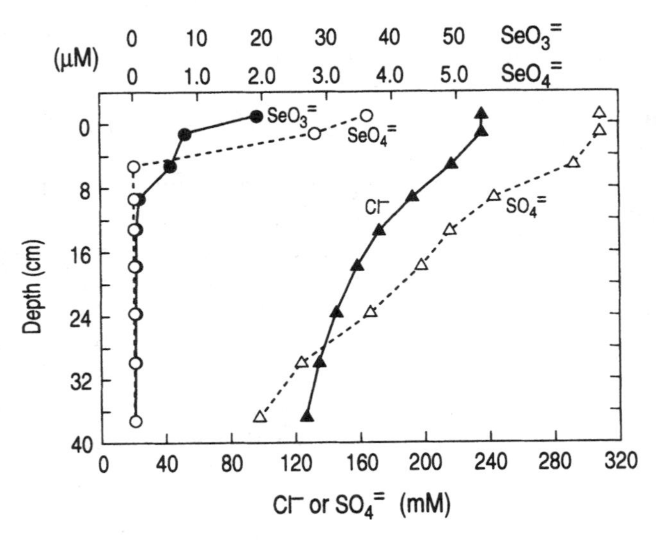

Figure 1. Dissolved selenate, selenite, sulfate, and chloride in the porewaters of sediments from an agricultural wastewater evaporation pond located in the San Joaquin Valley, Calif. Reprinted from reference 55 with permission of the American Society for Microbiology.

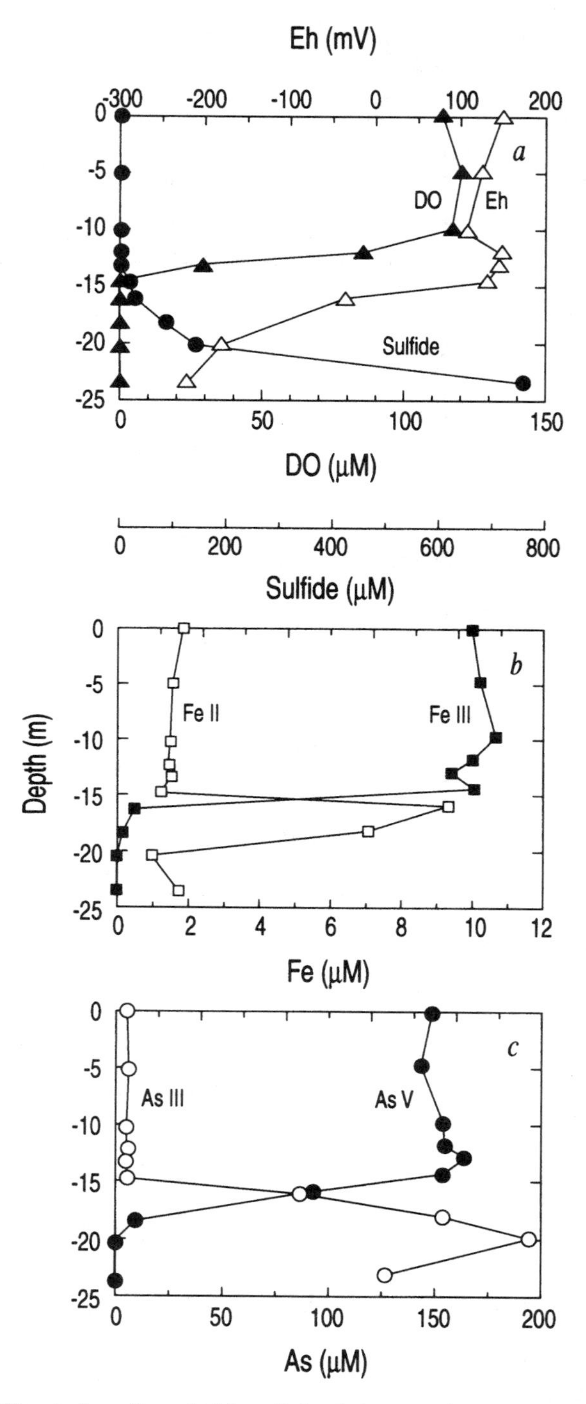

Figure 2. Chemical gradients in Mono Lake during a period of meromixis in the 1980s. Reprinted from reference 40, with permission of the publisher.

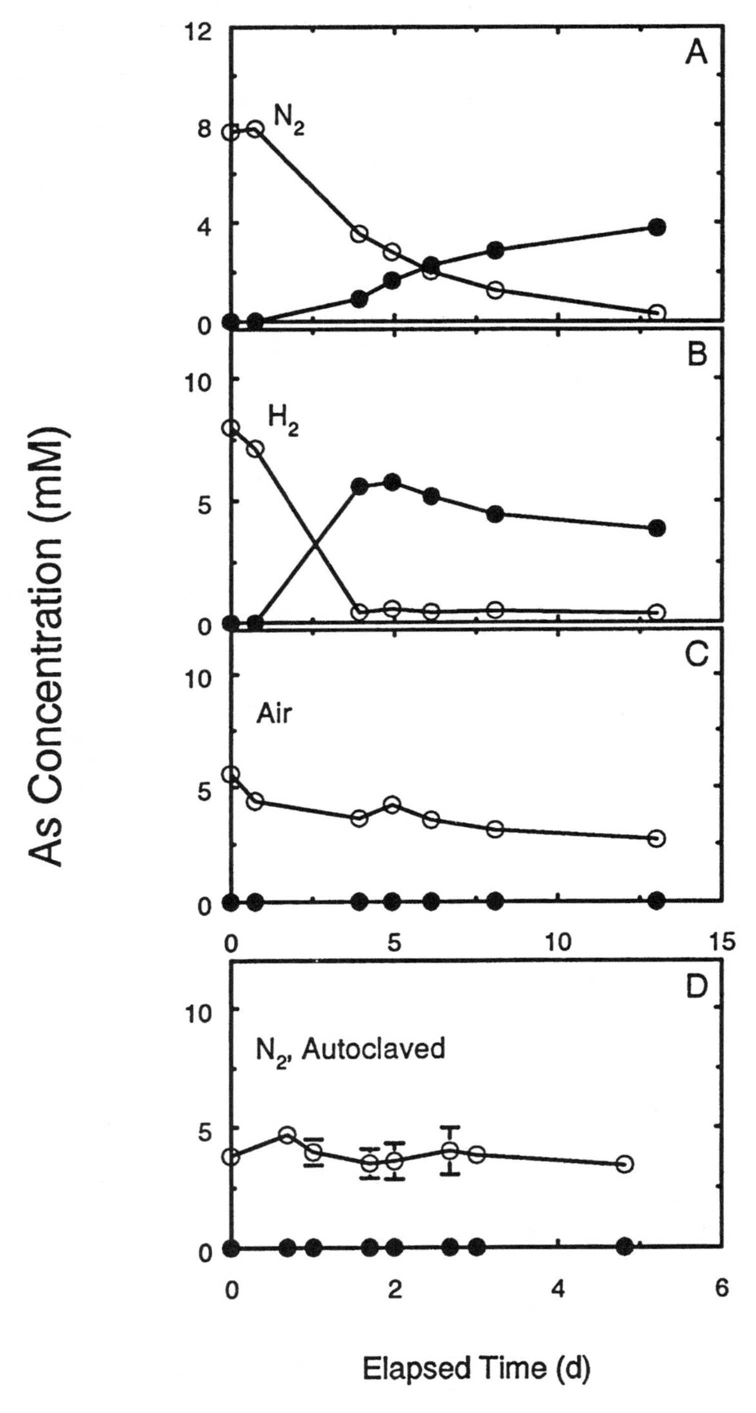

Figure 3. Arsenate reduction in estuarine sediments incubated under an atmosphere of nitrogen (A), hydrogen (B), air (C), or nitrogen with autoclaved sediments (D). Symbols: ○, As(V); ●, As(III). Reprinted from reference 12 with permission of the American Society for Microbiology.

For As(V) reduction to As(III), the role of anaerobic bacteria can be demonstrated in experiments with live versus heat-killed sediments and by the fact that reduction is inhibited by an air atmosphere whereas it is stimulated by the electron donor H_2 (Fig. 3). Similarly, dissolution of As(III) in sediments from iron-arsenate oxides is enhanced by bacteria (1). Other evidence includes As(V) reduction inhibition by chloramphenicol, dinitrophenol, cyanide, and tungstate (12), as well as by formalin (2). Tungstate is an antagonist for molybdenum-containing enzymes. However, naturally occurring reducing agents such as sulfide can chemically reduce As(V) to As(III), resulting in the formation of minerals such as orpiment (As_2S_3). The occurrence of orpiment in sulfidic hot springs formations was attributed to abiotic mechanisms (13).

Newman et al. (48), however, isolated the sulfate-reducing bacterium *Desulfotomaculum auripigmentum* from an arsenic-contaminated tanning waste site. The bacterium formed orpiment in culture by mediating biological reduction of As(V) followed by the reduction of sulfate during growth (49). Therefore, the biological generation of orpiment and other reduced minerals of As(III) is also possible, and their occurrence in nature cannot easily be attributed to either biotic or abiotic mechanisms. Dowdle et al. (12), however, noted that high concentrations of As(V) or its reduction product As(III) (~10 mM) inhibited sulfate reduction in sediment slurries. Rittle et al. (59) observed the removal of lower arsenic levels (~1 to 2 mM) while sulfate reduction was occurring in lake sediments by apparent formation of an arsenic-iron sulfide solid phase.

For Se(VI) reduction, the line of evidence supporting a biological mechanism is even stronger than it is for As(V). This is because Se(VI) does not readily undergo chemical reduction under physiological conditions of pH and temperature. Thus, chemical reduction does not occur in heat-killed sediments from an agricultural drain (Fig. 4) or even when heat-killed estuarine sediments are incubated under extremely reducing conditions by imposing an H_2 atmosphere and adding sterile HS^- ions to the mileu (55). The elimination of Se(VI) reduction by autoclaving was not limited to these two observations. In a broad survey of selenate reductase activity, sediments from chemically disparate environments were assayed for their ability to reduce 25 μM Se(VI) (67). The recovery of ^{75}Se counts in the pellet and CS_2-soluble fractions proved that the product of Se(VI) reduction was Se(0) (Fig. 5). A total of 11 sediment types were examined, and the gradients ranged from freshwater to extremely hypersaline (e.g., saturation) and from neutral pH to highly alkaline conditions (e.g., pH 9.8). In 10 of the 11 cases, no activity occurred in autoclaved sediments whereas live materials demonstrated clear reduction of ^{75}Se(VI) to ^{75}Se(0). The exception was for a very saline and alkaline brine, in which chemical precipitation of $Na_2{}^{75}SeO_4$ in heat-killed controls could not be distinguished from formation of ^{75}Se(0) in the live samples. It is commonly argued by skeptics that autoclaving by itself does not constitute sufficient evidence for a biotic mechanism, because the heat may have destroyed any naturally occurring chemical catalyst in addition to the targeted microbes. However, it is unlikely that such a hypothetical catalyst would have been so broadly distributed and heat sensitive under the range of conditions examined in the above study.

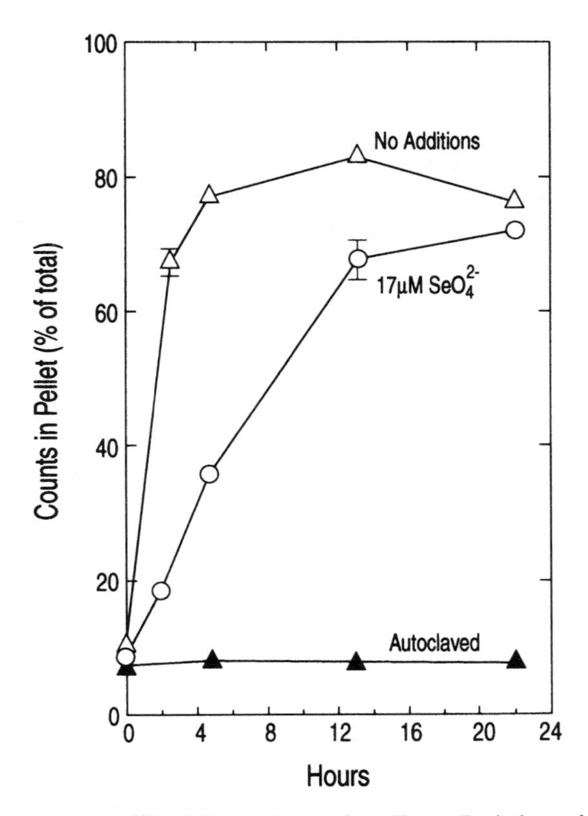

Figure 4. Reduction of ^{75}Se(VI) by sediments from Hunter Drain located in western Nevada. Counts in pellet indicate the formation of ^{75}Se(0). Sediments were incubated at ambient Se(VI) concentrations of ~0.5 μM with no additions or with addition of 17 μM unlabeled Se(VI) or were heat killed and incubated with ambient levels of Se(VI). Reprinted from reference 54 with permission of the publisher.

The finding that certain Fe(II,III) oxides ("green rust"), when added to sediments, can carry out the reduction of nitrite (20) led Myneni et al. (45) to present evidence for the reduction of Se(VI) to Se(0) by green rust in an artificially constituted laboratory system. Although Myneni et al. did not work with sediments, they extrapolated their laboratory results to soils and sediments by noting that their rate constants (k = ~0.01 h^{-1}) were similar to those reported by various groups working with soils and sediments (23, 65, 71). Myneni et al. (45) concluded that green rust must be an important mechanism for Se(VI) reduction in natural systems. Such logic is flawed, because agreement of k values constitutes only circumstantial evidence; it does not provide proof that the underlying mechanism responsible for reduction in sediments is abiotic as opposed to bacterial. We now review a further body of evidence, in addition to that obtained with the heat-killed controls cited above, which demonstrates that Se(VI) reduction in suboxic sediments is attributable to bacteria.

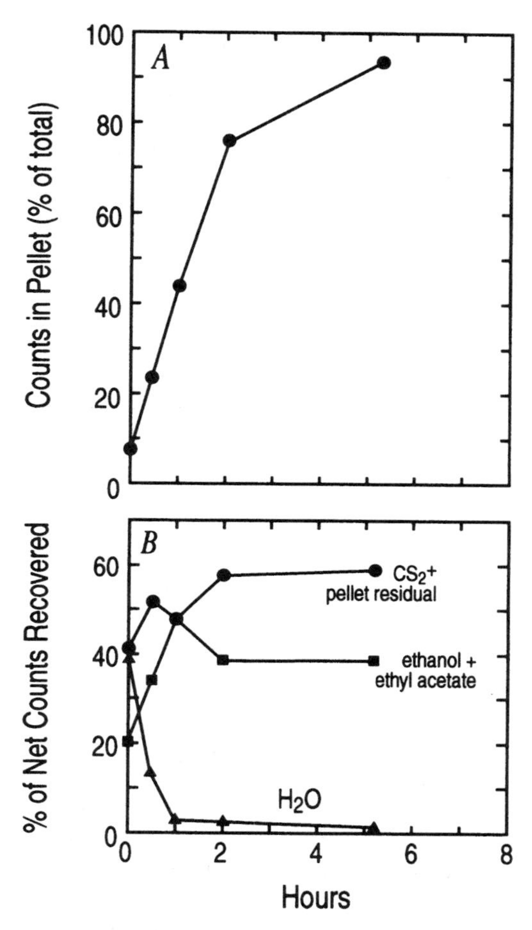

Figure 5. (A) Reduction of ^{75}Se(VI) to solid ^{75}Se(V) by sediments taken from Massie Slough in western Nevada. (B) Recovery of counts into various solvent fractions. Reprinted from reference 67 with permission of the American Society for Microbiology.

There are several additional experimental results which prove that Se(VI) reduction is a bacterial process. First, total inhibition of the reduction of 0.5 mM Se(VI) in estuarine sediment slurries was achieved by addition of 20 mM tungstate or chromate but not molybdate (55), and substantial inhibition of 0.5 mM Se(VI) reduction by tungstate was also achieved with freshwater sediments (67). Such results are consistent with the functioning of a molybdenum-containing enzyme carrying out the reduction of Se(VI) (9). Schröder et al. (60) demonstrated that the membrane-associated selenate reductase of *Thauera selenatis* contains a molybdenum cofactor. Second, the reduction of Se(VI) in sediments from a number of different sources displayed Michaelis-Menten kinetics characteristic of an enzymatic process (Fig. 6), and the apparent K_m values ranged from 8 to 720 μM (67).

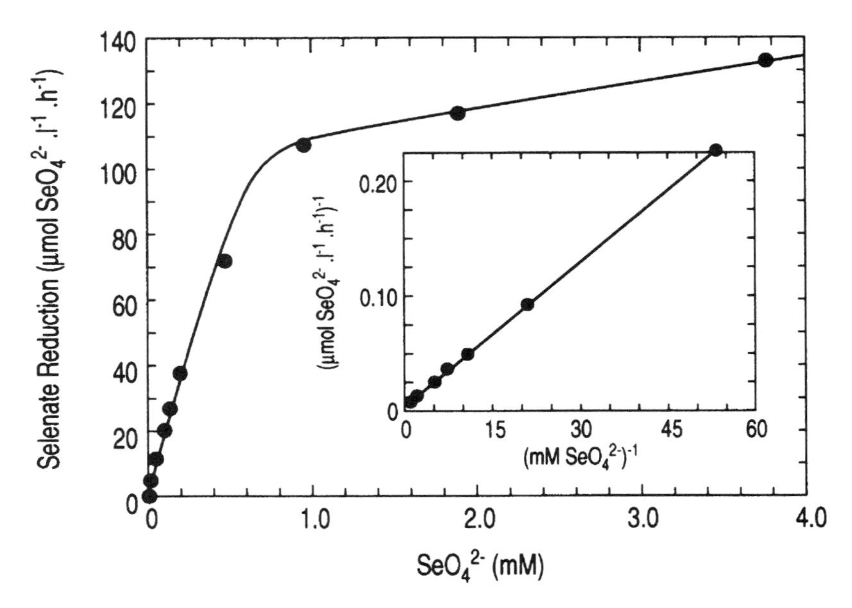

Figure 6. Michaelis-Menten kinetics displayed by selenate reduction in sediments from Massie Slough. Reprinted from reference 67 with permission of the American Society for Microbiology.

Third, Se(VI) reduction in estuarine sediment slurries is speeded by inclusion of common substrates of respiratory anaerobes (e.g., H_2, acetate, and lactate) in the milieu, and the oxidation of acetate to CO_2 demonstrates a concentration dependence on selenate but not on molybdate (55) (Fig. 7). A similar concentration-dependent oxidation of acetate can be achieved with As(V) as the electron acceptor (Fig. 8). The above results can be achieved only by a bacterial dissimilatory reduction of Se(VI), not one of which is driven by abiotic chemistry.

In fieldwork conducted with freshly recovered samples from the San Joaquin Valley (54) and from western Nevada (53), selenate reduction was assayed under in situ conditions at ambient Se(VI) concentrations by using the radiotracer ^{75}Se(VI). In these experiments, k values were often $\sim$1 h^{-1}, or about 100-fold faster than those reported by Myneni et al. (45) as well as in the highly selective literature they cited. Figure 9 shows depth profiles of Se(VI) reduction taken at three Se-impacted locations in western Nevada, in the vicinity of the Carson Sink. Rate constants for the surface samples at these sites ranged between $\sim$0.3 and 2.0 h^{-1} and declined with depth. We attribute this high activity to the fact that these were fresh sediments taken from organic-rich environs (e.g., marshes, lake bottoms, and drains), which typically harbor dense bacterial populations. The studies cited by Myneni et al. (45) were conducted with stored soils or sediments of apparent low organic content, which would explain their lower rate constants. In natural systems, rate constants for sediments can be quite variable. In a survey of 11 sediment types amended with 25 mM Se(VI), Steinberg and Oremland (67) obtained rate constants which ranged from a low of 0.0025 h^{-1} for the sediments from the extreme envi-

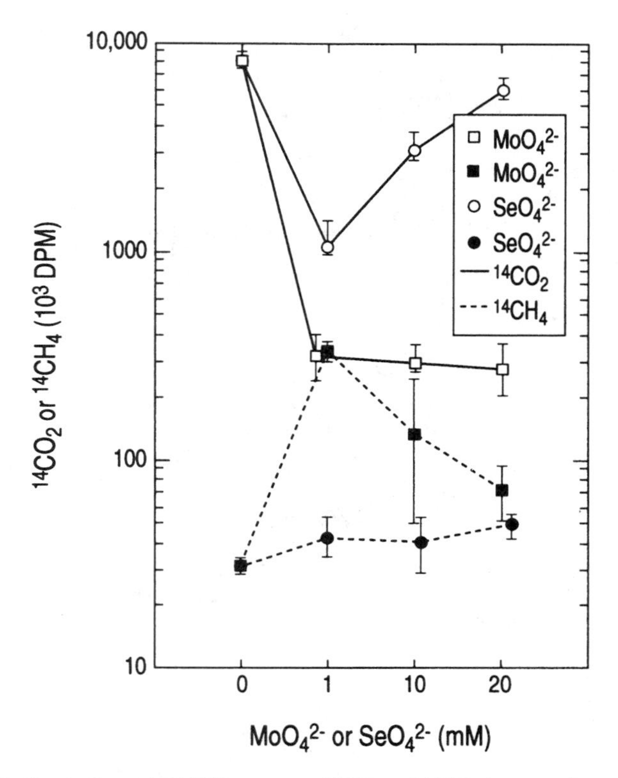

Figure 7. Metabolism of [2-^{14}C]acetate to $^{14}CH_4$ and $^{14}CO_2$ by estuarine sediments incubated with 10 mM sulfate, with sulfate plus 1, 10, or 20 mM molybdate, or with sulfate plus 1, 10, or 20 mM selenate. Reprinted from reference 55 with permission of the American Society for Microbiology.

ronment of Mono Lake (salinity, 90 g/liter; pH 9.8) to 1.0 h^{-1} for Massie Slough, a selenium-contaminated freshwater marsh.

The E_0' for the Se(VI)/Se(IV) couple is +475 mV (Table 1), while that for environmentally relevant forms of iron, such as the $FeOOH/FeCO_3$ couple, is about 0.0 mV (75). If we assume that significant Fe(II,III) oxides are present in sediments (an unproven assumption), thermodynamic considerations dictate that all the Se(VI) must be reduced before significant Fe(II) can be formed. Hence, Fe(II,III) oxides can be present only beneath the region where Se(VI) reduction occurs, and sediment diagenesis would result in a spatial separation that precludes any significant Se(VI) reduction by abiotic mechanisms. Such a separation occurs, for example, between selenate reduction and sulfate reduction (54).

The final point is that there are now a number of reports of natural environments for microbially mediated selenate reduction (42, 51, 53–55) and arsenate reduction (1, 2, 12, 56, 61). The number of bacterial species known to respire selenate and arsenate continues to increase. These species and the several new dissimilatory enzymes which have been described are discussed in the following sections.

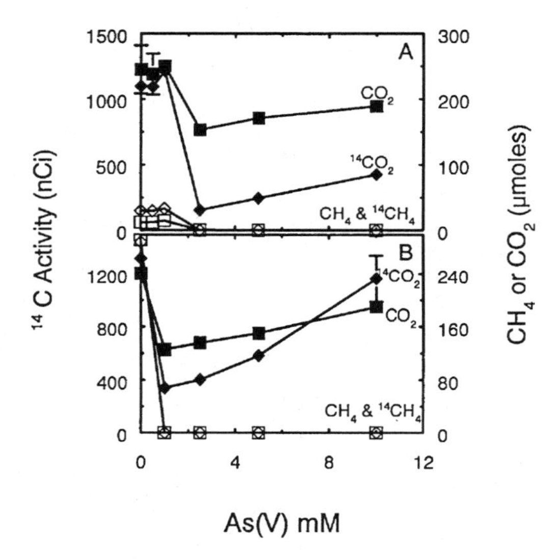

Figure 8. Metabolism of [2-^{14}C]acetate to $^{14}CH_4$ and $^{14}CO_2$ by anoxic sediments from salt marsh (A) or freshwater lake (B) sources in the presence of different concentrations of arsenate. Reprinted from reference 12 with permission of the American Society for Microbiology.

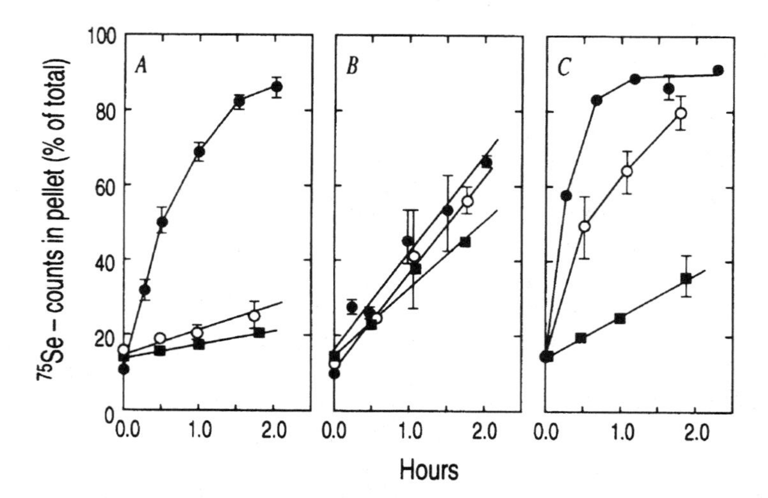

Figure 9. Selenate reduction activity in depth profiles taken from three sites in western Nevada: (A) South Lead Lake; (B) Hunter Drain; and (C) Massie Slough. Symbols: ●, 0 to 5 cm; ○, 5 to 10 cm; ■, 10 to 15 cm. Reprinted from reference 53 with permission of the American Society for Microbiology.

Table 1. Standard potentials, free energies, and molar growth yields of three species of selenate- and/or arsenate-reducing bacteria

Couple	E_0' (mV)	G (kJ/subst)	G (kJ/elect)	Y_M (g/mol)
S. barnesii (lactate as substrate)				
Se(VI)/Se(IV)	+475[a]	−343	−86	11.5
NO_3^-/NO_2^-	+425	−231	−58	7.1
As(V)/As(III)	+250	−140	−23	5.3
T. selenatis (acetate as substrate)				
Se(VI)/Se(IV)	+475	−575	−72	57
NO_3^-/NO_2^-	+425	−548	−69	55
D. auripigmentum (lactate as substrate)				
As(V)/As(III)	+250	−172	−29	5.5

[a]From reference 39.

DIVERSITY OF SELENATE- AND ARSENATE-REDUCING BACTERIA

Dissimilatory selenate reduction does not appear to be limited to any particular group of prokaryotes. Enrichment cultures (41) and several species of bacteria are capable of reducing submillimolar levels of Se(VI) to Se(0); these include *Wolinella succinogenes* (73), *Desulfovibrio desulfuricans* (72), *Pseudomonas stutzeri* (30), and *Enterobacter cloacae* (31). These organisms do not achieve respiratory growth on Se(VI), and the benefit gained from reducing Se(VI) is not clear and may be an incidental phenomenon. Not all strains of *P. stutzeri* are capable of Se(VI) reduction (66). *D. desulfuricans* can also reduce Se(VI) to Se(−II), but this process does not appear to have much importance in anoxic sediments (76). This reduction of selenate might be attributed to other reductases that have a broad range substrate specificity. For example, the membrane-associated nitrate reductases of *Escherichia coli* appear to be able to reduce Se(VI), as indicated by the oxidation of reduced methyl viologen (4).

Two species of selenate-respiring bacteria isolated from freshwater selenium-contaminated sediments have been well characterized. *Thauera selenatis*, a gram-negative motile rod originally classified as a pseudomonad was isolated on mineral medium with acetate as the electron donor (38). *T. selenatis* is a facultative anaerobe and belongs to the beta subclass of the *Proteobacteria* (37). It demonstrated molar growth yields (Y_M) of 57 and 55 g (dry weight)/mol of acetate (Table 1) for growth with selenate and nitrate as electron acceptors (36), and physiological studies suggest that these two anions are reduced by separate reductases (57). Its selenate reductase has been recently purified and is discussed in detail in the next section.

The second organism, *Sulfurospirillum barnesii* strain SES-3, was isolated from an acetate enrichment (66) and uses lactate as an electron donor for the reduction of a number of compounds including selenate and nitrate compounds (52). SES-3, a motile, gram-negative, curved organism (Fig. 10A and B) has only recently been assigned to the genus *Sulfurospirillum*. Earlier reports describing its ability to grow on iron (29, 34) and preliminary biochemical characterization (68) used the proposed name *Geospirillum barnesii*. The description of several new species of vibrioid microaerophilic sulfur bacteria belonging to the epsilon subgroup of the

Proteobacteria, including *Sulfurospirillum deleyianum* and *S. arcachonense*, suggested, however, that strain SES-3 was a member of this new genus (15). Phylogenic analysis using 16S rRNA has confirmed this (69) (Fig. 11B). The Y_Ms for SES-3 growth on selenate and nitrate were 11.5 and 7.1 g/mol of lactate, respectively (Table 1), and physiological studies also suggest that the reductases for selenate and nitrate are separate enzymes (52). This latter point has been borne out by the different cyctochrome contents of membrane fractions grown on selenate or nitrate (68). *S. barnesii* grows on Se(VI) by reducing it quantitatively to Se(IV) with the oxidation of lactate to acetate plus CO_2 (Fig. 12), although cell suspensions completely reduce Se(VI) to Se(0). The reason for this disparity is probably due to the toxicity of accumulated Se(IV) to growing cells. Growth on nitrate is by its dissimilatory reduction to ammonium rather than by denitrification (52). Preliminary growth experiments indicate that *Aeromonas hydrophila* represents a third species of freshwater *Proteobacteria* from the gamma subgroup which can respire selenate (26).

Two novel species of gram-positive selenium-respiring bacteria have been isolated from Mono Lake: the spore-forming *Bacillus arsenicoselenatis* E-1H (Fig. 10 C and D) and *Bacillus selenitireducens* MLS-10 (Fig. 10E and F). Both isolates are moderate halophiles as well as alkaliphiles, and both exhibit maximal growth between pH 9 and 11 (70). Strain E-1H grows by reducing Se(VI) to Se(IV) (Fig. 13), while strain MLS-10 grows by reduction of Se(IV) to Se(0) (Fig. 14). Both isolates grow by reduction of As(V) to As(III). The closest relative of E-1H, based on 16S rRNA analysis, is *Bacillus alcalophilus*, but the sequence similarity is distant enough to warrant a new species (Fig. 11A).

There are also several species of organisms known to grow by dissimilatory arsenate reduction. They, too, are not confined to any particular group of bacteria and are distributed throughout the domain *Bacteria*. Among these are several species of selenate-respiring bacteria that also have the ability to grow on arsenate. One such organism is *S. barnesii*. It reduces As(V) to As(III) (Fig. 15) but does not reduce As(III), as shown with As(V)-grown washed cell suspensions (28). The Y_M for growth on As(V) is 5.3 g/mol of lactate (Table 1). Three other species are capable of growth on As(V). These include strain MIT-13 (2), now classified as *Sulfurospirillum arsenophilum* (69), *Desulfotomaculum auripigmentum* (48, 49), and *Chrysiogenes arsenatis* (35). *S. arsenophilum* strain MIT-13 is a gram-negative, vibrioid, microaerobic sulfur-reducing bacterium closely related to SES-3 (2, 69). It was isolated from arsenic-contaminated watershed sediments in eastern Massachusetts (2). *D. auripigmentum* strain OREX-4 is a low-G+C, gram-positive organism which was isolated from surface lake sediments, also in eastern Massachusetts (48, 49). It differs significantly from both *S. barnesii* and *S. arsenophilum* in cell wall structure (i.e., it is gram positive) and its ability to use sulfate as a terminal electron acceptor (48). It can grow on the arsenic-containing mineral scorodite [$(FeAsO_4)\cdot 2H_2O$] and produces the arsenic sulfide mineral orpiment both intra- and extracellularly when grown in the presence of both arsenate and sulfate (48, 49). *C. arsenatis* strain BAL-1T, a strict anaerobe isolated from gold mine wastewater at Ballarat gold fields in Australia, represents its own deeply branching lineage in the domain *Bacteria* (35). Their phylogenic affiliations (except for *C. arsenatis*) are shown in Fig. 11A.

BIOCHEMISTRY OF Se(VI) AND As(V) REDUCTION

The biological reduction of selenate and arsenate occurs for a number of reasons. In general, these are assimilation, regulation of reducing equivalents, detoxification, and dissimilation. Each is discussed in turn below.

The toxic effects of arsenate (uncoupling of oxidative phosphorylation) and arsenite (destabilization of proteins) might make it seem highly unlikely that arsenic is actively assimilated or incorporated into cellular constituents. However, there are several reports of arsenic sequestration and use by marine animals (crustaceans, fish, elasmobranchs) and algae (14, 17, 18). Arsenobetaine is a structural analogue of the osmolyte glycine betaine, with As substituting for the N in the latter compound. Arsenobetaine can be transported into the cells of haloalkaliphilic bacteria isolated from Mono Lake (7) and may play a role in the maintenance of the osmotic balance of these cells. Arsenobetaine can be degraded by bacteria, resulting in the liberation of inorganic arsenic (19). Methylated arsenic compounds are also important in aquatic environments (3, 58) and have been found in polysaccharides from marine algae (62).

In contrast to arsenic, selenium is an essential element found in antioxidant proteins and thus is readily assimilated. It is found in selenocysteine-containing proteins (22), as well as selenoenzymes such as the nicotinic acid hydroxylase from *Clostridium barkeri* (10) and a formate dehydrogenase from *Methanococcus vannielii* (cited in reference 25). Recently, a selenium-containing enzyme was found in a virus (63). However, these Se-containing proteins do not contribute fundamentally to the cycling of Se in nature.

The second mechanism of reduction involves the regulation of reducing equivalents. In organisms such as *Rhodobacter sphaeroides*, intracellular redox potential is maintained through the continuous dumping of electrons (43, 44). During anaerobic growth in the light, a pool of reducing equivalents is built up, which eventually affects photosynthesis. One mechanism by which this pool is diminished is through the reduction of heavy-metal oxyanions. Although the exact mechanism has not been worked out, Moore and Kaplan (43) have identified a membrane-bound, $FADH_2$-dependent metal reductase. A side effect of this process is that the cells are resistant to high levels of heavy metal oxides ($Cr_2O_7^{2-}$, Rh_2O_3, Eu_2O_5, TeO_4^{2-}, TeO_3^{2-}) including selenate and arsenate (44). Based on what is known for tellurium, the reduced metal or metalloid may actually be precipitated in the cell membrane.

Arsenic resistance is widespread among clinical isolates of *Escherichia coli*, *Klebsiella pneumoniae*, *Pseudomonas aeruginosa*, and *Staphylococcus aureus* (46). This resistance is conferred by plasmids, two of which, R773 from *E. coli* (16, 50) and pI258 from *S. aureus* (24), have been well characterized. Plasmid R773 contains four genes, *arsR*, *arsA*, *arsB*, and *arsC* (16). Plasmid pI258 contains three genes, *arsR*, *arsB*, and *arsC* (24). In both cases, it is the *arsC* gene that encodes a soluble 13-kDa polypeptide that reduces As(V) to As(III). Arsenite is then exported

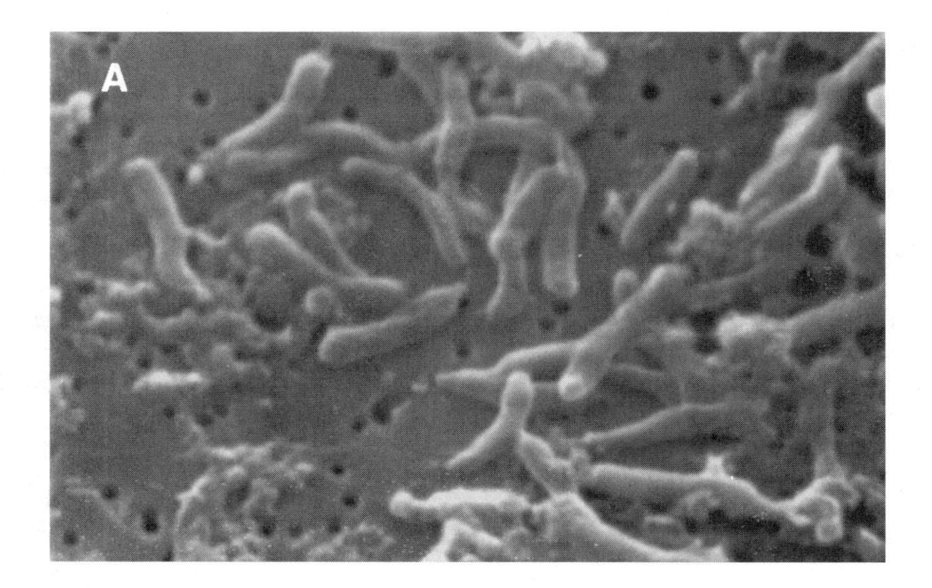

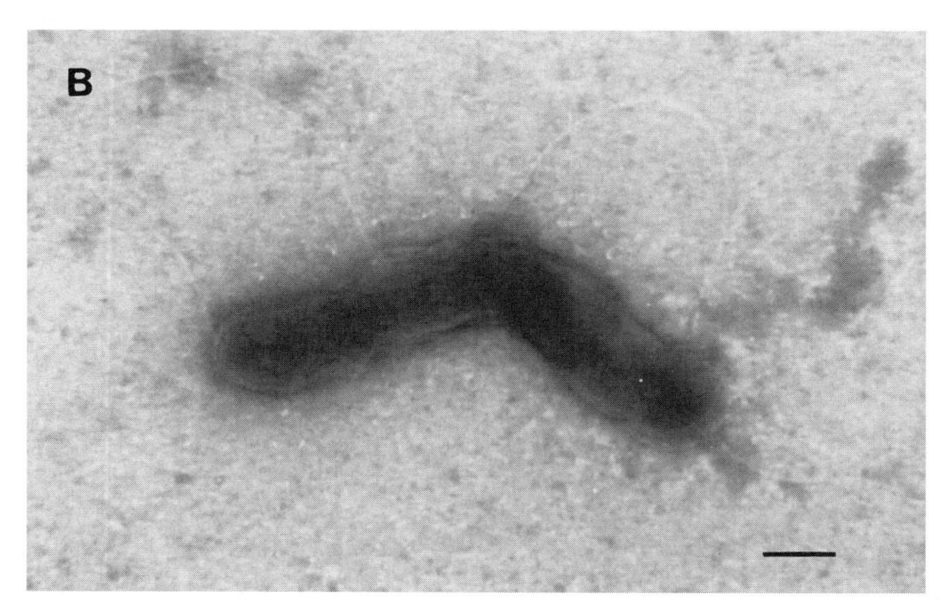

Figure 10. Electron micrographs of the selenium-reducing bacteria *S. barnesii* strain SES-3 (A and B), strain E-1H from Mono Lake (C and D), and strain MLS-10 from Mono Lake (E and F). (A, C, and E) Scanning electron microscopy by J. Switzer Blum, A. Burns, and R. S. Oremland (unpublished). (B, D, and F) Transmission electron microscopy by J. F. Stolz. Bars, 0.5 μm. Extracellular ball-like particles in panel E are elemental selenium as determined from X-ray energy-dispersive spectrometry done in association with the scanning electron microscopy.

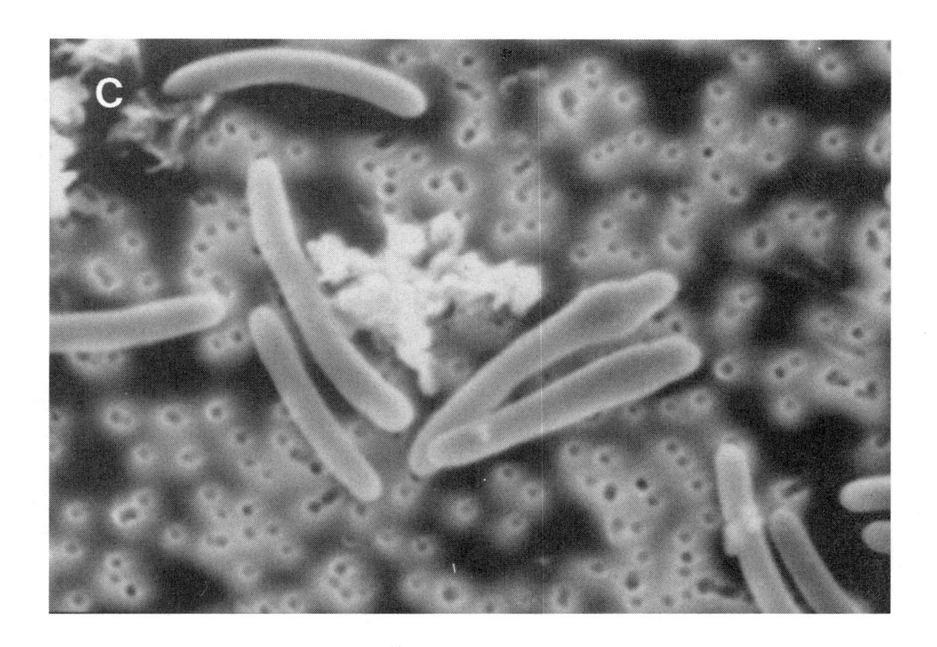

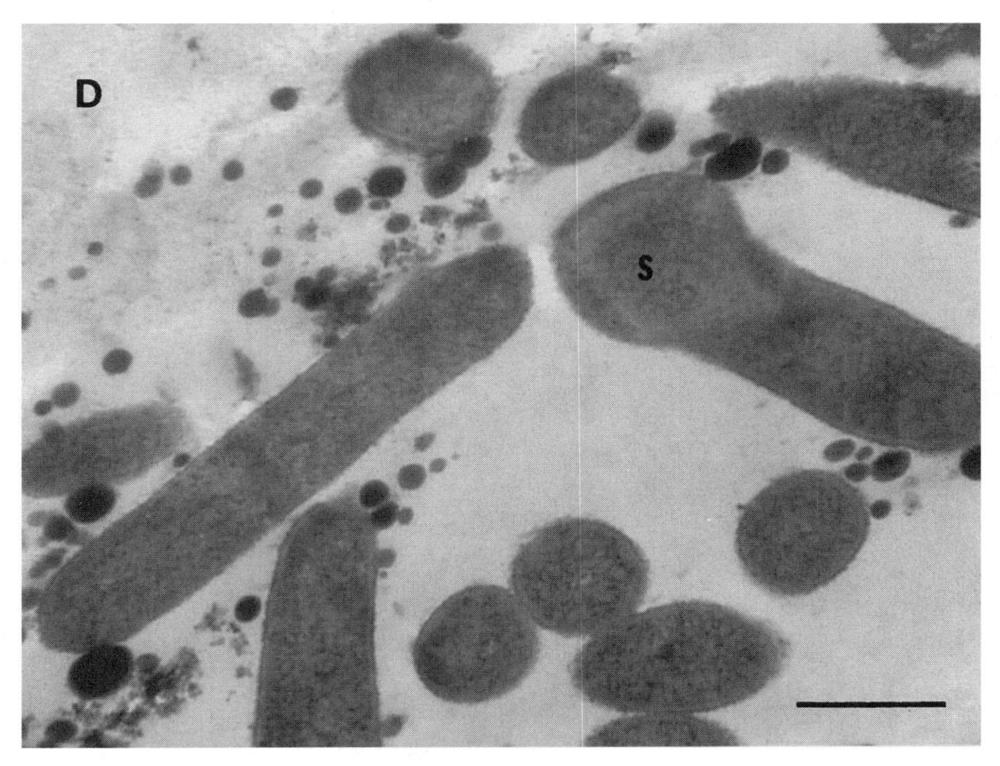

Figure 10. *Continued*

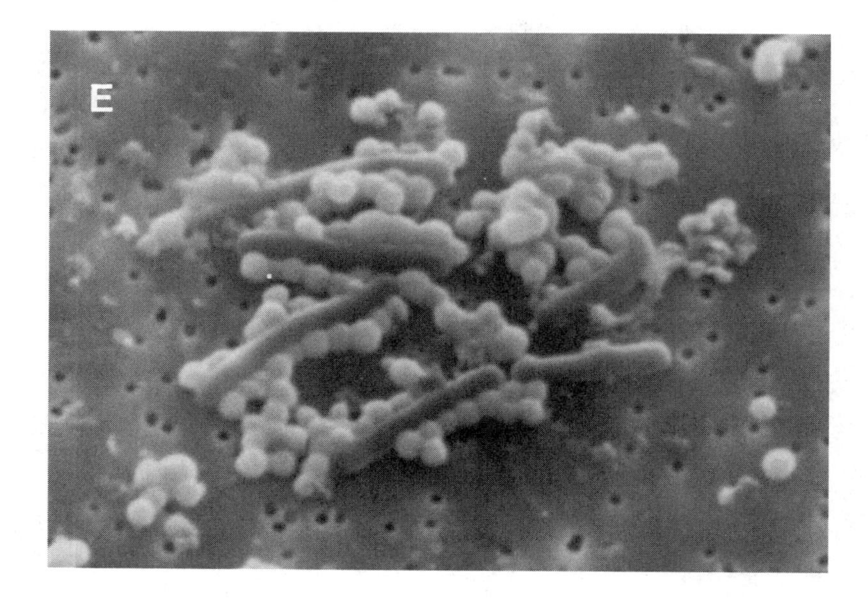

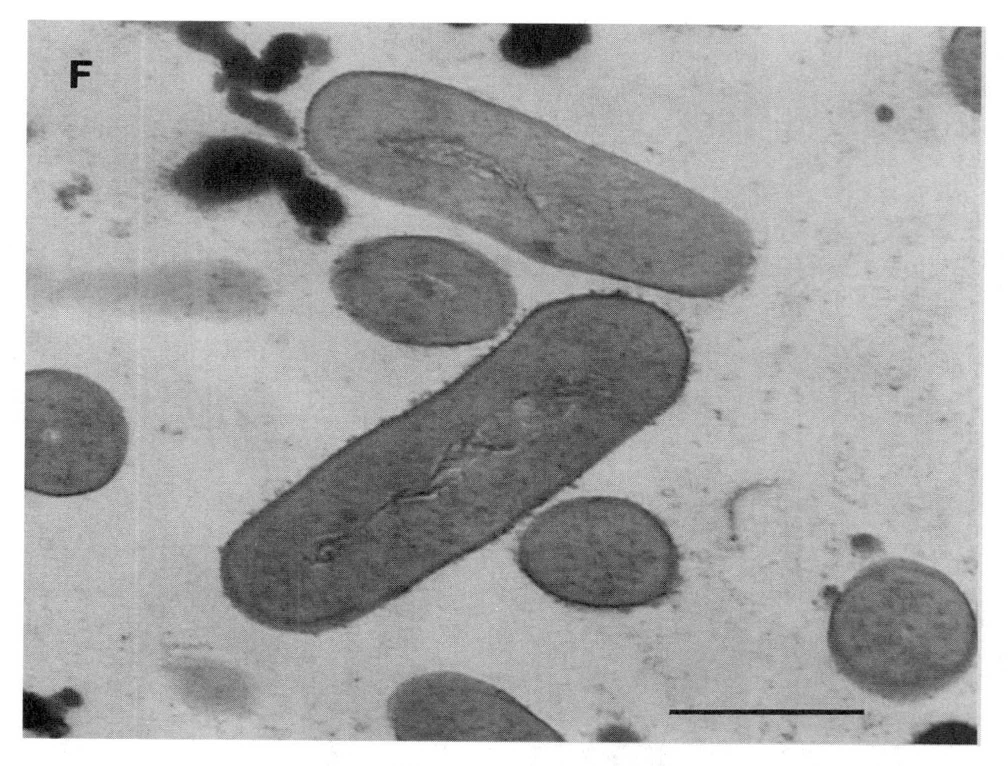

Figure 10. *Continued*

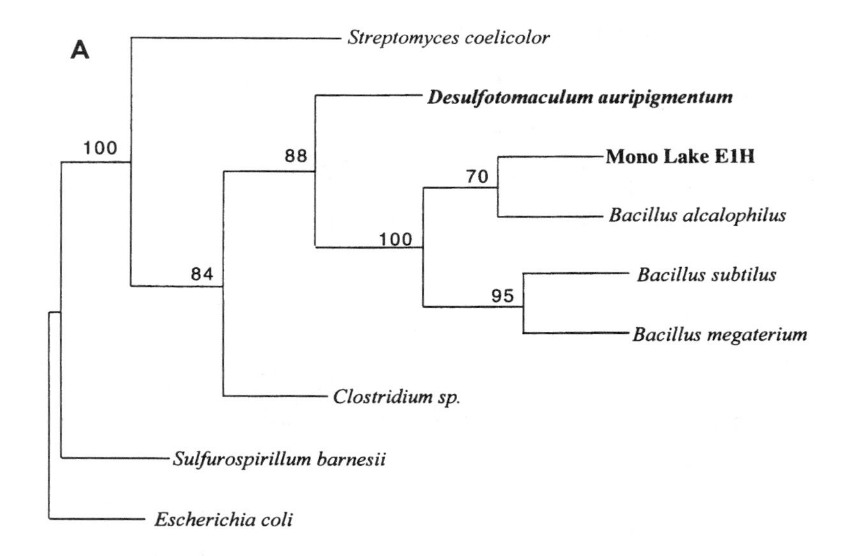

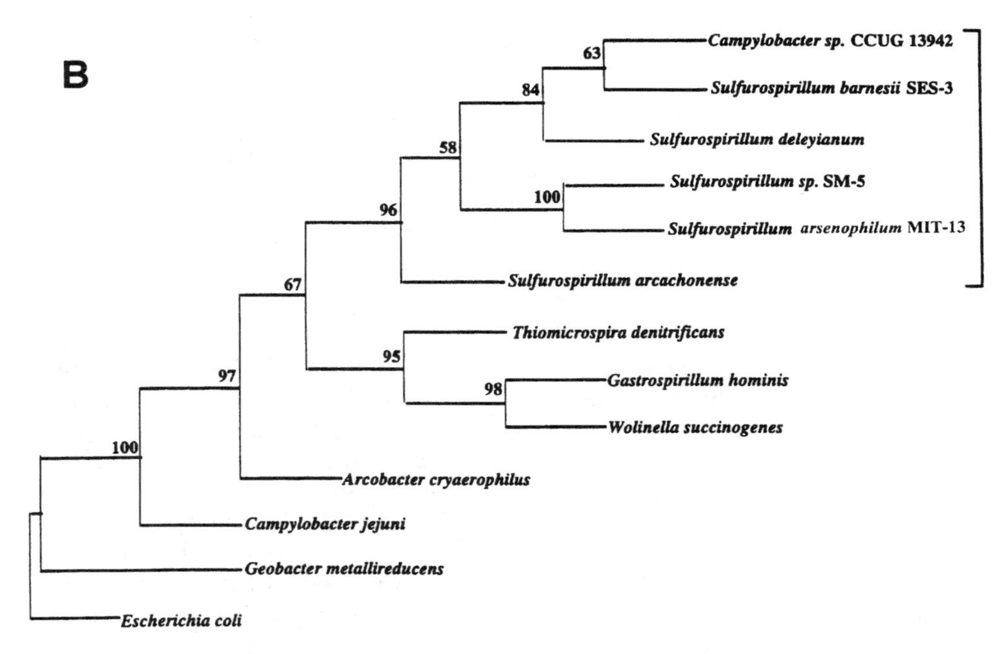

Figure 11. Phylogenic trees based on 16S rRNA sequence data using maximum-parsimony analysis of the gram-positive arsenate- and selenate-reducing bacterium E1-H (A) and the gram-negative arsenate- and selenate-reducing bacterium *S. barnesii* strain SES-3 and the arsenate-reducing bacterium *S. arsenophilus* strain MIT-13 (B).

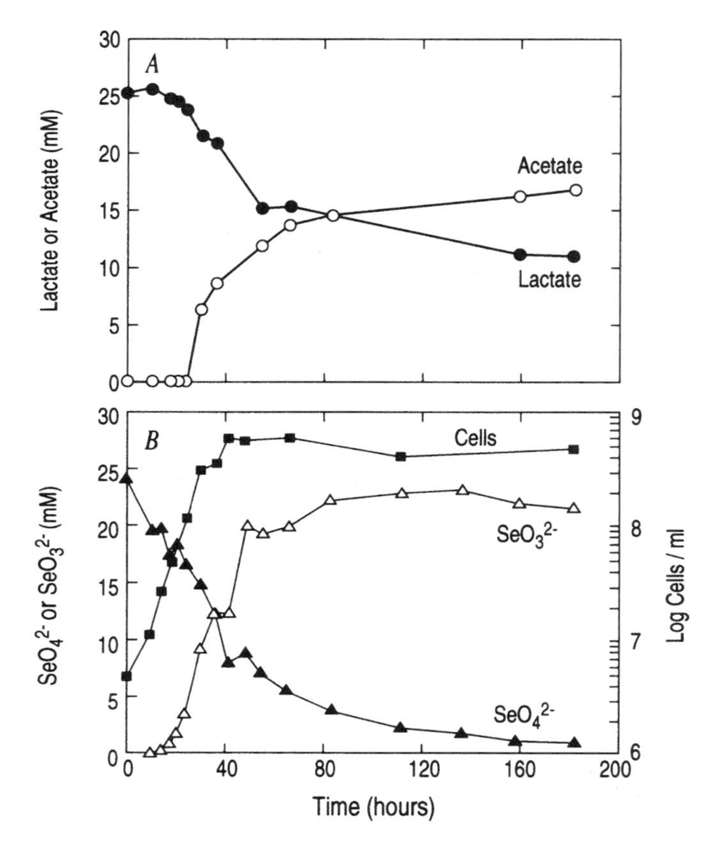

Figure 12. Growth of *S. barnesii* strain SES-3 with Se(VI) as the electron acceptor. Reprinted from reference 52 with permission of the American Society for Microbiology.

out of the cell by either ATP-driven or chemiosmotic transport systems (64). Diorio et al. (11) have also discovered a functional chromosomal *ars* operon homolog. Based on Southern analyses, they found evidence for other chromosomal *ars* operon homologs in *Shigella sonnei*, *Citrobacter freundii*, *Enterobacter cloacae*, *Salmonella enterica* serovar Arizonae, *Erwinia carotovora*, *Klebsiella pneumoniae*, and *Pseudomonas aeruginosa*, which suggested to them that chromosomal arsenate resistance is widespread within the *Enterobacteriaceae* (11). Although reduction of As(V) by resistant microbes has been relatively well studied (6), direct evidence for the involvement of these types of bacteria in the biogeochemical cycling of arsenic in nature is lacking.

The last mechanism is dissimilation, where the reduction of selenate and arsenate is coupled to the oxidation of an organic substrate such as lactate or acetate. The reduction is carried out by a terminal reductase, which is typically associated with a cytochrome. To date, one selenate reductase and one arsenate reductase have been purified and characterized. The dissimilatory selenate reductase of *Thauera*

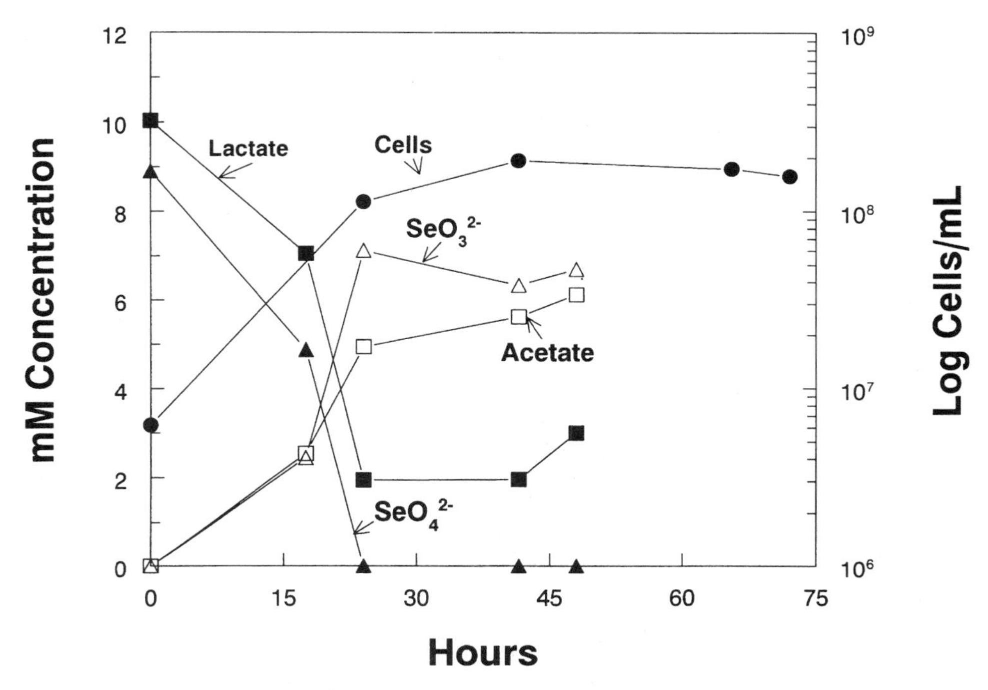

Figure 13. Growth of strain E-1H in Mono Lake water with Se(VI) as the electron acceptor. From J. Switzer Blum, A. Burns Bindi, and R. S. Oremland, unpublished data.

selenatis is a trimeric enzyme localized in the periplasm (60). The complex has an apparent molecular mass of 180 kDa with an α subunit of 96 kDa, a β subunit of 40 kDa, and a γ subunit of 23 kDa. The reductase contains molybdenum, iron, and acid-labile sulfur and has an apparent K_m of 16 μM. One of the subunits, presumably the 23-kDa protein, is a cytochrome *b* with a difference spectrum with absorbance maxima at 558, 528, and 424 nm. The enzyme is very substrate specific, reducing selenate only to selenite and unable to use nitrate, nitrite, chlorate, or sulfate.

Selenate reduction by *Sulfospirillum barnesii* is apparently different from that by *T. selenatis*. In whole cells selenate can be reduced all the way to elemental selenium (52), and selenate reductase activity has been localized in the membrane fraction (68). Although a *b*-type cytochrome has been detected in the membrane fraction from selenate-grown cells, its difference spectrum is different from that detected in *T. selenatis*, with absorbance maxima at 554, 523, and 422 nm (68). The apparent substrate specificity of the selenate reductase is much broader in that membrane fractions from cells grown on selenate had appreciable activity for nitrate, thiosulfate, and fumarate, even though components of these reductive pathways were not detectable (68).

Krafft and Macy (27) reported the characteristics of the purified arsenate reductase from *Chrisiogenes arsenatis*. It has two subunits (87 and 29 kDa); contains

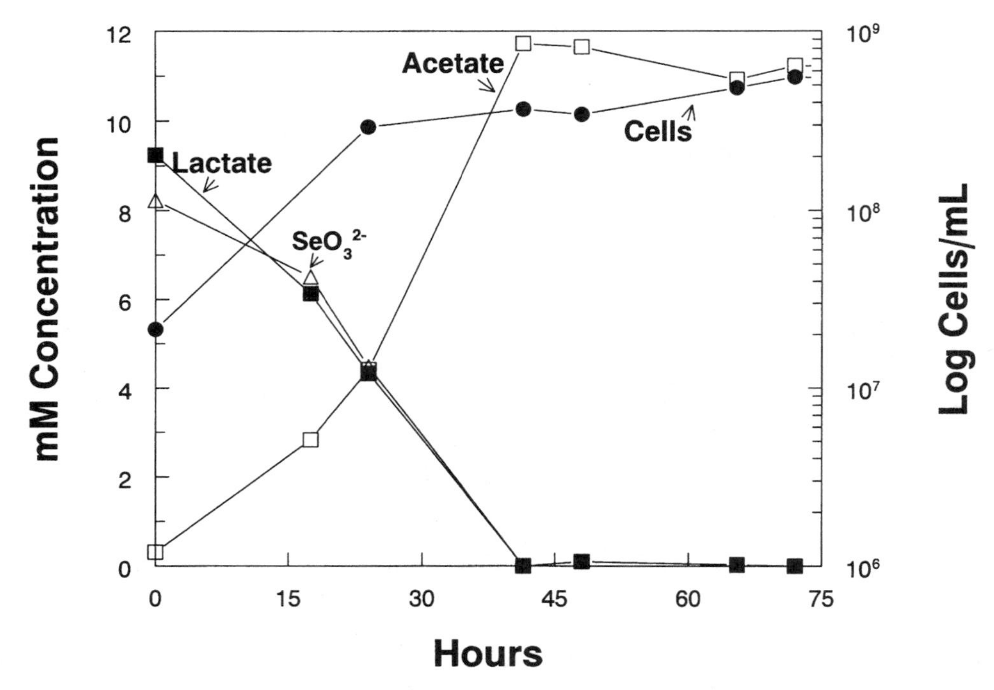

Figure 14. Growth of strain MLS-10 in Mono Lake water with Se(IV) as the electron acceptor. From J. Switzer Blum, A. Burns Bindi, and R. S. Oremland, unpublished data.

molybdenum, iron, sulfur, and zinc as cofactors; and is specific for arsenate. It has a K_m of 300 μM and is located in the periplasm of the organism. The arsenate reductase from *S. barnesii* has also recently been purified and characterized (D. K. Newman, R. S. Oremland, P. Dowdle, F. M. M. Morel, and J. F. Stolz, submitted for publication). Unlike the plasmid-encoded cytoplasmic arsenate reductases of *E. coli* and *S. aureus*, this enzyme is strongly membrane bound and is a dissimilatory enzyme. A trimeric complex, with a calculated molecular mass of about 120 kDa, it has an α subunit of 65 kDa, a β subunit of 31 kDa, and a γ subunit of 22 kDa. A *b*-type cytochrome with a difference spectrum with absorbance maxima at 554, 520, and 416 nm was also detected in the membrane fraction. Although a complete metals analysis has yet to be done (i.e., to test for the presence of molybdenum), evidence for Fe-S prosthetic groups has been found. The enzyme is able to couple the reduction of As(V) to As(III) to the oxidation of methyl viologen and has an apparent K_m of 200 μM. Arsenate reductase activity is inhibited by arsenite, phosphate, molybdate, and nitrate. NADH can also serve as an electron donor, but FADH has no effect.

S. barnesii is a versatile organism that is able to use nitrate, thiosulfate, fumarate, and iron in addition to selenate and arsenate as terminal electron acceptors. Initial studies of the cytochrome content and enzyme activity strongly suggested that there are separate pathways for As(V), Se(VI), nitrate, fumarate, and thiosulfate reduc-

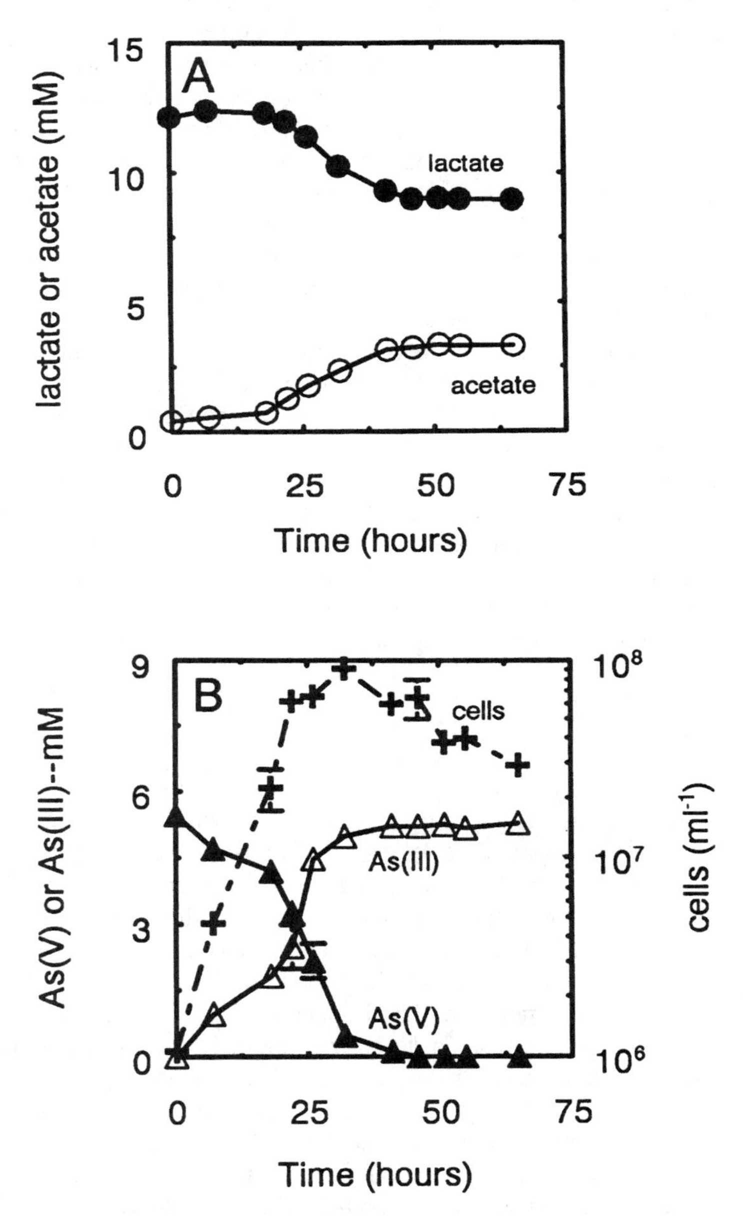

Figure 15. Growth of *S. barnesii* strain SES-3 with As(V) as the electron acceptor. Reprinted from reference 28 with permission of the American Society for Microbiology.

tion, i.e., that there are separate terminal reductases and cytochromes associated with each terminal electron acceptor. At least three different *b*-type and two different *c*-type cytochromes have been detected. In addition to the two membrane-bound *b*-type cytochromes described above that are associated with arsenate and selenate reduction, fumarate-grown cells also contain a membrane-bound *b*-type cytochrome (556, 520, and 420 nm). A membrane-bound 61.1-kDa *c*-type cytochrome, with absorbance maxima at 552, 523, and 423 nm, has been detected only in cells grown on nitrate. Similarly, only a membrane-bound 70.2-kDa *c*-type cytochrome has been found in thiosulfate-grown cells (68).

Using methyl viologen as an artificial electron donor, the substrate specificity of the enzymes expressed in the cells grown with different electron acceptors was tested (68). It was found that methyl viologen could be used to detect not only nitrate and nitrite reductase activity but also selenate, selenite, arsenate, fumarate, and thiosulfate reductase activity. The results showed that membrane fractions from cells grown on nitrate, fumarate, selenate, and thiosulfate had the ability to reduce all the substrates to a certain degree. This activity, however, could be attributed to the specific terminal reductase for the substrate on which the cells had been grown. That is, the nitrate reductase expressed in nitrate-grown cells also exhibited activity for selenate (40% of the nitrate activity), thiosulfate (9% of the nitrate activity), and fumarate (6% of the nitrate activity). Thus, it is tempting to propose that the the terminal reductases and cytochromes involved in nitrate, selenate, fumarate, and thiosulfate reduction in *S. barnesii* are under separate regulatory control.

CONCLUSIONS

The investigation into the number of species of microorganisms capable of dissimilatory arsenate and selenate reduction has only begun, and yet a great diversity has already been demonstrated. Species of gram-positive bacteria, alkaliphilic gram-positive bacteria, and epsilon and gamma *Proteobacteria*, as well as a microbe belonging to its own deeply branching lineage in the domain *Bacteria*, have been identified. The realization that arsenate and selenate are indeed suitable electron acceptors and are readily available in both natural and contaminated environments suggests that even more unrelated species will be discovered. The initial biochemical studies also suggest that there may be different pathways for selenate and arsenate reduction, with specific terminal reductases and cytochromes. Nonetheless, there are some similarities. Although the selenate reductase from *T. selenatis* is periplasmic and the arsenate reductase from *S. barnesii* are membrane bound, they both are trimeric enzymes associated with a *b*-type cytochrome. It also seems likely that the active site in all these enzyme is a molybdopterin. This last point will remain moot until more arsenate and selenate reductases are purified and characterized.

Acknowledgments. We thank M. Haggblom and M. Sylvester for their constructive comments on the manuscript.

REFERENCES

1. **Ahmann, D., L. R. Krumholz, H. F. Hemond, D. R. Lovley, and F. M. M. Morel.** 1997. Microbial mobilization of arsenic from sediments of the Aberjona Watershed. *Environ. Sci. Technol.* **31:**2923–2930.
2. **Ahmann, D., A. L. Roberts, L. R. Krumholz, and F. M. M. Morel.** 1994. Microbe grows by reducing arsenic. *Nature* **371:**750.
3. **Anderson, L. C. D., and K. W. Bruland.** 1991. Biogeochemistry of arsenic in natural waters: the importance of methylated species. *Environ. Sci. Technol.* **25:**420–427.
4. **Avazeri, C., R. J. Turner, J. Pommier, J. H. Weiner, G. Giordano, and A. Vermeglio.** 1997. Tellurite reductase activity of nitrate reductase is responsible for the basal resistance of *Escherichia coli* to tellurite. *Microbiology* **143:**1181–1189.
5. **Azcue, J. M., and J. O. Nriagu.** 1994. Arsenic: historical perspectives, p. 1–16. *In* J. O. Nriagu (ed.), *Arsenic in the Environment.* I. *Cycling and Characterization.* John Wiley & Sons, Inc., New York, N.Y.
6. **Cervantes, C., G. Ji, J. L. Ramirez, and S. Silver.** 1994. Resistance to arsenic compounds in microorganisms. *FEMS Microbiol. Rev.* **15:**355–367.
7. **Ciulla, R. A., M. R. Diaz, B. F. Taylor, and M. F. Roberts.** 1997. Organic osmolytes in aerobic bacteria from Mono Lake, an alkaline, moderately hypersaline environment. *Appl. Environ. Microbiol.* **63:**220–226.
8. **Cullen, W. R., and K. J. Reimer.** 1989. Arsenic speciation in the environment. *Chem. Rev.* **89:** 713–764.
9. **Deng, M., T. Moureaux, and M. Caboche.** 1989. Tungstate, a molybdate analog inactivating nitrate reductase, deregulates the expression of the nitrate reductase structural gene. *Plant Physiol.* **91:**304–309.
10. **Dilworth, G. L.** 1982. Properties of the selenium-containing moiety of nicotinic acid hydroxylase from *Clostridium barkeri. Arch. Biochem. Biophys.* **219:**30–38.
11. **Diorio, C., J. Cai, J. Marmor, R. Shinder, and M. S. DuBow.** 1995. An *Escherichia coli* chromosomal *ars* operon homolog is functional in arsenic detoxification and is conserved in gram-negative bacteria. *J. Bacteriol.* 177:2050–2056.
12. **Dowdle, P. R., A. M. Laverman, and R. S. Oremland.** 1996. Bacterial reduction of arsenic(V) to arsenic(III) in anoxic sediments. *Appl. Environ. Microbiol.* **62:**1664–1669.
13. **Eary, L. E.** 1992. The solubility of amorphous As_2S_3 from 25 to 90°C. *Geochim. Cosmochim. Acta* **56:**2267–2278.
14. **Edmonds, J. S., and K. A. Francesconi.** 1988. The origin of arsenobetaine in marine animals. *Appl. Organomet. Chem.* **2:**283–295.
15. **Finster, K., W. Liesack, and B. J. Tindall.** 1997. *Sulfurospirillum arcachonense* sp. nov., a new microaerophilic sulfur-reducing bacterium. *Int. J. Syst. Bacteriol.* **47:**1212–1217.
16. **Gladysheva, T. B., K. L. Oden, and B. P. Rosen.** 1994. Properties of the arsenate reductase of plasmid R 773. *Biochemistry* **33:**7288–7293.
17. **Hanaoka, K. and S. Tagawa.** 1985. Isolation and identification of arsenobetaine as a major water-soluble arsenic compound from muscle of blue pointer *Isurus oxyrhincus* and whitetip shark *Carcarhinus longimanus. Bull. Jpn. Soc. Fish. Sci.* **51:**681–685.
18. **Hanoaka, K., T. Fujita, M. Matsuura, S. Tagawa, and T. Kaise.** 1987. Identification of arsenobetaine as a major arsenic compound in muscle of two demersal sharks, shortnose dogfish *Squalus brevirostris* and starspotted shark *Mustelus manazo. Comp. Biochem. Physiol.* **86B:**681–682.
19. **Hanaoka, K., S. Tagawa, and T. Kaise.** 1992. The degradation of arsenobetaine to inorganic arsenic by sedimentary microorganisms. *Hydrobiologia* **235/236:**623–628.
20. **Hansen, H. C. B., O. K. Borggaard, and J. Sørensen.** 1994. Evaluation of the free energy of formation of Fe(II)-Fe(III) hydroxide-sulphate (green rust) and its reduction of nitrite. *Geochim. Cosmochim. Acta* **58:**2599–2608.
21. **Haygarth, P. M.** 1994. Global importance and global cycling of selenium, p. 1–28. *In* W. T. Frankenberger, Jr., and S. Benson (ed.), *Selenium in the Environment.* Marcel Dekker, Inc., New York, N.Y.

22. **Heider, J., and A. Boeck.** 1994. Selenium metabolism in micro-organisms. *Adv. Microbiol. Physiol.* **35:**71–109.
23. **Jayaweera, G. R., and J. W. Biggar.** 1996. Role of redox potential in chemical transformations of selenium in soils. *Soil Sci. Soc. Am. J.* **60:**1056–1063.
24. **Ji, G., E. A. E. Garber, L. G. Armes, C.-M. Chen, J. A. Fuchs, and S. Silver.** 1994. Arsenate reductase of *Staphylococcus aureus* plasmid pI258. *Biochemistry* **33:**7294–7299.
25. **Kletzin, A., and M. W. W. Adams.** 1996. Tungsten in biological systems. *FEMS Microbiol. Rev.* **18:**5–63.
26. **Knight, V., and R. P. Blakemore.** 1998. Reduction of diverse electron acceptors by *Aeromonas hydrophila*. *Arch. Microbiol.* **169:**239–248.
27. **Krafft, T., and J. M. Macy.** 1998. Purification and characterization of the respiratory arsenate reductase of *Chrysiogenes arsenatis*. *Eur. J. Biochem.* **255:**647–653.
28. **Laverman, A. M., J. Switzer Blum, J. K. Schaefer, E. J. P. Phillips, R. R. Lovley, and R. S. Oremland.** 1995. Growth of strain SES-3 with arsenate and other diverse electron acceptors. *Appl. Environ. Microbiol.* **61:**3556–3561.
29. **Lonergan, D. J., H. Jenter, J. D. Coates, T. Schmidt, and D. R. Lovley.** 1996. Phylogeny of dissimilatory Fe(III)-reducing bacteria. *J. Bacteriol.* **178:**2402–2408.
30. **Lortie, L., W. D. Gould, S. Rajan, R. G. L. McCready, and K-J. Cheng.** 1992. Reduction of selenate and selenite by a *Pseudomonas stutzeri* isolate. *Appl. Environ. Microbiol.* **58:**4043–4044.
31. **Losi, M. E., and W. T. Frankenberger, Jr.** 1997. Reduction of selenium oxyanions by *Enterobacter cloacae* SLD1a-1: isolation and growth of the bacterium and its expulsion of selenium particles. *Appl. Environ. Microbiol.* **63:**3079–3084.
32. **Losi, M. E., and W. T. Frankenberger, Jr.** 1997. Bioremediation of selenium in soil and water. *Soil. Sci.* **162:**692–702.
33. **Lovley, D. R.** 1993. Dissimilatory metal reduction. *Annu. Rev. Microbiol.* **47:**263–290.
34. **Lovley, D. R., J. D. Coates, D. A. Saffarini, and D. J. Lonergan.** 1997. Dissimilatory iron reduction, p. 187–215. *In* G. Winkelman and C. J. Carrano (ed.), *Iron and Related Transition Metals in Microbial Metabolism*. Harwood Academic Press, Zurich, Switzerland.
35. **Macy, J. M., K. Nunan, K. D. Hagen, D. R. Dixon, P. J. Harbour, M. Cahill, and L. I. Sly.** 1996. *Chrysiogenes arsenatis*, gen nov., sp. nov., a new arsenate-respiring bacterium isolated from gold mine wastewater. *Int. J. Syst. Bacteriol.* **46:**1153–1157.
36. **Macy, J. M., and S. Lawson.** 1993. Cell yield (Y_M) of *Thauera selenatis* grown anaerobically with acetate plus selenate or nitrate. *Arch. Microbiol.* **160:**295–298.
37. **Macy, J. M., S. Rech, G. Auling, M. Dorsch, E. Stackenbrandt, and L. I. Sly.** 1993. *Thauera selenatis* gen. nov., sp. nov., a member of the beta subclass of *Proteobacteria* with a novel type of anaerobic respiration. *Int. J. Syst. Bacteriol.* **43:**135–142.
38. **Macy, J. M., T. A. Michel, and D. A. Kirsch.** 1989. Selenate reduction by a *Pseudomonas* species: a new mode of anaerobic respiration. *FEMS Microbiol. Lett.* **61:**195–198.
39. **Madigan, M. T., J. M. Martinko, and J. Parker.** 1997. *Brock Biology of Microorganisms*. Prentice-Hall Inc., Upper Saddle River, N.J.
40. **Maest, A. S., S. P. Pasilis, L. G. Miller, and D. K. Nordstrom.** 1992. Redox geochemistry of arsenic and iron in Mono Lake, California, USA, p. 507–511. *In* Y. K. Kharaka and A. S. Maest (ed.), *Water-Rock Interaction*. A. A. Balkema, Rotterdam, The Netherlands.
41. **Maiers, D. T., P. L. Wichlacz, D. L. Thompson, and D. F. Bruhn.** 1988. Selenate reduction by bacteria from a selenium-rich environment. *Appl. Environ. Microbiol.* **54:**2591–2593.
42. **Masscheleyn, P. H., R. D. Delaune, and W. H. Patrick, Jr.** 1990. Transformations of selenium as affected by sediment oxidation-reduction potential and pH. *Environ. Sci. Technol.* **24:**91–96.
43. **Moore, M. D., and S. Kaplan.** 1992. Identification of intrinsic high-level resistance to rare-earth oxides and oxyanions in members of the class *Proteobacteria*: characterization of tellurite, selenite, and rhodium sesquioxide reduction in *Rhodobacter sphaeroides*. *J. Bacteriol.* **174:**1505–1514.
44. **Moore, M. D., and S. Kaplan.** 1992. Members of the family Rhodospirillaceae reduce heavy-metal oxyanions to maintain redox poise during photosynthetic growth. *ASM News* **60:**17–23.
45. **Myneni, S. C., T. K. Tokunaga, and G. E. Brown, Jr.** 1997. Abiotic selenium redox transformations in the presence of Fe(II,III) oxides. *Science* **278:**1106–1109.

46. **Nakahara, H., T. Ishikawa, Y. Sarai, and I. Kondo.** 1977. Frequency of heavy-metal resistance in bacteria from inpatients in Japan. *Nature* **266:**165–167.
47. **Newman, D. K., D. Ahmann, and F. M. M. Morel.** 1998. A brief review of dissimlatory arsenate reduction. *Geomicrobiol. J.* **15:**255–268.
48. **Newman, D. K., E. K. Kennedy, J. D. Coates, D. Ahmann, D. J. Ellis, D. R. Lovley, and F. M. M. Morel.** 1997. Dissimilatory arsenate and sulfate reduction in *Desulfotomaculum auripigmentum*, sp. nov. *Arch. Microbiol.* **165:**380–388.
49. **Newman, D. K., T. J. Beveridge, and F. M. Morel.** 1997. Precipitation of arsenic trisulfide by *Desulfotomaculum auripigmentum. Appl. Environ. Microbiol.* **63:**2022–2028.
50. **Oden, K. L., T. B. Gladysheva, and B. P. Rosen.** 1994. Arsenate reduction mediated by the plasmid-encoded *arsC* protein is coupled to glutathione. *Mol. Microbiol.* **12:**301–306.
51. **Oremland, R. S.** 1994. Biogeochemical transformations of selenium in anoxic environments, p. 389–420. *In* W. T. Frankenberger, Jr. and S. Benson (ed.), *Selenium in the Environment*. Marcel Dekker, Inc., New York, N.Y.
52. **Oremland, R. S., J. Switzer Blum, C. W. Culbertson, P. T. Visscher, L. G. Miller, P. Dowdle, and R. S. Oremland.** 1994. Isolation, growth, and metabolism of an obligately anaerobic, selenate-respiring bacterium, strain SES-3. *Appl. Environ. Microbiol.* **60:**3011–3019.
53. **Oremland, R. S., N. A. Steinberg, T. S. Presser, and L. G. Miller.** 1991. In situ selenate reduction in the agricultural drainage systems of western Nevada. *Appl. Environ. Microbiol.* **57:**615–617.
54. **Oremland, R. S., N. A. Steinberg, A. S. Maest, L. G. Miller, and J. T. Hollibaugh.** 1990. Measurement of in situ rates of selenate removal by dissimilatory bacterial reduction in sediments. *Environ. Sci. Technol.* **24:**1157–1164.
55. **Oremland, R. S., J. T. Hollibaugh, A. S. Maest, T. S. Presser, L. G. Miller, and C. W. Culbertson.** 1989. Selenate reduction to elemental selenium by anaerobic bacteria in sediments and culture: biogeochemical significance of a novel, sulfate-independent respiration. *Appl. Environ. Microbiol.* **55:**2333–2343.
56. **Peterson, M. L., and R. Carpenter.** 1983. Biogeochemical processes affecting total arsenic and arsenic species distributions in an intermittently stratified fjord. *Mar. Chem.* **12:**295–321.
57. **Rech, S. A., and J. M. Macy.** 1992. The terminal reductases for selenate and nitrate respiration in *Thauera selenatis* are two distinct enzymes. *J. Bacteriol.* **174**:7316–7320.
58. **Reimer, K. J. and J. A. J. Thompson.** 1988. Arsenic speciation in marine interstitial water. The occurrence of organoarsenicals. *Biogeochemistry* **6:**211–237.
59. **Rittle, K. A., J. I. Drever, and P. J. Colberg.** 1995. Precipitation of arsenic during sulfate reduction. *Geomicrobiol. J.* **13:**1–12.
60. **Schröder, I., S. Rech, T. Krafft, and J. M. Macy.** 1997. Purification and characterization of the selenate reductase from *Thauera selenatis. J. Biol. Chem.* **272:**23765–23768.
61. **Seyler, P., and J. M. Martin.** 1989. Biogeochemical processes affecting total arsenic and arsenic species distribution in a permanently stratified lake. *Environ. Sci. Technol.* **23:**1258–1263.
62. **Shibata, Y., and M. Morita.** 1988. A novel, trimethylated arseno-sugar isolated from the brown alga *Sargassum thunbergii. Agric. Biol. Chem.* **52:**1087–1089.
63. **Shisler, J. L., T. G. Senkevich, M. J. Berry, and B. Moss.** 1998. Ultraviolet-induced cell death blocked by a selenoprotein from a human dermatotropic poxvirus. *Science* **279:**102–105.
64. **Silver, S., G. Ji, S. Broer, S. Dey, D. Dou, and B. P. Rosen.** 1993. Orphan enzyme or patriarch of a new tribe: arsenic resistance ATPase of bacterial plasmids. *Mol. Microbiol.* **8:**637–642.
65. **Sposito, G., A. Yang, R. H. Neal, and A. Mackzum.** 1991. Selenate reduction in alluvial soil. *Soil Sci. Soc. Am. J.* **55:**1597–1602.
66. **Steinberg, N. A., J. Switzer Blum, L. Hochstein, and R. S. Oremland.** 1992. Nitrate is a preferred electron acceptor for growth of freshwater selenate-respiring bacteria. *Appl. Environ. Microbiol.* **58:** 426–428.
67. **Steinberg, N. A, and R. S. Oremland.** 1990. Dissimilatory selenate reduction potentials in a diversity of sediment types. *Appl. Environ. Microbiol.* **56:**3550–3557.
68. **Stolz, J. F., T. Gugliuzza, J. Switzer Blum, R. Oremland, and F. M. Murillo.** 1997. Differential cytochrome content and reductase activity in *Geospirillum barnesii* strain SES-3. *Arch. Microbiol.* **167:**1–5.

69. **Stolz, J. F., D. J. Ellis, J. Switzer Blum, D. Ahmann, D. R. Lovley, and R. S. Oremland.** 1999. *Sulfurospirillum barnesii* sp. nov., *Sulfurospirillum arsenophilus* sp. nov., and the *Sulfurospirillum* clade in the Epsilon *Proteobacteria*. *Int. J. Syst. Bacteriol.* **49:**1177–1180.
69a. **Stolz, J. F., and R. S. Oremland.** 1999. Bacterial respiration of arsenic and selenium. *FEMS Microbiol. Rev.* **23:**615–627.
70. **Switzer Blum, J., A. Burns Bindi, J. Buzzelli, J. F. Stolz, and R. S. Oremland.** 1998. *Bacillus arsenicoselenatis*, sp. nov., and *Bacillus selenitireducens* sp. nov.: two haloalkaliphiles from Mono Lake, California, that respire oxyanions of selenium and arsenic. *Arch. Microbiol.* **171:**19–30.
71. **Tokunaga, T. K., I. J. Pickering, and G. E. Brown, Jr.** 1996. Selenium transformations in ponded sediments. *Soil Sci. Soc. Am. J.* **60:**781–790.
72. **Tomei, F. A., L. L. Barton, C. L. Lemanski, T. G. Zocco, N. H. Fink, and L. O. Sillerud.** 1995. Transformation of selenate and selenite to elemental selenium by *Desulfovibrio desulfuricans*. *J. Ind. Microbiol.* **14:**329–336.
73. **Tomei, F. A., L. L. Barton, C. L. Lemanski, and T. G. Zocco.** 1992. Reduction of selenate and selenite to elemental selenium by *Wolinella succinogenes*. *Can. J. Bacteriol.* **38:**1328–1333.
74. **Velinsky, D. J., and G. A. Cutter.** 1991. Geochemistry of selenium in a coastal salt marsh. *Geochim. Cosmochim. Acta* **55:**179–191.
75. **Zehnder, A. J. B., and W. Stumm.** 1988. Geochemistry and biogeochemistry of anaerobic habitats, p. 1–38. *In* A. J. B. Zehnder (ed.), *Biology of Anaerobic Microorganisms*. John Wiley & Sons, Inc., New York, N.Y.
76. **Zehr, J. P., and R. S. Oremland.** 1987. Reduction of selenate to selenide by sulfate-respiring bacteria: experiments with cell suspensions and estuarine sediments. *Appl. Environ. Microbiol.* **53:** 1365–1369.
77. **Zhang, Y., and J. M. Moore.** 1996. Selenium fractionation and speciation in a wetland system. *Environ. Sci. Technol.* **30:**2613–2619.
78. **Zhang, Y., and J. M. Moore.** 1996. Controls on selenium distribution in wetland sediment, Benton Lake, Montana. *Water Air Soil Pollut.* **97:**323–340.

Environmental Microbe-Metal Interactions
Edited by Derek R. Lovley

Chapter 10

Microbial Reduction of Chromate

Yi-Tin Wang

Chromium is one of the most widely used metals in industry. It is used to manufacture stainless steel, jet aircraft, automobiles, hospital equipment, and mining equipment. In contrast to most metals, chromium is usually soluble under oxidizing conditions, and only limited removal can be achieved by conventional precipitation processes. Both trivalent chromium, Cr(III), and hexavalent chromium, Cr(VI), exist in various bodies of water in a variety of forms. Based on the theoretical equilibrium redox and acid-base chemistry, Cr(VI) appears to be the predominant dissolved, stable chromium species in aqueous system at natural pH range of 6.5 to 8.5 (31). However, under redox conditions commonly found in natural aquatic systems, Cr(III) is the most stable valence state. Trivalent chromium is not expected to significantly migrate in natural systems because it readily precipitates as Cr(III) minerals or is removed by adsorption (24).

Chromium has been designated a priority pollutant by the U.S. Environmental Protection Agency. Hexavalent chromium has long been considered to be much more toxic than trivalent chromium. In addition to being carcinogenic and mutagenic in animals and humans (11, 25), Cr(VI) at 10 mg/kg of body weight may result in liver necrosis, nephritis, and death in humans. Lower doses can cause gastrointestinal tract irritation and lung cancer (21a). Trivalent chromium, on the other hand, is less likely to cause mutation. A Cr(III) concentration 1,000 times higher than Cr(VI) is required to obtain the same mutation frequency (20). In addition, Cr(III) is an essential trace element necessary for glucose and lipid metabolism and for the utilization of amino acids. It also has been reported to be important in the prevention of mild diabetes and atherosclerosis in humans (33).

Current treatment techniques for chromium-containing wastes generally involve aqueous reduction of Cr(VI) to Cr(III) by using a reducing agent at lowered pH with subsequent adjustment of the solution pH to near-neutral ranges to precipitate the less soluble Cr(III) (10, 13, 30).

Yi-Tin Wang • Department of Civil Engineering, University of Kentucky, Lexington, KY 40506.

Biological transformation of Cr(VI) to Cr(III) has only recently been realized. In such a process, Cr(VI) may serve as an electron acceptor for the oxidation of organic compounds.

Cr(VI)-REDUCING BACTERIA

Many facultative anaerobes are capable of reducing Cr(VI) to Cr(III) under appropriate conditions. As shown in Table 1, chromium-reducing bacteria belong to a variety of genera such as *Achromobacter*, *Aeromonas*, *Agrobacterium*, *Bacillus*, *Desulfovibrio*, *Enterobacter*, *Escherichia*, *Micrococcus*, and *Pseudomonas*. Most are facultative anaerobes and are widespread in nature. Earlier studies assumed that Cr(VI) was reduced to Cr(III) because the medium changed from yellow to white (14, 18, 34). Cr(V) was detected as an intermediate in *Pseudomonas ambigua*, indicating that the reduction of Cr(VI) to Cr(III) is at least a two-step reaction (32). Recently, direct evidence has been provided that Cr(VI) is completely and quantitatively transformed to Cr(III) by *Escherichia coli* (26), *Bacillus* sp. (38), and *Agrobacterium radiobacter* (19).

Although chromium-reducing bacteria are widespread, high cell densities are generally required for significant Cr(VI) reduction to occur. Cell suspensions of *Desulfovibrio vulgaris* ATCC 29579 at 1.1 mg/ml buffer was needed to completely reduce 470 μM Cr(VI) in 100 min (21). The rate of Cr(VI) reduction increased with increasing cell density, as observed with *E. coli* ATCC 33456 (28), *Pseudo-*

Table 1. Microbial populations that transform Cr(VI) to Cr(III)

Organism	Substrate(s)	Redox condition
Achromobacter eurydice	Acetate, glucose	Anaerobic
Aeromonas dechromatica	Galactose, mannose, melibiose, sucrose, fructose, lactose, cellobiose, arabinose, mannitol dulcitol, sorbitol, glycerol	Anaerobic
Agrobacterium radiobacter	Glucose, fructose, maltose, lactose, mannitol, glycerol	Aerobic
Bacillus cereus	Acetate, glucose	Anaerobic
Bacillus sp.	Glucose	Aerobic
Bacillus subtilis	Acetate, glucose	Anaerobic
Desulfovibrio vulgaris	Hydrogen	Anaerobic
Enterobacter cloacae	Acetate, glycerol, glucose	Anaerobic
Escherichia coli	Acetate	Anaerobic
Escherichia coli ATCC 33456	Glucose, acetate, propionate	Aerobic and anaerobic
Micrococcus roseus	Acetate, glucose	Anaerobic
Pseudomonas aeruginosa	Acetate, glucose	Anaerobic
Pseudomonas dechromaticans	Peptone, glucose	Anaerobic
Pseudomonas chromatophila	Ribose, fructose, fumarate, lactate, acetate, succinate, butyrate, glycerol, ethylene glycol	Anaerobic
Pseudomonas ambigua G-1	Nutrient broth	Aerobic
Pseudomonas fluorescens LB 300	Glucose	Aerobic
Pseudomonas putida PRS 2000		

monas fluorescens LB 300 (36), *Bacillus* sp. (36), *A. radiobacter* EPS-916 (19), and *Enterobacter cloacae* HO1 (34). However, the specific rate of Cr(VI) reduction (normalized by cell mass) did not necessarily follow the same pattern. The specific rate of Cr(VI) reduction by *E. coli* was higher at relatively lower cell densities, with a maximum of 86 mg of Cr/h/g (dry weight) of cells observed at a cell density of 3×10^8 cells/ml (28).

INFLUENCE OF ENVIRONMENTAL FACTORS ON Cr(VI) REDUCTION

Electron Donor

Chromium-reducing bacteria may utilize a variety of organic compounds as electron donors for chromium reduction (Table 1). However, the majority of the known electron donors are natural aliphatic compounds, mainly low-molecular-weight carbohydrates, amino acids, and fatty acids. Hydrogen may serve as the electron donor in *D. vulgaris* (21). Enhanced Cr(VI) reduction was observed in several cultures, including *P. fluorescens* ATCC 27663, ATCC 31483, and ATCC 17573, *E. coli* ATCC 15489, and *E. cloacae* ATCC 529, ATCC 29893, and ATCC 35930, in growth media containing yeast extract or nutrient broth but not in minimal salts medium containing a sole carbon source (36). Yeast extract or nutrient broth alone may reduce Cr(VI), presumably by organic compounds with sulfhydryl groups, especially in the absence of oxygen (36).

Although Cr(VI) reduction was observed during the growth phase, cell growth is not necessarily required for Cr(VI) reduction. Resting cells of *P. fluorescens* (2) and *E. coli* (28) reduced Cr(VI) at the same rate as in the growth media. Cr(VI) reduction was also noted with resting cells of *D. vulgaris* (21) and *A. radiobacter* (19).

Cr(VI) Concentration

The rate of Cr(VI) reduction depends on the concentration of Cr(VI). The time required for complete reduction increased progressively as the initial concentration of Cr(VI) increased in cultures of *E. cloacae* (34), *E. coli* (28), *P. fluorescens* (36), and *Bacillus* sp. (36). The rate of Cr(VI) reduction may not be inhibited by high levels of Cr(VI) during the early phase of reduction. The initial specific rate of Cr(VI) reduction by *E. coli*, as normalized by cell dry weight, was higher at a higher initial Cr(VI) concentration (30). However, the opposite trend was observed with *E. cloacae* cultures, in which lower initial rates of Cr(VI) reduction were obtained with higher initial Cr(VI) concentrations (16).

Dissolved Oxygen

Bacterial reduction of Cr(VI) may occur both aerobically and anaerobically (Table 1). Under aerobic conditions, organisms may reduce Cr(VI) through the action of a soluble reductase (SR), using NADH or endogenous electron reserves as an electron donor. Organisms may also reduce Cr(VI) under anaerobic conditions via the mediation of either a soluble reductase, a membrane-bound reductase, or both, with the possible involvement of cytochrome *b*, *c*, and *d* (Fig. 1). The aerobic activity of Cr(VI) reduction is generally associated with soluble proteins, with NADH as an electron donor required to drive the reaction or to provide enhanced activity (14, 26, 36; Y. Ishibashi, M. Beck, C. Cervantes, and S. Silver, Abstr. 89th Annu. Meet. Am. Soc. Microbiol. 1989, p. 361, 1989). In the absence of added electron donors, chromium-reducing organisms may utilize endogenous reserves for the reduction of Cr(VI) through the activity of soluble reductase (15, 26, 36). However, the physiological functions of the electron flow to Cr(VI) through the soluble reductase have not been thoroughly examined. Under anaerobic conditions, Cr(VI) serves as a terminal electron acceptor through the respiratory chains of *E. cloacae* (35), *E. coli* (26), and *D. vulgaris* (21). Recent studies have also implicated the membrane-bound respiratory chain in the transfer of reducing equivalents to Cr(VI) through cytochrome *c* in *E. cloacae* (34) and cytochromes *b* and *d* in *E. coli* (26). In the absence of oxygen, the soluble-reductase activity may mediate electron transport to Cr(VI), as observed in *E. coli* (26) and *D. vulgaris* (21), in which the cytochrome c_3 in the soluble protein fraction of *D. vulgaris* was needed for Cr(VI) reduction. However, there is no evidence to show that electron transport to Cr(VI) yields energy to support anaerobic growth of Cr(VI)-reducing strains.

Although chromium-reducing bacteria are widespread, only two species, *A. radiobacter* EPS-916 (19) and *E. coli* ATCC 33456 (26), have been reported to reduce Cr(VI) in liquid media both aerobically and anaerobically. However, these organisms reduced Cr(VI) better under anaerobic conditions than under aerobic conditions. *A. radiobacter* EPS-916 actively reduced 0.05 mM chromate while growing aerobically but reduced up to 0.15 mM chromate under anaerobic conditions (19). Cr(VI) reduction by *E. coli* ATCC 33456 was found to be repressed by dissolved

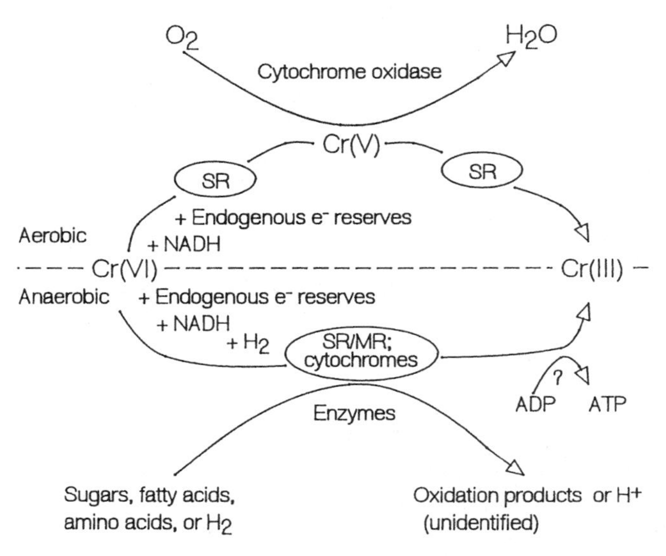

Figure 1. Mechanisms of Cr(VI) reduction in bacteria. SR, soluble reductase; MR, membrane-bound reductase.

oxygen, with an apparent uncompetitive inhibition behavior of oxygen observed. The maximum specific Cr(VI) reduction rate, *k*, decreased from 0.5 mmol/h/g (dry weight) of cells under anaerobic conditions to 0.27 mmol/h/g (dry weight) in the presence of oxygen (26). From a thermodynamic point of view, oxygen is the preferred final electron acceptor compared to Cr(VI) since higher energy levels will be obtained by cells through the electron transport systems involving cytochromes. This may explain why lower rates of Cr(VI) reduction were always noted in the presence of oxygen.

Other Electron Acceptors

Inhibition of sulfate and nitrate on Cr(VI) reduction has not been reported for aerobic cultures. Sulfate up to 1 mM and nitrate at 200 μM had no effect on chromate reduction either with whole cells or with cell-free supernatant fluid of *Pseudomonas putida* (15). Sulfate at 10 mM and nitrate at 16 mM did not affect Cr(VI) reduction by whole cells and cell extract of *Bacillus* sp. (36). In anaerobic cultures, the rate of Cr(VI) reduction by *E. coli* was not affected by up to 83 mM sulfate and 129 mM nitrate (28). Sulfate concentrations as high as 50 mM did not inhibit Cr(VI) reduction in *D. vulgaris* (21). However, Cr(VI) reduction activity of *E. cloacae* HO1 under anaerobic conditions was reduced to 68% in the presence of 25 μM $ZnSO_4$ and to 84% in the presence of 5 mM $NaNO_3$ (16).

pH and Temperature

The optimal pH and temperature for Cr(VI) reduction generally coincide with the optimal growth conditions of cells. Cr(VI) reduction by *E. cloacae* was observed at pH 6.0 to 8.5 and at temperatures of 20 to 40°C, with pH and temperature optima ranging from 7.0 to 7.8 and 30 to 37°C, respectively (16). Cr(VI) reduction by *E. coli* was observed within a pH range of 3 to 8 and a temperature range of 10 to 45°C, and the observed maximum initial specific rate of Cr(VI) reduction occurred at pH 7 and about 36°C. The rate data were found to fit the Arrhenius equation closely (28). For another Cr(VI)-reducing species, a *Bacillus* sp., the optimal conditions for Cr(VI) reduction were pH 7 and 30°C (36).

Redox Potential

The best range of redox potential for Cr(VI) reduction has not yet been conclusively established. Using washed, resting cells of *A. radiobacter* pregrown under different carbon and energy sources, the rate of Cr(VI) reduction was shown to be greater in cell suspensions at −240 than at −198 mV (19). When Cr(VI) was introduced into *E. coli* cultures with a redox potential greater than −140 mV, no reduction of Cr(VI) was observed within the first 1 h (12). However, in *B. subtilis* cultures, even with a high initial redox potential of about +250 mV, Cr(VI) reduction occurred after 1 hr of incubation as the redox potential decreased (12). Cr(VI) reduction by *Bacillus* sp. occurred over a wide range of redox potentials. Cr(VI) reduction began almost immediately at the beginning of the incubation in cultures with a redox potential of about +250 mV and continued at a constant rate throughout the incubation, despite a rapid drop in the redox potential to a low of −500 mV after 48 h, when the growth phase was evident (36).

Low redox potential generated by microbial activity was shown not to affect the reduction of Cr(VI), as demonstrated by using *E. coli* ATCC 33456 (28). *E. coli* ATCC 33456 was inoculated to a very low cell concentration (10^4 cells/ml) and incubated anaerobically with glucose as the sole carbon source and Cr(VI) in the form of CrO_4^{2-} as a potential electron acceptor. The anaerobic culture of *E. coli* ATCC 33456 quickly lowered the redox potential to below −400 mV despite a very low initial cell density. However, no significant Cr(VI) reduction was noted even at such a low redox potential. Cr(VI) reduction, on the other hand, occurred in the aerobic culture with a high initial cell density (10^{10} cells/ml). Cr(VI) reduction continued even after the redox potential of the aerobic culture had increased to above +150 mV from a low of about −500 mV after 6 h of incubation. In addition, a rapid decrease of the redox potential in the aerobic culture, from above +200 mV to below −500 mV after 6 h of incubation, did not seem to significantly enhance Cr(VI) reduction compared to that found in the positive redox potential range.

Inhibition by Metals

Cr(VI)-reducing organisms are susceptible to several heavy metals. Cr(VI) reduction by *E. cloacae* was completely inhibited by 0.5 mM Zn^{2+} and was reduced to 70% by 0.5 mM Cu^{2+} (17). Cr(VI) reduction by *E. coli* was reduced to 80 and 84% of control levels by 0.8 mM Zn^{2+} and 3 mM Cu^{2+}, respectively, while up to 0.2 mM Cd^{2+} and 0.1 mM Pb^{2+} did not exhibit inhibition (28). Strong inhibition of Cr(VI) reduction by Hg^{2+} and Ag^+ in *P. putida* was characterized as noncompetitive, with a K_I of 20 μM observed for both Hg^{2+} and Ag^+ (15). However, at 100 μM, a range of 11 metals including Zn^{2+} and Cu^{2+} had no effect on Cr(VI) reduction by *D. vulgaris* (21). The reduced form of Cr(VI), Cr^{3+}, was found not to inhibit Cr(VI) reduction by *P. putida* at 200 μM (21).

Inhibition by Phenolic Compounds

In addition to heavy metals, phenolic compounds have been found as cocontaminants in chromium-containing wastes and chromium-polluted sites. Inhibition of Cr(VI) reduction by phenolic compounds was observed with a few Cr(VI)-reducing organisms. Phenol and *p*-cresol at 5 mM each and 2-chlorophenol at 2 mM severely inhibited both Cr(VI) reduction and cell growth of *P. fluorescens* LB 3000 (36). The toxic effect of three phenolic compounds on Cr(VI) reduction by *E. coli* ATCC 33456 was recently assessed by examing the initial specific Cr(VI) reduction rate (26). Anaerobic cultures of *E. coli* ATCC 33456 were found to be more susceptible to phenolic compounds than were aerobic cultures, with a 50% decrease in Cr(VI) reduction activity being caused by *p*-cresol, 2-chlorophenol, and phenol at 9, 12, and 56 mM, respectively, in anaerobic cultures and 15, 20, and 82 mM, respectively, in aerobic cultures.

Cr(VI) REDUCTION CAPACITY AND Cr(VI) TOXICITY

The toxicity of Cr(VI) toward Cr(VI) reduction may be illustrated by the finite capacity of cells. Based on a Cr(VI) reduction experiment performed at a range of Cr(VI) concentrations, a finite Cr(VI) reduction capacity ranged from 21 to 65 mg of Cr(VI)/liter was observed in a *Bacillus* sp., *P. fluorescens* LB 300, and *E. coli* (38). These finite capacities were simulated well with an enzyme-based kinetic model (38). The ability of this model to analyze Cr(VI) reduction was further demonstrated with two other strains, *D. vulgaris* and *P. ambigua,* reported in the literature. The model illustrates an important characteristic of microbial Cr(VI) reduction. The rate and extent of Cr(VI) reduction in batch cultures are dependent upon the finite Cr(VI) reduction capacity of the cells, which is governed by the initial cell density regardless of subsequent growth. During Cr(VI) reduction, both increases and decreases in cell density were observed. For the observed cell density decrease during Cr(VI) reduction, the finite capacity of Cr(VI) reduction in batch cultures may be the result of termination of metabolic activity due to Cr(VI) toxicity. This phenomenon may include Cr(VI) reduction in *E. coli* and a *Bacillus* sp., in which bacterial cell density was observed to continuously decrease during the course of Cr(VI) reduction.

The oxidizing power of Cr(VI) constitutes its toxic effects toward bacterial cells. In addition to the enzymatic mechanism of Cr(VI) reduction, Cr(VI) may directly interact with various intracellular reducing agents such as sulfhydryl groups inside cells once it has penetrated the cells through a specific energy-dependent transport system (1, 23). Destruction of the reducing powers in bacterial cells may inhibit the metabolism of bacterial cells and thus result in a decrease and eventual termination of Cr(VI) reduction. The toxic effects of Cr(VI) toward bacterial cells are largely attributed to metabolic conversion of Cr(VI) to Cr(III) rather than Cr(VI) or its transformation product Cr(III) (3, 9). Hence, the rate of Cr(VI) reduction by batch cultures decreases progressively with continued Cr(VI) reduction and ceases when the finite Cr(VI) reduction capacity becomes exhausted. However, other mechanisms may be responsible for Cr(VI) reduction in the Cr(VI)-reducing strain *E. cloacae* HO1. *E. cloacae* reduced Cr(VI) with an increasing rate as the cell density decreased progressively during Cr(VI) reduction at initial Cr(VI) concentrations ranging from 0.46 to 0.85 mM (39).

During Cr(VI) reduction, continuous cell growth or utilization of substrate may occur, as was the case with *P. fluorescens* LB 300 and *P. ambigua* (41). However, such metabolic activity or cell growth did not result in increased or even prolonged Cr(VI) reduction. Instead, the rate of Cr(VI) reduction in these cultures decreased progressively and eventually ceased. This phenomenon may be attributed to the mutagenic effects of Cr(VI) toward bacterial cells. Mutations in bacterial cells as a result of DNA damage due to cellular interactions with Cr(VI) have been frequently observed (16, 23). DNA damage influences the genes governing critical aspects of cell metabolism (9) and thus may lead to infidelity of cell replication, which results in mutated descendants of the affected cells. Mutant cells may be more resistant to Cr(VI), as demonstrated by continuous utilization of glucose in LB 300 culture or by significant cell growth in *P. ambigua* culture. However, Cr(VI) resistance is not correlated with Cr(VI) reduction (4, 6). Resistance to Cr(VI) in mutant cells was considered to be associated with decreased Cr(VI) transport due to chromosomal mutations (4). In turn, the decrease in Cr(VI) transport may inhibit Cr(VI) reduction by the mutant cells. Therefore, the mutagenic effects of Cr(VI) on cells lead to progressive inactivation of bacterial cells for Cr(VI) reduction as Cr(VI) is reduced. Consequently, the growth of bacterial cells may play an insignificant role in Cr(VI) reduction because of the mutagenic effects of Cr(VI).

The finite Cr(VI) reduction capacity also suggests that energy yields from Cr(VI) reduction may not be conserved as biochemical energy for cell growth under both aerobic and anaerobic conditions. This is indicated by the kinetic model, which assumes a continuous reduction of active cells during Cr(VI) reduction, and is also supported by the absence of a correlation between Cr(VI) reduction and cell growth. Therefore, Cr(VI) reduction may occur as a result of cometabolism or a side activity of bacteria. This hypothesis is supported by the model analysis of Cr(VI) reduction data and is in agreement with the previous conclusion derived from an analysis of the enzymatic characteristics of Cr(VI) reduction (6, 29). However, the model has not been tested in the field, where many environmental factors such as pH, temperature, dissolved-oxygen concentration and cocontaminants may influence the rate of Cr(VI) reduction (27, 36, 37).

SIMULTANEOUS REMOVAL OF Cr(VI) AND ORGANIC POLLUTANTS

The electron donors known for Cr(VI) reduction are generally limited to nontoxic aliphatic compounds, mainly low-molecular-weight carbohydrates, amino acids, and fatty acids. A variety of aromatic contaminants, including phenol, 2-chlorophenol, *p*-cresol, 2,6-dimethylphenol, 3,5-dimethylphenol, benzene, and toluene, can be metabolized as the sole carbon source for chromium reduction in a defined coculture consisting of a chromium reducer, *E. coli* 33456, and a non-chromium-reducing species, *P. putida* DMP 1 (29). This finding is significant since these organic contaminants are present together with chromium in a number of industrial wastes and subsurface environment (2, 5). Cr(VI) reduction was correlated with phenol degradation, and both the rate and extent of Cr(VI) reduction and phenol degradation were significantly influenced by the population composition of the coculture. In the coculture, *P. putida* oxidized phenol using molecular oxygen as the electron acceptor. Metabolites formed from phenol degradation were then used by *E. coli* for growth and served as the electron donor for Cr(VI) reduction since *E coli* does not utilize phenol for energy and growth. The oxidation of phenol by *P. putida* initiated the energy flow of the coculture and served as a primary energy source for both strains, while Cr(VI) reduction occurred only as a result of cometabolism.

Simultaneous Cr(VI) reduction and phenol degradation was also demonstrated under anaerobic conditions in a mixed culture consisting of phenol degraders and Cr(VI) reducers (35a). The anaerobic consortium served as the principal phenol-degrading organisms, while *E. coli* was the Cr(VI)-reducing species. Similar to the *P. putida-E. coli* coculture, Cr(VI) reduction by *E. coli* in the anaerobic consortium depends on metabolites formed from phenol degradation by the anaerobic consortium. Simultaneous reduction of Cr(VI) and degradation of other organic pollutants including benzene, naphthalene, trichloroethylene, 2-chlorophenol, 2,4-dichlorophenol, and chloroform was also noted in the anaerobic mixed culture.

ENVIRONMENTAL SIGNIFICANCE

Recent work has revealed the potential of using Cr(VI)-reducing microorganisms for detoxifying Cr(VI)-contaminated environments or for treating Cr(VI)-containing wastes, even if the biochemical mechanisms of Cr(VI) reduction are not yet fully understood. The finding that biological reduction of Cr(VI) occurs near the neutral pH and over a moderate temperature range implies that no costly chemical reagents and extensive energy inputs are required. However, the need for added carbon sources and inhibition by heavy metals, the most likely cocontaminants, may render biological treatment of industrial effluents less efficient (22). A potential solution for the carbon source requirement is to use defined cocultures or mixed cultures for simultaneous removal of Cr(VI) and organic pollutants. Another challenge to be overcome before the application of biological treatment for Cr(VI) is Cr(VI) toxicity, which may lead to cell inactivation and loss of Cr(VI) reduction capacity (27, 39). Recent success in using a biofilm reactor for continuous reduction of Cr(VI) may have shed some light on methods for the biological treatment of Cr(VI)-containing wastes (7, 8).

REFERENCES

1. **Alvarez-Cohen, L., and P. L. McCarty.** 1991. A cometabolic biotransformation model for halogenated aliphatic compounds exhibiting product toxicity. *Environ. Sci. Technol.* **25:**1381–1387.
2. **Beszedits, S.**, 1988. Chromium removal from industrial wastewaters, p. 232–263. *In* O. Nriagu and E. Nieboer (ed.), *Chromium in the Natural and Human Environments.* John Wiley & Sons, Inc., New York, N.Y.
3. **Bianchi, V., L. Celotti, and G. Lanfranchi.** 1983. Genetic effects of chromium compounds. *Mutat. Res.* **117:**279–300.
4. **Bopp, L. H., and H. L. Ehrlich.** 1988. Chromate resistance and reduction in *Pseudomonas fluorescens* strain LB300. *Arch. Microbiol.* **150:**426–431.
5. **Canter, L. W., and R. C. Knox.** 1986. *Ground Water Pollution Control.* Lewis Publishers, Inc., Chelsea, Mich.
6. **Cervantes, C., and S. Silver.** 1992. Plasmid chromate resistance and chromate reduction. *Plasmid* **27:**65–71.
7. **Chirwa, E. M. N., and Y. T. Wang.** 1997. Hexavalent chromium reduction by *Bacillus* sp. in a packed-bed bioreactor. *Environ. Sci. Technol.* **31:**1446–1451.
8. **Chirwa, E. M. N., and Y. T. Wang**. 1997. Biological reduction of hexavalent chromium by *Pseudomonas fluorescens* LB 300 in a fixed-film reactor. *J. Environ. Eng.* **123:**760–766.
9. **DeFlora, S., V. Bianchi, and A. G. Levis.** 1984. Distinctive mechanisms for interaction of hexavalent omium and trivalent chromium with DNA. *Toxicol. Environ. Chem.* **8:**287–294.
10. **Eary, L. E., and D. Ral.** 1988. Chromate removal from aqueous wastes by reduction with ferrous ion. *Environ. Sci. Technol.* **22:**972–977.
11. **Enterline, P. E.** 1974. Respiratory cancer among chromate workers. *J. Occup. Med.* **16:**523–526.
12. **Gvozdyak, P. I., N. F. Mogilevich, A. F. Rylskii, and N. I. Grishchenko.** 1986. Reduction of hexavalent chromium by collection strains of bacteria. *Mikrobiologiya* **55:**962–965.
13. **Heller, R. J., and C. H. Roy.** 1986. Hexavalent chromium reduction. *Plating Surf. Finishing* **73:** 22.

14. **Horitsu, H., S. Futo, Y. Miyazawa, S. Ogai, and K. Kawai.** 1987. Enzymatic reduction of hexavalent chromium by hexavalent chromium tolerant *Pseudomonas ambigua* G-1. *Agric. Biol. Chem.* **51:**2417–2420.
15. **Ishibashi, Y., C. Cervantes, and S. Silver.** 1990. Chromium reduction in *Pseudomonas putida. Appl. Environ. Microbiol.* **56:**2268–2270.
16. **Komori, K., A. Rivas, K. Toda, and H. Ohtake.** 1989. Biological removal of toxic chromium using an *Enterobacter cloacae* strain that reduces chromate under anaerobic conditions. *Biotechnol. Bioeng.* **35:**951–954.
17. **Komori, K., P. C. Wang, K. Toda, and H. Ohtake.** 1989. Factors affecting chromate reduction in *Enterobacter cloacae* strain HO1. *Appl. Microbiol. Biotechnol.* **31:**567–570.
18. **Lebedeva, E. V., and N. N. Lyalikova.** 1979. Reduction of crocoite by *Pseudomonas chromatophila* sp. nov. *Mikrobiologiya* **48:**517–522.
19. **Llovera, S., R. Bonet, M. Simon-Pujol, and F. Congregado.** 1993. Chromate reduction by resting cells of *Agrobacterium radiobacter* EPS-916. *Appl. Environ. Microbiol.* **59:**3516–3518.
20. **Lofroth, G., and B. N. Ames.** 1978. Mutagenicity of inorganic compounds in *Salmonella typhimurium*: arsenic, chromium and selenium. *Mutat. Res.* **53:**65–66.
21. **Lovley, D. R., and E. J. P. Phillips.** 1994. Reduction of chromate by *Desulfovibrio vulgaris* and its c_3 cytochrome. *Appl. Environ. Microbiol.* **60:**726–728.
21a. **Mancuso, T. F.** 1975. Consideration of chromium as an industrial carcinogen, p. 343–356. *In* T. C. Hutchinson (ed.), *Proceedings of the International Conference on Heavy Metals in the Environment.* Toronto Institute of Environmental Studies, Toronto, Canada.
22. **Ohtake, H., E. Fujii, and T. Toda.** 1990. Reduction of toxic chromate in an industrial effluent by use of a chromate-reducing strain of *Enterobacter cloacae. Environ. Technol.* **11:**663–668.
23. **Petrilli, F. L., and S. DeFlora.** 1977. Toxicity and mutagenicity of hexavalent chromium on *Salmonella typhimurium. Appl. Environ. Microbiol.* **33:**805–809.
24. **Richard, F. C., and A. C. M. Bourg.** 1991. Aqueous geochemistry of chromium: a review. *Water Res.* **25:**807–816.
25. **Roe, F. J. C., and R. L. Carter.** 1969. Chromium carcinogenesis: calcium chromate as a potent carcinogen from the subcutaneous tissues of the rat. *Br. J. Cancer* **23:**172–176.
26. **Shen, H., and Y. T. Wang.** 1993. Characterization of enzymatic reduction of hexavalent chromium by *Escherichia coli* ATCC 33456. *Appl. Environ. Microbiol.* **59:**3771–3777.
27. **Shen, H., and Y. T. Wang.** 1994. Modeling hexavalent chromium reduction in *Escherichia coli* 33456. *Biotechnol. Bioeng.* **43:**293–300.
28. **Shen, H., and Y. T. Wang.** 1994. Biological reduction of chromium by *E. coli. J. Environ. Eng.* **120:**560–572.
29. **Shen, H., and Y. T. Wang.** 1995. Simultaneous chromium reduction and phenol degradation in a coculture of *Escherichia coli* ATCC 33456 and *Pseudomonas putida* DMP-1. *Appl. Environ. Microbiol.* **61:**2754–2758.
30. **Smillie, R. H., K. Hunter, and M. Loutit.** 1981. Reduction of chromium(VI) by bacterially produced hydrogen sulphide in a marine environment. *Water Res.* **15:**1351–1354.
31. **Stumm, W., and J. J. Morgan.** 1980. *Aquatic Chemistry*, 2nd ed. John Wiley & Sons, Inc., New York, N.Y.
32. **Suzuki, T., N. Miyata, H. Horitsu, K. Kawai, K. Takamizawa, Y. Tai, and M. Okazaki.** 1992. NAD(P)H-dependent chromium(VI) reductase of *Pseudomonas ambigua* G-1: a Cr(VI) intermediate is formed during the reduction of Cr(VI) to Cr(III). *J. Bacteriol.* **174:**5340–5345.
33. **Towhill, L. E., C. R. Shriner, J. S. Drury, A. S. Hammons, and J. W. Holleman.** 1978. *Reviews of the Environmental Effects of Pollutants.* III. *Chromium.* EPA 600/1-78-023. U.S. Environmental Protection Agency, Washington, D.C.
34. **Wang, P. C., T. Mori, K. Komori, M. Sasatsu, K. Toda, and H. Ohtake.** 1989. Isolation and characterization of an *Enterobacter cloacae* strain that reduces hexavalent chromium under anaerobic conditions. *Appl. Environ. Microbiol.* **55:**1665–1669.
35. **Wang, P. C., K. Toda, H. Ohtake, I. Kusaka, and I. Yabe.** 1991. Membrane-bound respiratory system of *Enterobacter cloacae* strain HO1 grown anaerobically with chromate. *FEMS Microbiol. Lett.* **78:**11–16.

35a. **Wang, Y. T., and E. M. Chirwa.** 1997. Simultaneous removal of chromium and organic pollutants by an anaerobic consortium of bacteria, p. 187–196. *In Proceedings of the 29th Mid-Atlantic Industrial and Hazardous Waste Conference.* Technomic Publishing Co., Lancaster, Pa.
36. **Wang, Y. T., and C. Xiao.** 1995. Factors affecting hexavalent chromium reduction in pure cultures of bacteria. *Water Res.* **29:**2467–2474.
37. **Wang, Y. T., and H. Shen.** 1995. Bacterial reduction of hexavalent chromium. *J. Ind. Microbiol.* **14:**159–163.
38. **Wang, Y. T., and H. Shen.** 1997. Modelling Cr(VI) reduction by pure bacterial cultures. *Water Res.* **31:**727–732.
39. **Yamamoto, K., J. Kato, T. Yano, and H. Ohtake.** 1993. Kinetics and modeling of hexavalent chromium in *Enterobacter cloacae. Biotechnol. Bioeng.* **41:**129–133.

Environmental Microbe-Metal Interactions
Edited by Derek R. Lovley

Chapter 11

Influence of Fungi on the Environmental Mobility of Metals and Metalloids

Geoffrey M. Gadd and Jacqueline A. Sayer

In the terrestrial environment, fungi are of fundamental importance as decomposer organisms and plant symbionts (mycorrhizas), playing important roles in carbon mineralization and other biogeochemical cycles (152). They are often dominant under acidic conditions, where toxic metals may be speciated into mobile forms (94), and in soil they can comprise the largest pool of biomass (including other microorganisms and invertebrates) (91). This, combined with their branching filamentous explorative growth habit and high surface-area-to-mass ratio, ensures that fungus-metal interactions are an integral component of major environmental cycling processes. Metals and their derivatives can interact with fungi in various ways depending on the metal species, organism, and environment, while fungal metabolic activities can also influence speciation and mobility (Fig. 1) (41). Certain mechanisms may mobilize metals into forms available for cellular uptake and leaching from the system, e.g., complexation with organic acids, other metabolites, and siderophores (37). Metals may also be immobilized by, e.g., sorption onto cell components or exopolymers, transport, and intra- and extracellular sequestration or precipitation (Fig. 2) (44, 93, 126). The relative importance of such apparently opposing phenomena of solubilization and immobilization is a key component of the biogeochemical cycles for toxic metals, whether indigenous or introduced into a given location, and is a fundamental determinant of fungal growth, physiology, and morphogenesis (Fig. 1) (94, 122, 159). Furthermore, several processes are relevant to environmental bioremediation (127, 159). This chapter seeks to highlight the physicochemical and biochemical mechanisms by which fungi can interact with and transform toxic metal species between soluble and insoluble forms, the significance of these processes in the environment, and their potential for use in bioremediation.

Geoffrey M. Gadd and Jacqueline A. Sayer • Department of Biological Sciences, University of Dundee, Dundee DD1 4HN, Scotland.

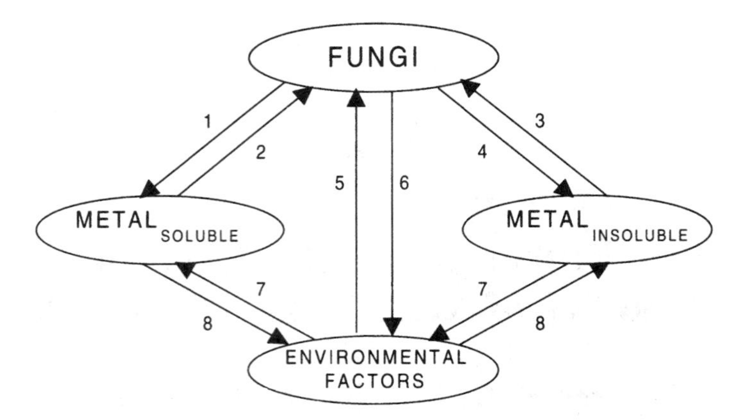

Figure 1. Simple model for the biogeochemical significance of metal and metalloid transformations by fungi. Their influence in effecting changes in metal solubility is emphasized, as well as the influence of environmental factors on these processes and on fungal growth, morphogenesis, and physiology. The relative balance between the processes will depend on the environment, organism(s), and interactions with other organisms including animals, plants and anthropogenic activities. 1, Metal solubilization by, e.g., heterotrophic leaching, siderophores, metabolite excretion including organic acids and H^+, redox reactions, methylation, and biodegradation of organometal(loid)s. 2, Effects of soluble metal species on fungi and metal immobilization by, e.g., biosorption, transport, intracellular sequestration and compartmentation, redox reactions, precipitation, and crystallization. 3, Effects of insoluble metal species on fungi, particulate adsorption, and entrapment by polysaccharide and/or mycelial network. 4, Metal immobilization by, e.g., precipitation, crystallization, or reduction. 5, Influence of environmental factors, e.g., pH, O_2, CO_2, nutrients, salinity, toxic metals, and other pollutants, on fungal growth, metabolism, and morphogenesis. 6, Influence of fungal activities on the environment, e.g., alterations in pH, O_2, CO_2, and redox potential; depletion of nutrients; and enzyme and metabolite excretion. 7 and 8, Environmental factors which direct the equilibrium between soluble and insoluble metal species towards metal mobilization (step 7) or metal immobilization (step 8) (44, 94, 159).

SOLUBILIZATION

Mechanisms of Metal Solubilization

Fungal solubilization of insoluble metal compounds, including certain oxides, phosphates, sulfides, and mineral ores, occurs by several mechanisms. Solubilization can occur by protonation of the anion of the metal compound, decreasing its availability to the cation, with the proton-translocating ATPase of the plasma membrane and the production of organic acids being sources of protons (41, 47, 74, 94, 129). In addition, organic acid anions are frequently capable of soluble complex formation with metal cations, thereby increasing mobility (159). Such complexation is dependent on the relative concentrations of the anions and metals in solution, pH, and the stability constants of the various complexes (27). A further mechanism of metal solubilization is the production of low-molecular-weight iron-chelating siderophores which solubilize iron(III). Siderophores are the most com-

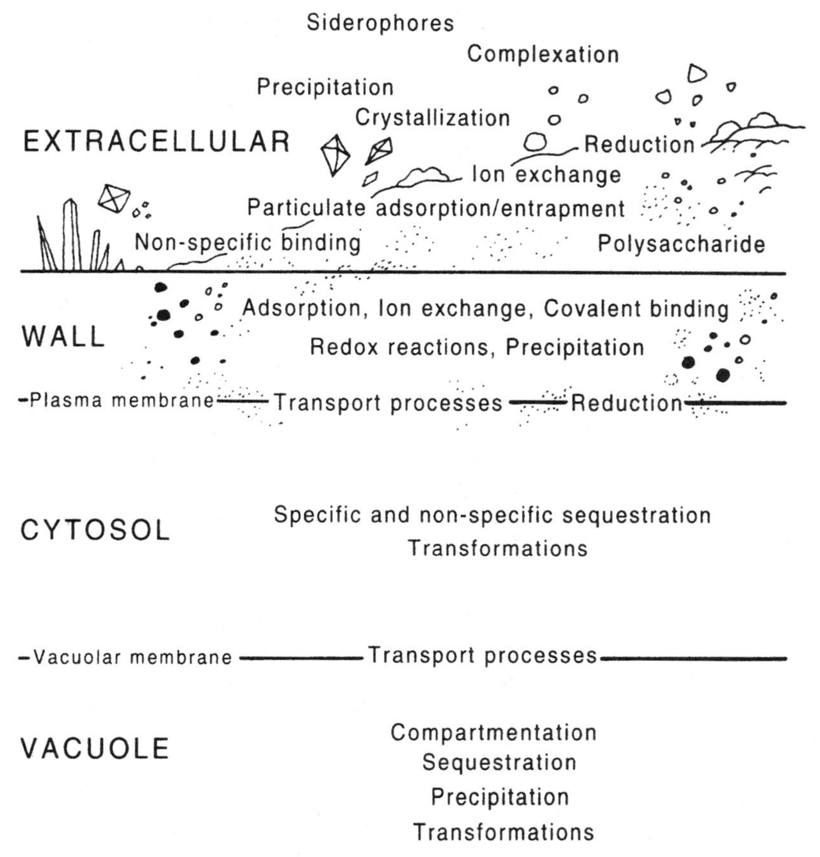

Figure 2. Mechanisms and cellular location of key fungal transformations of metals and metalloids. The list of interactions is not exhaustive, and considerable differences may occur between different species and strains. The location of some processes, especially certain sequestration and transformation reactions, is still uncertain, and this diagram does not include the possible involvement of other organelles, e.g., mitochondria, endoplasmic reticulum, and nucleus, in metal homeostasis and compartmentation.

mon means of acquisition of iron by bacteria and fungi and are effective in a wide range of soils, including calcareous soils. The most common fungal siderophore is ferrichrome (24).

A simple method of screening fungi for the solubilization of insoluble metal compounds is based on observing clear zones of solubilization around colonies growing on solid medium amended with the desired insoluble metal compound (18, 129). Using this method, *Aspergillus niger* was found to be capable of solubilizing a wide range of insoluble metal compounds, including CdS, $Cu_3(PO_4)_2$, nickel phosphate, manganese sulfide (J. A. Sayer and G. M. Gadd, unpublished work), and metal-bearing mineral ores including cuprite (CuS), rhodochrosite $[Mn(CO_3)_x]$

(128), gypsum ($CaSO_4{\cdot}2H_2O$) (57, 58), and pyromorphite (130). The incidence of metal-solubilizing ability among natural soil fungal communities appears to be relatively high; in one study, approximately 1/3 of the isolates tested were able to solubilize at least one of $Co_3(PO_4)_2$, ZnO, or $Zn_3(PO_4)_2$ and approximately 1/10 were able to solubilize all three (129).

Environmental Significance of Metal Solubilization by Fungi

Solubilization of insoluble metal compounds is an important but unappreciated aspect of fungal physiology for the release of anions, such as phosphate, and essential metal cations into forms available for intracellular uptake and into biogeochemical cycles. Most phosphate fertilizers are applied in a solid form (e.g., calcium phosphate), which has to be solubilized before it is available to plants and other organisms. In fact, increased phosphate uptake by mycorrhizal plants is believed to be due to the high phosphate-solubilizing ability of the mycorrhizal partner (81), and in Canada, a formulation containing spores of *Penicillium bilaii* has been registered for application to crops since it is believed that the phosphate-solubilizing ability of the fungus will improve crop productivity (25). As well as phosphate, soil fungi including mycorrhizas may increase inorganic-nutrient availability to plants and other microorganisms by increasing the mobility of essential metal cations and other anions, e.g., sulfate (58). It is significant that root infection is favored in nutrient-deficient soils, and this promotes increased growth of the host plant. Read (124) has proposed that mycorrhizas provide host plants with growth-limiting resources at the stages in their growth when they are most required. Solubilization may therefore be important for the release of essential metal ions.

Solubilization of insoluble toxic metal compounds in the environment can have adverse effects if potentially toxic metal ions are released into soil and/or water systems from metal-contaminated locations. Metal-citrate complexes are highly mobile and are not readily degraded, with degradation depending on the type of complex formed rather than the toxicity of the metal cation involved. Hence, the presence of citric acid in the terrestrial environment will leach potentially toxic metals from soil (38). In fact, it has been suggested that certain microbiological solubilization processes, including proton efflux, are included in the safety assessment of waste repositories (3). Some rock phosphate fertilizers contain cadmium, and solubilization of these to release the phosphate could also release Cd, increasing its availability to the soil biota (84). Metal-contaminated mining areas are often revegetated to help stabilize soils and reduce runoff and wind erosion. The production of organic acids and other metabolites by rhizosphere microorganisms in such locations can also influence metal mobility, especially when the soil microflora has not been completely restored (8).

IMMOBILIZATION

Metal immobilization by fungi may be metabolism independent, occurring whether the biomass is dead or alive, or metabolism dependent, comprising processes which sequester, precipitate, internalize, or transform the metal species and may involve both organic and inorganic extracellular metabolites (41, 94).

Physicochemical Mechanisms of Metal Immobilization

Fungal cell walls are complex macromolecular structures that consist predominantly of chitins, chitosans, and glucans but also contain other polysaccharides, proteins, lipids, and pigments, e.g. melanin (112). This variety of structural components ensures that many different functional groups are able to bind metal ions to various degrees depending on their chemical proclivities. Such uptake of metals by fungal biomass normally corresponds to models with a multiplicity of non-equivalent, heterogeneous binding sites (5, 6, 15, 29, 143). The strength and extent of bonding are dependent on the electrostatic interactions between the metal ions and the ligand groupings, and this is itself mediated by factors such as metal speciation, ionic radius, electronegativity, ligand ionic charge, and accessibility. Adsorptive preferences have frequently been shown to be related to the Irving-Williams scale of ion-ligand attraction (69) as well as to the Lewis theory of acid-base interactions (16, 66). Loci on fungal cell walls can further act as precipitation nuclei, with precipitation of metals occurring in and around cell wall components, a phenomenon particularly evident for actinides such as uranium and thorium (52, 146, 147). Passive uptake of metals is also highly pH dependent, with adsorption of cationic species often decreasing with external pH as protons compete more strongly with the metal ions for sorbing sites (21, 35, 49, 50).

Physiological Mechanisms of Metal Immobilization

Transport and Intracellular Fate

Many metals are essential for fungal growth and metabolism, and mechanisms exist for the acquisition of metals such as Na, Mg, K, Ca, Mn, Fe, Co, Ni, Cu, and Zn from the external environment by the action of transport systems of varyious specificities (41). Inessential toxic metals, although generally of low abundance in the environment unless redistributed by anthropogenic activities, can often compete with physiologically essential ions for such transport systems. Cesium, for example, competes for K^+ transport systems and can substitute for K^+ in K^+-regulated enzymes, with deleterious results (4, 113, 114).

Most work on metal ion transport in fungi has concerned K^+ and Ca^{2+}, in view of their importance in fungal growth, metabolism, and differentiation, although the transport of other essential metal species is now receiving renewed attention (33, 54, 77). Work carried out at low micronutrient metal concentrations has clearly demonstrated the existence of multiple and high-affinity metal transport systems in unicellular and filamentous fungi (Fig. 3) (28, 33, 64, 111, 116, 134, 158, 163, 164). *Saccharomyces cerevisiae* possesses a high-affinity Mn^{2+} transport system (K_m = 0.3 μM) that is functional at low Mn^{2+} concentrations (25 to 1,000 nM) and is of low specificity, being inhibited by Mg^{2+}, Co^{2+}, Zn^{2+}, and Cd^{2+} to various extents. At higher concentrations of $MnCl_2$ (5 to 200 μM), a transport system of lower affinity (K_m = 63 μM) is available to the cell. Although the manganese transport system of *S. cerevisiae* is relatively nonspecific, transport of manganese

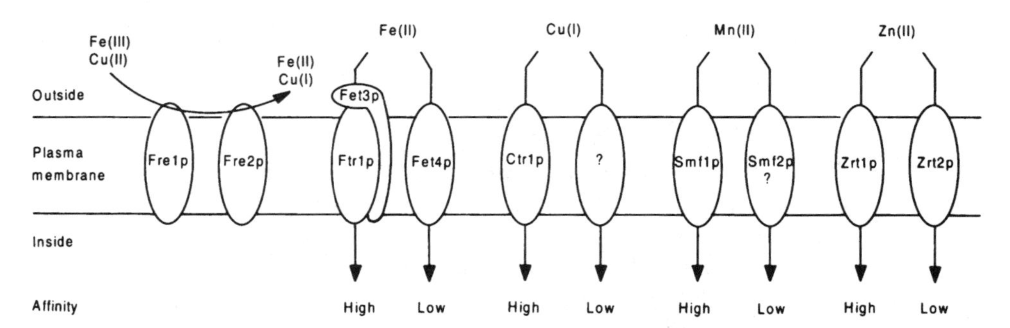

Figure 3. Some metal ion transport sytems of *S. cerevisiae.* The separate high-affinity transport systems ensure the accumulation of essential metals that are present at low external concentrations. Note that iron uptake has two phases: reduction by Fe(III) reductases of Fe(III) to Fe(II) followed by cellular entry of Fe(II) (*S. cerevisiae* does not produce siderophores, in contrast to many other fungi [133]). In the high-affinity copper uptake system, the *FRE1* gene product reduces Cu(II) to Cu(I) before transport of Cu(I). ?, incomplete characterization. The K_m values for the high- and low-affinity systems for Fe(II), Mn, and Zn are 0.15 and 30 μM, 0.3 and 60 μM, and 1 and 10 μM, respectively (33, 54, 163, 164). Adapted from reference 33 with permission of the American Society for Microbiology.

could still occur even when there was an excess of competing divalent cations such as Mg^{2+} (54). This is important since organisms must be able to acquire essential "trace" divalent cations even when there is an excess of other divalent cations in their external environment. In general terms, divalent cations appear to enter cells as a result of the electrochemical gradient ($\Delta\tilde{\mu}_{H^+}$) generated by the activity of the plasma membrane H^+-ATPase (12, 41, 73, 125), although there is some evidence that the transmembrane electrochemical K^+ gradient ($\Delta\tilde{\mu}_{K^+}$) may also be important (102, 120). While the existence of a Ca^{2+}-ATPase in fungi has been described (92), there is no evidence of a comparable mechanism for other essential divalent cations, although divalent-cation efflux mechanisms have frequently been proposed (73, 100, 102).

It should be noted that for potentially toxic inessential and essential metals, transport analysis can be complicated by toxic symptoms, which include membrane disruption and also compartmentation in organelles such as the vacuole (51). Even low concentrations of toxic metals can affect the structural integrity of the cell membrane and associated functions, which can alter the kinetics of uptake. Progressive inhibition of H^+ extrusion could result in deenergization of the plasma membrane, while increased membrane permeability results in K^+ efflux (12, 157, 158). For both Zn^{2+} and Mn^{2+}, apparent low-affinity transport coincided with such effects in *S. cerevisiae*, which suggesting that altered kinetics were a result of toxicity (54, 157, 158). In fact, if affinity constants are compared for a range of potentially toxic divalent cations, e.g., Zn^{2+}, Cu^{2+}, Cd^{2+}, Ni^{2+}, and Co^{2+}, it seems that when the concentration range used is high, the calculated affinity of the transport system is low (51, 77).

A number of intracellular fates are possible for transported metals; these include sequestration by metal-binding molecules and compartmentation in organelles such as the vacuole. In addition, intracellular metal concentrations may be regulated by transport, including efflux mechanisms (41, 87). Such mechanisms are involved in normal metal homeostasis within cells but also play a role in detoxification of potentially toxic metals.

The yeast cell vacuole is involved in a number of essential functions within the cell, including macromolecular degradation, storage of metabolites, and cytosolic pH homeostasis (76). In addition, the vacuole plays an important role in the regulation of cytosolic metal ion concentrations both for essential metabolic functions and for the detoxification of potentially toxic metal ions (41, 43, 156). Metals preferentially sequestered by the vacuole include the divalent cations Mn^{2+} (54, 105), Fe^{2+} (11), Zn^{2+} (158), Co^{2+} (156), Ca^{2+}, and Sr^{2+} (13, 41, 103, 105), and Ni^{2+} (70) and the monovalent cations K^{+}, Li^{+}, and Cs^{+} (104, 113). Transport of these metal cations across the vacuolar membrane occurs by cation-proton exchange driven by the vacuolar transmembrane electrochemical pH gradient, which is energized by the vacuolar H^{+}-ATPase (99, 101, 105). The absence of a vacuole or a functional vacuolar H^{+}-ATPase in *S. cerevisiae* is associated with increased sensitivity and a largely decreased capacity of the cells to accumulate Zn, Mn, Co, and Ni (121), which are known to be detoxified mainly in the vacuole (41, 70).

For Cu and Cd, intracellular detoxification appears to depend predominantly on sequestration in the cytosol by induced metal-binding molecules (63, 87, 110, 123). These include low-molecular-weight cysteine-rich proteins (metallothioneins [MT]) and peptides derived from glutathione (phytochelatins) (68, 87, 89, 110, 160). The latter peptides have the general structure of $(\gamma\text{Glu-Cys})_n$-Gly, where the number of γGlu-Cys repeating units may extend up to 11 (110). In *Schizosaccharomyces pombe*, the value of n ranges from 2 to 5, while in *S. cerevisiae*, only an n_2 isopeptide has been observed (87). As well as being termed phytochelatins, the most widely used trivial name, such peptides are known as cadystins and metal γ-glutamyl peptides, although the chemical structure, $(\gamma EC)_nG$, is a more precise description. Although $(\gamma EC)_nG$ induction has been reported with a wide variety of metal ions, including Ag, Au, Hg, Ni, Pb, Sn, and Zn, metal binding has been shown for only a few, primarily Cd and Cu (110). For Cd, two types of complexes exist in *S. pombe* and *Candida glabrata*. A low-molecular-weight complex consists of $(\gamma EC)_nG$ and Cd, whereas a higher-molecular-weight complex also contains acid-labile sulfide (96, 110). The $(\gamma EC)_nG$-Cd-S^{2-} complex has a greater stability and higher Cd-binding capacity than does the low-molecular-weight complex and has a structure consisting of a CdS crystallite core and an outer layer of $(\gamma EC)_nG$ peptides (26). The higher binding capacity of the sulfide-containing complex confers a greater degree of tolerance to Cd (110). Evidence has also been presented for subsequent vacuolar localization of $(\gamma EC)_nG$-Cd-S^{2-} complexes in *S. pombe* (107–109), illustrating a dynamic link between cytosolic sequestration and vacuolar compartmentation. Although the main function of *S. cerevisiae* MT is cellular copper homeostasis, induction and synthesis of MT as well as amplification of MT genes lead to enhanced copper resistance in both *S. cerevisiae* and *C. glabrata* (87).

Production of MT has been detected in both Cu- and Cd-resistant strains of *S. cerevisiae* (68, 145). However, it should be noted that other determinants of tolerance, e.g., transport phenomena and sulfide precipitation (51, 68, 162), also occur in these and other organisms, while some organisms, e.g., *Kluyveromyces lactis*, are not capable of MT or $(\gamma EC)_nG$ synthesis (87). In *S. cerevisiae*, changes in amino acid pools can occur in response to nickel exposure, with the formation of vacuolar nickel-histidine complexes being proposed as a survival mechanism (70). Negligible work has been carried out on MT or $(\gamma EC)_nG$ peptides in filamentous fungi (41, 55, 65). Other mechanisms for metal immobilization within cells include precipitation by, e.g., reduction, sulfide production, and association with polyphosphate (40, 41).

Extracellular Metal-Binding Molecules

A diverse range of specific (see above) and nonspecific metal-binding compounds are produced by fungi, some of these are associated with exterior surfaces and/or released into the environment. Specific metal-binding compounds may be produced in response to the levels of metals present in the environment. The best-known extracellular metal-binding compounds are siderophores, which are low-molecular-mass ligands (500 to 1,000 Da) possessing a high affinity for Fe(III) (98). They scavenge for Fe(III) and complex and solubilize it, making it available to the organisms. Although produced primarily as a means of obtaining iron, siderophores are also able to bind other metals such as magnesium, manganese, chromium(III), gallium(III), and radionuclides such as plutonium(IV) (10). Nonspecific extracellular metal-binding compounds range in size from, e.g., organic acids and alcohols to macromolecules such as polysaccharides, and they all may affect metal bioavailability and toxicity (48). Extracellular polymeric substances are produced by many fungi, as well as bacteria and algae, and may bind significant amounts of potentially toxic metals (56). Extracellular polysaccharides may bind charged metal species and may adsorb or entrap particulate matter such as metal sulfides and oxides (34, 150).

Oxalate Production

Organic acids are released into the soil by both plant roots and fungal hyphae, with citric and oxalic acids being the most commonly reported (36, 72). However, while most metal citrates are highly mobile, the production of oxalic acid by fungi provides a means of immobilizing soluble metal ions, or complexes, as insoluble oxalates (20, 47) (Table 1) thus decreasing bioavailability and conferring tolerance (126). Colonies of *Aspergillus niger* form oxalate crystals when grown on agar amended with a wide range of metal compounds including insoluble phosphates of Ca, Cd, Co, Cu, Mn, Sr, and Zn (126) and powdered metal-bearing minerals (57, 58, 128) (Fig. 4). Morphological examination of fungus-produced oxalate crystals and comparison, where possible, with chemically synthesized oxalates has sometimes shown clear differences in form (126, 151).

Environmental Significance of Metal Immobilization

Although clay minerals are the major metal-sorbing components in the soil environment, fungi play an important role in several contexts. Fungi can be efficient

Table 1. Solubility products of some metal oxalates[a]

Metal	Metal oxalate	Temp (°C)	Solubility product
Barium	$BaC_2O_4 \cdot 3\frac{1}{2}H_2O$	18	1.62×10^{-7}
	$BaC_2O_4 \cdot 2H_2O$	18	1.2×10^{-7}
	$BaC_2O_4 \cdot \frac{1}{2}H_2O$	18	2.1×10^{-7}
Cadmium	CdC_2O_4	18	1.53×10^{-8}
Calcium	$CaC_2O_4 \cdot H_2O$	18	1.78×10^{-9}
	$CaC_2O_4 \cdot H_2O$	25	2.57×10^{-9}
Copper	CuC_2O_4	25	2.87×10^{-8}
Iron	FeC_2O_4	25	2.1×10^{-7}
Lead	PbC_2O_4	18	2.74×10^{-11}
Magnesium	$MgC_2O_4 \cdot 2H_2O$	18	8.57×10^{-5}
Strontium	$SrC_2O_4 \cdot H_2O$	18	5.61×10^{-8}
Zinc	$ZnC_2O_4 \cdot 2H_2O$	18	1.35×10^{-9}

[a] The solubility product is the product of the molar concentrations of the ions in a saturated solution (20).

sorbents of metal ions over a wide range of pH values, and although they may take up less metal per unit dry weight than clay minerals do, they are more efficient sorbents per unit surface area (93). Furthermore, the capacity of the fungi for metal sorption varies less with changing pH than does that of clay minerals, probably due to the heterogeneity of the ligands in the biomass (94). It seems likely that fungi play a more significant role in metal speciation and mobility in the terrestrial environment than has previously been supposed (78, 79, 82, 94). It should also be appreciated that the morphology of filamentous fungi allows interconnection of hyphae and continuity of cytoplasm, enabling fungi to cross nutrient-poor regions by supplying the hyphae in these regions with nutrients translocated from the part of the mycelium in a nutritionally replete locality. This ability of fungi to translocate nutrients (and organelles, oxygen, and metabolites) may also be important in the translocation of metal species: continuity of cytoplasm means that the cytosol as well as metal-laden organelles may be free to move, allowing their translocation and concentration in specific regions of the fungal thallus, including fruiting bodies (23, 30, 60, 62). Elevated metal and radionuclide concentrations, particularly radiocesium, commonly occur in the fruiting bodies of basidiomycetous fungi during growth in polluted environments (2, 41, 45, 62, 153). In fact, it has been concluded that the fungal component of soil can immobilize the total Chernobyl radiocesium fallout received in upland grasslands (31), although grazing of fruit bodies by animals may lead to radiocesium transfer along the food chain (7). Metal immobilization by fungi may also be significant to plant productivity. Although there is wide variation in responses between different mycorrhizal symbioses, the amelioration of metal phytotoxicity by mycorrhizas has been frequently described (22). The sequestering ability of the fungal partner may effectively reduce metal bioavailability adjacent to the plant root systems and reduce or prevent translocation to the plant (14, 55). This relationship, though, might not always be beneficial, since it is possible that fungal accumulation of toxic metals will increase the apparent metal concentration in the roots if soil physicochemical conditions alter or when the fungi die and degrade or otherwise release the accumulated metals (94).

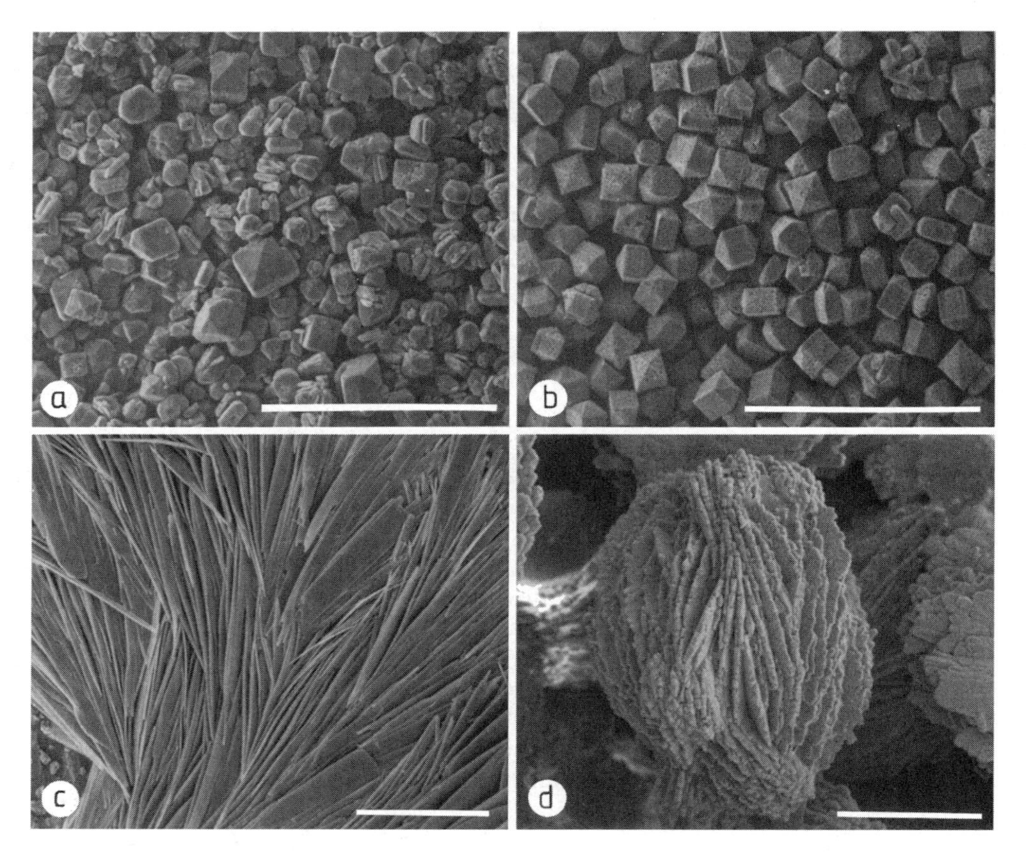

Figure 4. Scanning electron micrographs of metal oxalate crystals (126, 128, 129, 159). (a) Strontium oxalate produced by *A. niger* grown on strontium nitrate-containing malt extract agar. (b) Chemically synthesized strontium oxalate crystals made by allowing 100 mM oxalic acid to diffuse from wells cut in strontium nitrate-containing malt extract agar; crystals were purified from the agar surrounding the wells. (c and d) Manganese oxalate crystals produced by *A. niger* grown on malt extract agar amended with the manganese-containing mineral rhodochrosite (c) or manganese phosphate (d). Bars, 100 μm.

The formation of oxalates containing potentially toxic metal cations may provide a mechanism whereby oxalate-producing fungi can tolerate environments containing high concentrations of toxic metals. Most metal oxalates are insoluble, some exceptions being Na, K, Li, and Fe oxalates (47, 135). Copper oxalate (moolooite) has been observed around hyphae growing on wood treated with copper as a preservative (97, 137). The copper appeared on the surface of the wood and around hyphae as copper oxalate, which was reported to be nontoxic because of its insolubility. Copper oxalate has also been observed in lichens growing on copper-rich rocks, where it is thought that the precipitation of copper oxalate could be a detoxification mechanism; up to 5% copper (dry weight) was fixed in the lichen thallus as copper oxalate (117, 118).

In contrast to the formation of toxic metal oxalates, however, there are many reports of the formation of calcium oxalate (whewellite [calcium oxalate monohydrate] and weddellite [calcium oxalate dihydrate]). Oxalate, ubiquitous in the terrestrial environment, can reach concentrations of 10^{-6} to 10^{-3} M in the soil (1). Calcium oxalate crystals are often found around free-living hyphae and around mycorrhizal roots, where they are thought to play a major role in Ca detoxification (71, 80), since high concentrations of free Ca^{2+} can be toxic (41, 43). Some fungi, such as brown rot fungi, often secrete oxalic acid along with cell wall-degrading enzymes, and this is thought to assist the solubilization of pectin in membranes in the middle lamellae (61). Calcium oxalate is resistant to further solubilization, with only a few aerobic bacteria and fungi and anaerobic bacteria of the gastrointestinal tract known to degrade these oxalates (1, 95).

TRANSFORMATION

Mechanisms of Transformation

Several species of fungi, including unicellular and filamentous forms, can transform metals, metalloids, and organometallic compounds by reduction, methylation, and dealkylation, these are processes of environmental importance, since transformation of a metal or metalloid may modify its mobility and toxicity (42). For example, methyl derivatives of selenium are volatile and less toxic than inorganic forms, while reduction of metalloid oxyanions such as selenite to amorphous selenium results in immobilization and detoxification (94).

Reduction

There are several reports of metal and metalloid reduction by fungi. Reduction of Ag(I) to Ag(0) during growth on $AgNO_3$-containing media results in blackened colonies with metallic Ag(0) precipitated in and around cell walls (75). Both enzymatic and nonenzymatic Cu(II)-reducing systems have been purified from *Debaryomyces hansenii* cell walls, with the enzyme being involved in the control of Cu(II) uptake (154, 155). In fact, it is now known that reductive processes, e.g., of Cu(II) to Cu(I) and Fe(III) to Fe(II), are integral prerequisites for high-affinity transport of these metals (33, 77, 83). Reduction of Hg(II) to Hg(0) by fungi has also been demonstrated (161), but in contrast to bacteria, there has been little detailed characterization of this system. The ability of fungi to reduce metalloids is perhaps more clearly demonstrated. Reduction of selenate [Se(VI)] and selenite [Se(IV)] to elemental selenium can be catalyzed by numerous fungal species, resulting in red colonies (119). Both extracellular and intracellular deposition of Se(0) has been demonstrated (59). Tellurite (TeO_3^{2-}) reduction to Te(0) results in black or dark grey colonies (133).

Methylation

The biological methylation (biomethylation) of metalloids has been demonstrated in filamentous fungi and yeasts, and this frequently results in their volatilization (42). Arsenic and selenium have received most attention, with the biochemical pathway for the fungal production of trimethylarsine from arsenite being suggested initially by Challenger (19). Subsequent studies have shown that several fungi, such as *Gliocladium roseum*, *Candida humicola*, and a *Penicillium* sp., can convert monomethylarsonic acid to trimethylarsine (67, 138). The pathway for arsenic methylation involves the transfer of methyl groups as carbonium (CH_3^+) ions by *S*-adenosylmethionine (42, 138). Numerous fungi can convert both selenite (SeO_3^{2-}) and selenate (SeO_4^{2-}) to methyl derivatives such as dimethylselenide [$(CH_3)_2Se$] and dimethyldiselenide [$(CH_3)_2Se_2$] (17, 141, 142). Inorganic forms of Se (SeO_3^{2-} and SeO_4^{2-}) appear to be methylated more rapidly than organic forms, such as Se-containing amino acids, with the mechanism for Se methylation appearing to be similar to that for arsenic (42). There are few detailed studies on fungal biomethylation of other metals and metalloids. Mercury biomethylation by fungal species has been reported (161), while there is evidence of dimethyltelluride [$(CH_3)_2Te$] and dimethylditelluride [$(CH_3)_2Te_2$] production from both TeO_3^{2-} and TeO_4^{2-} by a *Penicillium* sp. (67).

Dealkylation

Organometallic compounds can be degraded by fungi, either by direct biotic action (enzymes) or by facilitation of abiotic degradation, for instance by alteration of pH and excretion of metabolites. Organotin compounds, such as tributyltin oxide and tributyltin naphthenate, may be degraded to mono- and dibutyltins by fungal action, with inorganic Sn(II) being the ultimate degradation product (9, 106). Organomercury compounds may be detoxified by conversion to Hg(II) by fungal organomercury lyase, with the Hg(II) being subsequently reduced to Hg(0) by mercuric reductase, a system broadly analogous to that found in mercury-resistant bacteria (139). Trimethyllead degradation has been demonstrated in an alkyllead-tolerant yeast (85) and in the wood decay fungus *Phaeolus schweintzii* (86).

Significance of Fungus-Metal Interactions for Bioremediation

As described above, many species of fungi are able to remove, or leach, metals ("heterotrophic leaching") from industrial wastes and by-products, low-grade ores (18), and metal-bearing minerals (32, 149), a process relevant to metal recovery and recycling and/or bioremediation of contaminated solid wastes (18, 159). Most biotechnological processes for the winning of metals and metalloids from ores by solubilization processes involve chemoautotrophic bacteria, particularly members of the genus *Thiobacillus*. However, the pH of growth media may be increased by many industrial metal-containing wastes, e.g., filter dust from metal processing, and most *Thiobacillus* spp. cannot solubilize metals effectively above pH 5.5, with an optimum of pH 2.4 (18). Although fungi need a source of carbon and aeration, they can solubilize metals at higher pH values and so could perhaps become more important when leaching with bacteria is not possible. However, fungi may not readily be used in situ, and their use in specific bioreactors would be envisaged. Leaching of metals with fungi can be very effective, although a high level of organic acid production may have to be maintained. For example, laboratory-scale leaching of Ni and Co from low-grade laterite ore has been carried out using strains of *Aspergillus* and *Penicillium* (149). In one study, 55 to 60% of the available Ni was leached when the fungi were grown in the presence of the ore and 70% was leached at high temperature (95°C) by application of metabolic products obtained from cultivation of the fungi at 30°C in a glucose- and sucrose-containing medium (148). Heterotrophic leaching by fungi can occur as a result of several processes, including the production of siderophores (for iron), but in most fungal strains, leaching occurs mainly by the production of organic acids (18, 126). The pH of nonregulated *A. niger* cultures can fall to values between 1.5 and 2.0 due to high citric acid production, since the optimal pH for citric acid production is below 3.5 (132). Organic acid production can be manipulated by changes in the growth medium, for example, a deficiency of manganese (less than 10^{-8} M) leads to the production of large amounts of citric acid by *A. niger* (90), with the concentrations of citric acid produced industrially by this fungus being capable of reaching 600 mM (88). Oxalic acid production can also be manipulated to yield concentrations of up to 200 mM on low-cost carbon sources (135). A strain of *Penicillium simplicissimum* which was isolated from a metal-contaminated site developed the ability to produce citric acid (>100 mM) in the presence of metal contaminants and has been used successfully to leach Zn from insoluble ZnO contained in industrial filter dust (39, 131). Culture filtrates from *A. niger* have also been used to leach Cu, Ni, and Co from copper converter slag (136). If necessary, metal-citrate complexes could eventually be degraded for ultimate metal recovery (37). Another possible application or effect of fungal metal solubilization is the removal of unwanted contaminants such as phosphates (18), while interaction of leaching technologies with biosorption is also a possibility (41, 144).

Regarding metal immobilization, fungi and their by-products have received considerable attention as possible biosorbent materials for metal-contaminated aqueous solutions. Their suitability is supported by the ease with which they are grown and their availability as waste products of industrial processes, e.g., *A. niger* (citric acid production) and *S. cerevisiae* (brewing). Many studies have shown their efficacy in sorbing metal contaminants as either living or dead biomass, in pelleted whole-cell or dissembled forms, and as mobile or immobilized sorbents in batch and continuous processes (144). They are also amenable to removal of the loaded metals by acids, alkalis, or chelating agents, retaining a high percentage of their original sorption capacity after repeated regenerative stages (41, 53). However, although there have been several attempts to commercialize biosorption using microbial biomass, success has been limited, primarily due to competition with commercially produced ion-exchange media (127).

The ability of fungi, along with bacteria, to transform metals and metalloids has been utilized in the bioremediation of contaminated land and water. Selenium methylation results in volatilization, a process which has been used to remove selenium from the San Joaquin Valley and Kesterson Reservoir, Calif., by evaporation pond management and primary-pond operation (140, 142). Incoming Se-contaminated drainage water was evaporated to dryness, and the process was repeated until the sediment Se concentration approached 100 mg kg^{-1} on a dry-weight basis. The volatilization process was then optimized until the selenium concentration in the sediment fell to acceptable limits, using parameters such as carbon source, moisture, temperature, and aeration (140). Fungal organometal transformations may also be envisaged for the removal of alkylleads or organotins from water (46, 85, 86).

Acknowledgments. G.M.G. gratefully acknowledges financial support from the BBSRC (GR/J48214, 94/SPC02812) and NATO (Envir.Lg.950387 Linkage Grant) for some of the work described in this chapter.

REFERENCES

1. **Allison, M. J., S. L. Daniel, and N. A. Cornick.** 1995. Oxalate degrading bacteria, p. 131–168. *In* S. R. Khan (ed.), *Calcium Oxalate in Biological Systems.* CRC Press, Inc., Boca Raton, Fla.
2. **Anderson, P., C. M. Davidson, D. Littlejohn, A. M. Ure, C. A. Shand, and M. V. Cheshire.** 1997. The translocation of caesium and silver by fungi in some Scottish soils. *Commun. Soil Sci. Plant Anal.* **28:**635–650.
3. **Arter, H. E., K. W. Hanselmann, and R. Bachofen.** 1991. Modelling of microbial-degradation processes—the behaviour of microorganisms in a waste repository. *Experientia* **47:**542–548.
4. **Avery, S. V.** 1995. Caesium accumulation by microorganisms—uptake mechanisms, cation competition, compartmentalization and toxicity. *J. Ind. Microbiol.* **14:**76–84.
5. **Avery, S. V., and J. M. Tobin.** 1992. Mechanisms of strontium uptake by laboratory and brewing strains of *Saccharomyces cerevisiae*. *Appl. Environ. Microbiol.* **58:**3883–3889.
6. **Avery, S. V., and J. M. Tobin.** 1993. Mechanisms of adsorption of hard and soft metal-ions to *Saccharomyces cerevisiae* and influence of hard and soft anions. *Appl. Environ. Microbiol.* **59:**2851–2856.
7. **Bakken, L. R., and R. A. Olsen.** 1990. Accumulation of radiocaesium in fungi. *Can. J. Microbiol.* **36:**704–710.
8. **Banks, M. K., A. P. Schwabb, G. R. Fleming, and B. A. Herrick.** 1994. Effects of plants and soil microflora on leaching of zinc from mine tailings. *Chemosphere* **29:**1691–1699.
9. **Barug, G.** 1981. Microbial degradation of bis(tributyltin) oxide. *Chemosphere* **10:**1145–1154.
10. **Birch, L., and R. Bachofen.** 1990. Complexing agents from microorganisms. *Experientia* **46:**827–834.
11. **Bode, H.-P., M. Dumschat, S. Garotti, and G. F. Fuhrmann.** 1995. Iron sequestration by the yeast vacuole. A study with vacuolar mutants of *Saccharomyces cerevisiae. Eur. J. Biochem.* **228:**337–342.
12. **Borst-Pauwels, G. W. F. H.** 1981. Ion transport in yeast. *Biochim. Biophys. Acta* **650:**149–156.
13. **Borst-Pauwels, G. W. F. H.** 1989. Ion transport in yeast including lipophilic ions. *Methods Enzymol.* **174:**603–616.
14. **Bradley, R., A. J. Burt, and D. J. Read.** 1981. Mycorrhizal infection and resistance to heavy metals. *Nature* **292:**335–337.
15. **Brady, J. M., and J. M. Tobin.** 1994. Adsorption of metal-ions by *Rhizopus arrhizus* biomass—characterization studies. *Enzyme Microb. Technol.* **16:**671–675.
16. **Brady, J. M., and J. M. Tobin.** 1995. Binding of hard and soft metal-ions to *Rhizopus arrhizus* biomass. *Enzyme Microb. Technol.* **17:**791–796.
17. **Brady, J. M., J. M. Tobin, and G. M. Gadd.** 1996. Volatilization of selenite in aqueous medium by a *Penicillium* species. *Mycol. Res.* **100:**955–961.
18. **Burgstaller, W., and F. Schinner.** 1993. Leaching of metals with fungi. *J. Biotechnol.* **27:**91–116.
19. **Challenger, F.** 1945. Biological methylation. *Chem. Rev.* **36:**15–61.
20. **Chang, J. C.** 1993. Solubility product constants, p. 39. *In* D. R. Lide (ed.), *CRC Handbook of Chemistry and Physics.* CRC Press, Inc., Boca Raton, Fla.
21. **Collins, Y. E., and G. Stotzky.** 1992. Heavy metals alter the electrokinetic properties of bacteria, yeasts and clay minerals. *Appl. Environ. Microbiol.* **58:**1592–1600.
22. **Colpaert, J. V., and K. K. Van Tichelen.** 1996. Mycorrhizas and environmental stress, p. 109–128. *In* J. C. Frankland, N. Magan, and G. M. Gadd (ed.), *Fungi and Environmental Change.* Cambridge University Press, Cambridge, United Kingdom.
23. **Connolly, J. H., and J. Jellison.** 1997. Two-way translocation of cations by the brown rot fungus *Gloeophyllum trabeum. Int. Biodeterior. Biodegrad.* **39:**181–188.
24. **Crichton, R. R.** 1991. *Inorganic Biochemistry of Iron Metabolism.* Ellis Horwood, Chichester, United Kingdom.

25. **Cunningham, J. E., and C. Kuiack.** 1992. Production of citric and oxalic acids and solubilization of calcium phosphates by *Penicillium bilaii*. *Appl. Environ. Microbiol.* **58:**1451–1458.
26. **Dameron, C. T., R. N. Reese, R. K. Mehra, A. R. Kortan, P. J. Carrol, M. L. Steigerwald, L. E. Brus, and D. R. Winge.** 1989. Biosynthesis of cadmium sulfide quantum semiconductor crystallites. *Nature* **338:**596–597.
27. **Denêvre, O., J. Garbaye, and B. Botton.** 1996. Release of complexing organic acids by rhizosphere fungi as a factor in Norway Spruce yellowing in acidic soils. *Mycol. Res.* **100:**1367–1374.
28. **de Rome, L., and G. M. Gadd.** 1987. Measurement of copper uptake in *Saccharomyces cerevisiae* using a Cu^{2+}-selective electrode. *FEMS Microbiol. Lett.* **43:**283–287.
29. **de Rome, L., and G. M.Gadd.** 1987. Copper adsorption by *Rhizopus arrhizus*, *Cladosporium resinae* and *Penicillium italicum*. *Appl. Microbiol. Biotechnol.* **26:**84–90.
30. **Dighton, J., and G. Terry.** 1996. Uptake and immbilization of caesium in UK grassland and forest soils by fungi, following the Chernobyl accident, p. 184–200. *In* J. C. Frankland, N. Magan, and G. M. Gadd (ed.), *Fungi and Environmental Change*. Cambridge University Press, Cambridge, United Kingdom.
31. **Dighton, J., G. M. Clint, and J. Poskitt.** 1991. Uptake and accumulation of ^{137}Cs by upland grassland soil fungi: a potential pool of Cs immobilization. *Mycol. Res.* **95:**1052–1056.
32. **Drever, J. I., and L. L. Stillings.** 1997. The role of organic acids in mineral weathering. *Colloids Surf.* **120:**167–181.
33. **Eide, D., and M. L. Guerinot.** 1997. Metal ion uptake in eukaryotes. *ASM News* **63:**199–205.
34. **Flemming, H.-K.** 1995. Sorption sites in biofilms. *Water Sci. Technol.* **32:**27–33.
35. **Fourest, E., and J.-C. Roux.** 1992. Heavy metal biosorption by fungal mycelial by-products, mechanisms and influence of pH. *Appl. Microbiol. Biotechnol.* **37:**399–403.
36. **Fox, T. R., and N. B. Comerford.** 1990. Low-molecular weight organic acids in selected forest soils of the southeastern USA. *Soil Sci. Soc. Am. J.* **54:**1139–1144.
37. **Francis, A. J.** 1994. Microbial transformations of radioactive wastes and environmental restoration through bioremediation. *J. Alloys Compounds* **213/214:**226–231.
38. **Francis, A. J., C. J. Dodge, and J. B. Gillow.** 1992. Biodegradation of metal citrate complexes and implications for toxic metal mobility. *Nature* **356:**140–142.
39. **Franz, A., W. Burgstaller, B. Muller, and F. Schinner.** 1993. Influence of medium components and metabolic inhibitors on citric acid production by *Penicillium simplicissimum*. J. Gen. Microbiol. **139:**2101–2107.
40. **Gadd, G. M.** 1990. Fungi and yeasts for metal binding, p. 249–275. *In* H. L. Ehrlich and C. L. Brierley (ed.), *Microbial Mineral Recovery*. McGraw-Hill Book Co., New York, N.Y.
41. **Gadd, G. M.** 1993. Interactions of fungi with toxic metals. *New Phytol.* **124:**25–60.
42. **Gadd, G. M.** 1993. Microbial formation and transformation of organometallic and organometalloid compounds. *FEMS Microbiol. Rev.* **11:**297–316.
43. **Gadd, G. M.** 1995. Signal transduction in fungi, p. 183–210. *In* N. A. R. Gow and G. M. Gadd (ed.), *The Growing Fungus*. Chapman & Hall, Ltd., London, United Kingdom.
44. **Gadd, G. M.** 1996. Influence of microorganisms on the environmental fate of radionuclides. *Endeavour* **20:**150–156.
45. **Gadd, G. M.** 1997. Roles of microorganisms in the environmental fate of radionuclides. *CIBA Found. Symp.* **203:**94–108.
46. **Gadd, G. M.** Microbial interactions with tributyltin compounds: detoxification, accumulation, environmental fate and effects. *Sci. Total Environ.*, in press.
47. **Gadd, G. M.** 1999. Fungal production of citric and oxalic acid: importance in metal speciation, physiology and biogeochemical processes. *Adv. Microb. Physiol.* **41:**47–92.
48. **Gadd, G. M., and A. J. Griffiths.** 1978. Microorganisms and heavy metal toxicity. *Microb. Ecol.* **4:**303–317.
49. **Gadd, G. M., and A. J. Griffiths.** 1980 Influence of pH on copper uptake and toxicity in *Aureobasidium pullulans*. *Trans. Br. Mycol. Soc.* **75:**91–95.
50. **Gadd, G. M., and C. White.** 1985. Copper uptake by *Penicillium ochro-chloron*: influence of pH on toxicity and demonstration of energy-dependent copper influx using protoplasts. *J. Gen. Microbiol.* **131:**1875–1879.

51. **Gadd, G. M., and C. White.** 1989. Heavy metal and radionuclide accumulation and toxicity in fungi and yeasts, p. 19–38. *In* R. K. Poole and G. M. Gadd (ed.), *Metal-Microbe Interactions.* IRL Press, Oxford, United Kingdom.
52. **Gadd, G. M., and C. White.** 1990. Biosorption of radionuclides by yeast and fungal biomass. *J. Chem. Technol. Biotechnol.* **49:**331–343.
53. **Gadd, G. M., and C. White.** 1992. Removal of thorium from simulated acid process streams by fungal biomass: potential for thorium desorption and reuse of biomass and desorbent. *J. Chem. Technol. Biotechnol.* **55:**39–44.
54. **Gadd, G. M., and O. S. Lawrence.** 1996. Demonstration of high-affinity Mn^{2+} uptake in *Saccharomyces cerevisiae*—specificity and kinetics. *Microbiology* **142:**1159–1167.
55. **Galli, U., H. Schuepp, and C. Brunold.** 1994. Heavy metal binding by mycorrhizal fungi. *Physiol. Plant.* **92:**364–368.
56. **Geesey, G., and L. Jang.** 1990. Extracellular polymers for metal binding, p. 223–275. *In* H. L. Ehrlich and C. L. Brierley (ed.), *Microbial Mineral Recovery.* McGraw-Hill Book Co., New York, N.Y.
57. **Gharieb, M. M., and G. M. Gadd.** 1999. Influence of nitrogen source on the solubilization of natural gypsum ($CaSO_4 \cdot 2H_2O$) and the formation of calcium oxalate by different oxalic and citric acid-producing fungi. *Mycol. Res.* **103:**473–481.
58. **Gharieb, M. M., J. A. Sayer, and G. M. Gadd.** 1998. Solubilization of natural gypsum ($CaSO_4 \cdot 2H_2O$) and the formation of calcium oxalate by *Aspergillus niger* and *Serpula himantiodes. Mycol. Res.* **102:**825–830.
59. **Gharieb, M. M., S. C. Wilkinson, and G. M. Gadd.** 1995. Reduction of selenium oxyanions by unicellular, polymorphic and filamentous fungi: cellular location of reduced selenium and implications for tolerance. *J. Ind. Microbiol.* **14:**300–311.
60. **Gray, S. N., J. Dighton, and D. H. Jennings.** 1996. The physiology of basidiomycete linear organs. 3. Uptake and translocation of radiocaesium within differentiated mycelia of *Armillaria* spp. growing in microcosms and in the field. *New Phytol.* **132:**471–482.
61. **Green, F., C. A. Clausen, T. A. Kuster, and T. L. Highley.** 1995. Induction of polygalacturonase and the formation of oxalic acid by pectin in brown rot fungi. *World J. Microbiol. Biotechnol.* **11:** 519–524.
62. **Haselwandter, K., and M. Berreck.** 1994. Accumulation of radionuclides in fungi, p. 259–277. *In* G. Winkelmann and D. R. Winge (ed.), *Metal Ions in Fungi.* Marcel Dekker, Inc., New York, N.Y.
63. **Hayashi, Y., and N. Mutoh.** 1994. Cadystin (phytochelatin) in fungi, p. 311–337. *In* G. Winkelmann and D. R. Winge (ed.), *Metal Ions in Fungi.* Marcel Dekker, Inc., New York, N.Y.
64. **Hockertz, S., J. Schmid, and G. Auling.** 1987. A specific transport system for manganese in the filamentous fungus *Aspergillus niger. J. Gen. Microbiol.* **133:**3513–3519.
65. **Howe, R., R. L. Evans, and S. W. Ketteridge.** 1997. Copper binding proteins in ectomycorrhizal fungi. *New Phytol.* **135:**123–131.
66. **Hughes, M. N., and R. K. Poole.** 1991. Metal speciation and microbial growth—the hard (and soft) facts. *J. Gen. Microbiol.* **137:**725–734.
67. **Huysmans, K. D., and W. T. Frankenberger.** 1991. Evolution of trimethylarsine by a *Penicillium* sp. isolated from agricultural evaporation pond water. *Sci. Total Environ.* **105:**13–28.
68. **Inouhe, M., M. Sumiyoshi, H. Tohoyama, and M. Joho.** 1996. Resistance to cadmium ions and formation of a cadmium-binding complex in various wild-type yeasts. *Plant Cell Physiol.* **37:**341–346.
69. **Irving, H., and R. J. P. Williams.** 1948. Order of stability of metal complexes. *Nature* **162:**746–747.
70. **Joho, M., M. Inouhe, H. Tohoyama, and T. Murayama.** 1995. Nickel resistance mechanisms in yeasts and other fungi. *J. Ind. Microbiol.* **14:**164–168.
71. **Jones, D., W. J. McHardy, M. J. Wilson, and D. Vaughan.** 1992. Scanning electron microscopy of calcium oxalate on mantle hyphae of hybrid larch roots from a farm forestry experimetal site. *Micron Microsc. Acta* **23:**315–317.
72. **Jones, D. L., and L. V. Kochian.** 1996. Aluminium-organic acid interactions in acid soils. *Plant Soil* **182:**221–228.

73. **Jones, R. P., and G. M. Gadd.** 1990. Ionic nutrition of yeast—physiological mechanisms involved and implications for biotechnology. *Enzyme Microb. Technol.* **12:**402–418.
74. **Karamushka, V. I., J. A. Sayer, and G. M. Gadd.** 1996. Inhibition of H^+ efflux from *Saccharomyces cerevisiae* by insoluble metal phosphates and protection by calcium and magnesium: inhibitory effects a result of soluble metal cations? *Mycol. Res.* **100:**707–713.
75. **Kierans, M., A. M. Staines, H. Bennett, and G. M. Gadd.** 1991. Silver tolerance and accumulation in yeasts. *Biol. Metals* **4:**100–106.
76. **Klionsky, D. J., P. K. Herman, and S. D. Emr.** 1990. The fungal vacuole: composition, function and biogenesis. *Microbiol. Rev.* **54:**266–292.
77. **Kosman, D. J.** 1994. Transition metal ion uptake in yeasts and filamentous fungi, p. 1–38. *In* G. Winkelmann and D. R. Winge (ed.), *Metal Ions in Fungi.* Marcel Dekker, Inc., New York, N.Y.
78. **Krantz-Rulcker, C., B. Allard, and J. Schnurer.** 1993. Interactions between a soil fungus, *Trichoderma harzianum* and IIB metals—adsorption to mycelium and production of complexing metabolites. *Biometals* **6:**223–230.
79. **Krantz-Rulcker, C., B. Allard, and J. Schnurer.** 1996. Adsorption of IIB metals by 3 common soil fungi—comparison and assessment of importance for metal distribution in natural soil systems. *Soil Biol. Biochem.* **28:**967–975.
80. **Lapeyrie, F., G. A. Chilvers, and C. A. Bhem.** 1987. Oxalic acid synthesis by the mycorrhizal fungus *Paxillus involutus. New Phytol.* **106:**139–146.
81. **Lapeyrie, F., J. Ranger, and D. Vairelles.** 1991. Phosphate solubilizing activity of ectomycorrhizal fungi *in vitro. Can. J. Bot.* **69:**342–346.
82. **Ledin, M., C. Krantz-Rulcker, and B. Allard.** 1996. Zn, Cd and Hg accumulation by microorganisms, organic and inorganic soil components in multicompartment systems. *Soil Biol. Biochem.* **28:**791–799.
83. **Lesuisse, E., and P. Labbe.** 1994. Reductive iron assimilation in *Saccharomyces cerevisiae*, p. 149–178. *In* G. Winkelmann and D. R. Winge (ed.), *Metal Ions in Fungi.* Marcel Dekker, Inc., New York, N.Y.
84. **Leyval, C., T. Surtiningish, and J. Berthelin.** 1993. Mobilization of P and Cd from rock phosphates by rhizosphere microorganisms (phosphate dissolving bacteria and ectomycorrhizal fungi). *Phosphorus Sulfur Sil.* **77:**133–136.
85. **Macaskie, L. E., and A. C. R. Dean.** 1987. Trimethyllead degradation by an alkyllead-tolerant yeast. *Environ. Technol. Lett.* **8:**635–640.
86. **Macaskie, L. E., and A. C. R. Dean.** 1990. Trimethyl lead degradation by free and immobilized cells of an *Arthrobacter* sp. and by the wood decay fungus *Phaeolus schweintzii. Appl. Microbiol. Biotechnol.* **38:**81–87.
87. **Macreadie, I. G., A. K. Sewell, and D. R. Winge.** 1994. Metal ion resistance and the role of metallothionein in yeast, p. 279–310. *In* G. Winkelmann and D. R. Winge (ed.), *Metal Ions in Fungi.* Marcel Dekker, Inc., New York, N.Y.
88. **Mattey, M**. 1992. The production of organic acids. *Crit. Rev. Biotechnol.* **12:**87–132.
89. **Mehra, R. K., and D. R. Winge.** 1991. Metal ion resistance in fungi: molecular mechanisms and their related expression. *J. Cell. Biochem.* **45:**30–40.
90. **Meixner, O., H. Mischack, C. P. Kubicek, and M. Rohr.** 1985. Effect of manganese deficiency on plasma-membrane lipid composition and glucose uptake in *Aspergillus niger. FEMS Microbiol. Lett.* **26:**271–274.
91. **Metting, F. B.** 1992. Structure and physiological ecology of soil microbial communities, p. 3–25, *In* F. B. Metting (ed.), *Soil Microbial Ecology, Applications and Environmental Management.* Marcel Dekker, Inc., New York, N.Y.
92. **Miller, A. J., G. Vogg, and D. Sanders.** 1990. Cytosolic calcium homeostasis in fungi: roles of plasma membrane transport and intracellular sequestration of calcium. *Proc. Natl. Acad. Sci. USA* **87:**9348–9352.
93. **Morley, G. F., and G. M. Gadd.** 1995. Sorption of toxic metals by fungi and clay minerals. *Mycol. Res.* **99:**1429–1438.
94. **Morley, G. F., J. A. Sayer, S. C. Wilkinson, M. M. Gharieb, and G. M. Gadd.** 1996. Fungal sequestration, solubilization and transformation of toxic metals, p. 235–256. *In* J. C. Frankland, N.

Magan, and G. M. Gadd (ed.), *Fungi and Environmental Change.* Cambridge University Press, Cambridge, United Kingdom.

95. **Morris, S. J., and M. F. Allen.** 1994. Oxalate metabolizing microorganisms in sagebrush steppe soil. *Biol. Fertil. Soils* **18:**255–259.
96. **Murasugi, A., C. Wada, and Y. Hayashi.** 1983. Occurrence of acid labile sulfide in cadmium binding peptide 1 from fission yeast. *J. Biochem.* **93:**661–664.
97. **Murphy, R. J., and J. F. Levy.** 1983. Production of copper oxalate by some copper tolerant fungi. *Trans. Br. Mycol. Soc.* **81:**165–168.
98. **Neilands, J. B.** 1981. Microbial iron compounds. *Annu. Rev. Biochem.* **50:**715–731.
99. **Nelson, N., C. Beltran, F. Supek, and H. Nelson.** 1992 Cell biology and evolution of proton pumps. *Cell. Physiol. Biochem.* **2:**150–158.
100. **Nieuwenhuis, B. J. W. M., A. G. M. Weijers, and G. W. F. H. Borst-Pauwels.** 1981. Uptake and accumulation of Mn^{2+} and Sr^{2+} in *Saccharomyces cerevisiae. Biochim. Biophys. Acta* **649:**83–88.
101. **Ohsumi, Y., and Y. Anraku.** 1983. Calcium transport driven by a proton motive force in vacuolar membrane vesicles of *Saccharomyces cerevisiae. J. Biol. Chem.* **41:**17–22.
102. **Okorokov, L. A.** 1985. Main mechanisms of ion transport and regulation of ion concentrations in the yeast cytoplasm, p. 463–472. *In* I. S. Kulaev, E. A. Dawes, and D. W. Tempest (ed.), *Environmental Regulation of Microbial Metabolism.* Academic Press, Ltd., London, United Kingdom.
103. **Okorokov, L. A.** 1994. Several compartments of *Saccharomyces cerevisiae* are equipped with Ca^{2+} ATPase(s). *FEMS Microbiol. Lett.* **117:**311–318.
104. **Okorokov, L. A., L. P. Lichko, and I. S. Kulaev.** 1980. Vacuoles: main compartments of potassium, magnesium and phosphate in *Saccharomyces carlsbergensis* cells. *J. Bacteriol.* **144:**661–665.
105. **Okorokov, L. A., T. V. Kulakovskaya, L. P. Lichko, and E. V. Polorotova.** 1985. H^+/ion antiport as the principal mechanism of transport systems in the vacuolar membrane of the yeast *Saccharomyces carlsbergensis. FEBS Lett.* **192:**303–306.
106. **Orsler, R. J., and G. E. Holland.** 1982. Degradation of tributyltin oxide by fungal culture filtrates. *Int. Biodeterior. Bull.* **18:**95–98.
107. **Ortiz, D. F., D. F. Kreppel, D. M. Speiser, G. Scheel, G. McDonald, and D. W. Ow.** 1992. Heavy-metal tolerance in the fission yeast requires an ATP-binding cassette-type vacuolar membrane transporter. *EMBO J.* **11:**3491–3499.
108. **Ortiz, D. F., T. Ruscitti, K. F. McCue, and D. W. Ow.** 1995. Transport of metal-binding peptides by HMT1, a fission yeast ABC-type vacuolar membrane protein. *J. Biol. Chem.* **270:**4721–4728.
109. **Ow, D. W.** 1993. Phytochelatin-mediated cadmium tolerance in *Schizosaccharomyces pombe. In Vitro Cell. Dev. Biol. Plant* **29P:**213–219.
110. **Ow, D. W., D. F. Ortiz, D. M. Speiser, and K. F. McCue.** 1994. Molecular genetic analysis of cadmium tolerance in *Schizosaccharomyces pombe*, p. 339–359. *In* G. Winkelmann and D. R. Winge (ed.), *Metal Ions in Fungi.* Marcel Dekker, Inc., New York, N.Y.
111. **Parkin, M. J., and I. S. Ross.** 1986. The specific uptake of manganese in the yeast *Candida utilis. J. Gen. Microbiol.* **132:**2155–2160.
112. **Peberdy, J. F.** 1990. Fungal cell walls—a review, p. 5–30. *In* P. J. Kuhn, A. P. J. Trinci, M. J. Jung, M. W. Coosey, and L. E. Copping (ed.), *Biochemistry of Cell Walls and Membranes in Fungi.* Springer-Verlag KG, Berlin, Germany.
113. **Perkins, J., and G. M. Gadd.** 1993. Accumulation and intracellular compartmentation of lithium ions in *Saccharomyces cerevisiae. FEMS Microbiol. Lett.* **107:**255–260.
114. **Perkins, J., and G. M. Gadd.** 1993. Caesium toxicity, accumulation and intracellular localization in yeasts. *Mycol. Res.* **97:**717–724.
115. **Perkins, J., and G. M. Gadd.** 1996. Interactions of Cs^+ and other monovalent cations (Li^+, Na^+, K^+, Rb^+, NH_4^+) with K^+-dependent pyruvate-kinase and malate-dehydrogenase from the yeasts *Rhodotorula rubra* and *Saccharomyces cerevisiae. Mycol. Res.* **100:**449–454.
116. **Pilz, F., G. Auling, D. Stephan, U. Rau, and F. Wagner.** 1981. A high affinity Zn^{2+} uptake system controls growth and biosynthesis of an extracellular, branched β-1,3-β-1,6-glucan in *Sclerotium rolfsii* ATCC 15205. *Exp. Mycol.* **15:**181–192.
117. **Purvis, O. W.** 1984. The occurrence of copper oxalate in lichens growing on copper sulphide-bearing rocks in Scandinavia. *Lichenologist* **16:**197–204.

118. **Purvis, O. W., and C. Halls.** 1996. A review of lichens in metal-enriched environments. *Lichenologist* **28:**571–601.
119. **Ramadan, S. E., A. A. Razak, Y. A. Youssef, and N. M. Sedky.** 1988. Selenium metabolism in a strain of *Fusarium. Biol. Trace Elem. Res.* **18:**161–170.
120. **Ramos, S., P. Pena, E. Valle, L. Bergillos, F. Parra, and P. S. Lazo.** 1985. Coupling of protons and potassium gradients in yeast, p. 351–357. *In* I. S. Kulaev, E. A. Dawes, and D. W. Tempest (ed.), *Environmental Regulation of Microbial Metabolism.* Academic Press, Ltd., London, United Kingdom.
121. **Ramsay, L. M., and G. M. Gadd.** 1997. Mutants of *Saccharomyces cerevisiae* defective in vacuolar function confirm a role for the vacuole in toxic metal ion detoxification. *FEMS Microbiol. Lett.* **152:**293–298.
122. **Ramsay, L. M., J. A. Sayer, and G. M. Gadd.** 1998. Stress responses of fungal colonies towards toxic metals, p. 178–200. *In* N. A. R. Gow, G. D. Robson and G. M. Gadd (ed.), *The Fungal Colony.* Cambridge University Press, Cambridge, United Kingdom.
123. **Rauser, W. E.** 1995. Phytochelatins and related peptides. *Plant Physiol.* **109:**1141–1149.
124. **Read, D. J.** 1991. Mycorrhizas in ecosystems. *Experientia* **47:**376–391.
125. **Sanders, D**. 1990. Kinetic modelling of plant and fungal membrane transport systems. *Annu. Rev. Plant Physiol. Plant Mol. Biol.* **41:**77–107.
126. **Sayer, J. A., and G. M. Gadd.** 1997. Solubilization and transformation of insoluble inorganic metal compounds to insoluble metal oxalates by *Aspergillus niger. Mycol. Res.* **101:**653–661.
127. **Sayer, J. A., C. White, T. A. M. Bridge, and G. M. Gadd.** Metals and metalloids. *In* H. Eccles (ed.), *Bioremediation: Sustainable Technology for the Twenty-First Century*, in press. Taylor & Francis, London, United Kingdom.
128. **Sayer, J. A., M. Kierans, and G. M. Gadd.** 1997. Solubilization of some naturally-occurring metal-bearing minerals, limescale and lead phosphate by *Aspergillus niger. FEMS Microbiol. Lett.* **154:**29–35.
129. **Sayer, J. A., S. L. Raggett, and G. M. Gadd.** 1995. Solubilization of insoluble metal compounds by soil fungi: development of a screening method for solubilizing ability and metal tolerance. *Mycol. Res.* **99:**987–993.
130. **Sayer, J. A., J. D. Cotter-Howells, C. Watson, S. Hillier, and G. M. Gadd.** 1999. Lead mineral transformation by fungi. *Curr. Biol.* **9:**691–694.
131. **Schinner, F., and W. Burgstaller.** 1989. Extraction of zinc from an industrial waste by a *Penicillium* sp.. *Appl. Environ. Microbiol.* **55:**1153–1156.
132. **Schrickz, J. M., M. J. H. Raedts., A. H. Southamer, and H. W. van Versveld.** 1994. Organic acid production by *Aspergillus niger* in recycling culture analysed by capillary electrophoresis. *Anal. Biochem.* **231:**175–181.
133. **Smith, D. G.** 1974. Tellurite reduction in *Schizosaccharomyces pombe. J. Gen. Microbiol.* **83:**389–392.
134. **Starling, A. P., and I. S. Ross.** 1991. Uptake of zinc by *Penicillium notatum. Mycol. Res.* **95:**712–714.
135. **Strasser, H., W. Burgstaller, and F. Schinner.** 1994. High yield production of oxalic acid for metal leaching purposes by *Aspergillus niger. FEMS Microbiol. Lett.* **119:**365–370.
136. **Sukla, L. B., R. N. Kar, and V. Panchanadikar.** 1992. Leaching of copper converter slag with *Aspergillus niger* culture filtrate. *Biometals* **5:**169–172.
137. **Sutter, H.-P., E. B. G. Jones, and O. Walchi.** 1984. Occurrence of crystalline hyphal sheaths in *Poria placenta* (Fr.) Cke. *J. Inst. Wood Sci.* **10:**19–23.
138. **Tamaki, S., and W. T. Frankenberger.** 1992. Environmental biochemistry of arsenic. *Rev. Environ. Contam. Toxicol.* **124:**79–110.
139. **Tezuka, T., and Y. Takasaki.** 1988. Biodegradation of phenylmercuric acetate by organomercury-resistant *Penicillium* sp. MR-2. *Agric. Biol. Chem.* **52:**3183–3185.
140. **Thompson-Eagle, E. T., and W. T. Frankenberger.** 1992. Bioremediation of soils contaminated with selenium, p. 261–309. *In* R. Lal and B. A. Stewart (ed.), *Advances in Soil Science.* Springer-Verlag, New York, N.Y.
141. **Thompson-Eagle, E. T., W. T. Frankenberger, and U. Karlson.** 1989. Volatilization of selenium by *Alternaria alternata. Appl. Environ. Microbiol.* **55:**1406–1413.

142. **Thompson-Eagle, E. T., W. T. Frankenberger, and K. E. Longley.** 1991. Removal of selenium from agricultural drainage water through soil microbial transformations, p. 169–186. *In* A. Dinar and D. Zilberman (ed.), *The Economics and Management of Water and Drainage in Agriculture*. Kluwer Academic Publishers, New York, N.Y.
143. **Tobin, J. M., D. G. Cooper, and R. J. Neufeld.** 1990. Investigation of the mechanism of metal uptake by denatured *Rhizopus arrhizus* biomass. *Enzyme Microb. Technol.* **12:**591–595.
144. **Tobin, J. M., C. White, and G. M. Gadd.** 1994. Metal accumulation by fungi—applications in environmental biotechnology. *J. Ind. Microbiol.* **13:**126–130.
145. **Tohoyama, H., M. Inouhe, M. Joho, and T. Murayama.** 1995. Production of metallothionein in copper-resistant and cadmium-resistant strains of *Saccharomyces cerevisiae. J. Ind. Microbiol.* **14:** 126–131.
146. **Tsezos, M., and B. Volesky.** 1982. The mechanism of uranium biosorption by *Rhizopus arrhizus. Biotechnol. Bioeng.* **24:**385–401.
147. **Tsezos, M., and B. Volesky.** 1982. The mechanism of thorium biosorption by *Rhizopus arrhizus. Biotechnol. Bioeng.* **24:**955–969.
148. **Tzeferis, P. G.** 1994. Leaching of a low-grade hematitic laterite ore using fungi and biologically produced acid metabolites. *Int. J. Miner. Proc.* **42:**267–283.
149. **Tzeferis, P. G., S. Agatzini, and E. T. Nerantzis.** 1994. Mineral leaching of non-sulphide nickel ores using heterotrophic micro-organisms. *Lett. Appl. Microbiol.* **18:**209–213.
150. **Vieira, M. J., and L. F. Melo.** 1995. Effect of clay particles on the behaviour of biofilms formed by *Pseudomonas fluorescens. Water Sci. Technol.* **32:**45–52.
151. **Vivier, H., B. Marcant, and M.-N. Pons.** 1994. Morphological shape characterization: application to oxalate crystals. *Part. Part. Syst. Char.* **11:**150–155.
152. **Wainwright, M.** 1988. Metabolic diversity of fungi in relation to growth and mineral cycling in soil—a review. *Trans. Br. Mycol. Soc.* **90:**159–170.
153. **Wainwright, M., and G. M. Gadd.** 1997. Fungi and industrial pollutants, p. 85–97. *In* D. T. Wicklow and B. E. Soderstrom (ed.), *The Mycota.* V. *Environmental and Microbial Relationships.* Springer-Verlag UG, Berlin, Germany.
154. **Wakatsuki, T., S. Hayakawa, T. Hatayama, T. Kitamura, and H. Imahara.** 1991. Solubilization and properties of copper reducing enzyme systems from the yeast cell surface in *Debaromyces hansenii. J. Ferment. Bioeng.* **72:**79–86.
155. **Wakatsuki, T., S. Hayakawa, T. Hatayama, T. Kitamura, and H. Imahara.** 1991. Purification and some properties of copper reductase from cell surface of *Debaromyces hansenii. J. Ferment. Bioeng.* **72:**158–161.
156. **White, C., and G. M. Gadd.** 1986. Uptake and cellular distribution of copper, cobalt and cadmium in strains of *Saccharomyces cerevisiae* cultured on elevated concentrations of these metals. *FEMS Microbiol. Ecol.* **38:**277–283.
157. **White, C., and G. M. Gadd.** 1987. Inhibition of H^+ efflux and induction of K^+ efflux in yeast by heavy metals. *Toxic. Assess.* **2:**437–444.
158. **White, C., and G. M. Gadd.** 1987. The uptake and cellular distribution of zinc in *Saccharomyces cerevisiae. J. Gen. Microbiol.* **133:**727–737.
159. **White, C., J. A. Sayer, and G. M. Gadd.** 1997. Microbial solubilization and immobilization of toxic metals: key biogeochemical processes for treatment of contamination. *FEMS Microbiol. Rev.* **20:**503–516.
160. **Wu, J. S., H. Y. Sung, and R. J. Juang.** 1995. Transformation of cadmium-binding complexes during cadmium sequestration in fission yeast. *Biochem. Mol. Biol. Int.* **36:**1169–1175.
161. **Yannai, S., I. Berdicevsky, and L. Duek.** 1991. Transformations of inorganic mercury by *Candida albicans* and *Saccharomyces cerevisiae. Appl. Environ. Microbiol.* **57:**245–247.
162. **Yu, W., R. A. Farrell, D. J. Stillman, and D. R. Winge.** 1996. Identification of SLF1 as a new copper homeostasis gene involved in copper sulfide mineralization in *Saccharomyces cerevisiae. Mol. Cell. Biol.* **16:**2464–2472.
163. **Zhao, H., and D. Eide.** 1996. The yeast *ZRT1* gene encodes the zinc transporter of a high affinity uptake system induced by zinc limitation. *Proc. Natl. Acad. Sci. USA* **93:**2454–2458.
164. **Zhao, H., and D. Eide.** 1996. The *ZRT2* gene encodes the low affinity zinc transporter in *Saccharomyces cerevisiae. J. Biol. Chem.* **271:**232031–23210.

Environmental Microbe-Metal Interactions
Edited by Derek R. Lovley

Chapter 12

Bacterial Surface-Mediated Mineral Formation

Gordon Southam

Bacteria are ubiquitous prokaryotic organisms in surface and subsurface environments (4, 106, 131). They exhibit tremendous molecular and metabolic diversity (7, 125), which can even arise within a single microenvironment (47, 48, 138). In natural systems, bacteria exploit a wide range of redox reactions that possess negative $\Delta G^{\circ\prime}$ to support metabolism (125). These reactions often affect the solubility of metals.

Bacteria can react with soluble heavy metals by binding and precipitating these metal ions on their surfaces, producing fine-grained minerals. This precipitation is attributable to the physical and chemical nature of the bacterial cell envelope and can be promoted by dissimilatory metabolic activity (see chapters 1, 2, and 9). Mineral-forming bacteria occur as individual cells (159), as complex-particle associations of cells (69) including sediments (54), and as biofilms (57, 100, 159). Natural environments where biomineralization occurs include freshwater (56, 59, 68, 99, 100), seawater (34, 44), sediments (24), zones of groundwater discharge (47, 48, 53), mine tailings (19, 63–66, 157, 158), acid mine drainage environments (20, 54, 57, 58, 73, 77), pyritic soils (88, 170), bog iron deposits (35), and hydrothermal systems (52, 81, 98, 150). The inherent ability of bacteria to precipitate metals from solution has been exploited for the reclamation of metals from contaminated water (86).

In these mineral-forming environments, certain bacteria are more extensively mineralized than are other adjacent bacteria, allowing the nonmineralized bacteria to persist in these metal-stressed systems (159). Mineral precipitation reduces the cell surface area available for nutrient uptake. Consequently, those mineralized cells are starved and die, and their proton motive force dissipates. Without the competition from hydronium ions within the cell envelope, mineralization proceeds to completion, reducing the concentration of soluble, toxic heavy metals in the surrounding microenvironment and thereby promoting the survival of the remaining

Gordon Southam • Department of Biological Sciences, Northern Arizona University, Flagstaff, AZ 86011-5640.

nonmineralized bacteria. The high surface-to-volume ratio of bacteria facilitates their growth. Because access to nutrients is based on diffusion, the greater this ratio, the greater the diffusion of nutrients into cells (10, 96, 134, 135). In terms of mineralization, high surface-to-volume ratios also provide a tremendous surfacial biomineralization potential compared with eukaryotic microorganisms. For example, the eukaryotic algae typically possess a surface-to-volume ratio 1/10 that of bacteria. The implications of the differing mineralization potentials are twofold. First, bacteria can precipitate more metal than could a comparable measure of eukaryotic biomass, and second, in a mineral-precipitating environment, the ability of bacteria to precipitate more metal increases the probability that the bacteria will survive. Therefore, from a population perspective, mineral formation represents a resistance mechanism to the presence of toxic soluble heavy metals.

Anthropogenically or naturally 'contaminated' ecosystems containing toxic heavy metals will select for and enrich metal-resistant bacteria (146, 147, 163). One resistance mechanism which is a direct response to the presence of toxic heavy metals is the ability of bacteria to form membrane porters which function by transporting soluble metals out of the cell (41, 152, 153). In this system, plasmid-encoded membrane porters often confer resistance to a variety of metals. For example, the *Alcaligenes eutrophus* plasmid pMOL28 confers resistance to nickel, cobalt, mercury, and chromate while its plasmid pMOL30 confers resistance to cadmium, zinc, cobalt, mercury, and copper (126, 127). Since these bacterial porters do not change the stability (i.e., the redox properties) of the metal or produce stable organometallic complexes (i.e., the metal remains soluble), other bacteria in these metal-contaminated systems will have to cope with the presence of the toxic metal. A second resistance mechanism involves the bacterial formation of proteinaceous metal-chelating compounds as a direct response to a metal, e.g., cadmium (78, 144), copper (94), nickel (67), and zinc (137, 144). These metal chelators prevent the metal from entering the cell, which reduces metal bioavailablity and can prevent mineral formation (67). A third resistance mechanism is the ability of bacteria to reduce the concentration of some toxic compounds by producing volatile metal forms. Examples include the formation of methyl-mercury by *Desulfovibrio desulfuricans* LS (26), dimethyl selenide by *Alternaria alternata* (168), and elemental selenium by dissimilatory selenate reduction [SeO_4^{2-} to Se^0] (129). Despite these elaborate resistance mechanisms, mineral formation still occurs in all metal-containing systems.

This chapter focuses on the role of bacterial cell surfaces in catalyzing biomineralization processes and in promoting microfossil formation. Previously unpublished micrographs are provided as representative samples of bacterium-mineral precipitates that are commonly observed in natural systems. Ultrathin sections were prepared by conventional embedding (72) without the addition of osmium tetroxide or uranyl acetate as heavy-metal fixatives and contrasting agents. The electron density in these ultrathin sectioned samples as well as in the whole mounts presented in this chapter was provided by the naturally immobilized metals.

CELL ENVELOPE STRUCTURE AND CHEMISTRY

Prokaryotes, excluding bacteria which do not possess cell walls (e.g., *Mycoplasma*, *Thermoplasma*, and *Methanoplasma*), possess a variety of surfaces which can interact with soluble metals in the environment. The domains *Bacteria* and the *Archaea* represent the two major bacterial groups that have been identified based on 16S rRNA phylogenetic studies (130, 183, 184). The domain *Bacteria* is divided into the gram-positive or gram-negative groups based on cell envelope structure and chemistry (although gram-variable organisms also exist). The *Archaea*, like the *Bacteria*, have demonstrated both positive- and negative-staining characteristics in the Gram reaction. However, the *Archaea* have different chemistry (91), antigenicity (30, 31), and ultrastructure (90, 92, 101) from the *Bacteria*.

Gram-positive bacteria possess a thick (typically 15- to 25-nm), peptidoglycan-containing cell wall that provides the framework to which the secondary polymers (teichoic acids or teichuronic acids) attach (9). Peptidoglycan, the major shape-determining structure for the organism, consists of repeating β(1–4)-linked *N*-acetylglucosamine–*N*-acetylmuramic acid dimers. The *N*-acetylmuramic acid residues, which possess short peptide stems (4 or 5 amino acids), may be covalently bound via these peptide stems to other muramic acid residues on neighboring strands. This results in a three-dimensional macromolecule shaped like the bacterium (11). These carbohydrate matrices and peptide stems possess carboxylate groups, which dominate the charge density of this structure (17, 18, 43). Teichoic acids are composed of polyalcohol-based chains joined by phosphodiester linkages (e.g., polyglycerol-phosphate-in *Bacillus subtilis* 168). The phosphate moieties confer a net negative electrical charge upon this polymer. Teichuronic acids are generally found under phosphate-limiting conditions and are composed of uronic acid polymers, which also possess anionic reactive sites (12).

One morphotype of the gram-positive *Archaea* resembles the gram-positive *Bacteria* cell envelope in possessing a single homogeneous layer external to the plasma membrane. In the *Archaea*, the envelope consists of either an *N*-acetyltalosaminuronic acid–*N*-acetylglucosamine peptidoglycan-like polymer termed pseudomurein (found in *Methanobacterium* spp.). A second morphotype is found in *Methanosarcina* spp., which possess a proteinaceous layer plus a nonsulfated heteropolysaccharide. Shape varies among the gram-positive *Archaea*, with *Methanobacterium* spp. growing as rod-shaped organisms and *Methanosarcina* spp. growing as irregularly shaped cells in packets enclosed by the heteropolysaccharide matrix.

Gram-negative bacterial cell envelopes are structurally and chemically more complicated than those of the gram-positive bacteria. External to the plasma membrane is a thin (2- to 3-nm) layer of peptidoglycan contained within a periplasm possibly having a gel-like consistency (80). The periplasm is bounded by an outer membrane which consists of a lipopolysaccharide (LPS)-phospholipid-protein mosaic in which the LPS and phospholipid occur on opposing membrane faces of the bilayer (9). The LPS is anchored to the outer membrane via its lipid moiety and extends its polysaccharide chains outward from the bacterial surface (105). The outer membrane is often cemented to the peptidoglycan via salt bridging or covalent bonding of the outer membrane proteins. One class of these proteins forms hydrophilic pores or channels (see reference 9 for more details). The LPS and the peptidoglycan of the gram-negative bacteria possess a net electronegative charge, which allows interaction with soluble cations (14, 50, 84, 85).

Additional wall layers (e.g., capsules, S layers, or sheaths) can exist external to the gram positive or gram-negative cell envelopes described above. Commonly produced by bacteria in natural systems, the capsule is a highly hydrated, amorphous assemblage of polysaccharides or polypeptides that are chemically linked to the cell surface and may extend up to 1 μm from the cell (10). Because of its carboxylate groups, the capsule usually possesses a net negative charge, and it may also possess additional anionic reactive groups due to the presence of phosphate moieties in polysaccharide chains. The highly hydrated nature of the bacterial capsule and its cell surface location allow for extensive interaction between capsular material and soluble metal cations (32, 70, 120). In the study by Kidambi et al. (93), alginate, a copper-chelating uronic acid-containing capsule, was produced by *Pseudomonas syringae* in response to the presence of copper. A capsule has also been synthesized by a *Rhizobium sp.* in response to the presence of manganese (5).

S layers are paracrystalline cell surface assemblages consisting of protein or glycoprotein with p2, p4, or p6 symmetry (154, 155). They self-assemble and associate with the underlying wall though noncovalent interaction (16, 103, 104). S layers are located external to the peptidoglycan-based cell wall of the gram-positive bacteria, the LPS of gram-negative bacteria, the pseudomurein of gram-positive archaea (154), or the cell membrane of gram-negative archaea (8, 9, 14, 82, 91, 101, 102, 104, 154, 155). S layers generally have acidic pI values and thus exhibit a net negative charge. They are often the site of intersubunit salt bridging and subunit-cell surface interaction, which neutralize the acidic groups (103). The cell envelope of gram-negative archaea (e.g., *Methanococcus* spp.) consists of an S layer as the sole envelope component external to the plasma membrane. Most of these S-layer envelopes are sensitive to treatment with the detergent sodium dodecyl sulfate, which suggests that covalent intermolecular linkages do not exist within these structures. The lack of covalent linkages in *Methanococcus* spp. is remarkable, since ionic bonding forces, hydrogen bonding, and hydrophobic interactions are rarely strong enough to resist cellular turgor pressure. It also makes one wonder about the initiation of a constriction annulus for division, which must be developed through ionic and hydrophobic forces between S-layer subunits.

Sheaths, which surround chains of cells, are less frequently encountered. In the bacteria, they are typically recalcitrant structures composed of homo- or heteropolymers and usually remain intact even after cell degradation. The role of these sheath structures in the enzymatic precipitation of iron and manganese is described later. Sheaths in the archaea are covalently linked proteinaceous structures that are found on *Methanospirillum hungatei* and *Methanosaeta concilii*. In these species, the sheath overlies a proteinaceous layer (based on the staining properties of standard embedded material [160]) and the plasma membrane is the innermost envelope structure. *M. hungatei* grows as chains of cells in filaments and possess an unusual cell spacer region composed of a combination of two types of S-layer discs (61, 160).

SURFACE CATALYSIS

The formation of many secondary minerals in natural as well as laboratory systems is catalyzed by microorganisms (113) (Table 1). These precipitation re-

Table 1. Secondary minerals known to form on microbial cell surfaces via passive interaction or as a consequence of microbial metabolism

Process	Minerals formed and process used (reference)
Surface catalysis (interaction between metallic cations and net negative surface charge on bacteria)	Iron-oxides (ferrihydrite, hematite, goethite) (57, 58)
	Hydroxy iron sulfates (136)
	Metal phosphates, phosphorite (15, 128)
	Fe- silicates; geothermal sediments (52)
	Fe-Al- silicates; sediment bacteria (54)
	Metal sulfides, e.g., millerite (54)
Mineral formation via dissimilatory metabolic activity and surface catalysis	Gypsum, calcite, magnesite, celestite, or strontianite (e.g., *Synechococcus* sp. [149, 165]); photosynthetic shift in [CO_3^{2-}] from uptake of bicarbonate from solution and release of hydroxyl anions
	Dolomite (178) and siderite (130, 131, 141); CO_2 flux from heterotrophic activity
	Iron oxide/oxyhydroxide (e.g., *Sphaerotilus* [142, 143] and *Leptothrix* [2]); enzymatic iron oxidation of Fe(II) to Fe(III)
	Manganese oxide (e.g., *Leptothrix* [22]; enzymatic manganese oxidation of Mn(II) to Mn (IV)
	Ferromanganese concretions (Fe-Mn oxides [45]); combination of above two systems
	Metal sulfides (mackinawite [FeS_{1-x}], millerite [NiS], chalchocite [CuS], sphalerite [ZnS]) (e.g., *Desulfotomaculum* [67]); dissimilatory sulfate reduction (173)
	Metal phosphates; phosphatase activity (114)
	$Cr(OH)_3$; chromium (VI) to chromium (III); dissimilatory metal reduction (87)
	UO_2; uranium (VI) to uranium (IV); dissimilatory metal reduction (112)

actions have been divided into two general categories: passive and active mineralization. Passive mineralization, or surface catalysis, is caused by the net negative charge on most bacterial cell surfaces (9) that nucleates the precipitation of metallic cations from solution. Active surface-mediated mineralization occurs either by the direct transformation of metals into unstable forms (71) or by the formation of metal-reactive by-products (67, 173).

Fundamental Aspects of Surface-Mediated Mineral Precipitation

Ion-exchanging reactions play an important role in the initiation of mineral formation through competition between hydronium ions, alkaline earth ions, and heavy metals for anionic reactive sites on bacterial surfaces (115, 116). An acid-base titration of the *B. subtilis* cell wall demonstrated that carboxyl, phosphoryl, and hydroxyl groups could potentially interact with soluble heavy metals (37, 38, 49). Under normal growth conditions, divalent cations (usually Ca^{2+} and Mg^{2+}) contribute to the stability of teichoic and teichuronic acid polymers (12), LPS (51), and S layers (17, 18). It is these cation-stabilized anionic sites that are replaced by heavy metals and serve as nucleation sites for the formation of minerals at the bacterial cell surface. Planktonic and sessile bacteria nucleate metals from solution, resulting initially in the precipitation of fine-grained minerals tens of nanometers in diameter. The nucleation of fine-grained minerals on bacterial cells can be observed by transmission electron microscopy (TEM) of bacterial whole mounts because the minerals are more electron dense than are the bacteria (Fig. 1).

Bacterial metal precipitation typically exceeds the stoichiometry expected per chemical reaction site within the cell envelope (122). Mineral formation results from neutralization of chemically reactive sites and proceeds via nucleation of additional metallic ions with these previously sorbed metals (17, 18). These critical nuclei, stabilized by the wall, are less prone to remobilization by dissolution, because the wall reduces the interfacial tension between the mineral nucleus and the bulk water phase. Mineral growth, then, is most active at the outer surface of the bacterium, where these nuclei are formed and where space constraints by the en-

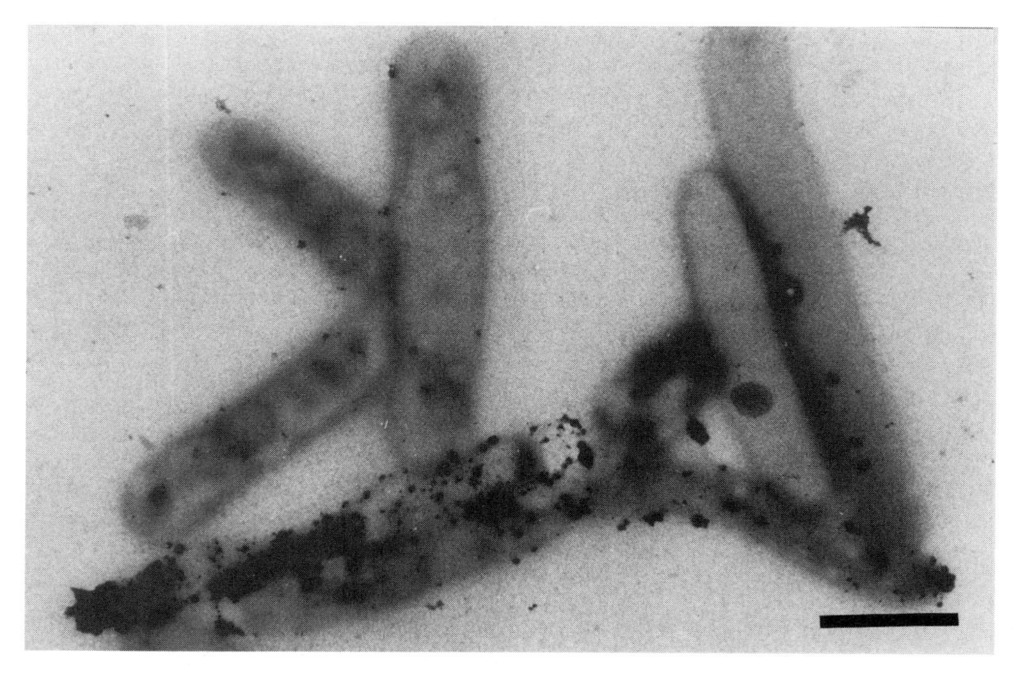

Figure 1. Unstained transmission electron micrograph of a water sample from the Golden Giant mine tailings pond (Hemlo gold region, Marathon, Ontario, Canada) prepared by drying an aliquot of water onto a Formvar-carbon coated 200 mesh Cu-TEM grid and examined using a Philips EM300 electron microscope (a whole mount). Not all of these bacteria have precipitated fine-grained iron-arsenic minerals on their surface. Elemental analysis was performed using a Philips EM400T electron microscope equipped with a LINK X-ray analyzer for energy-dispersive X-ray spectroscopy (data not shown). The immobilization of metal on bacterial cell surfaces occurs at distinct nucleation sites, resulting in the development of fine-grained minerals. Bar, 1 μm.

velope polymers do not inhibit metal precipitation. Mineral growth at these nucleation sites initially prevents a bacterium from being completely encrusted in metal. This is clearly important to an organism that obtains its nutrients by diffusion. The eventual, complete mineralization of microbial surfaces produces hollow minerals that are generally the size (i.e., on the order of micrometers) and shape of the cell (Fig. 2).

As metal precipitation proceeds in the planktonic phase, larger aggregates form containing insoluble particulate material and mineralized bacteria (179). When iron is present in the reaction system, walls first interact with one another to form visible flocs, which immobilize a variety of metals and then sediment by gravity (118). These iron oxyhydroxide wall matrices enhance the immobilization of metals such as Cu and Ni (57, 58). Cooperative binding between iron and other metals has also been observed for *Bacillus licheniformis*, a bacterium that contains an extensive anionic capsule (119). When capsule formation occurs in metal-contaminated systems, metal nucleation and precipitation within the capsule can protect the internal microcolony from the deleterious effects of mineralization (Fig. 3).

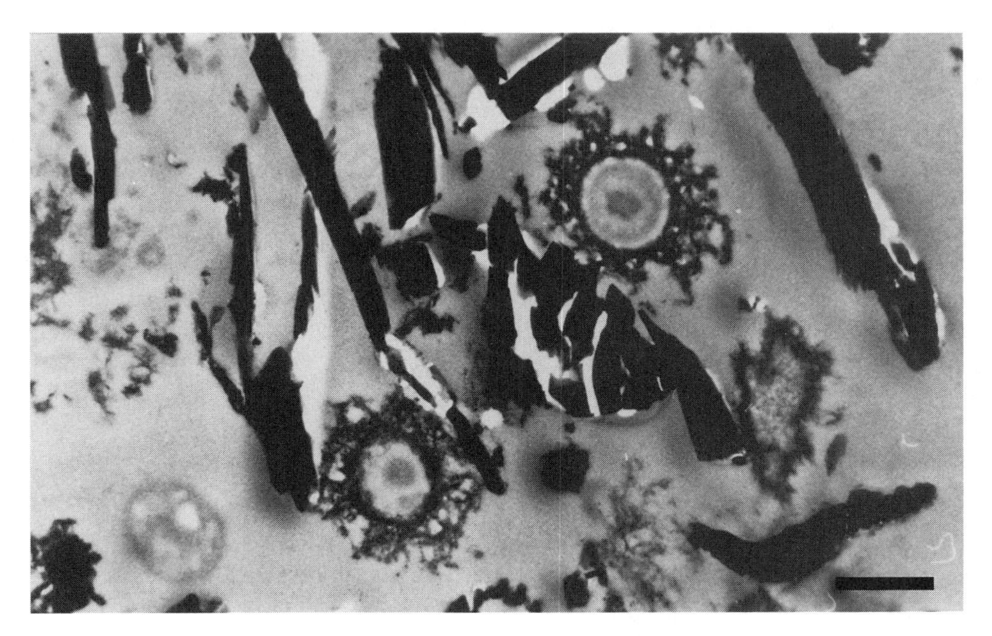

Figure 2. Unstained, ultrathin section transmisison electron micrograph of a mineralized biofilm from the Copper Rand mine tailings pond (Chibougamou, Quebec, Canada), revealing an unmineralized bacterium, mineralized bacteria, and inorganic particulate material that has been trapped by the biofilm. The nonmineralized cell is presumably viable, containing a hydronium "cloud" produced by its proton motive force. The proton motive force creates an acidic environment at the bacterial cell surface and prevents or limits heavy-metal binding, while the other two cells probably exhibited little or no metabolic activity at the time of sampling and are mineralized. Bar, 0.5 μm.

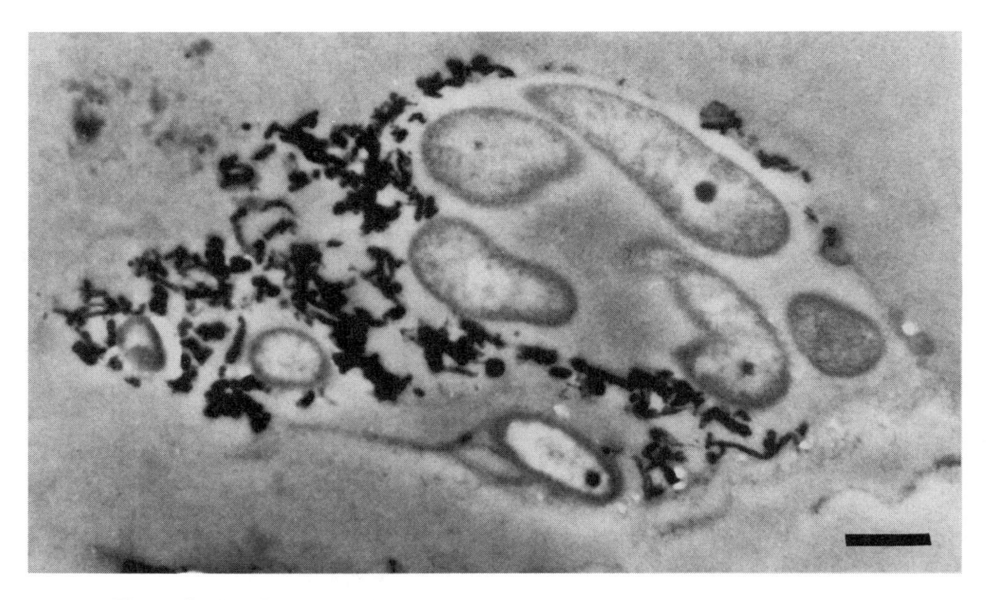

Figure 3. Unstained, ultrathin-section transmission electron micrograph of a mineralized biofilm from the Lemoine tailings pond (Chibougamau, Quebec, Canada), demonstrating how capsular material is capable of protecting a bacterial microcolony from mineralization. Bar, 0.5 μm.

In laboratory metal binding experiments, living *B. subtilis* cells bound less metal than did nonliving cells (177). These experiments determined that the membrane-induced proton motive force, which pumps protons into the wall fabric, reduces the metal binding ability of the cell walls, probably through competition of protons with metal ions for anionic wall sites. The metabolic activity and energized membranes of individual bacterial cells may help explain why one bacterium in a natural population is not mineralized whereas a second, adjacent bacterium might be extensively mineralized (Fig. 2) (56).

Reactive By-Products and Surface Catalysis

Cyanobacteria have the capacity to precipitate calcite from solution (132). The S layer of *Synechococcus*, a cyanobacterial species, is responsible for nucleating gypsum or calcite in a freshwater environment (151, 165, 166). In this system, calcite formation is mediated by an increase in alkalinity that is caused by photosynthetic activity. When provided with alternative cations, this photosynthetic S-layer system was capable of precipitating celestite and strontianite (149). The oxidation of iron (biological reaction, $2Fe^{2+} + \frac{1}{2}O_2 + 2H^+ \rightarrow 2Fe^{3+} + H_2O$ [33, 47, 48]; geochemical reaction, $Fe^{3+} + 3H_2O \rightarrow Fe(OH)_{3(S)} + 3H^+$) and manganese ($2Mn^{2+} + O_2 + 2H_2O \rightarrow MnO_{2(S)} + 4H^+$ [net reaction of two-step oxidation] [1, 3, 45, 46]) typically occurs within the aerobic/anaerobic interface, where the reduced metals encounter an oxidizing environment. Sheaths of *Sphaerotilus* (142,

143) and *Leptothrix* (2, 22, 33) are integral in oxidative enzyme-mediated precipitation of iron and manganese because they contain the enzymes responsible for metal oxidation and they nucleate the oxidized mineral precipitates.

Dissimilatory metal-reducing bacteria are best known for their ability to utilize minerals for their electron acceptors, resulting in the solubilization of Mn and Fe oxides and any coprecipitated base metals (108–110, 123–125). Dissimilatory metal reduction is also responsible for the bacterial cell surface precipitation of other metals, e.g., chromium(VI) to chromium(III) [$Cr(OH)_{3(s)}$] (87, 111) and uranium(VI) to uranium(IV) ($UO_{2(s)}$) (71, 112). While iron reduction can also result in the formation of magnetite (112) or siderite (29, 131, 141), reduced iron and other base metals are commonly precipitated on dissimilatory sulfate-reducing bacterium (SRB) cell surfaces as sulfides (107). In a related system, a bacterial phosphatase released inorganic phosphate, catalyzing the formation of metal phosphates that nucleated on the bacterial cell surfaces, perhaps with previously immobilized metals (114). These dissimilatory processes, combined with bacterial surface catalysis are responsible for cell surface mineral formation in these systems.

Iron sulfide is the most common metal sulfide deposit attributed to biogenic activity (23, 83, 170). A core prerequisite to biogenic sulfide formation is the presence of sulfate. The basic biochemical (equation 1) and geochemical (equation 2) reactions mediated by dissimilatory SRB (172, 173) are as follows:

$$2CH_2O + SO_4^{2-} \rightarrow H_2S + 2HCO_3^- \quad (1)$$

$$Fe^{2+} + H_2S \rightarrow FeS_{(s)} + 2H^+ \quad (2)$$

The immobilization of FeS on the SRB surface is promoted by a combination of the ionic interaction of Fe^{2+} with the anionic cell surface polymers and biogenic H_2S (Fig. 4) (67, 121). While the precipitates found on bacterial surfaces represent the early stages in authigenic mineral formation (13, 54), in SRB systems, mineral transformations continue after the initial metal immobilization has occurred, resulting in the formation of pyrite. Compared to purely abiotic processes, the bacterially mediated transformation was shown to be more efficient in transforming FeS into FeS_2 (42). Also, in a radiolabeled organic-sulfur tracer study using TEM autoradiography, partitioning of the radiolabel from the inside of the bacteria to the cell surface has been demonstrated.

Since FeS is an important precursor to pyrite formation (139, 140, 181, 182), the precipitation of a thin layer of FeS ($\sim$25 μm [67]) would promote its reactivity with bacterially released H_2S, thus also promoting the formation of FeS_2. This, combined with the observation that pyrite nucleation is kinetically favored over crystal growth (139, 140), is the reason that most of the biogenic FeS_2 forms as a bilayer on the surface of the SRB (Fig. 5). The minute size (micrometer range) of SRB is a key feature in this process and confers a high surface area for the uptake of nutrients into cells (135), the precipitation of FeS (67), and the diagenesis of FeS to pyrite.

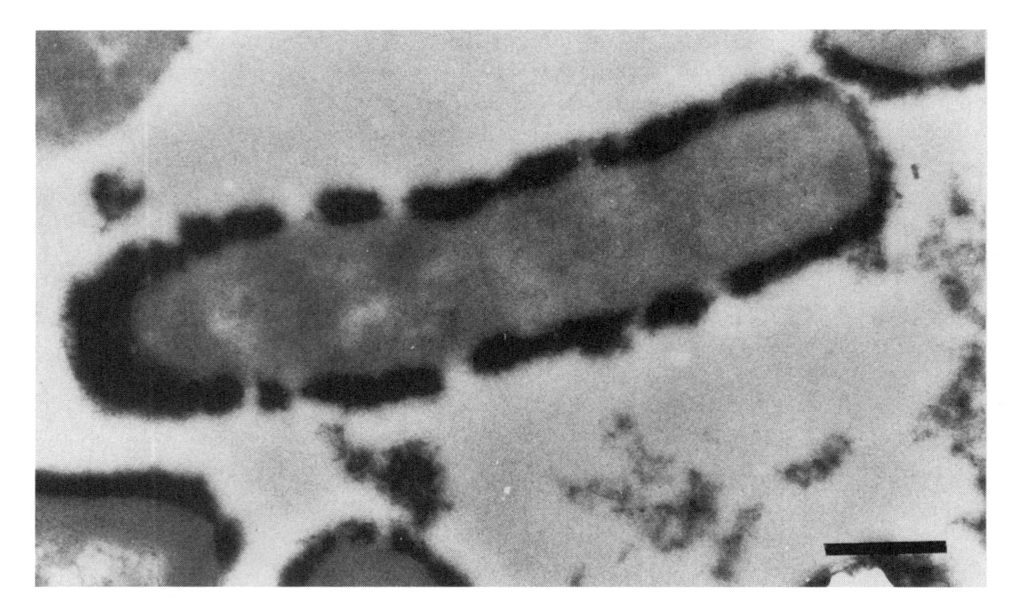

Figure 4. Unstained ultrathin-section transmission electron micrograph of a *Desulfotomaculum* sp. which has been cultured in the presence of 100 ppm of Fe^{2+}, resulting in the precipitation of amorphous FeS (reference 67 and data not shown). FeS precipitation at the cell surface is caused by the presence of HS^-, released as a by-product of SRB metabolism, presumably forming an HS^- rich microenvironment around the individual SRB. Even in active surface catalysis systems, the surface of the bacteria has an uneven distribution of minerals. Bar, 0.5 μm.

GEOLOGICAL CONSEQUENCES

In aquatic systems, mineral growth cross-links individual bacteria, forming flocs. Aggregate formation increases the floc density and can cause settling in the water column (118), promoting the transfer of once soluble metals into the sediment. The immobilization of soluble heavy metals by microbial mechanisms could potentially produce sediments rich in these metals (54). The enrichment of base metals in sediments was recognized by Timperley and Allan (169), who advocated the examination of sediments, presumably containing metal-encrusted bacteria, during exploration geochemistry to determine the presence of anomalous metal concentrations.

Metal precipitates on bacterial surfaces are generally hydrous, amorphous aggregates and become crystalline minerals only by lithification. However, poorly crystallized phases may also reorder and become more crystalline with time (e.g., by solution redeposition or solid-state transformation). Citing geological timescales, the precipitates found on bacterial surfaces represent the early stages in authigenic mineral formation (13). In a low temperature (100°C) bacterium-mineral diagenesis study, Beveridge et al. (15) found that bacteria contribute to the formation of crystalline metal phosphates and polymeric metal-coated organic residues by accelerating the formation of authigenic mineral phases.

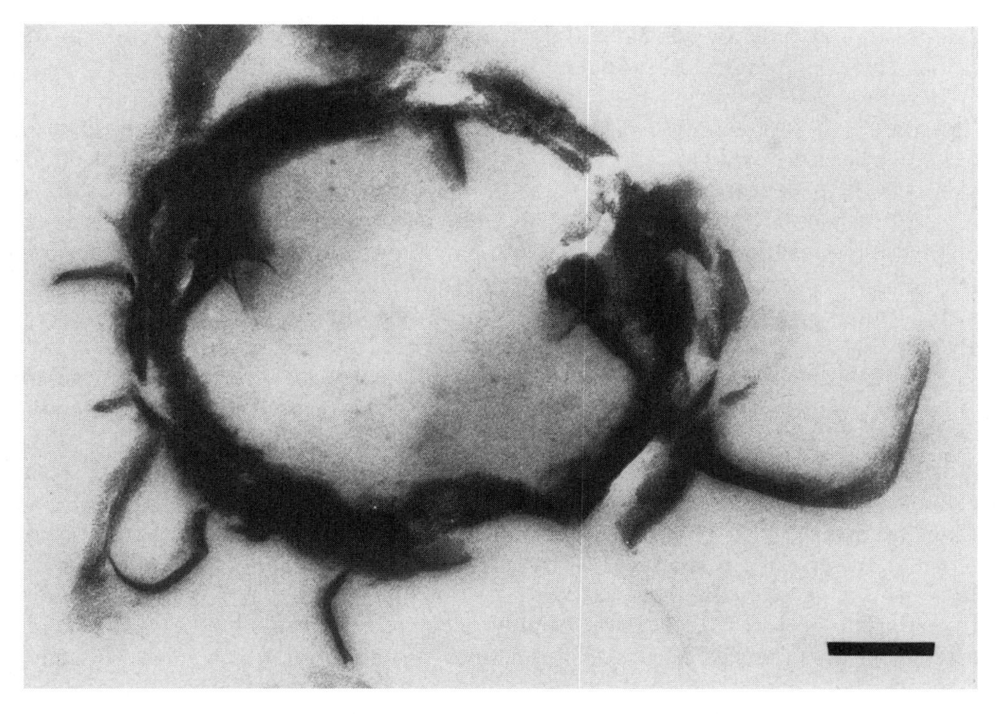

Figure 5. Unstained ultrathin-section transmission electron micrograph of an SRB culture/diagenesis system that has been incubated at 21°C for 6 months. Bacterial diagenesis promoted the nucleation of iron disulfide as a bilayer on the inner and outer surfaces of the SRB, representing the earliest and dominant stage of bacterial mineral diagenesis. Bar, 100 nm.

The formation of bacterial microfossils results from the cell surface immobilization of soluble heavy metals (biomineralization) by passive ionic interactions or by the formation and release of chemically reactive metabolic by-products that also promote the nucleation of metals on bacterial surfaces. Metal-encrusted cell surfaces are resistant to remobilization and are typically the only component of the cell that is preserved for, possibly, as long as several billion years. The size and shape of the microfossils are determined by bacterial morphology, which includes spherical, rod-shaped, filamentous, vibriod, helical, and stalked structures. The identification of bacterial microfossils as hollow mineral assemblages using ultrathin-section TEM requires preservation of the original biomineralization phenomenon. Preservation may occur via the immobilization of metals by bacteria (demonstrated in Fig. 1 to 3) in a geochemically and biogeochemically stable environment. However, for long-term preservation, silicification of bacteria, often in association with iron, results in the formation of extremely stable microfossils (52, 55, 97, 98, 164, 174–176, 180). In their simplest form, bacterial microfossils consist of only a mineralized cell envelope, which preserves the original bacterial morphology (Fig. 2). Even in the rare cases when both the cell envelope and the cytoplasm are mineralized, the cell envelope can be easily differentiated from the cytoplasm, preserving the original cell morphology.

In sedimentary environments, particulate complexes are instrumental in crystal formation and in the growth of authigenic minerals such as sulfides, phosphates, oxides, and carbonates (17, 18, 39, 76). Also, bacterium-metal-clay aggregates (tens to hundreds of micrometers in size) are less prone than are mineralized bacteria to remobilize metals (62). This is probably due to the extremely low surface-area-to-volume ratios of the metal-laden flocs compared to that of an individual bacterium. In sediments, the decomposition of organic matter does not typically include organics (most probably bacterial cell envelopes) that have bound heavy metals (17, 18, 40).

Since bacterial microfossils can be formed within several weeks and bacteria are ubiquitous (7, 125), bacterial microfossils do not have to result from the syngenetic formation of the fossil in the enclosed rock as recommended by Schopf and Walter (148). Bacterial fossil formation within naturally occurring current Earth systems can occur in any region that contains (or contained as some point in time) liquid water up to ~110°C, the upper limit of bacterial life (161). Therefore, the occurrence of bacteria and microfossils is not limited to sedimentary rock of scientifically established age, e.g., bacteria associated with fracture flow within the Columbia River basalt (131, 162). To put it simply, a microfossil must be of true biogenic origin and must be indigenous to the natural system from which it was obtained and must not be a contaminant of sampling.

Although individual bacteria are extremely small, bacterial mineral formation can result in the formation of geologically significant carbonates (6), iron-silica deposits (6, 27, 28, 133), sulfides (21, 23, 170), and phosphorites (25, 36, 128, 156, 185). From a geochemical perspective, the contribution of bacteria to the formation of ancient deposits is often inferred by the enrichment of stable light isotopes by microbial mineral-forming processes (145). Biogenic activity can also be implicated when thermodynamic constraints do not permit particular inorganic processes to occur. Stratum-bound and stratiform metal sulfides that formed at moderate to low geological temperatures (<200°C) are considered to be of biogenic origin because the geochemical precipitation of iron sulfides from ferrous iron, sulfate, and organic matter requires temperatures of ≥250°C (171).

The bacteriological processes (*Bacteria* and *Archaea*) that have produced these geologically recognized deposits are still occurring today. Cyanobacteria have been implicated in open-ocean whiting events (167) via the same photosynthesis-generated alkalinity and bacterial surface-active nanoenvironment described by Thompson et al. (166). Bacteria are actively involved in the precipitation of silica in ambient and hydrothermal current Earth systems (54, 60, 79, 89, 97–99), conceivably through the precipitation of silicic acid as amorphous silica (174) and dissolution and reprecipitation to chalcedony. Mineral formation in hypersaline systems has not received much attention, presumably because of the extremely high concentration of cations (e.g., Na^+) that would compete with anionic reactive groups on bacteria. However, Kobluk and Crawford (95) have described a modern hypersaline organic mud- and gypsum-dominated system containing microbialites. The formation of biogenic metal sulfides is enriched downstream from acid mine drainage systems (73). This bacterial mineralization process is being exploited for bioremediation of acid mine drainage (74, 75, 117).

Whether mineral formation results from surface catalysis or a combination of dissimilatory processes and surface catalysis, microfossils in current and ancient geological systems would not have been preserved if bacterial surfaces were not important to their preservation.

Acknowledgments. The Hemlo tailings pond samples were collected while working on an Ontario Geological Survey project (Geoscience Research grant 418). The Lemoine and Copper Rand mine drainage samples were collected during an environmental impact assessment for Westminer Canada Ltd. The TEM, SEM and EDS were performed in the NSERC Regional STEM facility located in the Department of Microbiology, University of Guelph, Guelph, Ontario, Canada, and in the Electron Microscopy facility located in the Department of Biological Sciences, Northern Arizona University.

Special thanks to C. Southam for assistance with preparation of this chapter.

REFERENCES

1. **Adams, L. F., and W. C. Ghiorse.** 1986. Physiology and ultrastructure of *Leptothrix discophera* SS-1. *Arch. Microbiol.* **145:**126–135.
2. **Adams, L. F., and W. C. Ghiorse.** 1987. Characterization of extracellular Mn^{2+}-oxidizing activity and isolation of an Mn^{2+}-oxidizing protein from *Leptothrix discophora* SS-1. *J. Bacteriol.* **169:** 1279–1285.
3. **Adams, L. F., and W. C. Ghiorse.** 1988. Oxidation state of Mn in the Mn-oxide produced by *Leptothrix discphora* SS-1. *Geochim. Cosmochim. Acta* **52:**2073–2076
4. **Amy, P. S., D. L. Haldeman, D. Ringelberg, D. H. Hall, and C. Russell.** 1992. Comparison of identification systems for classification of bacteria isolated from water and endolithic habitats within the deep subsurface. *Appl. Environ. Microbiol.* **58:**3367–3373.
5. **Appanna, V. D., and C. M. Preston.** 1987. Manganese elicits the synthesis of a novel exopolysaccharide in an arctic *Rhizobium. FEBS Lett.* **215:**79–82.
6. **Barghoorn, E. S., and S. A. Tyler.** 1965. Microorganisms from the Gunflint Chert. *Science* **147:** 563–577.
7. **Barns, S. M., and S. A. Nierzwicki-Bauer.** 1997. Microbial diversity in ocean, surface and subsurface environments. *Rev. Mineral.* **35:**35–79.
8. **Baumeister, W., I. Wildhaber, and B. M. Phipps.** 1989. Principles of organization in eubacterial and archaebacterial surface proteins. *Can. J. Microbiol.* **35:**215–227.
9. **Beveridge, T. J.** 1981. Ultrastructure, chemistry, and function of the bacterial wall. *Int. Rev. Cytol.* **72:**229–317.
10. **Beveridge, T. J.** 1988. The bacterial surface: general considerations towards design and function. *Can. J. Microbiol.* **34:**363–372.
11. **Beveridge, T. J.** 1989. Role of cellular design in bacterial metal accumulation and mineralization. *Annu. Rev. Microbiol.* **43:**147–171.
12. **Beveridge, T. J., C. W. Forsberg, and R. J. Doyle.** 1982. Major sites of metal binding in *Bacillus licheniformis* walls. *J. Bacteriol.* **150:**1438–1448.
13. **Beveridge, T. J., and W. S. Fyfe.** 1985. Metal fixation by bacterial cell walls. *Can. J. Earth Sci.* **22:**1893–1898.
14. **Beveridge, T. J., and S. F. Koval.** 1981. Binding of metals to cell envelopes of *Escherichia coli* K-12. *Appl. Environ. Microbiol.* **42:**325–335
15. **Beveridge, T. J., J. D. Meloche, W.S. Fyfe, and R. G. E. Murray.** 1983. Diagenesis of metals chemically complexed to bacteria: laboratory formation of metal phosphates, sulfides, and organic condensates in artificial sediments. *Appl. Environ. Microbiol.* **45:**1094–1108.
16. **Beveridge, T. J., and R. G. E. Murray.** 1976. Reassembly in vitro of the superficial wall components of *Spirillum putridiconchylium. J. Ultrastruct. Res.* **55:**105–118.
17. **Beveridge, T. J., and R. G. E. Murray.** 1976b. Uptake and retention of metals by cell walls of *Bacillus subtilis. J. Bacteriol.* **127:**1502–1518.

18. **Beveridge, T. J., and R. G. E. Murray.** 1980. Sites of metal deposition in the cell wall of *Bacillus subtilis*. *J. Bacteriol.* **141:**876–887.
19. **Bhatti, T. M., J. M. Bigham, L. Carlson, and O. H. Tuovinen.** 1993. Mineral products of pyrrhotite oxidation by *Thiobacillus ferrooxidans*. *Appl. Environ. Microbiol.* **59:**1984–1990.
20. **Bigham, J. M., U. Schuertmann, L. Carlson, and E. Murad.** 1990. A poorly crystallized oxy-hydroxy sulfate of iron formed by bacterial oxidation of Fe (II) in acid mine waters. *Geochim. Cosmochim. Acta* **54:**2743–2758.
21. **Birnbaum, S. J., and J. W. Wireman.** 1985. Sulfate-reducing bacteria and silica solubility: a possible mechanism for evaporite diagenesis and silica precipitation in banded iron formations. *Can. J. Earth Sci.* **22:**1904–1909.
22. **Boogerd, F. C., and J. P. M. deVrind.** 1987. Manganese oxidation by *Leptothrix discophora*. *J. Bacteriol.* **169:**489–494.
23. **Bubela, B., and J. A. McDonald.** 1969. Formation of banded sulphides: metal ion separation and precipitation by inorganic and microbial sulphide sources. *Nature* **221:**465–466.
24. **Burne, R. V., and L. S. Moore.** 1987. Microbialites: organosedimentary deposits of benthic microbial communities. *Palaios* **2:**241–254
25. **Chauhan, D. S.** 1979. Phosphate-bearing stromatolites of the Precambrian Aravalli phosphate deposits of the Udaipur region, their environmental significance and genesis of phosphorite. *Precambr. Res.* **8:**95–126.
26. **Choi, S.-C., T. Chase and R. Bartha.** 1994. Metabolic pathways leading to mercury methylation in *Desulfovibrio desulfuricans* LS. *Appl. Environ. Microbiol.* **60:**4072–4077.
27. **Cloud, P. E.** 1973. Paleoecological significance of the banded iron-formation. *Econ. Geol.* **68:**1135–1143.
28. **Cloud, P. E., Jr., and G. R. Licari.** 1968. Microbiotas of the banded iron formations. *Proc. Natl. Acad. Sci. USA* **61:**779–786.
29. **Coleman, M. L., D. B. Hedrick, D. R. Lovley, D. C. White, and K. Pye.** 1993. Reduction of Fe(III) in sediments by sulfate-reducing bacteria. *Nature* **361:**436–438.
30. **Conway de Macario, E. H. König, and A. J. L. Macario.** 1986. Antigenic determinants distinctive of *Methanospirillum hungatei* and *Methanogenium cariaci* identified by monoclonal antibodies. *Arch. Microbiol.* **144:**20–24.
31. **Conway de Macario, E., M. J. Wolin, and A. J. L. Macario.** 1981. Immunology of archaebacteria that produce methane gas. *Science* **214:**74–75.
32. **Corpe, W.** 1964. Factors influencing growth and polysaccharide formation by strains of *Chromobacterium violaceum*. *J. Bacteriol.* **88:**1433–1437.
33. **Corstjens, P. L. A. M., J. P. M. de Vrind, P. Westbroek, and E. W. de Vrind-de Jong.** 1992. Enzymatic iron oxidation by *Leptothrix discophora*: identification of an iron-oxidizing protein. *Appl. Environ. Microbiol.* **58:**450–454.
34. **Cowen, J. P., and K. W. Bruland.** 1985. Metal deposits associated with bacteria: implications for Fe and Mn marine biogeochemistry. *Deep-Sea Res.* **32:**253–272.
35. **Crerar, D. A., G. W. Knox, and J. L. Means.** 1979. Biogeochemistry of bog iron in the New Jersey pine barrens. *Chem. Geol.* **24:**111–135.
36. **Dahanayake, K., and W. E. Krumbein.** 1985. Ultrastructure of a microbial mat generated phosphorite. *Miner. Depos.* **20:**260–265.
37. **Daughney, C. J., and J. B. Fein.** 1998. The effect of ionic strength on the adsorption of H^+, Cd^{2+}, Pb^{2+}, and Cu^{2+} by *Bacillus subtilis* and *Bacillus licheniformis*: a surface complexation model. *J. Colloid Interface Sci.* **198:**53–77.
38. **Daughney, C. J., J. B. Fein, and N. Yee.** 1998. A comparison of the thermodynamics of metal adsorption onto two common bacteria. *Chem. Geol.* **144:**161–176.
39. **Degens, E. T., and V. Ittekkot.** 1982. *In situ* metal-staining of biological membranes in sediments. *Nature* **298:**262–264.
40. **Degens, E. T., S. W. Watson, and C. C. Remsen.** 1970. Fossil membranes and cell wall fragments from a 7000-year-old Black Sea Sediment. *Science* **168:**1207–1208.
41. **Diels, L., and M. Mergeay.** 1990. DNA-probe mediated detection of new bacteria resistant to heavy metals. *Appl. Environ. Microbiol.* **56:**1485–1491.

42. **Donald, R., and G. Southam.** 1999. Low temperature anaerobic bacterial diagenesis of ferrous monosulfide to pyrite. *Geochim. Cosmochim. Acta* **63:**2019–2023.
43. **Doyle, R. J., T. H. Matthews, and U. N. Streips.** 1980. Chemical basis for selectivity of metal ions by the *Bacillus subtilis* cell wall. *J. Bacteriol.* **143:**471–480.
44. **Ehrlich, H. L.** 1975. The formation of ores in the sedimentary environment of the deep sea with microbial participation: the case for ferromanganese concretions. *Science* **119:**36–41.
45. **Ehrlich, H. L.** 1996. How microbes influence mineral growth and dissolution. *Chem. Geol.* **132:**5–9.
46. **Emerson, D., and W. C. Ghiorse.** 1992. Isolation, cultural maintenance, and taxonomy of a sheath-forming strain of *Leptothrix discophora* and characterization of manganese-oxidizing activity associated with the sheath. *Appl. Environ. Microbiol.* **58:**4001–4010.
47. **Emerson, D., and N. P. Revsbech.** 1994. Investigation of an iron-oxidizing microbial mat community located near Aarhus, Denmark: field studies. *Appl. Environ. Microbiol.* **60:**4022–4031.
48. **Emerson, D., and N. P. Revsbech.** 1994. Investigation of an iron-oxidizing microbial mat community located near Aarhus, Denmark: laboratory studies. *Appl. Environ. Microbiol.* **60:**4032–4038.
49. **Fein, J. B., C. J. Daughney, N. Yee, and T. Davis.** 1997. A chemical equilibrium model of metal adsorption onto bacterial surfaces. *Geochim. Cosmochim. Acta* **61:**3319–3328.
50. **Ferris, F. G., and T. J. Beveridge.** 1984. Binding of a paramagnetic metal cation to *Escherichia coli* K-12 outer membrane vesicles. *FEMS Microbiol. Lett.* **24:**43–46.
51. **Ferris, F. G., and T. J. Beveridge.** 1986. Physicochemical roles of soluble metal cations in the outer membrane of *Escherichia coli* K-12. *Can. J. Microbiol.* **32:**594–601.
52. **Ferris, F. G., T. J. Beveridge, and W. S. Fyfe.** 1986. Iron-silica crystallite nucleation by bacteria in a geothermal sediment. *Nature* **320:**609–611.
53. **Ferris, F. G., C. M. Fratton, J. P. Gerits, S. Schultze-Lam, and B. Sherwood-Lollar.** 1995. Microbial precipitation of a strontium calcite phase at a groundwater discharge zone near Rock Creek, British Columbia, Canada. *Geomicrobiol. J.* **13:**57–67.
54. **Ferris, F. G., W. S. Fyfe, and T. J. Beveridge.** 1987. Bacteria as nucleation sites for authigenic minerals in a metal-contaminated lake sediment. *Chem. Geol.* **63:**225–232.
55. **Ferris, F. G., W. S. Fyfe, and T. J. Beveridge.** 1988. Metallic ion activity by *Bacillus subtilis*: implications for the fossilization of microorganisms. *Geology* **16:**149–152.
56. **Ferris, F. G., W. S. Fyfe, T. Whitten, S. Schultze, and T. J. Beveridge.** 1989. Effect of mineral substrate hardness on the population density of epilithic microorganisms in two Ontario rivers. *Can. J. Microbiol.* **35:**744–747.
57. **Ferris, F. G., S. Schultze, T. C. Witten, W. S. Fyfe, and T. J. Beveridge.** 1989. Metal interaction with microbial biofilms in acidic and neutral pH environments. *Appl. Environ. Microbiol.* **55:**1249–1257.
58. **Ferris, F. G., K. Tazaki, and W. S. Fyfe.** 1989. Iron oxides in acid mine drainage environments and their association with bacteria. *Chem. Geol.* **74:**321–330.
59. **Ferris, F. G., J. B. Thompson, and T. J. Beveridge.** 1997. Modern freshwater microbialites from Kelly Lake, British Columbia, Canada. *Palaios* **12:**213–219.
60. **Ferris, F. G., R. G. Wiese, and W. S. Fyfe.** 1994. Precipitation of carbonate minerals by microorganisms: Implications for silicate weathering and the global carbon dioxide budget. *Geomicrobiol. J.* **12:**1–13.
61. **Firtel, M., G. Southam, T. J. Beveridge, M. H. Jericho, B. L. Blackford, P. J. Mulhern, and W. Xu.** 1992. Investigation of lattice surface layers by scanning probe microscopy, p. 243–256. *In* T. J. Beveridge and S. F. Koval (ed.), *Advances in Bacterial Paracrystalline Surface Layers.* Plenum Publishing Corp., New York, N.Y.
62. **Flemming, C. A., F. G. Ferris, T. J. Beveridge, and G. W. Bailey.** 1990. Remobilization of toxic heavy metals adsorbed to bacterial wall-clay composites. *Appl. Environ. Microbiol.* **56:**3191–3203.
63. **Fortin, D., and T. J. Beveridge.** 1997. Microbial sulfate reduction within sulfidic mine tailings: formation of diagenetic Fe-sulfides. *Geomicrobiol. J.* **14:**1–21.
64. **Fortin, D., and T. J. Beveridge.** 1997. Role of the bacterium, *Thiobacillus*, in the formation of silicates in acidic mine tailings. *Chem. Geol.* **141:**235–250.
65. **Fortin, D., B. Davis, and T. J. Beveridge.** 1996. Role of *Thiobacillus* and sulfate-reducing bacteria in iron biocycling in oxic and acidic mine tailings. *FEMS Microbiol. Ecol.* **21:**11–24.

66. **Fortin, D., B. Davis, G. Southam, and T. J. Beveridge.** 1995. Biogeochemical phenomena induced by bacteria within sulfidic mine tailings. *J. Ind. Microbiol.* **14:**178–185.
67. **Fortin, D., G. Southam, and T. J. Beveridge.** 1994. An examination of iron sulfide, iron-nickel sulfide and nickel sulfide precipitation by a *Desulfotomaculum* species: and its nickel resistance mechanisms. *FEMS Microbiol. Ecol.* **14:**121–132.
68. **Fortin, D., A. Tessier, and G. C. Leppard.** 1993. Characteristics of lacustrine iron oxyhydroxides. *Geochim. Cosmochim. Acta* **57:**4391–4404.
69. **Fukui, M., and S. Takii.** 1990. Colony formation of free-living and particle-associated sulfate-reducing bacteria. *FEMS Microbiol. Ecol.* **73:**85–90.
70. **Geesey, G. G., P. J. Bremer, J. J. Smith, M. Muegge, and L. K. Jang.** 1992. Two-phase model for describing the interactions between copper ions and exopolymers from *Alteromonas atlantica*. *Can. J. Microbiol.* **38:**785–793.
71. **Gorby, Y. A., and D. A. Lovley.** 1992. Enzymatic uranium precipitation. *Environ. Sci. Technol.* **26:** 205–207.
72. **Graham, L. L., and T. J. Beveridge.** 1990. Evaluation of freeze-substitution and conventional embedding protocols for routine electron microscopic processing of eubacteria. *J. Bacteriol.* **172:** 2141–2149.
73. **Gyure, R. A., A. Konopka, A. Brooks, and W. Doemel.** 1990. Microbial sulfate reduction in acidic (pH 3) strip-mine lakes. *FEMS Microbiol. Ecol.* **73:**193–202.
74. **Hammack, R. W., and H. W. Edenborn.** 1992. The removal of nickel from mine waters using bacterial sulphate reduction. *Appl. Microbiol. Biotechnol.* **37:**674–678.
75. **Hammer, D. A.** 1990. Constructed wetlands for acid water treatment—an overview of emerging technology, p. 381–394. *In* J. W. Gadsby, J. A. Malik, and S. J. Dau (ed.), *Acid Mine Drainage Designing for Closure.* BiTech Publications Ltd., Vancouver, Canada.
76. **Henrot, J., and R. K. Wieder.** 1990. Processes of iron and manganese retention in laboratory peat microcosms subjected to acid mine drainage. *J. Environ. Qual.* **19:**312–320.
77. **Herlihy, A. T., and A. L. Mills.** 1985. Sulfate reduction in freshwater sediments receiving acid mine drainage. *Appl. Environ. Microbiol.* **49:**179–186.
78. **Higham, D. P., and P. J. Sadler.** 1984. Cadmium-resistant *Pseudomonas putida* synthesizes novel cadmium proteins. *Science* **225:**1043–1046.
79. **Hinman, N. W., and R. F. Lindstrom.** 1996. Seasonal changes in silica deposition in hot spring systems. *Chem. Geol.* **132:**237–246.
80. **Hobot, J. A., E. Carlemam, W. Villiger, and E. Kellenberger.** 1984. The periplasmic gel: a new concept resulting from the reinvestigation of bacterial cell envelope ultrastructure by new methods. *J. Bacteriol.* **160:**143–152.
81. **Holm, N. G.** 1987. Biogenic influences on the geochemistry of certain ferruginous sediments of hydrothermal origin. *Chem. Geol.* **63:**45–57.
82. **Hovmöller, S., A. Sjogren, and D. N. Wang.** 1988. The structure of crystalline bacterial surface layers. *Prog. Biophys. Mol. Biol.* **51:**131–163.
83. **Howarth, R. W.** 1979. Pyrite: its rapid formation in a salt marsh and its importance in ecosystem metabolism. *Science* **203:**49–51.
84. **Hoyle, B., and T. J. Beveridge.** 1983. Binding of metallic ions to the outer membrane of *Escherichia coli*. *Appl. Environ. Microbiol.* **46:**749–752.
85. **Hoyle, B., and T. J. Beveridge.** 1984. Metal binding by the peptidoglycan sacculus of *Escherichia coli* K-12. *Can. J. Microbiol.* **30:**204–211.
86. **Hutchins, S. R., M. S. Davidson, J. A. Brierly, and C. L. Brierly.** 1986. Microorganisms in reclamation of metals. *Annu. Rev. Microbiol.* **40:**311–336.
87. **Ishibashi, Y., C. Cervantes, and S. Silver.** 1990. Chromium reduction in *Pseudomonas putida*. *Appl. Environ. Microbiol.* **56:**2268–2270.
88. **Ivarson, K. C., and M. Sojak.** 1978. Microorganisms and ochre deposits in field drains of Ontario. *Can. J. Soil Sci.* **58:**1–17.
89. **Jones, B., and R. W. Renaut.** 1996. Influence of thermophilic bacteria on calcite and silica precipitation in hot springs with water temperatures above 90°C: evidence from Kenya and New Zealand. *Can. J. Earth Sci.* **33:**72–83.

90. **Kandler, O.** 1982. Cell wall structures and their phylogenetic implications. *Zentbl. Bakteriol. Mikrobiol. Hyg. I Abt. Orig. C* **3:**149–160.
91. **Kandler, O., and H. König.** 1978. Chemical composition of the peptidoglycan-free cell walls of methanogenic bacteria. *Arch. Microbiol.* **118:**141–152.
92. **Kandler, O., and H. König.** 1985. Cell envelopes of archaebacteria, p. 413–457. *In* C. R. Woese and R. S. Wolfe (ed.), *The Bacteria,* vol. 8. Academic Press, Inc., New York, N.Y.
93. **Kidambi, S. P., G. W. Sundin, D. A. Palmer, A. M. Chakrabarty, and C. L. Bender.** 1995. Copper as a signal for alginate synthesis in *Pseudomonas syringae* pv. syringae. *Appl. Environ. Microbiol.* **61:**2172–2179.
94. **Kim, B.-K., T. D. Pihl, J. N. Reeve, and L. Daniels.** 1995. Purification of the copper response extracellular proteins secreted by the copper-resistant methanogen *Methanobacterium bryantii* BKYH and cloning, sequencing, and transcription of the gene encoding these proteins. *J. Bacteriol.* **177:**7178–7185.
95. **Kobluk, D. R., and D. R. Crawford.** 1990. A modern hypersaline organic mud- and gypsum-dominated basin and associated microbialites. *Palaios* **5:**134–148.
96. **Koch, A. L.** 1996. What size should a bacterium be? A question of scale. *Annu. Rev. Microbiol.* **50:**317–348.
97. **Konhauser, K.** 1998. Diversity of bacterial iron mineralization. *Earth Sci. Rev.* **43:**91–121.
98. **Konhauser, K., and F. G. Ferris.** 1996. Diversity of iron and silica precipitation by microbial mats in hydrothermal waters, Iceland: implications for Precambrian iron formations. *Geology* **24:** 323–326.
99. **Konhauser, K. O., W. S. Fyfe, F. G. Ferris, and T. J. Beveridge.** 1993. Metal sorption and mineral precipitation by bacteria in two Amazonian river systems: Rio Solimoes and Rio Negro, Brazil. *Geology* **21:**1103–1106.
100. **Konhauser, K. O., S. Schultze-Lam, F. G. Ferris, W. S. Fyfe, F. J. Longstaffe, and T. J. Beveridge.** 1994. Mineral precipitation by epilithic biofilms in the Speed River, Ontario, Canada. *Appl. Environ. Microbiol.* **60:**549–553
101. **König, H.** 1988. Archaebacterial cell envelopes. *Can. J. Microbiol.* **34:**395–406.
102. **König, H., and K. O. Stetter.** 1986. Studies on archaebacterial S-layers. *Syst. Appl. Microbiol.* **7:** 300–309.
103. **Koval, S. F.** 1988. Paracrystalline surface arrays on bacteria. *Can. J. Microbiol.* **34:**407–414.
104. **Koval, S. F., and R. G. E. Murray.** 1986. The superficial protein arrays on bacteria. *Microbiol. Sci.* **3:**357–361.
105. **Lam, J. S., L. L. Graham, J. Lightfoot, T. Dasgupta, and T. J. Beveridge.** 1992. Ultrastructural examination of lipopolysaccharides of *Pseudomonas aeruginosa* strains and their isogenic mutants by freeze-substitution. *J. Bacteriol.* **174:**7159–7167.
106. **Lovley, D. R., and F. H. Chapelle.** 1995. Deep subsurface microbial processes. *Rev. Geophys.* **33:** 365–381.
107. **Lovley, D. R., and M. J. Klug.** 1983. Sulfate reducers can outcompete methanogens at freshwater sulfate concentrations. *Appl. Environ. Microbiol.* **45:**187–192.
108. **Lovley, D. R., and E. J. P. Phillips.** 1986. Organic matter mineralization with the reduction of ferric iron in anaerobic sediments. *Appl. Environ. Microbiol.* **51:**683–689.
109. **Lovley, D. R., and E. J. P. Philips.** 1987. Competitive mechanisms for inhibitions of sulfate reduction and methane production in the zone of ferric iron reduction in sediments. *Appl. Environ. Microbiol.* **53:**2636–2641.
110. **Lovley, D. R., and E. J. P. Phillips.** 1988. Novel mode of microbial energy metabolism: organic carbon oxidation coupled to dissimilatory reduction of iron or manganese. *Appl. Environ. Microbiol.* **54:**1472–1480.
111. **Lovley, D. R., and E. J. P. Phillips.** 1994. Reduction of chromate by *Desulfovibrio vulgaris* and its c_3 cytochrome. *Appl. Environ. Microbiol.* **60:**726–728.
112. **Lovley, D. R., E. J. P. Phillips, Y. A. Gorby, and E. R. Landa.** 1991. Microbial reduction of uranium. *Nature* **350:**413–416.
113. **Lowenstam, H. A.** 1981. Minerals formed by organisms. *Nature* **211:**1126–1131.

114. **Macaskie, L. E., A. C. R. Dean, A. K. Cheetham, R. J. B. Jakeman, and A. J. Skarnulis.** 1987. Cadmium accumulation by a *Citrobacter* sp.: the chemical nature of the accumulated metal precipitate and its location on the bacterial cells. *J. Gen. Microbiol.* **133:**539–544.
115. **Mann, S.** 1988. Molecular recognition in biomineralization. *Nature* **332:**119–124.
116. **Marquis, R. E., K. Mayzel, and E. L. Carstensen.** 1976. Cation exchange in cell walls of gram-positive bacteria. *Can. J. Microbiol.* **22:**975–982.
117. **Maree, J. P., and W. F. Strydom.** 1987. Biological sulphide removal from industrial effluent in an upflow packed bed reactor. *Water Resour. Res.* **19:**141–146.
118. **Mayers, I. T., and T. J. Beveridge.** 1989. The sorption of metals to *Bacillus subtilis* walls from dilute solutions and simulated Hamilton Harbour (Lake Ontario) water. *Can. J. Microbiol.* **35:**764–770.
119. **McLean, R. J. C., D. Beauchemin, L. Clapham, and T. J. Beveridge.** 1990. Metal-binding characteristics of the gamma-glutamyl capsular polymer of *Bacillus licheniformis* ATCC 9945. *Appl. Environ. Microbiol.* **56:**3671–3677.
120. **Mittelman, M. W., and G. G. Geesey.** 1985. Copper-binding characteristics of exopolymers from a freshwater sediment bacterium. *Appl. Environ. Microbiol.* **49:**846–851.
121. **Mohagheghi, A., D. M. Updegraff, and M. B. Goldhaber.** 1985. The role of sulfate-reducing bacteria in the deposition of sedimentary uranium ores. *Geomicrobiol. J.* **4:**153–173.
122. **Mullen, M. D., D. C. Wolf, F. G. Ferris, T. J. Beveridge, C. A. Flemming, and G. W. Bailey.** 1989. Bacterial sorption of heavy metals. *Appl. Environ. Microbiol.* **55:**3143–3149.
123. **Myers, C. P., and K. H. Nealson.** 1988. Bacterial manganese reduction and growth with manganese oxide as the sole electron acceptor. *Science* **240:**1319–1321.
124. **Nealson, K. H., and C. R. Myers.** 1992. Microbial reduction of manganese and iron: new approaches to carbon cycling. *Appl. Environ. Microbiol.* **58:**439–443.
125. **Nealson, K. H., and D. A. Stahl.** 1997. Microorganisms and biogeochemical cycles: what can we learn from layered microbial communities? *Rev. Miner.* **35:**5–34.
126. **Nies, D. H.** 1992. Resistance to cadmium, cobalt, zinc and nickel in microbes. *Plasmid* **27:**17–28.
127. **Nies, D. H., and S. Silver.** 1989. Plasmid-determined inducible efflux is responsible for resistance to cadmium, zinc and cobalt in *Alcaligenes eutrophicus. J. Bacteriol.* **171:**896–900.
128. **O'Brien, G. W., J. R.Harris, A. R. Milnes, and H. H. Veeh.** 1981. Bacterial origin of East Australian continental margin phosphorites. *Nature* **294:**442–444.
129. **Oremland, R. S., J. T. Hollibaugh, A. S. Maest, T. S. Presser, L. S. Miller, and C. W. Culbertson.** 1989. Selenate reduction to elemental selenium by anaerobic bacteria in sediments and culture: biogeochemical significance of a novel sulfate-independent respiration. *Appl Environ. Microbiol.* **55:**2333–2343.
130. **Pace, N. R.** 1997. A molecular view of microbial diversity and the biosphere. *Science* **276:**734–740.
131. **Pedersen, K., and S. Ekendahl.** 1990. Distribution and activity of bacteria in deep granitic groundwaters of southeastern Sweden. *Microb. Ecol.* **20:**37–52.
132. **Pentecost, A., and J. Bauld.** 1988. Nucleation of calcite on the sheaths of cyanobacteia using a simple diffusion cell. *Geomicrobiol. J.* **6:**129–135.
133. **Pflug, H. D., and H. Jaeschke-Boyer.** 1979. Combined structural and chemical analysis of 3,800-Myr-old microfossils. *Nature* **280:**483–486.
134. **Pirie, N. W.** 1973. On being the right size. *Annu. Rev. Microbiol.* **27:**119–131.
135. **Purcell, E.** 1977. Life at low Reynold's number. *Am. J. Phys.* **45:**3–11.
136. **Ramsay, B., J. Ramsay, M. deTremblay, and C. Chavarie.** 1988. A method for the quantification of bacterial protein in the presence of jarosite. *Geomicrobiol. J.* **6:**171–177.
137. **Remacle, J., and C. Vercheval.** 1991. A zinc-binding protein in a metal-resistant strain, *Alcaligenes eutrophus* CH34. *Can. J. Microbiol.* **37:**875–877.
138. **Revsbech, N. P., and B. B. Jorgensen.** 1986. Microelectrodes: their use in microbial ecology. *Adv. Microb. Ecol.* **9:**293–353.
139. **Rickard, D.** 1997. Kinetics of pyrite formation by the H_2S oxidation of iron (II) monosulfide in aqueous solutions between 25 and 125°C: the rate equation. *Geochim. Cosmochim. Acta* **61:**115–134.

140. **Rickard, D., and G. W. Luther III.** 1997. Kinetics of pyrite formation by the H_2S oxidation of iron (II) monosulfide in aqueous solutions between 25 and 125°C: the mechanism. *Geochim. Cosmochim. Acta* **61:**135–147.
141. **Roden, E. E., and D. R. Lovley.** 1993. Dissimilatory Fe(III)-reduction by the marine microorganism *Desulfuromonas acetoxidans*. *Appl. Environ. Microbiol.* **59:**734–742.
142. **Rodgers, S. R., and J. J. Anderson.** 1976. Measurement of growth and iron deposition in *Sphaerotilus discophorus*. *J. Bacteriol.* **126:**257–263.
143. **Rodgers, S. R., and J. J. Anderson.** 1976. Role of iron deposition in *Sphaerotilus discophorus*. *J. Bacteriol.* **126:**264–271.
144. **Sakamoto, K., M. Yagasaki, K. Kirimura, and S. Usami.** 1989. Resistance acquisition of *Thiobacillus thiooxidans* upon cadmium and zinc ion addition and formation of ion-binding and zinc ion-binding proteins exhibiting metallothionein-like properties. *J. Ferment. Bioeng.* **67:**266–273.
145. **Schidlowski, M., J. M. Hayes, and I. R. Kaplan.** 1983. Isotopic inferences of ancient biochemistries: carbon, sulfur, hydrogen and nitrogen, p. 149–186. *In* J. W. Schopf (ed.), *Earth's Earliest Biosphere. Its Origin and Evolution.* Princeton University Press, Princeton, N.J.
146. **Schmidt, T., and H. G. Schlegel.** 1989. Nickel and cobalt resistance of various bacteria isolated from soil and highly polluted domestic and industrial wastes. *FEMS Microbiol. Ecol.* **62:**315–328.
147. **Schmidt, T., R.-D. Stoppel, and H. G. Schlegel.** 1991. High-level nickel resistance in *Alcaligenes xylosoxydans* 31A and *Alcaligenes eutrophicus* KT02. *Appl. Environ. Microbiol.* **57:**3301–3309.
148. **Schopf, J. W., and M. R. Walter.** 1983. Archaean microfossils: new evidence of ancient microbes, p. 214–239. *In* J. W. Schopf (ed.), *Earth's Earliest Biosphere. Its Origin and Evolution.* Princeton University Press, Princeton, N.J.
149. **Schultze-Lam, S., and T. J. Beveridge.** 1994. Nucleation of celestite and strontianite on a cyanobacterial S-layer. *Appl. Environ. Microbiol.* **60:**447–453.
150. **Schultze-Lam, S., F. G. Ferris, K. O. Konhauser, and R. G. Wiese.** 1995. *In situ* silicification of an Icelandic hot spring microbial mat: implications for microfossil formation. *Can. J. Earth Sci.* **32:**2021–2026.
151. **Schultze-Lam, S., G. Harauz, and T. J. Beveridge.** 1992. Participation of a cyanobacterial S-layer in fine-grain mineral formation. *J. Bacteriol.* **174:**7971–7981.
152. **Sensfuss, C., and H. G. Schlegel.** 1988. Plamid pMOL28-encoded resistance to nickel is due to specific efflux. *FEMS Microbiol. Lett.* **55:**295–298.
153. **Silver, S., and M. Walderhaus.** 1992. Gene regulation of plasmid- and chromosome-determined inorganic ion transport in bacteria. *Microbiol. Rev.* **56:**195–228.
154. **Sleytr, U. B., and P. Messner.** 1983. Crystalline surface layers on bacteria. *Annu. Rev. Microbiol.* **37:**311–339.
155. **Sleytr, U. B., and P. Messner.** 1988. Crystalline surface layers in procaryotes. *J. Bacteriol.* **170:**2891–2897.
156. **Soudry, D., and Y. Champtier.** 1983. Microbial processes in Negev phosphorites (southern Israel). *Sedimentology* **30:**411–423.
157. **Southam, G., and T. J. Beveridge.** 1992. Enumeration of thiobacilli with pH-neutral and acidic mine tailings and their role in the development of secondary mineral soil. *Appl. Environ. Microbiol.* **58:**1904–1912.
158. **Southam, G., and T. J. Beveridge.** 1993. Examination of lipopolysaccharide (O-antigen) populations of *Thiobacillus ferrooxidans* from two mine tailings. *Appl. Environ. Microbiol.* **59:**1283–1288.
159. **Southam, G., F. G. Ferris, and T. J. Beveridge.** 1995. Mineralized bacterial biofilms in sulfide tailings and in acid mine drainage systems, p. 148–170. *In* H. M. Lappin-Scott and J. W. Costerton (ed.), *Microbial Biofilms.* Cambridge University Press, Cambridge, United Kingdom.
160. **Southam, G., G. D. Sprott, and T. J. Beveridge.** 1992. Paracrystalline layers of *Methanospirillum hungatei*. p. 129–142. *In* T. J. Beveridge and S. F. Koval (ed.), *Advances in Bacterial Paracrystalline Surface Layers.* Plenum Publishing Corp., New York, N.Y.
161. **Stetter, K. O., G. Fiala, G. Huber, R. Huber, and A. Segerer.** 1990. Hyperthermophilic microorganisms. *FEMS Microbiol. Rev.* **75:**117–124.
162. **Stevens, T. O., J. P McKinley, and J. K. Fredrickson.** 1993. Bacteria associated with deep, alkaline, anaerobic groundwaters in southeast Washington. *Microb. Ecol.* **25:**35–50.

163. **Stoppel, R.-D., and H. G. Schlegel.** 1995. Nickel-resistant bacteria from anthropogenically nickel-polluted and naturally nickel-percolated ecosystems. *Appl. Environ. Microbiol.* **61:**2276–2285.
164. **Tazaki K., F. G. Ferris, R. G. Wiese, and W. S. Fyfe.** 1992. Iron and graphite associated with fossil bacteria in chert. *Chem. Geol.* **95:**313–325.
165. **Thompson, J. B., and F. G. Ferris.** 1990. Cyanobacterial precipitation of gypsum, calcite, and magnesite from natural alkaline lake water. *Geology* **18:**995–998.
166. **Thompson, J. B., F. G. Ferris, and D. A. Smith.** 1990. Geomicrobiology and sedimentology of the mixolimnion and chemocline in Fayetteville Green Lake, New York. *Palaios* **5:**52–75.
167. **Thompson, J. B., S. Schultze-Lam, T. J. Beveridge, and D. J. Des Marais.** 1997. Whiting events: biogenic origin due to the photosynthetic activity of cyanobacterial picoplankton. *Limnol. Oceanogr.* **42:**133–141.
168. **Thompson-Eagle, E. T., W. T. Frankenburger, Jr., and U. Karlson.** 1989. Volatilization of selenium by *Alternaria alternata*. *Appl. Environ. Microbiol.* **55:**1406–1413.
169. **Timperley, M. H., and R. J. Allan.** 1974. The formation and detection of metal dispersion halos in organic lake sediments. *J. Geochem. Explor.* **3:**167–190.
170. **Trafford, B. D., C. Bloomfield, W. I. Kelso, and G. Pruden.** 1973. Ochre formation in field drains in pyritic soils. *J. Soil Sci.* **24:**453–460.
171. **Trudinger, P. A., L. A. Chambers, and J. W. Smith.** 1985. Low-temperature sulphate reduction: biological versus abiological. *Can. J. Earth Sci.* **22:**1910–1918.
172. **Tuttle, J. H., P. R. Dugan, C. B. MacMillan, and C. I. Randles.** 1968. Microbial dissimilatory sulfur cycle in acid mine water. *J. Bacteriol.* **100:**594–602.
173. **Tuttle, J. H., C. B. Dugan, and C. I. Randles.** 1969. Microbial sulfate reduction and its potential utility as an acid mine water pollution abatement procedure. *Appl. Microbiol.* **17:**297–302.
174. **Urrutia, M., and T. J. Beveridge.** 1993. Mechanism of silicate binding to the bacterial cell wall in *Bacillus subtilis*. *J. Bacteriol.* **175:**1936–1945.
175. **Urrutia, M. M., and T. J. Beveridge.** 1993. Remobilization of heavy metals retained as oxyhydroxides or silicates by *Bacillus subtilis* cells. *Appl. Environ. Microbiol.* **59:**4323–4329.
176. **Urrutia, M. M., and T. J. Beveridge.** 1994. Formation of fine-grained silicate minerals and metal precipitates by a bacterial surface (*Bacillus subtilis*) and the implications in the global cycling of silicon. *Chem. Geol.* **116:**261–280.
177. **Urrutia, M., M. Kemper, R. Doyle, and T. J. Beveridge.** 1992. The membrane-induced proton motive force influences the metal binding ability of *Bacillus subtilis* cell walls. *Appl. Environ. Microbiol.* **58:**3837–3844.
178. **Vasconcelos, C., J. A. McKenzie, S. Bernasconi, D. Grujic, and A. J. Tien.** 1995. Microbial mediation as a possible mechanism for natural dolomite formation at low temperatures. *Nature* **377:**220–222.
179. **Walker, S. G., C. A. Flemming, F. G. Ferris, T. J. Beveridge, and G. W. Bailey.** 1989. Physicochemical interaction of *Escherichia coli* cell envelopes and *Bacillus subtilis* cell walls with two clays and ability of the composite to immobilize heavy metals from solution. *Appl. Environ. Microbiol.* **55:**2976–2984.
180. **Walter, M. R.** 1983. Archean stromatolites: evidence of Earth's earliest benthos, p. 187–213. *In* J. W. Schopf (ed.), *Earth's Earliest Biosphere. Its Origin and Evolution.* Princeton University Press, Princeton, N.J.
181. **Wilkin, R. T., and H. L. Barnes.** 1996. Pyrite formation by reactions of iron monosulfides with dissolved inorganic and organic sulfur species. *Geochim. Cosmochim. Acta* **60:**4167–4179.
182. **Wilkin, R. T., and H. L. Barnes.** 1997. Formation processes of framboidal pyrite. *Geochim. Cosmochim. Acta* **61:**323–339.
183. **Woese, C. R.** 1987. Bacterial evolution. *Microbiol. Rev.* **51:**221–271.
184. **Woese, C. R., and G. E. Fox.** 1977. Phylogenetic structure of the prokaryotic domain: the primary kingdoms. *Proc. Natl. Acad. Sci. USA* **74:**5088–5090.
185. **Youssef, M. I.** 1965. Genesis of bedded phosphates. *Econ. Geol.* **60:**590–600.

Environmental Microbe-Metal Interactions
Edited by Derek R. Lovley

Chapter 13

Bioremediation of Radionuclide-Containing Wastewaters

Jon R. Lloyd and Lynne E. Macaskie

The release of radionuclides into the environment is a subject of intense public concern. Although significant quantities of radionuclides were released as a consequence of nuclear weapons testing in the 1950 and 1960s and via accidental release, e.g., from Chernobyl in 1986, the major burden of anthropogenic environmental radioactivity is from the controlled discharge of process effluents produced by industrial activities allied to the generation of nuclear power.

Radionuclide-containing wastes are produced at all steps in the nuclear fuel cycle. Wastes produced vary considerably, from low-level, high-volume radioactive effluents produced during uranium mining to the intensely radioactive plant, fuel, and liquid wastes produced from reactor operation and fuel reprocessing. All waste streams do, however, pose a potential threat to the environment and require treatment prior to release. Although chemical treatment methods are established for all waste types (27, 191), considerable recent interest has been generated in biological treatment strategies (59, 67, 75, 76, 131, 138). New technologies will be required to achieve more stringent targets set by regulatory bodies, and it is envisaged that biological approaches may play a role alongside existing chemical processes in meeting lower release limits.

This chapter highlights the key steps in the nuclear fuel cycle where biological treatment strategies may replace or augment existing chemical processes. The mechanisms of microbial interactions with key radionuclides in the wastes are discussed alongside the possible antagonistic effects of other organic and inorganic species copresented in solution. Although emphasis is placed on the development of "end-of-pipe" treatments, the application of biological agents in the detoxification of already polluted ecosystems via in situ bioremediation is also highlighted.

Jon R. Lloyd • Department of Microbiology, University of Massachusetts, Amherst, MA 01003. ***Lynne E. Macaskie*** • School of Biological Sciences, The University of Birmingham, Edgbaston, Birmingham B15 2TT, United Kingdom.

NUCLEAR FUEL CYCLE

A simplified scheme showing the major steps in the nuclear fuel cycle is shown in Fig. 1. Natural uranium contains only small amounts (0.7%) of fissile ^{235}U. Enrichment to 3% ^{235}U is required for use in power generation, although much of the fissile isotope (1.1%) remains in the spent fuel. The aim of fuel reprocessing is to recover, for reuse, the uranium from other actinides and fission products. This is particularly true for advanced "breeder" reactors, which aim to produce at least as much fissile material as is consumed, with future technology aiming to allow the selective recovery of plutonium as an additional fuel. All steps in the cycle, from mining to fuel enrichment and reprocessing and, ultimately, reactor decontamination, generate waste contaminated with radionuclides. The characteristic activity and background matrices of the wastes are different at each stage, however, and dictate the approach that is adopted for treatment, be it chemical or biological. Although the environmental burden imposed by activities associated with the nuclear fuel cycle must be emphasized, it should be noted that some radionuclides, for example ^{210}Po, occur at naturally elevated concentrations in some groundwaters and may also require treatment (39).

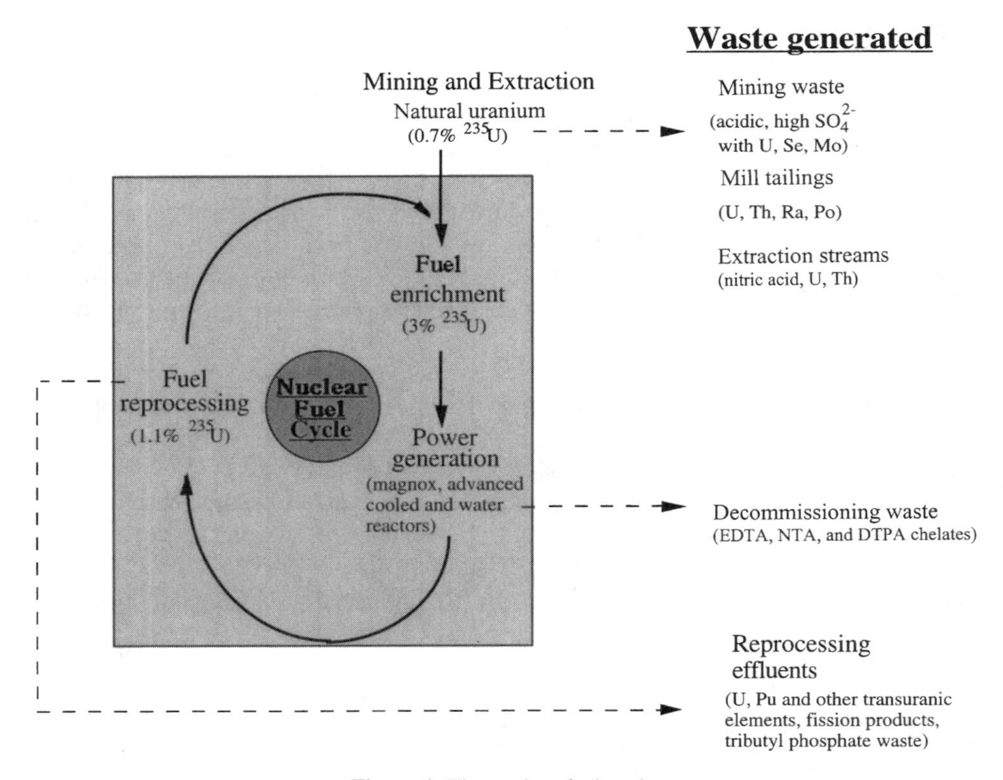

Figure 1. The nuclear fuel cycle.

Wastes from the Mining and Extraction of Uranium

Alongside uranium, mining wastes may contain high concentrations of sulfate as well as lower concentrations of other metals such as molybdenum and selenium. Sulfate, the major anionic pollutant, may be produced from the oxidation of pyrite (FeS_2) by chemolithotrophic bacteria, e.g., *Thiobacillus ferrooxidans* and *T. thiooxidans*, yielding sulfuric acid, which is required to solubilize uranium in low-grade ores (31). The accompanying decrease in pH (to pH 1 to 2), and liberation of ferric iron may have a negative effect on metal biosorption (see "Biosorption" below), which is optimal at pH 3.5 to 5.0 and is inhibited by Fe^{3+} (153). An alternative approach to the treatment of acidic mine effluents is to use sulfate-reducing bacteria to condition the waste stream, effectively reversing the action of the sulfur-oxidizing bacteria. Acidophilic and acidotolerant sulfate-reducing bacteria, recently discovered in mine tailings and effluents (66, 103) and sediments proximal to acidic hot springs in Yellowstone National Park (J. R. Lloyd, unpublished observations), can utilize sulfate as an electron acceptor during anaerobic growth. During sulfate reduction, the pH of the bulk solution is raised toward neutral, with dissolved heavy metals being precipitated as highly insoluble sulfides. The main mechanism of the pH increase is thought to be the conversion of a strong acid (H_2SO_4), which is completely dissociated in dilute solution, to a weak acid (H_2S) (256). Although the use of neutrophilic sulfate-reducing bacteria is established for metal- and sulfate-contaminated wastewaters at pH > 4.0 (14, 256) (see "Biomineralization via microbially generated ligands" below), the potential of acidophilic communities to treat acid waste streams has yet to be realized.

Two types of waste remain following the extraction of uranium from ores. Mill tailings, which are the crushed ore residues remaining after uranium extraction, contain the daughter elements of both ^{235}U and ^{238}U. Dominating the inventory of this type of waste are ^{230}Th and ^{226}Ra as well as Po and Pb. Recent studies have emphasized the poor stability of the solid waste if conditions are not maintained to inhibit microbial metabolism. Soluble Ra is often coprecipitated with $BaSO_4$ by the addition of $BaCl_2$ to the sulfate-rich tailing effluents. The resulting $(Ba,Ra)SO_4$ can be utilized by sulfate-reducing bacteria, releasing H_2S, Ba^{2+}, and Ra^{2+} (152). Dissimilatory iron-reducing bacteria, e.g., *Shewanella putrefaciens*, also release dissolved ^{226}Ra from the tailings where Ra is precipitated using $Fe(OH)_3$ (113).

Aqueous extraction streams are typically highly acidic from the use of nitric acid in uranium recovery procedures. The studies by Gadd and White (78) and Tsezos and Volesky (236, 237) have demonstrated the removal of thorium from acidic solution by microbial biomass. In addition to problems associated with solutions at low pH, the presence of high concentrations of nitrate can inhibit biological treatment of uranium (239, 264) and technetium (119). The study by Robinson et al. (K. G. Robinson, R. Ganesh, G. D. Reed, and D. A. Kucsmas, Proc. Water Environ. Fed. 67th Annu. Conf. Expos., 1994) noted, however, that up to 50 g of nitrate liter^{-1} had little effect on uranium reduction and precipitation by a sulfate-reducing bacterium (*Desulfovibrio desulfuricans*). In cases where nitrate removal is required prior to metal treatment, several recent studies have reported the development of denitrification processes to treat high-nitrate loadings in industrial wastewaters (174, 181). Pitt et al. (182) and Shumate and Strandberg (199) have also demonstrated the removal of nitrate from water cocontaminated with uranium.

Tributyl phosphate (TBP), which is added to facilitate uranium extraction and separation from thorium, may also be present in extraction streams and may affect cellular integrity via permeabilization of cellular membranes (R. E. Dick and L. E. Macaskie, unpublished observations). Finally, carbonate, which is added to "scrub" TBP for reuse, forms very strong soluble complexes with uranium (89), and the tightness of the uranium-bicarbonate complex has been implicated in inhibition of uranium biosorption at pH 8.0 (34). More recent work, however, has demonstrated that treatment of carbonate-extracted U(VI) is feasible using metal-reducing organisms, e.g., *D. desulfuricans* (179). Pretreatment of some soil types with hydrogen peroxide was required to remove organic carbon that inhibited the formation of insoluble U(IV) particles larger than 0.2 μm (179).

Wastes from Fuel Reprocessing

Fission reactions in the fuel material during nuclear-power generation leads to the production of medium-weight elements including radioiodine, noble gases, and rare earth elements. Neutron capture results in the formation of the high-mass transuranic elements and their decay products, including various isotopes of plutonium, americium, and neptunium. These products, together with residual uranium, dominate the waste inventory from nuclear power-generating facilities and are subsequently separated from fuel assemblies. In most examples, U and Pu are removed using a variant of the PUREX process, with TBP as an extractant and HNO_3 as a salting-out agent (208). The intensely radioactive high-level liquid waste remaining is not amenable to biological treatment and is converted to solid waste, via calcination or vitrification, for long-term storage (166). Finally, a solution of TBP in hydrocarbon solvent, e.g., odorless kerosene, is used to recover the uranium and plutonium, generating a low-level residual aqueous waste containing fission products, high concentrations of nitrate and other reagents used during extraction, and some residual uranium and TBP. TBP degradation by strains of *Pseudomonas* (194), by an *Acinetobacter* sp. (204), and within 3 days by a mixed culture containing seven *Pseudomonas* spp. (216) was reported. Immobilized biomass harnessed phosphate production from TBP hydrolysis to the effective bioprecipitation of uranium from aqueous test solutions (214) and acid mine wastewater (215). The involvement of a plasmid was implicated in TBP hydrolysis (216).

Wastes from Decontamination of Reactors and Plant

Structural materials including fuel cladding and reactor walls can be activated by neutrons and gamma rays during nuclear-power generation. Subsequent decontamination of surfaces during the decommissioning of reactors can be achieved using chelating agents which form selective and strong complexes with radionuclides. The aminopolycarboxylic chelating agents, EDTA, diethylenetriamine pentaacetic acid (DTPA), and nitrilotriacetic acid (NTA) are frequently used and form stronger complexes with actinides than do soil humic acids, preventing immobilization of the radionuclides in soil (138). Indeed, EDTA-enhanced nuclide migration has been implicated at both the Oak Ridge National Laboratory and Maxey Flats radioactive-waste burial sites (46, 157). EDTA is biodegraded slowly in the environment (28, 217) by two possible pathways (22), the first via the stepwise removal of acetate groups to leave ethylenediamine and the second via NTA-aldehyde with the removal of an iminodiacetate, a glycine, and then an ammonium group. Although EDTA degradation by mixed cultures has been reported (169), few studies have been performed using named axenic cultures, with the exception of that by Lauff et al. (116), who isolated an *Agrobacterium radiobacter* strain that was able to degrade the ammonium ferric complex of EDTA at concentrations in excess of 5 mM. This organism was, however, unable to degrade nickel-, cobalt-, or copper-EDTA complexes. Palumbo et al. (175) reported the degradation of the Co-EDTA complex but only if the bound Co was first displaced by Fe to give biodegradable Fe-EDTA. The interest in Co arises from the generation of ^{60}Co as an important fission product, while the related group VIII metal ^{59}Ni is produced as an activation product of the hardware of fuel assemblies. Until recently, no organism was described which could degrade the EDTA complexes of Co and Ni, but a mixed population was enriched over 2 years which grew at the expense of the EDTA complexes of Cd, Cu, Co and Ni (147). Subsequent tests confirmed a loss of EDTA from the medium in parallel to biomass growth (213), and by incorporation of inorganic phosphate as a metal precipitant, both EDTA and Fe were removed in a flowthrough reactor containing biofilm immobilized on shale particles (213).

For nuclear decontamination operations, the order of preference of the major chelating agents is EDTA > NTA > citrate, in terms of the relative strengths of their metal complexes and hence the degree of clean-up that can be achieved. The resistance of EDTA to chemical and biological attack has tended to preclude its widespread use. In contrast, more information is available on the biochemistry of NTA biodegradation, with initial oxidation of the chelating agent to iminodiacetic acid being catalyzed by a soluble oxygen-, NADH-, and Mg^{2+}-dependent monooxygenase (48). To our knowledge, no workers have reported the biodegradation of EDTA, NTA, or DTPA chelated to uranium or transuranic elements. This is in contrast to studies on metal-citrate biodegradation by Francis and coworkers. Citric acid, a naturally occurring organic complexing agent, has also found use in the decontamination of nuclear reactors (67). Several bidentate metal-citrate complexes are biodegradable (37, 38, 105), but a notable exception is the binuclear uranium-citrate complex, which was degraded by cell extracts of *Pseudomonas fluorescens* but not by intact cells, implying that biodegradation of the chelated ligand was limited by lack of transport into the cell. The uranyl-citrate complex is, however, rapidly decomposed when exposed to light (56), permitting precipitation of uranium as uranium trioxide (56, 68, 69). In other studies (212), remediation of Co- and Ni-citrate complexes was achieved by *Pseudomonas putida* and *Pseudomonas aeruginosa*, respectively, with the liberated metal being removed by addition of phosphate per se or provided as a precipitant ligand biogenically (see below).

RADIONUCLIDE-MICROBE INTERACTIONS

Microorganisms can interact with radionuclides via several mechanisms, some of which may be used as the basis of potential bioremediation strategies. The major types of interaction are summarized in Fig. 2. In addition to the mechanisms outlined, accumulation of radionuclides by plants (phytoremediation) warrants special attention but is beyond the scope of this review. Instead, the reader is referred to the work of Salt et al. (197), who give a detailed description of the use of plants for (i) phytoextraction (the use of metal-accumulating plants to remove toxic metals from soil), (ii) rhizofiltration (the use of plant roots to remove toxic metals from polluted waters), and (iii) phytostabilization (the use of plants to eliminate the bioavailability of toxic metals in soils). The role of the microbiota associated with the plant root system, in radionuclide accumulation by plant tissue, remains relatively poorly studied. In this connection, the use of reed bed technologies and artificial wetlands for bioremediation (87, 88; M. Kalin, Proc. Int. Symp. Biohydrometall., 1989) should also be mentioned, since these, too, would point to the involvement of rhizosphere microorganisms.

Biosorption

The term "biosorption" is used to describe the metabolism-independent sorption of heavy metals and radionuclides to biomass. It encompasses both adsorption (the accumulation of substances at a surface or interface) and absorption (the almost uniform penetration of atoms or molecules of one phase forming a solution with a second phase) (77). Both living and dead biomasses are capable of biosorption, and ligands involved in metal binding include carboxyl, amine, hydroxyl, phos-

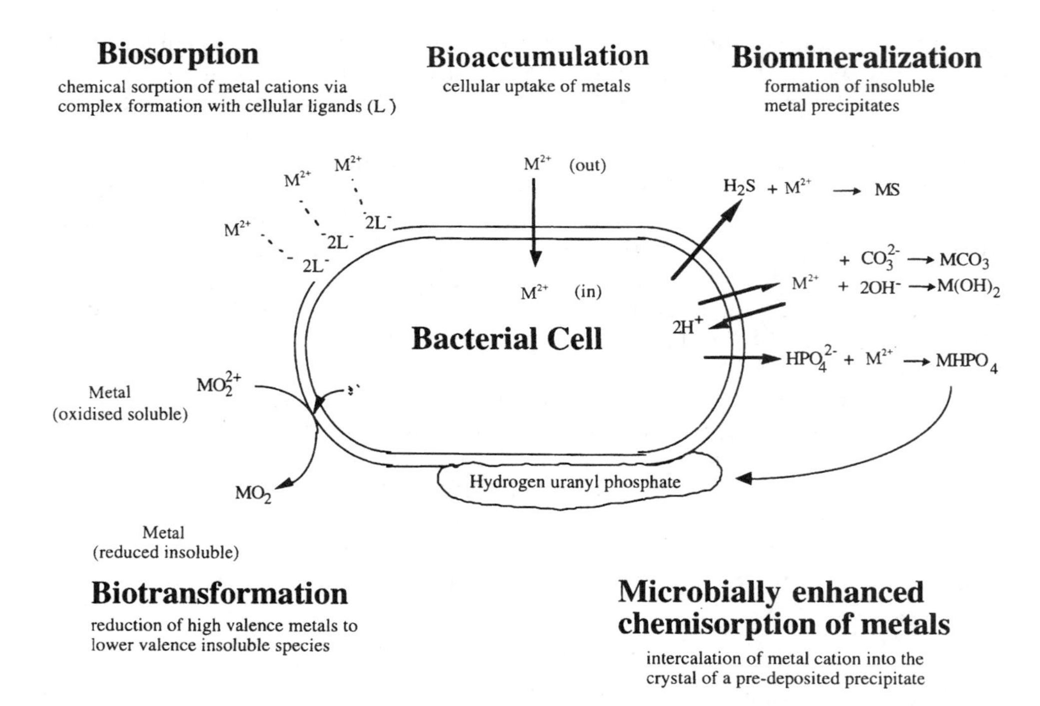

Figure 2. Mechanisms of radionuclide-microbe interactions.

phate, and sulfhydryl groups. Biosorption of metals has been reviewed extensively (156, 219, 244–246), and the present review notes only some salient points and recent developments of interest. Volesky and Holan (246) give an excellent overview of metal biosorption and a numerical assessment of uranium and thorium biosorption along with some more recent data. Beveridge et al. (24), in addition to overviewing metal-microbe interactions, give a very useful guide to the various methods which are now available for studying these and a good introduction to the problems of metal speciation, which always inflence the data and their interpretation and which make accurate comparisons between the reports of different laboratories very difficult indeed (190). However, some generalizations are possible.

Dead biomass often sorbs more metal than its live counterpart does (33, 247) and thus may be particularly suited to the treatment of highly toxic radioactive wastes. In many cases, the sorbing biomass (biosorbent) comprises the waste from another process, adding considerably to the economic attractiveness, even though the absolute biomass capacity may be less than that of a more attractive biomass that has to be grown for the purpose. This is not always the case; Avery and Tobin (8) noted that a laboratory strain of *Saccharomyces cerevisiae* removed less strontium from solution than did a brewery strain, possibly attributable to properties of the extracellular polymer that are desirable for good flocculation of the yeast after fermentation. Examples of the application of waste biomasses to uranium biosorption are *S. cerevisiae* (247) and filamentous fungi such as *Penicillium* sp. and *Aspergillus* sp. (156). The use of filamentous fungi and yeast biomasses has also been described by Kapoor and Viraraghavan (106) and Simmons et al. (200), respectively. Postharvest treatments, including the powdering of dry biomass (218) and the application of detergents (195), have been reported to improve biosorption by exposing additional metal-binding sites. Autoclaving after fermentation can also affect the biomass surface properties (200), as can washing (192). New approaches to the improvement of metal uptake can include the application of electrical pulses, which increased the initial rate of binding of uranyl ions to yeast biomass, and its capacity, from 70 to 140 mg of uranium/g of biomass (40), a value comparable to that of the filamentous fungi (246).

Biosorption is generally rapid and unaffected over modest temperature ranges, and in many cases it can be described by isotherm models such as the Langmuir and Freundlich isotherms (245, 246). Gadd and White (77), however, noted that more complex interactions are difficult to model because the adsorption of solutes by solids is affected by factors including diffusion, heterogeneity of the surface, and pH. An additional isotherm, the Brunauer-Emmett-Teller (BET) isotherm, which assumes multilayer binding at constant energy, has been used to describe metal biosorption (3, 53). This model assumes that one layer need not necessarily be completely filled before another is begun. Further insight is offered by Andres et al. (3) who summarize: "each adsorption layer of the BET model can be reduced to Langmuir behavior with homogeneous surface energy, in contrast to the adsorption energy requirements of the Freundlich isotherm." Other studies on biosorption of the uranyl ion have used complex multistage kinetic approaches to model biosorption (225, 252). The uranyl ion is a rather special case. High uptake by *Rhizopus arrhizus* and other fungal biomasses has been reported (up to approximately

20% of the biomass dry weight) (77, 245, 246). Here, the sorbing species is the chitin component of the fungal cell wall; the UO_2^{2+} ion was proposed to coordinate initially with the amine nitrogen of chitin, and then with hydroxylate, and the hydroxylated uranyl species was proposed to translocate to deeper wall layers, leaving the original coordination site free for the next uranyl ion (228, 238). Intact fungal cell walls are preferable; isolated chitin sorbs less UO_2^{2+} than does native biomass (228, 238), and Guibal et al. (91), working on *Mucor miehei* (a waste product of agroindustries, also produced as a fermentation by-product and sorbing uranyl comparably to *R. arrhizus* at pH 4 to 5 according to the BET isotherm [91]), suggested that the actual mechanism of uranyl deposition is more complex than originally suggested. Observations made for biomass sorption of uranyl differed from those of chitosan or, indeed, of chitin, suggesting the participation of additional functional groups (90), which is in accordance with the greater sorbing capacity of native biomass (see above).

Ultimately, however, the amount of residual metal remaining in solution at equilibrium is governed by the stability constant of the metal-ligand complex (138), and the only way to change the equilibrium position is to modify the binding ligand to one which has a greater binding affinity for the given metal or to transform the metal from a poorly sorbing species to one which has a higher ligand-binding affinity, e.g., by a change of metal valence (see below).

Metabolism-Dependent Bioaccumulation

Energy-dependent metal uptake has been demonstrated for most physiologically important metal ions, and some radionuclides enter the cell as chemical "surrogates" using these transport systems. Monovalent-cation transport, for example K^+ uptake, is linked to the plasma membrane-bound H^+-ATPase via the membrane potential and is therefore affected by factors that inhibit cell energy metabolism. These include the absence of substrate, anaerobiosis, incubation at low temperatures, and the presence of respiratory inhibitors such as cyanide (255). The requirement for metabolically active cells may therefore limit the practical application of this mode of metal uptake to the treatment of radioisotopes with low toxicity and radioactivity. For example, in the study by White and Gadd (255), increasing metal concentrations inhibited H^+ pumping, potentially deenergizing the cell membrane and reducing cation uptake. Although the presence of multiple transport mechanisms of differing affinities may cause added complications, metal influx frequently conforms to Michaelis-Menten kinetics (30).

Once in the cell, toxic metals and, potentially, radionuclides may be sequestered by cysteine-rich metallothioneins (96, 241) or, in fungi, compartmentalized into the vacuole (77, 171). In this context, it should be emphasized that the uptake of higher-mass radionuclides e.g., the actinides, into microbial cells has been reported sporadically and remains poorly characterized. Sections of the gram-negative bacterium *Citrobacter* sp. accumulating UO_2^{2+} showed deposition of needle-like deposits of uranium on the inner and outer membranes of the cells but not in the cytoplasm (101), suggesting the permeability of the outer membrane to UO_2^{2+}, in accordance with an earlier study using a *Pseudomonas* sp. (151). An earlier study (206) showed cell surface uptake of uranyl ion by the yeast *S. cerevisiae* and was attributed to biosorption. Uptake of uranium by *P. aeruginosa* was also reported (206) but was localized to the cytoplasm by electron microscopy. Uptake was rapid ($<$10 s) and was not affected by metabolic inhibitors. Cellular detection of actinides was also implied by the induction of chelating agents by *P. aeruginosa* in response to uranyl ion or thorium (188), but this, in itself, is not evidence for cellular penetration. More recent studies (247) seem to confirm the intracellular accumulation of uranium by *S. cerevisiae* cells; uranium was identified using energy-dispersive X-ray microanalysis.

Enzymatically Catalyzed Biotransformations

Microorganisms can catalyze the transformation of toxic metals to less soluble or more volatile forms. For example, the microbial reduction of Cr(VI) to Cr(III), Se(VI) to Se(0), V(V) to V(III), and Au(III) to Au(0) results in metal precipitation under physiological conditions (130). In many cases the high-valence metal can be used as an electron acceptor under anoxic conditions. Biomethylation may also increase the volatility of metals, with the methylation of mercury, cadmium, lead, tin, selenium, and tellurium being recorded (61). While the microbial reduction and precipitation of U(VI) (134), Tc(VII) (119, 122), and Np(V) (125a) have been demonstrated, biomethylation of radionuclides has received little attention. The high instability of alkylated actinides (15) may in part explain this observation.

In common with some examples of bioaccumulation (see "Metabolism-Dependent Bioaccumulation" above and "Biomineralization via Microbially Generated Ligands" below), enzymatic biotransformations of radionuclides can be described by Michaelis-Menten kinetics. This allows, for example, the description of a flowthrough bioreactor using an integrated form of the Michaelis-Menten equation (140, 145, 148, 222, 265). It is therefore possible to predict the degree of biomass loading or the operational temperature needed to maintain metal removal at a given efficiency within the constraints set by the reactor volume available, the background ionic matrix, and the flow rates required (140), e.g., for enzymatically catalyzed biomineralization of U (140, 145, 148, 265) and La (222, 264) and for bioreduction of Tc (120).

Another advantage of using a single-enzyme-mediated transformation over, for example, energy-dependent processes such as bioaccumulation is that nongrowing or even nonliving biomass with enzymatic activity may be utilized to treat radiotoxic effluents, yielding a waste material with a low organic content. The high radioresistance of several enzymes, potentially useful for the bioremediation of radionuclides, has been confirmed (120; L. F. Strachan, M. R. Tolley, and L. E. Macaskie, Proc. 201st Meet. Am. Chem. Soc., 1991). For many biological reductions, e.g., U(VI) or Tc(VII) reduction, simple, cheaply available electron donors such as hydrogen, acetate, and formate can be supplied, negating the requirement for cofactor regeneration.

Since radionuclides are usually not the natural substrate for an enzyme, for example, the hydrogenase-mediated reduction and precipitation of Tc(VII) (121, 124) or U(VI) reduction by Fe(III)-reducing (134, 227) and sulfate-reducing bacteria (128, 135, 136, 240), new mutagenic techniques for "directed evolution," such as PCR-based DNA shuffling (47), may potentially prove useful in altering the kinetic constants of enzymes to improve process efficiency. Such improvements must, however, be considered alongside regulatory constraints on the use of genetically modified organisms.

Biomineralization via Microbially Generated Ligands

Microorganisms are able to precipitate metals and radionuclides as carbonates and hydroxides via plasmid-borne resistance mechanisms, whereby proton influx countercurrent (antiport) to metal efflux results in localized alkalinization at the cell surface (55, 242). Alternatively, metals can precipitate with enzymatically generated ligands, e.g., sulfide (14) or phosphate (139, 144, 146; R. E. Dick, C. D. Boswell, and L. E. Macaskie, Proc. Int. Symp. Biohydrometall., 1995). The concentration of residual free metal at equilibrium is governed by the solubility product of the metal complex (e.g., 10^{-20} to 10^{-30} for the sulfides and phosphates, higher for the carbonates). Most of the metal should be removed from solution if an excess of ligand is supplied. This is difficult to achieve using chemical precipitation methods in dilute solutions; the advantage of microbial ligand generation is that high concentrations of ligand are achieved in juxtaposition to the cell surface and can provide nucleation foci for the rapid onset of metal precipitation; effectively, the metals are concentrated "uphill" against a concentration gradient. This was demonstrated using the gamma isotope ^{241}Am supplied at an input concentration of approximately 2.5 parts per billion; approximately 95% of the metal was removed as biomass-bound phosphate (146), with the use of gamma counting permitting detection at levels below those of most analytical methods. In many cases, the production of the ligand can also be fine-tuned by the application of Michaelis-Menten kinetics.

Sulfide precipitation, catalyzed by a mixed culture of sulfate-reducing bacteria, has been utilized to treat water cocontaminated by sulfate and zinc (14) and to treat soil leachate contaminated with sulfate alongside metal and radionuclides (T. Kearney, H. Eccles, D. Graves, and A. Gonzales, Proc. 18th Annu. Conf. Natl. Low-Level Waste Manage. Prog., 1996). Ethanol was used as the electron donor for the reduction of sulfate to sulfide in both examples. A more recent study has confirmed the potential of integrating the action of sulfur-cycling bacteria (257) employed by Kearney et al. (Proc. 18th Annu. Conf. Natl. Low-Level Waste Manage. Prog., 1996). In the first step of a two-stage process, sulfur-oxidizing bacteria were used to leach metals from contaminated soil via the generation of sulfuric acid, and in the second step, the metals were stripped from solution in an anaerobic bioreactor containing sulfate-reducing bacteria. The ubiquitous distribution of sulfate-reducing bacteria in acidic, neutral, and alkaline environments (185) suggests that they have the potential to treat a variety of effluents, while the ability of the organisms to metabolize a wide range of electron donors may also allow cotreatment of other organic contaminants. For example, growth of sulfate-reducing bacteria with TBP supplied as the sole electron donor for sulfate reduction has been demonstrated (R. A. P. Thomas, J. R. Lloyd, and L. E. Macaskie, unpublished observations).

Bioprecipitation of metal phosphates via hydrolysis of stored polyphosphate by *Acinetobacter* spp. is dependent upon alternating aerobic (polyphosphate synthesis) and anaerobic (polyphosphate hydrolysis and phosphate release) periods (Dick et al., Int. Symp. Biohydrometall., 1995). This obligately aerobic organism is fairly restricted in the range of carbon sources utilized, but the preferred substrates (acetate and ethanol) are widely and cheaply available. The best-documented organism for metal phosphate biomineralization, a *Citrobacter* sp., grows well on cheaply available substrates, and viable cells are not required for metal uptake, since this relies on hydrolytic cleavage of a supplied organic phosphate donor (144).

Microbially Enhanced Chemisorption of Heavy Metals

Microbially enhanced chemisorption of heavy metals (MECHM) is a generic term that describes a class of reactions whereby microbial cells first precipitate a biomineral of one metal (priming deposit). The priming deposit then acts as a nucleation focus, or host crystal, for the subsequent deposition of the metal of interest (target metal), acting to promote and accelerate target metal precipitation reactions (146, 147, 221). The priming deposit is made initially by the sulfide or phosphate biomineralization routes described above.

With the addition of Fe as the precipitant metal to sulfate-reducing bacteria producing H_2S, cell-bound FeS then acts as a sorbent for the target metals, including radionuclides (62, 249; J. H. P. Watson and D. C. Ellwood, Proc. Int. Conf. Control Environ. Problems Metal Mines, 1988). The use of FeS is notable because this provides the mechanism for rapid biomass separation from the liquor via the magnetic properties of Fe in a high-gradient magnetic separator (62, 249; Watson and Ellwood, Proc. Int. Conf. Control Environ. Problems Metal Mines, 1988).

Metal phosphate can also be used as the priming deposit, in two ways. $LaPO_4$ predeposited onto a metal-accumulating *Citrobacter* sp. via biogenic phosphate release can be used as the priming deposit for subsequent deposition of actinide phosphates (146) (see below). Also, by use of a priming deposit with an appropriate, well-defined cystalline lattice, the target metal can be intercalated within the host lattice, effectively by a mechanism of bioinorganic ion exchange, well established as a chemical process (45, 178, 186, 187) and best characterized biologically for Ni^{2+} removal into biomass-bound $HUO_2PO_4 \cdot nH_2O$ (29). Hydrogen uranyl phosphate (HUP) consists of sheets of uranyl phosphate ions separated by water molecules, creating a regular network of hydrogen bonds (45, 99). The overall outcome of this highly organized crystal lattice is a high mobility of protons in the interlamellar space (178), which gives rise to the ion-exchange intercalative property (99). Other than as a use for uranyl phosphate produced as solid waste from the remediation of radionuclide-containing wastewaters, there would be no market for this outside the nuclear industry, since uranium is both toxic and radioactive. Other metal phosphates such as zirconium phosphate (45) can perform similar ion exchange reactions, but attempts to substitute nontoxic zirconium or titanium as the priming metal in MECHM have met with only limited success, since these tetravalent metals are not removed in the correct crystal structure to act as a host lattice for intercalation; some posttreatment of the loaded biomass would be necessary, which would inactivate the enzyme performing the primary deposition reactions and rule out the possibility of biomass recycling (G. Basnakova and L. E. Macaskie, unpublished observations).

In MECHM, the hexavalent actinides (An) form a special case, since AnO_2^{2+} can substitute for UO_2^{2+} in the backbone of the lattice, displacing the uranyl ion to give a hybrid crystal (57). In theory, it would be possible to produce a lattice consisting of a uranyl phosphate carrier matrix with other actinides intercalated into the backbone and simple radionuclide cations intercalated within the lumen; such remediation of multiple species simultaneously has not yet been achieved but is the subject of ongoing research in the laboratory of one of us (L.E.M.).

TECHNICAL CHALLENGES ASSOCIATED WITH THE BIOTECHNOLOGICAL TREATMENT OF RADIOACTIVE WASTE

Although the application of biotechnology to radioactive-waste treatment has attracted much interest, studies have generally been confined to laboratory or pilot scale. Technical challenges associated with large-scale clean-up of highly complex wastes must be overcome prior to the full commercial realization of the technologies currently under consideration. The major technical challenges are summarized below.

Metal Bioremediation: a Question of Scale

All the different steps in the nuclear fuel cycle generate wastes of differing chemical composition. A common attribute, however, is the very large volume of waste produced. For example, Schmidt and Moffat quoted a value of 2.5×10^4 m^3 of wastewater discharged per day from tailings in the Elliot Lake area in Canada (J. W. Schmidt and D. Moffat, CNA Symp. Res. Relat. Radiol. Saf. Nuclear Fuel Cycles, 1979). Waste streams from nuclear reprocessing facilities and fuel rod storage ponds are also of a high volume, typically on the order of 1,000 m^3/day (138). Although the problems of operating bioprocesses at such high throughput volumes should not be understated, the long-term success of the water industry operating at a similar scale, albeit treating a very different type of waste, should be noted. Indeed, several metal treatment processes operating at high throughput volumes have also been reported (14, 160; J. L. Whitlock and G. R. Smith, Int. Symp. Biohydrometall., 1989).

In addition to newly generated process waters, sediments contaminated with previously generated waste must be treated. This also presents a problem of daunting scale, exemplified by that faced by the U.S. Department of Energy. By the early 1990s, the end of the Cold War heralded the shutdown of all nuclear-weapon production reactors in the United States, leaving a legacy of large volumes of contaminated water and sediments spread over 120 sites. This includes a staggering 475×10^9 gal of contaminated groundwater in 5,700 plumes, 75×10^6 m^3 of contaminated sediments, and 3×10^6 m^3 of leaking buried waste (154). Although "pump-and-treat" technologies are established for the decontamination of such sites, the scale and heterogeneity of the subsurface sediments to be treated may make such an approach uneconomical. The use of microorganisms able to solublize problematic contaminants from contaminated sediment could enhance extraction (for example, metal and radionuclide leaching by the use of sulfur-oxidizing bacteria (257; Kearney et al., Proc. 18th Annu. Conf. Natl. Low-Level Waste Manage. Prog, 1996). An alternative approach would be to use bacteria able to immobilize the radionuclides in situ, thus stabilizing the contaminated plume and preventing further migration into the groundwater. If either approach is adopted, engineering such a bioprocess at scale will be a considerable challenge that has yet to be attempted for radionuclides. A role for phytoremediation in such applications may also be possible and has been demonstrated in principle for radionuclide-contaminated sites (B. Ensley, personal communication); however, it suffers, along with other methods, from lack of specificity toward nuclides at low concentrations in competition with nonnuclide species.

Toxicity of Radioactive Waste

The acidic nature of some mining wastes may pose a problem to existing biotreatment processes, many of which utilize neutrophilic bacteria. Wastes from fuel processing and reprocessing can also be produced at extremes of pH, yielding high salinity waste after neutralization. Dilution of wastewater to a physiologically compatible composition is possible but will increase the volume of waste to be treated. Another alternative, the application of acidophilic, alkaliphilic, or halophilic microbes, has yet to be investigated.

The radiotoxicity of effluents containing high-activity radionuclides, e.g., ^{241}Pu, may also adversely effect the microbial component of a bioprocess. However, the large microbial gene pool may provide novel radioresistant organisms with very efficient DNA repair mechanisms. For example, bacteria of the genus *Deinococcus* are able to withstand the normally lethal mutagenic effects of DNA-damaging agents, particularly ionizing radiation (21). Recent studies have shown that *Deinococcus radiodurans* has the potential to reduce metals, including Fe(III)-NTA (H. M. Kostandarithes, M. J. Daly, and J. K. Fredrickson, Abstr. 99th Gen. Meet. Am. Soc. Microbiol. 1999, abstr. Q-11, 1999), and can act as a host for multicomponent enzymes for biodegradation (e.g., the toluene dioxygenase from *Pseudomonas putida* F1) (114), offering the potential to treat metals and organic contaminants in highly radioactive waste. It has also been noted that bacteria that have come into contact with industrial sources of radiation can have increased radioresistance (25), and such bacteria could also have future applications in nuclear waste remediation (25).

Most wastes from mining and fuel reprocessing contain a large background of nonfissile residual ^{238}U. The acquisition of heavy deposits of uranium around cells accumulating U(VI) as the phosphate or as reduced U(IV) oxide (see below) introduces the intriguing possiblity of a self-generating 'radiation shield' of ^{238}U nuclei, but little is known of the effectiveness of this to the bacteria at submicrometer to atomic distances, and this would also relate to the penetrating power of the radioactive emission of the target isotopes in a particular waste (e.g., nonpenetrating α-emissions of the transuranic elements, moderately penetrating β-emission of, e.g.,

^{99}Tc and ^{241}Pu, and highly penetrating γ-rays from isotopes such as ^{60}Co and ^{241}Am). Radioactivity lethal to cells may not inhibit bioremediation where only one or a few enzymatic steps are required; for example, the phosphatase activity of *Citrobacter* sp. accumulating uranyl phosphate was highly radioresistant (^{60}Co gamma source) after complete loss of cell viability (Strachan et al., Proc. 201st Meet. Am. Chem. Soc., 1991), while hydrogenase-mediated ^{99}Tc removal by *Escherichia coli* or *Desulfovibrio desulfuricans* (see below) continued to high loads of metal and very high local β-activity. Finally, other organic contaminants, e.g., TBP or odorless kerosene, may add to the toxicity of the effluent, could be used as energy sources, and, indeed, may have to be removed to convert the target metals into available forms (see above).

In general, there have been few attempts to develop versatile technologies for the treatment of aggressive wastes. Premuzic et al. have proposed several systems for the bioremediation of metal-contaminated geothermal wastes; the reader is referred to a recent paper (189) as a source document.

Competing Ions and Complexing Agents

The problems associated with treating wastes containing chelating agents such as EDTA and DTPA have already been highlighted. Such wastes already exist and are problematic (51). Anionic species may also be present at high concentration and can complex with radionuclides. For example, the use of nitric acid in fuel reprocessing dictates that high concentrations of nitrate (300 mM) be present in low-activity waste streams. The possible inhibitory effects of high-nitrate loadings on bioremediation have been outlined (see "Wastes from the Mining and Extraction of Uranium" and "Wastes from Fuel Reprocessing" above). Although nitrate can be coremoved via dissimilatory nitrate reduction for integrated waste treatment (see above), a preferred option, that of metal removal from a nitric acid background, is still not often achieved. Inactive metal species are also present at high concentrations; in some cases, iron is present at concentrations more than 4 orders of magnitude higher than those of target actinides (ferrous sulfamate is used as a reductant in fuel partition cycles). Nonfissile ^{238}U is also present at an excess of 2 orders of magnitude over Pu and Am (138). The need for specificity in the bioprocess, or a high-capacity system capable of removing all metals together, is evident.

Speciation and the Actinides

The actinides are among the principal radionuclides dominating nuclear-waste inventories and can exist in III, IV, V, and VI oxidation states. The presence of biomass or receiving waters can alter the oxidation state and therefore the solution chemistry of these elements. The latter is very complex due to redox instabilities (that are often light sensitive), ease of hydrolysis, and ligand-complexing properties of uranium and the transuranic elements (108). The trivalent actinides have a relatively simple solution chemistry. They have little tendency to form organic complexes or hydroxylated species in solution. For example, at physiological pH values, Am(III) is present as the Am^{3+} ion; a convenient surrogate is La^{3+}. The tetravalent actinides (An) are found only as the free cation at low pH. Under physiological conditions, hydroxylated forms exist as insoluble $An(OH)_4$ or soluble $An(OH)_5^-$ species. An(IV) form very strong complexes with ligands. Hydroxylation behaviour is suppressed when the ion is highly complexed. Th(IV) is a relatively innocuous stable surrogate for Pu(IV) and Np(IV); U(IV), like Np(IV), is easily oxidized in solution, but Np(IV) is stablilized in solutions containing complexing ligand (118). The chemical characteristics of the pentavalent actinides, for example, NpO_2^+, are similar to those of simple monovalent cations: low ligand-complexing capabilities and a high environmental mobility that warrant their consideration as high-priority pollutants. Although pentavalent species of all except Np(V) [and its decay product protactinium(V)] are unstable, Pu(V) can be the most common form in some natural waters (164). Finally, the hexavalent actinides, for example, the uranyl ion (UO_2^{2+}), also form strong complexes, although less strong than those of the tetravalent actinides. At neutral pH, the hexavalent actinides are predominantly hydroxylated. The complicated solution chemistry of the actinides is illustrated well by Pu, which can coexist in all four oxidation states and can give rise to many hydroxylated species in a single solution. It is therefore very difficult to predict the outcome of biological treatment of some radioactive waste, particularly in the presence of other redox-active metal species (e.g., Fe), radiolytic effects, UV-visible light, and complexing ligands.

The inclusion of a time variable, which may be important for long-term in situ applications and contaminated-site management, can bring further complications. For example, the decay chain from ^{241}Pu to ^{241}Am and then to ^{237}Np dictates that after 300 years the daughter elements will be radioactive species with very different solution chemistries, e.g., decay from mixed-valence Pu via Am(III) to highly mobile Np(V). Here, microbial activity could be used to alter or hold the oxidation state of the final radionuclide in the decay chain (^{237}Np) and affect its mobility and solubility by, for example, metal reduction (125a, 138; E. A. Bondietti, S. A. Reynolds, and M. H. Shanks, San Francisco Conf. 751105, 1976).

Potential Advantages of Bioremediation Strategies

Environmental biotechnology is developing rapidly, but the potential to treat radionuclide contamination via bioremediation has yet to be fully realized. Few comparisons have been done of the relative costs of bioremediation strategies with respect to other separative technologies. Two studies on biosorptive-metal treatment and one on bioprecipitation should, however, be cited. ATM-Bioclaim used 0.1-mm-diameter granules of modified biomass to treat metal-contaminated waste effluent (35). Cost analyses predicted that the ATM-Bioclaim process was significantly cheaper than chemical precipitation and ion exchange for primary and secondary (polishing) processes, respectively. Similar conclusions were drawn in the study of Jeffers et al. (100), who compared a biosorption process with chemical precipitation for treating acid mine drainage. Eccles (60) considered that the cost comparisons outlined in these two examples were far from rigorous but were indicative of the potential competitive advantages of biological treatment of metal- and radionuclide-contaminated water. In addition, several technical advantages associated with the use of biological agents were not considered in these examples.

Potential advantages that should be noted include the large gene pool and hence metabolic diversity of microorganisms, the ability of microorganisms to adapt to the most inhospitable environments, the operational flexibility with the potential to treat multiple contaminants in a single step, the potential for promoting metal transformations (e.g., oxidations-reductions and methylations-demethylations) , and the opportunity to enhance natural specificity or improve bioprocess efficiency using the tools of molecular biology. A study to evaluate critically the potential for enzymatic (phosphatase-mediated) remediation of uranium acid mine wastewater, using a biofilm immobilized on porous glass supports to accumulate uranyl phosphate, concluded that the cost of adding the substrate (glycerol-2-phosphate) to the feed was the single factor limiting economic viability (193). Future substitution of a cheap phosphate donor (e.g., TBP) could render this more economically attractive (193), while the efficiency of uranium removal was improved by cloning and expression of the appropriate gene (*phoN*, encoding the phosphatase activity responsible for biogenic HPO_4^{2-} production) in an innocuous *E. coli* host (20).

We are unaware of any cost analysis for the treatment of radionuclide-contaminated land using in situ bioremediation, but several potential advantages over more traditional chemical-based approaches should be highlighted. Although ex situ pump-and-treat techniques using chemical or biological agents are better understood and probably easier to control, in situ bioremediation offers the potential to treat radionuclides that are widely dispersed at low concentration or are otherwise inaccessible (154). An added advantage of attempting bioremediation in situ is the operational flexibility offered. Several different approaches are possible, depending on the geological setting and the indigenous organisms present. For example, intrinsic bioremediation relies on natural biological processes that may already be present at a contaminated site. If intrinsic bioremediation can be demonstrated by measuring the relevant metabolic activity at site and monitoring the size of the contaminated plume over time, more costly treatments may not be required. Where intrinsic bioremediation cannot be demonstrated, it may be possible to stimulate extant organisms through the delivery of a suitable limiting nutrient (e.g., an electron donor for metal reduction). This approach has been termed "biostimulation." Finally, the addition of microorganisms with the required activity can also be used to accelerate in situ bioremediation ("bioaugmentation"), and delivery of microorganisms into tightly packed sediments may be improved if the cultures are first stressed to generate "ultramicrobacteria" of reduced dimensions (36).

Specific interactions with radionuclides which may provide a useful basis in treatment strategies are now reviewed.

MICROBE-ACTINIDE INTERACTIONS

The application of microorganisms to actinide remediation has been reviewed in depth (7, 138); space does not permit extensive reiteration, and this chapter focuses on recent developments. Each actinide (An) is considered under the valence in which it is most common, but in general a strategy effective for a particular valence will apply to all actinides equally well, in addition to their more innocuous surrogates. Thus, studies have used La(III) for An(III) and Th(IV) and Zr(IV) for An(IV). There is no convenient surrogate for An(V), and even between the actinides the behavior of the An(V) species is not always consistent; for example, Np(V) forms soluble NpO_2^+ which does not hydrolyze extensively; in contrast Pa(V), uniquely, forms colloidal hydroxides and ligand complexes with phosphate (13, 110). This has been used to advantage in the separation of Np(V) from its Pa(V) daughter (125a).

Americium(III), Plutonium(III), and Curium(III)

The trivalent actinides, like the analogous lanthanides, pose no major problems for bioremediation. Sorption of ^{241}Am onto various algal biomasses was approximately 3-fold greater than for Pu, which was in turn more than 10-fold greater than for Np (65). However, gross-uptake comparisons are not meaningful since the Am was present at an initially higher concentration than the other metals, and even so, at the low concentration employed (83 pM), saturation may not have been reached. In another study using *E. coli* and a marine bacterium challenged with 0.24 nM Am (260), a diluting out of uptake occurred during growth, with a 20 to 30-fold increase as the cultures entered stationary phase, suggesting the production of higher-affinity ligand by the cells. Other studies using the bacterium *Aeromonas hydrophila* showed that an initial Am uptake was followed by an apparent desorption: initially 1×10^9 pCi g of cells^{-1} and later 1.8×10^3 pCi g^{-1} (84). It should be noted that none of these studies were targeted specifically toward bioremediation, and so there were no attempts to determine the maximum biomass loading. Biosorption of ^{241}Am by *Rhizopus arrhizus* was also reported (54). Am(III) uptake was 40-fold greater than that of Pu(IV) at pH 2, comparable at pH 4, and only 30% of that of Pu(IV) at pH 7. This is not attributable to the relative hydroxylation behavior of the two elements, because Am^{3+} uptake also fell with increasing pH from pH 2 to 7 (54). Removal of ^{241}Am (approximately 90%) from an initial solution of 299 nCi ml^{-1} by *Candida utilis* was reported, probably attributable to biosorption to cell-associated phosphate (249).

In a study of enzymatically mediated metal phosphate deposition (100% removal of ^{241}Am [146]) the total capacity of the cells was not achieved, but since $AmPO_4$ deposition would have continued with continuing phosphate supply, the final load would have been high; other tests using La showed extensive metal uptake over many hours of continuous flowthrough operation using immobilized cells (222). To saturate cell surface-biosorbing ligands, the cells were treated with La^{3+} before being challenged with Am^{3+}, and conclusive proof of enzymatically mediated uptake was obtained using a phosphatase-deficient mutant, which neither liberated HPO_4^{2-} nor accumulated Am (146).

Since the concentration of ^{241}Am in wastes and natural environments is likely to be very low this study was important not for ^{241}Am bioremediation per se but for its implications in the application of bioreduction. Np(III) is unlikely to be stable except under highly reducing conditions, but under appropriate circumstances Pu(IV) could be bioreduced to Pu(III) and efficient scavenging could then occur, as for Am(III). This bioreduction was suggested by Macaskie (138) and demonstrated by Rusin et al. (196) using iron-reducing bacteria, and easy scavenging of Pu(III) would be anticipated.

Uptake of Cm(III) has been little studied. Cm(III) in HCl carrier (100 ng ml^{-1} initial concentration in HCl [pH 2]) was partially removed by sterilized sediments but almost completely removed by natural sediments after 4 months (176). The mechanism of removal was not investigated.

Plutonium(IV), Thorium(IV), and Neptunium(IV)

Of all the actinides, the tetravalent forms should be the most amenable to bioremediation because of their high ligand-complexing abilities. However, this poses problems for bioremediation since chelating ligands in solution withold the metal from the biomass. In aqueous, uncomplexing solution, An(IV) is available only as the An^{4+} cation at low pH values (pH 2.0), where protonation of biomass ligands will also occur, effectively reducing the scope for biosorption. Despite this, several groups have reported biosorption of Th^{4+} onto fungal biomasses (78, 236, 237). The uptake values (236) were (milligrams of Th per gram [dry weight]) 90 (*Rhizopus arrhizus*), 61.5 (*Saccharomyces cerevisiae*), 65 (*Aspergillus niger*), 114 (*Penicillium italicum*), and 195 (*Penicillium chrysogenum*), with the data not giving a good fit to established biosorption models. It is difficult to formulate accurate models for these biosorption studies because if the localized pH is above that of the bulk solution, with hydrolysis of the metal the sorbed species will not be An^{4+} but a species carrying one to four hydroxyl groups. The application of biosorption to the bioremediation of Th^{4+} at acidic pH using *R. arrhizus* and *A. niger* with various bioreactor configurations was discussed in detail previously (254), and the reader is referred to this earlier work. In summary, static-bed designs and stirred beds gave low thorium removal, attributed to mixing constraints. In contrast, use of an airlift reactor allowed 90 to 95% removal of the thorium, followed by a rapid breakthrough at saturation. In these studies, the Th loading was 116 and 138 mg/g (dry weight) for *R. arrhizus* and *A. niger*, respectively (254). In a study of thorium biosorption at pH 1 using *Mycobacterium smegmatis*, Andres et al. (5) reported Th^{4+} loadings of approximately 4% of the biomass dry weight, a comparable value to that of UO_2^{2+} with that strain. Thorium removal was, as also reported by Yong and Macaskie (263, 267), promoted by the incorporation of NH_4^+, to form insoluble thorium ammonium phosphate (263, 267).

In the same study as mentioned above, Tsezos and Volesky (236) allowed Th(IV) to hydroxylate at pH 4 [forming colloidal $Th(OH)_4$]; this increased the uptake to 170 mg per g (dry weight) of *R. arrhizus* biomass. Biosorption of Pu(IV) by *R. arrhizus* was observed at pH 4 and 9, but *Gibberella* spp. did not perform biosorption (54). The capacity of the biomass at pH 9.4 (3 mg of Pu liter of $solution^{-1}$) was 0.8 mg g of $biomass^{-1}$. Pu sorption was reduced by approximately 30% and more than 80% in the presence of nitrate and acetate, respectively (54). An alternative approach to the removal of tetravalent Pu [$Pu(OH)_4$] employed *Pseudomonas aeruginosa* immobilized on a plasma-treated polypropylene web (211). This was intially used in a batch operation with nuclear plant wastewater containing 1.7 nCi of Pu. Up to 95% of the Pu was removed, but Pu removal was particle size dependent. The authors suggested that PuO_2 entrapment within the filaments occurred rather than true adsorption. The distinction between $Pu(OH)_4$ and PuO_2 is not well defined; in a given solution, these and intermediate species are all present, with the exact proportions dependent upon the concentration and pH (see reference 138 for discussion).

Other studies have demonstrated Th and Pu removal at neutral pH. Here, the tetravalent actinides were stabilized by extensive complexation with citrate. In addition to providing a realistic scenario for many cases (51), this poses a "worst case" scenario for the biomass, because the concentration of free An(IV) available for interaction with the biomass is very small. Initial tests done in parallel to the ^{241}Am study (above) showed that the metal phosphate-accumulating *Citrobacter* sp. removed 50% of the $^{238/239}$Pu from a 60 nM solution when phosphatase-containing cells were used at steady-state in a flowthrough system; none was removed by the corresponding phosphatase-deficient mutant (146). Removal of only 50% of the Pu would be unattractive for industrial bioremediation, and the problem was confirmed in a comparative study of the removal of La^{3+}, Th^{4+}, and $UO_2{}^{2+}$ as being attributable to the speciation problems suggested above; here, no Pu was removed when the concentration was increased to 400 μM (266). Further studies using a chemical model system (267) showed that Th was removed poorly, with only an amorphous deposit of thorium phosphate compared to the well-defined crystals of HUO_2PO_4 and $LaPO_4$ (identified by X-ray diffraction analysis). Incorporation of $NH_4{}^+$ promoted the precipitation of thorium phosphate; the solublity products of the ammonium actinide phosphates are much lower than those of their hydrogen phosphate counterparts (263, 267). Further, cochallenge of the cells with La and Th improved Th removal from a 300 μM solution to 90% (263), with analysis of the recovered precipitate showing a hybrid crystal of Th/La phosphate containing molar proportions of Th to La of 1:1 by proton-induced X-ray emission analysis. The toxicity of Pu precluded similar solid-state analysis, but it was predicted that incorporation of La should promote desolubilization of Pu according to a conceptual model formally stated as the MECHM hypothesis (see above). Accordingly, 100% of $^{239/241}$Pu was removed from a 400 μM solution with excess La^{3+} (262, 263) and 2 mM citrate against a high-nitrate background (300 mM) (the Pu was supplied in a carrier solution of 8 M HNO_3). As an alternative approach to MECHM, other tests used biomass preloaded with $LaPO_4$ and challenged with a solution of 39.5 nM $^{239/241}$Pu in a flowthrough column; here the removal was approximately 80% after 11 column volumes (10 mg of dry biomass per column) (17).

Most studies on An(IV) remediation have used very low concentrations of Pu, but one investigation utilized a 2-mg/ml stock solution of ^{239}Pu as the carbonate, introduced to natural sediments (176). All of the Pu was removed after 4 months, with only 34% of the removal being attributed to biosorption using heat-killed controls, implicating additional biochemical mechanisms.

In a novel study, Rusin et al. (196) provided evidence for microbially mediated reduction of Pu(IV) to Pu(III), but the Pu(III) reoxidized spontaneously. These authors made no attempt to trap the nascent Pu(III) species, but this should be possible given the easy bioremediation of An(III) (see above) and removal of Np(V) by a coupled bioreduction and bioprecipitation method (see below). Various other approaches have been attempted using biochemical chelating ligands either per se or modified to increase the specificity for An(IV), e.g., in the competing presence of Fe(III). As yet, none of these have realized their industrial potential for various reasons which are outside the scope of this review but are discussed by Macaskie (138). Biochemical ligands for selective Pu binding should be considered alongside chemical ligands which have been designed specifically for this purpose (170).

Neptunium(V) and Protactinium(V)

Np(V), as NpO_2^+, is especially difficult to remediate (see above), but Pa(V), in contrast to Np(V), forms both insoluble hydroxides and phosphates (13, 110) and so is relatively amenable. The study of Np is not trivial. ^{233}Pa is produced as a highly radioactive (β-emitting) daughter of ^{237}Np, small in mass but frustrating attempts to measure the removal of ^{237}Np by radioactive counting alone and also decoupling the major radioactivity (Pa component) from the mass (Np component), which complicates attempts to determine the radiotoxicity of ^{237}Np. Three methods are available for study of the fate of Np. It is possible to "clean" ^{237}Np by ion-exchange chromatography, but ^{233}Pa soon reappears. Alternatively, use can be made of the gamma tracer ^{239}Np, but the short half-life (2.35 days) of this isotope requires its rapid use. A third approach uses scintillation counting methods with α-β discrimination to separate the ^{237}Np (α) from ^{233}Pa (β) activity. A novel method of separation and visualization of isotopes was developed previously (122, 147; Lloyd et al., submitted), entailing paper chromatographic separation of the metal species followed by rapid counting of the radioactive spots using a PhosphorImager and image analysis software. Penetrating (β, γ) and nonpenetrating (α) activities are easily distinguished; in conjunction with scintillation counting, it is possible to do a single-step analysis of a given nuclide mixture (122, 125a, 147).

There have been few attempts to study the interaction of microorganisms with Np. Fisher et al. (65) reported negligible Np uptake by marine algae, and in the only detailed study performed prior to 1993, low uptakes (10 μg g [dry weight]$^{-1}$) were reported with *Pseudomonas aeruginosa*, *Streptomyces viridochromogenes*, *Scenedesmus obliquus*, and *Micrococcus luteus* (205). Even the metal phosphate-accumulating *Citrobacter* sp. which accumulated 100% of ^{241}Am and 50% of $^{238/239}$Pu (see above) was largely ineffective against Np (220).

According to the principle of MECHM (see above), the removal of Np could be promoted by priming the biomass with a metal phosphate such as $LaPO_4$, and, indeed, this gave more than 80% removal of Np over 11 column volumes from an initial concentration of 1 μM (17). However only the hexa- and tetravalent forms are amenable to the formation of a mixed crystal with $LaPO_4$ (see above), and the ultimate efficiency of MECHM against Np could be determined by the rate of the reestablishment of the disproportionation equilibrium

$$2Np(V) \leftrightarrow Np(VI) + Np(IV)$$

In summary, there is no easy way to remediate Np(V). An alternative approach (see above) is to harness the reductive capacity of some microorganisms to effect a valence change to Np(IV), under which conditions the tetravalent Np would be removed as for An(IV) above. Accordingly, a *Shewanella putrefaciens* strain previously reported to reduce Tc(VII) (122) and U(VI) (130, 134) reduced Np(V), and coremoval of the reduced species was accomplished by the concerted use of *S. putrefaciens* and the *Citrobacter* sp. to reduce and simultaneously desolubilize the Np as neptunium phosphate in the presence of NH_4^+ to promote Np(IV) precipitation (see above) (125a).

Uranium(VI), Plutonium(VI), and Neptunium(VI)

In contrast to the other actinides, the biological removal of UO_2^{2+} is well documented and removal of the other hexavalent actinide species would be similarly achieved. Although radiolytic oxidation of Pu(IV) to Pu(VI) in saline solution has been reported (262), such radioactivity could be incompatible with uptake mechanisms utilizing living cells. Microbial oxidation of Pu(IV), Np(IV), or Np(V) has never been attempted, although biooxidation of Fe, Mn and As is well documented (61). Intercalation of Np(VI) [but not Np(V)] into crystalline $HUO_2PO_4 \cdot 4H_2O$ is documented (57); a similar result would be expected for Pu(VI) and a coupling of an oxidative organism and a precipitative organism would provide an analogous solution to the bioreduction and bioprecipitation of Np(V) described above. Once in the hexavalent state, the other actinides would behave as UO_2^{2+}.

Most of the studies on U remediation have utilized biosorption by various biomasses; the literature is large, and no attempt is made to summarize it here other than to direct the reader to some recent reviews (see above) and to note significant advances since 1991.

Biosorption of U(VI)

Most studies on U(VI) removal have utilized biosorption of UO_2^{2+} onto various biomasses, with fungal biomass emerging as the leading technology, in spite of some promising early work using the actinomycetes *Streptomyces viridochromogenes* and *S. longwoodensis* (Table 1), which do not appear to have been developed further. To avoid unnecessary repetition, Table 1 summarizes the studies on uranyl biosorption which have achieved biomass loadings of in excess of 10% of the biomass dry weight and also some recent studies for comparison. No attempt is made to correct for different exposure conditions and solution ionic matrices; these discrepancies make objective comparisons very difficult (246). Table 1 does not attempt to distinguish between biosorption of UO_2^{2+} and the hydroxylated species which arise in noncomplexing solutions. The message from Table 1 is that biosorption technology for uranium removal is very well developed and, in some cases, has been applied to wastewater processing (see below). Since biosorption of uranium has been covered extensively in the literature and since biosorbents relate in general to structural, not metabolic, aspects of the biomass, this chapter notes only a few recent developments.

Biosorption by Fungal Biomasses

Fungal biosorbents are, in general, very well developed and in some cases have been implemented in wastewater treatment (233). The "traditional" fungal biosor-

Table 1. Biosorbents for U(VI)[a]

Biomass type	Organism	U uptake (% of dry wt)	Reference
Filamentous fungi	*Aspergillus niger*	21.5	261
		12	109
	Rhizopus arrhizus	18.0	225
		19.5	106
		22	248
	Penicillium sp.	8–17	269
	Penicillium chrysogenum	25	165
	Chaetomium distortium	27	109
	Trichiderma harzianum	26	109
	Alternaria tenulis	24	109
	Fusarium sp.	24	109
	Aspergillus amsta	22	109
	Penicillium herguei	20	109
	Zybgorenchus macrocarpus	13	109
	Aspergillus flavus	**4**	**92**
	Mucor meihei	**24**	**91**
Yeasts	*Saccharomyces cerevisiae*	10–15	206
		14	247
		24	109
		15–36[b]	**254**
		48–60[b]	**172**
	Kluyveromyces marxianus	**12–18**	**40**
	Talaromyces emersonii	**28**	**23**
Alga	*Chlorella regularis*	15	162
Actinomycetes	*Streptomyces longwoodensis*	44	74
	Streptomyces viridochromogenes	31	162
Bacteria	*Zoogloea ramigera*	38	168
	Pseudomonas aeruginosa	15	206
	***Pseudomonas* strain EPS 5028**	**15–20**	**184**
	Bacillus sp.	38	Cotoras et al.[c]
	Mycobacterium smegmatis	**4**	**3**
	Citrobacter sp.	900[d]	144
	Escherichia coli	ND[e]	20

[a] Most data are taken from reviews. Data in bold are from other studies published since 1992. With previously reviewed studies, only biomasses with U sorption in excess of 10% of the dry weight are considered.
[b] A mixture of biosorption and precipitation by biomass-associated fermentation waste products.
[c] E. Cotoras, P. Viedma, and J. Pimentel, Proc. Int. Biohydrometall. Symp. 1993.
[d] Not true biosorption but enzymatically promoted biomineralization (see text).
[e] Biomineralization using a cloned gene for enzymatically promoted biomineralization (see text). ND, the maximum capacity was not determined, but U uptake was better than for the *Citrobacter* sp.

bents are the filamentous fungi, where biosorption capacities of in excess of 20% of the dry weight are well documented. Biosorption is attributable to the chitin and chitosan components of fungal cell walls, but recent studies have suggested the participation of additional binding ligands (see above). Recent developments have utilized fungal biomasses produced as wastes from primary fermentation processes; brewery yeast is particularly applicable in terms of the availability of very large quantities, but in initial studies it showed inferior uptake to the filamentous fungi (Table 1). Volesky and May-Phillips (247) observed that uptake was improved after a longer contact time, attributable to intracellular uptake in addition to sorption to wall functional groups. Later studies (172, 254) have suggested that although yeast biomass per se is less effective, the brewery strains, unwashed, accumulate more uranyl ion due to surface-bound ligands, probably products of the fermentation process. Strictly, therefore, this process is a hybrid between biosorption and precipitation. An interesting recent study has used the thermophilic fungus *Talaromyces emersonii* CBS 814.70 to accumulate uranium to more than 28% of the dry weight (23). Bustard et al. (40) reported the stimulatory effect of an electric field on U uptake by yeast; such studies could usefully be extended to other sorbing biomasses. Applications of fungal biomass in environmental biotechnology have been reviewed recently (106, 219).

Biosorption by Algal Biomasses

Although *Chlorella* shows potential as a biosorbent material (Table 1), development of the technology seems to be limited. A biologically derived resin of algal biomass immobilized on silica gel beads (AlgaSORB) has been applied to the removal of uranium from contaminated groundwaters, e.g., by Feiler and Darnall (cited in reference 97).

Biosorption by Bacterial Biomasses

Several bacterial biomasses have been evaluated for uranium removal. Although the *Streptomyces* biomasses are promising, the literature does not appear to show any recent developments. *Mycobacterium smegmatis* has been evaluated recently (3–5). Although biosorption of uranium is less extensive than that described for other biomasses (Table 1), one possible advantage of this sorbent is a reported selectivity for Th (5), which is more difficult to remove than U (see above). Studies using ^{31}P nuclear magnetic resonance suggested that cellular phosphate groups play an important part in uranium biosorption (4, 5). Similarly, in addition to the biogenic phosphate ligand, which precipitates with uranyl ions (see below), phosphate groups of the lipid A component of the exocellular lipopolysaccharidic material of *Citrobacter* sp. were suggested by ^{31}P nuclear magnetic resonance to be involved in the formation of nucleating (priming) deposits of uranyl phosphate (142). For *M. smegmatis*, the phosphate was hypothesized to be derived from sugar phosphates or adenosine phosphates or even from cellular polyphosphate (5), which is known to occur in this organism (79) and the hydrolysis of which is the mechanism of metal accumulation by *Acinetobacter* spp. (L. E. Macaskie and R. E. Dick, 1993, United Kingdom patent application GB93/02330). Hydrolysis of polyphosphate (proposed under conditions of heat drying by Andres et al. [5]) should precede uranyl precipitation. This hypothesis was substantiated by the observation that in *Acinetobacter* spp., which are known to store large amounts of polyphosphate, hydrolysis of the latter and phosphate efflux can be coupled to the bioprecipitation of uranyl and other metal phosphates (Dick et al., Int. Symp. Biohydrometall., 1995; Macaskie and Dick, United Kingdom patent application, 1993).

Aside from these examples, most of the studies of uranyl accumulation by bacterial biomass have utilized *Pseudomonas* species following the demonstration of uranyl removal by *P. aeruginosa* in early work (206). A *Pseudomonas* strain accumulated uranyl ion to up to 5.5% of the bacterial dry weight (151), with the deposited metal being clearly seen as needle-like fibrils. Subsequent studies using polyacrylamide gel-immobilized cells (150) showed application in the removal of U from synthetic effluents, while a recent study has demonstrated the various immobilized-cell technologies that can be used with this biosorbent (97).

Microbially Enhanced Chemisorption of Heavy Metals

There are several examples where microbially generated mineral deposits can serve as the sorbent material for uranium and, by implication, other hexavalent actinide species. Ferric (and manganese) oxides are readily formed by a variety of organisms (61), but this has not been applied to nuclide sorption, even though nuclide sorption to ferric hydroxide flocs forms the basis of the commercial Enhanced Actinide Recovery Process at BNFL (Sellafield, United Kingdom). A major problem with this chemical approach is the formation of bulky wet sludge, which is difficult to separate and concentrate for final disposal. Microorganisms are used to produce high-quality biominerals for simultaneous sorption and separation. The geochemical literature provides good evidence for sorption of U(VI) to sulfide minerals (253), with concomitant partial reduction of U(VI) to U(IV) suggested by using complementary techniques of X-ray photoelectron spectroscopy, Auger electron spectroscopy, and Fourier transform infrared analysis.

Watson and Ellwood described a novel approach to metal sorption and separation of the sorbed species (62, 249; Watson and Ellwood, Proc. Int. Conf. Control Environ. Problems Metal Mines, 1988). Initial studies used anaerobic sulfate-reducing bacteria to deposit FeS, which was then used as sorbent for various metals. Biogenic FeS is a superior sorbent to "geochemical" FeS; it was produced as discrete intertwined fibrils of very high surface area (62), 10 to 100 times better as metal sorbent than chemically precipitated FeS (249). The loaded bacteria are separated effectively from solution using a high-gradient magnetic separator.

A second approach for uranium involved deposition of uranyl phosphate (see below) by using *Candida utilis* and *Bacillus subtilis* with magnetic separation, taking advantage of the strong paramagnetic property of uranium (249). Magnetic separation provides a novel alternative to the use of immobilized biomasses for metal removal. Biomass deposition of Pu by the phosphate route was also shown with *C. utilis* (98% removal from a 426 nCi liter^{-1} solution of ^{242}Pu [249]) and in the presence of a magnetic ion (wastes often contain an excess of Fe and U [7, 138]); easy separation would be anticipated.

Biomineralization of Hexavalent Actinides

Metal phosphates and sulfides are highly insoluble. Many metals can be precipitated via H_2S produced by the activities of sulfate-reducing bacteria, but the main focus of these organisms with respect to uranyl ions has been in the bioreduction of U(VI) (see below). The best-documented biomineralization system for U(VI) is its precipitation using inorganic phosphate generated by the mobilization of cellular polyphosphate (Dick et al., Int. Symp. Biohydrometall., 1995) or via the cleavage of a supplied organic phosphate donor ligand such as TBP (216) or glycerol-2-phosphate (for review, see references 138, 144, 146, and 147) via phosphatase activity localized in the periplasmic space (101) and exocellularly (141). Recent studies on the latter system have suggested that the production of exocellular enzyme together with phosphate-containing extracellular polymer enables effective resistance to metal toxicity and promotes metal phosphate biomineralization (141). First, incoming metal forms a complex with extracellular polymer-phosphate groups, which acts as the nucleation site. Subsequently, the "tethered" metal phosphate crystal grows at the expense of inorganic phosphate fed in via phosphatase activity, trapping more metal incoming from the bulk solution. This scenario provides the answer to several outstanding pertinent questions. Blockage by accumulated metal phosphate has never been seen; presumably, the crystals are effectively compartmentalized away from the enzyme and sites of substrate access. The second question relates to activity at low pH. The pH optimum of the enzyme is approximately 5 to 7.5 (102, 222), yet acidic mine waters containing U were successfully remediated at pH values of 3 to 4 (148). It is assumed that exocellular phosphate (and carboxyl) groups act as natural buffers for pH stasis.

The uranyl phosphate biomineralization system, although not likely to be ultimately economically attractive for the remediation of routine metal cations (193), could find application in radionuclide remediation, where the desirability of removal outweighs the cost of the glycerol-2-phosphate substrate. Uranium removal, as HUO_2PO_4 (144, 265), is usually >90% (e.g., from a 1 mM [238 ppm] solution), with sustained and steady-state activity to very high cellular loads (e.g., 900% of the biomass dry weight) (the experiments were terminated due to the high back pressure caused by column blockage by acumulated metal [140]). In practice, it is easy to regenerate the column by removing the uranium using, e.g., (bi)carbonate (145).

Removal of uranyl ions from solution was not completely efficient unless it was done with an excess of glycerol-2-phosphate, i.e., an excess of liberated phosphate, interpreted mathematically as an intrinsic inefficiency factor (145). The chemical literature predicts that $NaUO_2PO_4$ and $NH_4UO_2PO_4$ should be easier to remove, since their solubility products are several orders of magnitude lower than that of HUO_2PO_4, and, indeed, incorporation of NH_4^+ into the solution promoted uranyl removal (265). In practice, this means that the inefficiency factor due to the chemical and crystallization processes becomes smaller and the system tends more toward the behavior predicted by Michaelis-Menten kinetics relating to product (phosphate) release; i.e., it is the chemistry of the system and not the biochemistry that is limiting. This very important point is usually overlooked in attempts to compare chemical and biochemical processes and provides proof of the superiority of the latter. Furthermore, it is possible for a biochemical system to further overcome the constraints of chemistry by providing an excess of ligand locally, together with the template surface upon which to accelerate the nucleation of crystallization and, a local pH which prevents excessive protonation of the ligand group and hence the problems of increased metal phosphate solubility at low pH.

Bioreduction of U(VI) to U(IV)

From the above discussion, it follows that bacteria which produce sulfides should be able to reduce U(VI) by chemical processes, but this is still not clear. Earlier literature (159) suggested that aqueous sulfide species can reduce U(VI) to U(IV), with precipitation of the U(IV) oxide mineral uraninite. This is disputed by Lovley and Coates (131), who state that U(VI) is resistant to reduction by sulfide. This may reflect the different timescales of geochemical and biochemical processes. Regardless of the role of additional chemical redox reactions, the biological reduction of uranium (and, by implication, other hexavalent actinides) is now well established. The net result of reduction of U(VI) to U(IV) is the precipitation of insoluble UO_2 (uraninite) (86). Precipitation and entrapment of PuO_2 is also documented (263), but the proportion of colloidal plutonium hydroxides to the insoluble oxide is dependent on the local conditions (see above) and on the concentration. Several microbial systems are documented for U(VI) reduction anaerobically; soluble U(IV) oxidizes rapidly to U(VI) in oxic solution. Anaerobically, U(VI) is a terminal electron acceptor in lieu of others such as O_2, NO_3^-, Fe^{3+}, SO_4^{2-} or CO_2. Dissimilatory metal reduction (including U reduction) in bacteria has been reviewed extensively (129, 130). The microbial systems thus far reported to reduce U(VI) are listed in Table 2.

An early study established the biological reduction of U(VI) to U(IV) by a new organism, GS-15 (86, 134), later identified as *Geobacter metallireducens* (132), a representative of the family *Geobacteraceae* (127), a group of bacteria which normally reduce Fe(III) as the terminal electron acceptor. It was calculated that acetate oxidation coupled to U(VI) reduction by this organism could yield more than twice the energy that is available from Fe(III) reduction (134). In the same study, *Shewanella putrefaciens* similarly utilized U(VI), with H_2 as the sole electron donor, according to the following equation (134):

$$H_2 + U(VI) \leftrightarrow U(IV) + 2H^+$$

In each case, U(VI) was reduced only in the presence of actively metabolizing

Table 2. Microbial systems reported to reduce U(VI) to U(IV)

Organism	Comments	Reference(s)
Micrococcus lactilyticus	Cell extracts reduced U at the expense of molecular H_2; hydrogenase activity implicated but not proven	259
Geobacter metallireducens gen. nov. sp. nov. (formerly strain GS-15)	Strict anaerobe; couples reduction of U(VI) with oxidation of acetate or H_2 as electron donor; normally reduces Fe(III)	130, 132, 134
Shewanella putrefaciens	Facultative anaerobe; couples oxidation of H_2 to U(VI) reduction	130, 134
Desulfovibrio desulfuricans	Couples reduction of U(VI) to oxidation of H_2 via cytochrome c_3 activity	128, 240
Desulfovibrio vulgaris	Use of in vitro cytochrome c_3 coupled to hydrogenase for U(VI) reduction	136
Clostridium sp.	Strict anaerobe; bioreduction of U(VI) in wastes.	70, 71

cells. The resulting black precipitate, identified by X-ray diffraction analysis as UO_2 [U(IV), uraninite] (86), was stable to reoxidation in the groundwater samples remediated. However, the difficulty of culture of process-scale quantities of *Geobacter* makes this organism unattractive industrially, and subsequent studies concentrated on the use of *S. putrefaciens* and the sulfate-reducing bacteria, which also reduced U(VI) (128, 133, 135). Under comparable conditions the rates of U(VI) loss were similar for *S. putrefaciens* and *Desulfovibrio desulfuricans*, i.e., approximately 0.2 mmol/liter/h/mg of protein with lactate as the electron donor, while the rate for *G. metallireducens* was half of that value, with acetate instead of lactate (128, 133). The presence of sulfate did not affect the rate of U(VI) reduction, which is important for the application to remediation of wastewaters cocontaminated with sulfate. This was confirmed in a recent study (239), in which cells of *D. desulfuricans* were insensitive to up to 2,000 mg of sulfate per liter, but 50 mg of nitrate per liter inhibited formate- and lactate-dependent U(VI) reduction. Similarly, formate-dependent reduction of Tc(VII) by *D. desulfuricans* is also sensitive to low concentrations of nitrate; however, this was alleviated by the use of hydrogen as the electron donor for Tc(VII) reduction (125).

Using *D. desulfuricans*, U(VI) reduction was growth decoupled; attempts to grow the organism with U(VI) as the electron acceptor were unsuccessful, and prior exposure to O_2 (usually inhibitory to sulfate-reducing bacteria) did not affect the rate of U(VI) reduction (128). Selection of a suitable nongrowing strain is important, since substantial biomass growth within a remediation system produces undesirable biological oxygen demand; there is evidence that at least one sulfate-reducing bacterium, *Desulfotomaculum reducens*, can couple U(VI) reduction to growth (210).

In growth-decoupled processes, a single enzymatic step or pathway can catalyze metal reduction, and for *S. putrefaciens* and the sulfate-reducing bacteria, hydrogenase activity is implicated since they both use H_2 as an electron donor. Indeed, early studies using cell extracts of *Micrococcus lactilyticus* implicated, but did not prove, hydrogenase-mediated U(VI) reduction (259), and since hydrogenase alone did not reduce U(VI), another redox enzyme was also implicated (136).

In contrast to *D. desulfuricans*, U(VI) reductase activity could not be recovered from broken cell extracts of *G. metallireducens* or *S. putrefaciens* (135), but the U(VI) reductase activity of whole cells of *Desulfovibrio* is stable in whole and broken cells and tolerant to air; of the *Desulfovibrio* species known to reduce U(VI), the biochemistry of *D. vulgaris* has been studied most extensively. Using membrane-free preparations, preliminary studies suggested that 95% of the H_2-dependent U(VI) reduction was in the soluble fraction and the U(VI) reductase was attributable to an activity of cytochrome c_3, with hydrogen/hydrogenase acting as the reductant (136), a function that could be replaced by electrochemical reduction of cytochrome c_3 (167). Decoupling the system from an enzymatic supply of reducing power effectively gives the potential for an in vitro "electrobioreactor" similar to that described for nitrate reduction by Mellor et al. (158). Myers and Myers (161) highlighted the difficulties encountered by microorganisms when the solid forms of Fe and Mn oxides (larger than the cells themselves) were used as electron acceptors and postulated the need for electron transport chains in the outer membrane, previously considered unlikely. A recent study (161) confirmed the localization of *c*-type cytochromes to the outer membrane of *S. putrefaciens*, these are likely to be involved also in the reduction of U(VI), but this has not yet been addressed per se.

The above finding facilitates kinetic studies, since with whole cells it has been difficult to ascertain whether the observed K_m and V_{max} values (see references 202 and 240 for specimen data) are those of the enzymatic system per se or the uptake process for the target ion. Exocellular metal reduction explains the extracellular localization of reduced-metal precipitate observed when sulfate-reducing bacteria are reducing UO_2^{2+} (86) and also clearly delineates the metal-reducing and the sulfate-reducing activities, explaining why the latter has no inhibitory effect and further strengthening the case for the use of sulfate-reducing bacteria to remediation of sulfate-laden wastes.

An early study identified anaerobic biotransformations of uranium in actual wastes (70); later studies (71) showed clearly that this was attributable to reduction of U(VI) to U(IV) by *Clostridium* spp. that should be included in the portfolio of potentially useful organisms (Table 2). The use of *D. desulfuricans* with uranium mine drainage water and uranium-containing groundwaters was demonstrated (133). Here, the organisms were held within semipermeable membranes; nascent U(IV) oxide diffused outward to spatially separate the bioreduction and precipitation steps. Robinson et al. (Proc. Water Environ. Fed. 67th Annu. Conf. Expos., 1994) used a similar technique to confirm that excess sulfate (and nitrate) did not inhibit U(VI) reduction and also showed a decreased rate in U(VI) reduction by the entrapped versus the equivalent free cells, which was attributable to the rate of diffusion of uranium across the dialysis membrane. Recent studies have further shown the application of sulfate-reducing bacteria to minewater remediation (93) and waste disposal cell stabilization (26). In the former study, naturally occurring sulfate-reducing bacteria (including some strains capable of growth at pH 4) coupled the oxidation of methanol to the reduction of sulfate, sulfur, and nitrate, but metal reduction was not tested. The study made the assumption that biogenic sulfide would precipitate the metals Al, Cd, Cr, Cu, Pb, Ag, U, Ni, Fe, Mn, and Co, but in view of the sulfide-independent enzymatic U-reduction process (see above), similar cytochrome-mediated processes should be examined in the new (acid-tolerant) strains. Ethanol has also been used as an electron donor for U(VI) reduction by bacteria indiginous to contaminated groundwater (1), with sulfate-reducing bacteria again being implicated in U(VI) reduction. In this study, a comparatively slow sulfide-mediated reduction in controls suggested that enzymatic reduction was important under the experimental conditions.

INTERACTIONS WITH NONACTINIDES

Group I: Cesium Removal

Radiocesium is released in large quantities during the controlled discharge of low-level liquid wastes (nuclear fuel reprocessing streams and reactor cooling waters). For example, ^{137}Cs has accounted for 65% of all isotopes in low-level wastes from the reprocessing plant at Sellafield, resulting in 3×10^{16} to 4×10^{16} Bq of ^{137}Cs released into the Irish Sea up to 1990 (149, 155). Similarly high activities are reported in effluents from other facilities (138). Cs, in common with other alkali metals and unlike most other radionuclides of environmental significance, is a very weak Lewis acid with a low tendency to interact with ligands (98). It forms electrostatic (ionic) rather than covalent bonds with oxygen-donor ligands, binding only weakly to organic and inorganic ligands. Although the chemical toxicity of stable Cs is low, the long half-life of ^{137}Cs (30.1 years), in combination with high solubility, has led to concern over its release. Indeed, recent reports have shown that higher than expected levels of ^{137}Cs persisted in a mobile form in the environment following the Chernobyl accident in 1986, with accumulation of the Cs by microorganisms playing a role in passing the radionuclide to higher trophic levels in the food chain (9). The current method of treatment is by ion exchange using zeolites (aluminosilicates), e.g., clinoptilolite, offering high ion exchange selectivity, moderate resistance to radiolytic degradation, and low solubility. Some interference by K^+ and Na^+ is noted, however, with a stronger effect exerted by K^+, which has the more similar ionic radius to Cs. Specialist ion-exchange resins, including hexacyanoferrates, are required for high-salt effluents (126).

Microbial biosorbents have also been evaluated for low-salt effluents but exhibited low Cs uptake and offered no improvement over the use of zeolites (138). Uptake of Cs^+ by actively metabolizing microorganisms has proved more successful. Due to similarity of the K^+ and Cs^+ cations, both are taken up by the same metabolism-dependent transport systems. Indeed, broad-specificity alkaline earth metal uptake transporters have been reported in all microbial groups (9). The need for energy-dependent transport dictates that growing cells are required for Cs^+ uptake. Studies by Harvey and Patrick (94) and deRome and Gadd (53) have illustrated the superiority of batch over continuous-flowthrough systems for algal and fungal model systems, respectively. In the latter study, which utilized pelleted mycelial biomass of *Rhizopus arrhizus* and *Penicillium chrysogenum*, uptake by energy-independent processes was more important than energy-dependent uptake, suggesting that Cs^+ uptake could be related to the physical properties of the immobilized biomass.

Biochemistry of Cs Accumulation

Cs uptake by bacteria. *Escherichia coli* has several well-characterized K^+ uptake transport systems, which are required for enzyme activation and osmotic regulation. The low-specificity Kup (formerly TrkD) system has a 13.5-fold higher affinity for K^+ than for Cs^+ (with a V_{max} of 27 compared to 17 μmol min^{-1} g [dry weight] cells^{-1} [32]). A higher affinity for Rb than for Cs was also reported. *kup* mutants transported Cs^+ into the cell very slowly, suggesting that the other transport systems, Trk and Kdp, were not important in Cs^+ uptake. This distinction precludes the manipulation of an osmotic stress response to achieve Cs^+ uptake via the osmotically regulated Trk system. Also, wastewaters typically contain high concentrations of K^+, effectively limiting the commercial application of this technology. In a recent study, Appanna et al. (6) noted that Cs was immobilized with the cell pellet of *Pseudomonas fluorescens* as an insoluble gelatinous precipitate. The bacterium was grown with 5 mM ferric iron, and although the mode of uptake was not determined, X-ray fluorescence analyses indicated that the two test metals (Cs and Fe) were initially localized within the bacterial cell.

Cs uptake by cyanobacteria and eukaryotic algae. Two transport systems with different affinities for K^+ have been reported in cyanobacteria. The low-affinity K^+ transporter of *Anabaena variabilis* is probably solely responsible for Cs^+ uptake in this organism (11). Rb^+ and K^+ inhibited Cs^+ uptake by the estuarine eukaryotic alga *Chlorella salina*, implicating a similar, common transport system (12). In contrast, Singh et al. (201) demonstrated Cs^+ uptake by the ammonium-repressible/derepressible ammonium transport system of *Nostoc muscorum*.

Cs uptake by fungi. A Cs^+ uptake system is well characterized in the yeast *Saccharomyces cerevisiae*. The order of affinity for the K^+ transport system is $K^+ > Rb^+ > NH_4{}^+ > Cs^+ > Na^+ > Li^+ > Mg^{2+} > Ca^{2+}$ (30). Three sites are involved in translocation (52). A high-affinity activator site is utilized at low Cs^+ concentrations, with low-activity modifier and translocating substrate sites being active only at high concentrations of Cs^+. K^+ uptake is inhibited at very high Cs^+ concentrations. In *S. cerevisiae*, imported Cs is detoxified via accumulation in vacuoles, accounting for 90% of Cs taken up into the fungal biomass (177). Indeed, this would seem to be a general phenomenon in eukaryotes; Avery (9) postulated that enhanced uptake of Cs by eukaryotes compared to prokaryotes may be due to vacuolar sequestration.

Effect of Process Variables on Cs Uptake

Most studies on microbial Cs^+ accumulation have used relatively clean, well-defined laboratory solutions. Plato and Denovan (183), however, quantified ^{137}Cs uptake by *Chlorella pyrenoidosa* against a background of K^+ ions. Approximately 83 to 88% of the Cs^+ was removed if K^+ was present at concentrations between 50 to 375 μM. Removal was less than 20% if the concentration of K^+ was below or above these values, due to poor growth of the organism or K^+ competition. Other studies have shown, however, that the presence of excess K^+ does not inhibit Cs uptake in all examples. For example, Cs^+ uptake by *Synechocystis* strain PCC 6803 was unaffected by equimolar concentrations of K^+ (10). High concentrations of H^+ ions have also been shown to exert a deleterious effect on Cs^+ uptake; in the same study, bioaccumulation was enhanced at alkaline pH values (pH 10). Similar results have been obtained with yeast cells, where Cs^+ uptake was pH-independent at high pH but decreased as the pH of the growth medium dropped from 5.5 to 3.0 (177). Although some wastes are highly acidic, Cs uptake by acidophiles has yet to be investigated.

Strategies To Optimize Cs Uptake

Two recent studies have used autoradiography-based techniques to select for high-Cs^+-accumulating microbes growing on solid media (104, 223). *Rhodococcus erythropolis* was isolated in the former study, and a *Cladosporium* sp. and a *Bacillus* sp. were isolated in the latter study. All the organisms accumulated large amounts of Cs. An alternative approach is to manipulate the environmental conditions to stimulate Cs^+ uptake. Hyperosmotic shock has been proposed as one method to stimulate Cs^+ uptake and is most suited to estuarine organisms such as

Chlorella salina, which is able to cope with rapid changes in the osmolarity of its growth environment (12). The single system which is responsible for Cs^+ and K^+ uptake by this organism does not discriminate between the two cations, making it an ideal candidate for this form of metabolic manipulation. Indeed, a 28-fold increase in Cs^+ uptake was noted when *C. salina* cells were supplemented with 0.5 M NaCl. An additional advantage of the use of these organisms was the rapid loss of Cs when the cells were subjected to hypoosmotic shock by being resuspended in less than 50 mM NaCl. Genetic manipulation of the well-characterized monovalent cation uptake systems may also prove useful but has been little-explored, despite early reports of enhanced Cs^+ uptake by *E. coli* overexpressing the *trkD* gene which encoded the Kup transport system (32).

Group IIA: Radium Removal

Radium is present in dilute wastewaters from the uranium mining and milling industries and also in an unknown number of sites resulting from the historically widespread use of luminous paints containing Ra. The long half-life of ^{226}Ra (1,600 years) raises serious concerns about its release. The concentration of Ra^{2+} in wastewaters is typically in the picogram-per-liter range, in contrast to other cations such as Ca^{2+}, Mg^{2+} and Na^+, which are typically in the order of several milligrams per liter (231). Macaskie (138) calculated that for a biological treatment strategy to meet the current low permissible discharge limits, over 99% of the Ra presented must be removed against this high concentration of contaminating cations.

The biosorption of ^{226}Ra has been studied in detail. Tsezos and Keller (232) used equilibrium biosorption isotherms to quantify the Ra uptake capacity of various types of biomass and activated carbon. Maximum uptake occurred at pH 7.0, with reduced and negligible uptake at pH 4.0 and 2.0, respectively. The best absorbent was biomass from a municipal-wastewater-activated return sludge (40,000 nCi g^{-1}), which bound significantly more Ra than did commercially available activated carbon (3,600 nCi g^{-1}). With some biomass types, over 99% of the Ra was removed from solution. Interestingly, *Rhizopus arrhizus*, which accumulated substantial quantities of U (236), was a poor accumulator of Ra, indicating the presence of different mechanisms of uptake for U and Ra. A subsequent study (230) determined that radium biosorption was rapid, with equilibrium attained in 2 min (*Penicillium chrysogenum*) and 5 minutes (activated sludge). The rate of Ra adsorption was not affected by changes in pH (in the range from pH 4 to 9) or by the presence of competing ions including Ca^{2+}, Ba^+, Cu^{2+}, and Fe^{2+}. Biosorption was also similar in synthetic laboratory and actual waste solutions. Immobilized sludge from a municipal wastewater treatment plant was used for radium recovery from Elliot Lake uranium tailing streams (229, 231). The biomass was immobilized into 1-mm-diameter beads, containing not less than 90% dead biomass, using a proprietary technique. Mass transfer limitations inside the biosorbent were assumed to be responsible for the long time required for adsorption equilibrium (7 days, compared to 24 h for free biomass under similar conditions). The possible reuse of the biosorbent was also assessed (229, 231). Although bound Ra could be eluted and recovered by a range of acids and chelating agents, effective adsorption was possible only after a maximum of two elution cycles.

In a more recent study, Dwivedy and Mathur (58) reported the use of an *Arthrobacter* sp. for the precipitation of Mn and ^{226}Ra from neutralized uranium mill effluents. Yeast extract (0.005%) and sucrose (0.005%) were added as growth substrates, and 95% of the Ra was coprecipitated with 92% of the Mn after treatment (Mn was precipitated as $MnO_2 \cdot nH_2O$). The concentrations of the two elements were within the limits set by the International Committee of Radiological Protection (Mn, <0.5 mg liter^{-1}; Ra, < 3.0 pCi liter^{-1}). Anaerobic reduction of Mn(IV) is possible (163), however, and could potentially remobilize the ^{226}Ra. Liberation of Ra from stored tailings via the biological oxidation of pyrite (generating sulfuric acid) or microbial ferric iron reduction (113) is already documented. Similarly, a MECHM-type reaction could permit Ra to intercalate into biogenic $HUO_2PO_4 \cdot 4H_2O$; formation of solid metal phosphates can be problematic in nuclear fuel reprocessing, with phosphate arising via radiolysis of TBP (49); therefore, there is a precedent for this type of reaction to form $Ra(UO_2PO_4)_2$.

Group IIA: Strontium Removal

In common with other group IIA alkaline earth cations, e.g., Ca^{2+}, Ba^{2+}, and Ra^{2+}, Sr^{2+} readily forms complexes and insoluble precipitates. The metal is removed by zeolites, but Cs removal by nonspecific ion-exchange materials would be enhanced if the competing Sr^{2+} ion was removed first (138).

Strontium Carbonate Precipitation by Metabolizing Cells

Two recent studies have demonstrated the biomineralization of strontium by actively metabolizing microorganisms. *Pseudomonas fluorescens* grew in a minimal medium containing strontium complexed to citrate, which was supplied as the sole carbon source (2). Strontium was excluded from the cell and precipitated in the growth medium as crystalline strontium carbonate, as determined by X-ray fluorescence spectroscopy, X-ray diffraction spectrometry, and Fourier transform infrared spectroscopy. A threefold increase in the concentration of dissolved carbon dioxide in the growth medium implied a role for this gas in mineral deposition. Microbial activity was also implicated in the deposition of a strontium calcite phase at a groundwater discharge zone (64). The authors interpreted the deposition of the mineral to be the result of carbonate precipitation by epilithic cyanobacteria (driven by HCO_3^-/OH^- exchange during photosynthesis), with a Sr/Ca ratio that promoted the deposition of $SrCO_3$ in calcite (up to 1% Sr).

Strontium Uptake Applied to Bioremediation

Sr accumulation by a *Citrobacter* species has also been studied (143). Cells immobilized in polyacrylamide gel removed 47% of the Sr from 5.6 liters of flow over 8 days (60-ml columns; flow rate, 30 ml h^{-1}) as an insoluble phosphate mediated via phosphatase activity. This corresponded to a loading of 363 mg of Sr g (dry weight) of biomass^{-1} (138). Sr removal was poor at neutral pH and was maximal only at pH values above the pH optimum of the phosphatase. At alkaline pH and using a series of sequential columns, removal of Sr was increased to >90%

(143). Following on from work of Zajic and Chiu (269), who noted Sr uptake by a *Penicillium* sp. to 75 mg g of biomass^{-1}, Watson et al. (251) made a detailed study of Sr biosorption by various microorganisms. *Rhizopus* was the most effective organism tested overall, but *Micrococcus luteus* gave comparable uptake after 2 days (compared to 14 days of incubation with the former organism). Other studies by this group showed microbial Sr uptake to be instantaneous (maximum loading 1.2 mg per g of cells), with metal uptake being identical after 4 h for both free cells and cells immobilized in gelatin (250). This is in contrast to the work of Tsezos et al. (230), who observed considerable mass transfer limitations in the biosorption of Ra upon immobilization. Immobilized cells were used in both batch and flowthrough contactors. Almost complete removal was obtained in columns (52 ml) for 2,500 ml of flow, with breakthrough noted after 3,500 ml of flow. Sr was successfully desorbed by 0.3 M KCl, although gel stability problems were reported at high salt concentrations. deRome and Gadd (53) also studied the biosorption of strontium, along with uranium and cesium, by *Rhizopus arrhizus* and *Penicillium chrysogenum*. The uptake mechanism for the two organisms was different. In the former, uptake was metabolism independent. With the yeast, however, uptake was biphasic in the presence of glucose, with surface biosorption being followed by energy-dependent influx. The surface-bound metals were removed by elution with carbonate-bicarbonate solution or mineral acids. Biosorptive uptake of strontium by waste *Saccharomyces cerevisiae* (to 21% of the biomass dry weight) was also noted by Avery and Tobin (8).

Sr uptake by plant tissue has also been studied. Scott (198) noted adsorption of Sr by tobacco and tomato root tissue and recorded distribution coefficients of greater than 550. The adsorption isotherms fitted Langmuir-type expressions. The tomato root tissue was immobilized in carrageenan gel beads, packed in a 9-ml column, and challenged with Sr solution (10 ppm). Over 99% removal was recorded for the first 25 ml of eluant, with gradual breakthrough noted for the next 125 ml. Quantitative removal of the Sr was achieved by elution with 1 M KCl. Uptake by *Pinus ponderosa* and *P. radiata* has also been studied by Entry et al. (63). Seedlings inoculated with ectomycorrhizal fungi accumulated three to five times as much ^{90}Sr from contaminated soil than did seedlings without ectomycorrhizae.

Group IIIA: Yttrium Removal

^{90}Sr decays to the short-lived and relatively innocuous isotope ^{90}Y (half-life, 64 h). Biological interactions with Y have been little studied, although they should mirror those with La, which is bioaccumulated by several mechanisms. Both elements are stable as the trivalent cation, and immobilized cells of *Citrobacter* sp. accumulated Y^{3+} very rapidly in a similar manner to that described for La^{3+} (K. M. Bonthrone and L. E. Macaskie, unpublished observations). Lear and Oppenheimer (117) showed that ^{90}Sr and ^{90}Y were removed from seawater by marine microorganisms, following the early work of Spooner (203), which suggested that accumulation of ^{90}Y by marine algae is greater than that of ^{90}Sr. A more recent study quantified Y and Sc biosorption by *Saccharomyces cerevisiae* and *Aspergillus terreus* (107) and concluded that Al, Fe(III), and Ti affected the biosorption of both elements (particularly Y). Biosorption was maximal at high pH, and Y could be eluted by acid washing of the biomass. Tzesos et al. (234) also noted that yttrium biosorption by two pseudomonads was inhibited by the presence of uranium, suggesting that the latter exhibited a higher affinity for the binding loci on the biomass. In an earlier pilot-scale study, Tsezos et al. (233) reported that ytrrium was also displaced from immobilized *Rhizopus arrhizus* biomass by incoming uranium in a mine effluent. In addition to competing cations, the deleterious effects of organic chelating agents on biosorption were emphasized by the study of Sun et al. (207), who noted reduced uptake of rare earth elements (La, Gd, and Y) when supplied as citrate, NTA, or EDTA complexes.

Group IVA: Zirconium Removal

Zirconium isotopes arise from fission and activation reactions and also by corrosion of fuel cladding, but Zr is generally considered to be biologically inert (82). ^{95}Zr and ^{93}Zr (half lives, 64 days and 1.5×10^6 years, respectively) are present in some nuclear wastes; in a typical low-activity waste stream the concentration of ^{95}Zr is 2.2 μCi liter^{-1}, while in unreprocessed fuel the activity of ^{93}Zr is approximately 20% of that of ^{99}Tc (in becquerels [138]). Zirconium is a neglected element; its general chemistry is similar to that of Ti(IV) and in terms of bioremediation it suffers from the same constraints as Th(IV). There are few studies on Zr bioaccumulation. Garnham et al. (80) reported energy-independent, pH-dependent Freundlich-type biosorption of Zr by microalgae and cyanobacteria. Biosorption of ^{95}Zr *R. arrhizus* was observed; it was maximal at pH 2 and was reduced by approximately 40% at pH values of 4 and above (54). A direct comparison between Zr(IV) uptake and Pu(IV) uptake was not possible because the challenge concentrations were not identical, but the behavior of Zr(IV) should be contrasted with that of Pu(IV), where increased pH gave increased biosorption of $Pu(OH)_n$ (see above). Attempts to accumulate zirconium phosphate using the *Citrobacter* system met with limited success (16), mainly attributable to the amorphous nature of the deposit (as shown by X-ray diffraction [Basnakova and Macaskie, unpublished]). Zirconium phosphate exists in several amorphous and crystalline forms (243), and attempts to improve the crystallinity by various chemical treatments were unsuccessful (16). The simultaneous presence of UO_2^{2+} during Zr^{4+} bioprecipitation by *Citrobacter* did not promote removal of the latter, but a predeposited layer of HUO_2PO_4 promoted Zr^{4+} removal with 100% efficiency at pH 2 to 6, maintained over 38 column volumes before column saturation (16).

Group VA: Niobium Removal

The isotope ^{95}Nb was reported to be present at 8.8 μCi liter^{-1} in a typical low-activity waste steam (138). It is included here for completeness; the authors are not aware of any attempts to remediate ^{95}Nb biologically. Nb is a member of the same chemical group as vanadium and would be expected to behave similarly biologically. Vanadium is an essential element for some biochemical reactions (72), and therefore biochemical mechanisms for sequestration and management of V must exist. The vanadyl oxycation VO_2^+ would probably behave like NpO_2^+ with respect to bioremediation. Attempts to bioremediate VO_2^+ by bioreduction, as described for NpO_2^+ (125a) and UO_2^{2+} (134) were not pursued, although V(V) can be reduced by microbial action (259, 268). Studies on ^{95}Nb would be facilitated by the recent development of a method for separation and quantification of radionuclide species in mixed-valence solutions (122, 125a), but this has not yet been applied to investigation of the biology of Nb.

Group VIB: Polonium Removal

^{210}Po is the last member of the ^{238}U decay series, produced by the decay of ^{210}Pb via ^{210}Bi. Although the overall abundance of naturally occurring Po in crustal rocks is on the order of 3×10^{-10} ppm, concentrations of about 0.1 mg of Po per tonne (10^{-4} ppm) are recorded in uranium ores. Po has attracted recent attention, since it accounts for all α-emitting activity in some samples of groundwater from the Central Florida Phosphate District (39). Levels of over 1,000 dpm liter^{-1} were recorded in the absence of the radiogenic precursors. The ^{210}Po was considered to originate either from the naturally occurring phosphate rock of the area (high in uranium and daughter decay elements) or from phosphogypsum, which is a by-product of the wet-process manufacture of phosphoric acid and is high in ^{226}Ra, ^{210}Pb and ^{210}Po. ^{210}Po has also been identified as a troublesome isotope contaminating and fouling drilling equipment during oil exploration in areas rich in uranium ores (A. Clerkin, personal communication).

Little is known about the geochemistry of Po, but there is evidence that factors affecting the sulfur cycle may affect Po availability. Both elements are in the same group of the periodic table, occur simultaneously in groundwater, and are removed by bacteria under aerobic conditions. Indeed, Cherrier et al. (44) demonstrated the microbial uptake of Po into the cellular pool, indicating the potential for assimilative Po reduction and a biochemical role analogous to that of S. LaRock et al. (115) subsequently determined that aerobic bacteria were capable of releasing Po from gypsum, possibly via utilization of the sulfate in the mineral. Sulfate-reducing bacteria were also able to mobilize Po, but the Po was precipitated at higher concentrations of sulfide (approximately 10 μM and above). The use of sulfate-reducing bacteria has been proposed, therefore, as a method of removing Po from solution.

Group VIIA: Technetium Removal

Technetium-99, a fission product of ^{235}U, is long-lived (half-life, 2.1×10^5 years) and is a significant pollutant in waste streams. It is normally present in the heptavalent form as the pertechnetate anion (TcO_4^-), which is both highly soluble and mobile in the environment (258). Despite its artificial nature, the biological activity of Tc(VII) is high, since it acts as a sulfate analogue (43), and assimilation by plants facilitates its entry into the food chain (50). These factors, considered in combination, led Trabalka and Garten (224) to conclude that Tc may be the critical radionuclide in determining the long-term impact of the nuclear fuel cycle.

Uptake of the pertechnetate anion by biosorption is low (81). However, an alternative approach utilizing metal-reducing microorganisms has proved more successful. Although Tc(VII) is highly soluble, several low-valence oxides of Tc are insoluble (111). Recent studies have aimed, therefore, at characterizing the role of microorganisms in the reduction and precipitation of Tc. Henrot (95) showed that anaerobically (but not aerobically) grown mixed cultures of soil bacteria accumulated Tc and that addition of the sulfate-reducing bacteria *Desulfovibrio vulgaris* and *D. gigas* increased Tc removal by more than an order of magnitude. It was postulated, but not conclusively shown, that Tc reduction and removal were mediated by microbially generated H_2S. Pignolet et al. (180) considered that microbially derived H_2S also played a major role in Tc reduction and precipitation by anaerobic cultures containing mixed populations of microorganisms isolated from a marine sediment. In contrast to indirect (chemical) mechanisms of reduction, the authors suggested that Tc reduction and removal by oxygen-limited cultures of *Moraxella* and *Planococcus* spp. may be enzymatically catalyzed. Cells incubated at 21°C removed significantly more Tc from solution than did those incubated at 4°C, and negligible Tc was concentrated by heat-treated (tyndallized) cultures.

Lloyd and Macaskie (122) subsequently developed a PhosphorImager-based technique to monitor the microbial reduction of Tc(VII). Using this technique, direct enzymatic reduction of Tc(VII) by resting cells of the Fe(III)-reducing bacteria *Shewanella putrefaciens* and *Geobacter metallireducens* was demonstrated. Tc products were species specific; only soluble, reduced species were detected in supernatants from *S. putrefaciens*, but appreciable quantities of the radionuclide were precipitated by *G. metallireducens*, probably as a low-valence insoluble Tc oxide. X-ray absorbance spectroscopy studies have now confirmed that Tc(VII) is reduced to insoluble TcO_2 by a closely-related species, *Geobacter sulfurreducens* (J. R. Lloyd, V. A. Sole, C. Gaw, and D. R. Lovley, submitted for publication). Another potentially important mechanism by which Fe(III)-reducing bacteria can reduce Tc(VII) is via biogenic Fe(II), with microbially produced magnetite being a particularly efficient reductant (Lloyd et al., submitted).

More recent studies have demonstrated that the ability to reduce Tc(VII) is not restricted to iron-reducing bacteria. For example, anaerobic cultures of the enteric bacterium *E. coli* are able to couple the oxidation of formate or hydrogen to Tc(VII) reduction and precipitation, presumably also as TcO_2 (119, 120). Using physiological and molecular biological approaches, the enzyme responsible for Tc(VII) reduction was identified as the hydrogenase 3 component of the formate hydrogenlyase complex (119). The normal physiological role of the enzyme complex is to convert the fermentation product formic acid to hydrogen and CO_2, thus minimizing acidification of the medium during anaerobic growth. Anaerobically grown resting cells of *E. coli*, immobilized in a flowthrough hollow-fiber membrane bioreactor, have also been used to remove Tc(VII) from a challenge solution; formate or hydrogen was supplied as the electron donor for metal reduction (120). The biocatalyst was highly stable in the reactor, with no loss in activity noted during 200 h of continuous use.

Sulfate-reducing bacteria, which have well-documented metal-reducing activities (130), can also reduce Tc(VII) enzymatically. In a recent study, resting cells of *D. desulfuricans* also reduced and precipitated Tc (123), but the lack of available deletion mutants has hampered attempts to elucidate the mechanism of Tc(VII) reduction by this organism. In common with the *E. coli* system described above, the optimal electron donors for the bioreduction were identified as formate and hydrogen (124). Electron microscopy in combination with energy-dispersive X-ray microanalysis showed that the reduced Tc was associated with the periphery of the cell (124), while X-ray absorbance spectroscopy confirmed that the precipitated Tc was reduced from the heptavalent state (123). Under sulfate-reducing conditions (i.e., in the presence of high SO_4^{2-} concentrations), Tc was also precipitated, this time as an insoluble sulfide, but removal was less efficient than via enzymatic reduction (123). Subsequent studies have demonstrated that the specific rate of enzymatic reduction of Tc(VII) by *D. desulfuricans* is more than an order of magnitude greater than that by both a wild-type *E. coli* strain and a regulatory mutant with elevated hydrogenase 3 activities (produced via overexpression of the whole formate hydrogenlyase complex [125]). Hydrogen-dependent Tc(VII) reduction by *D. desulfuricans* is also insensitive to high concentrations of nitrate, which may be present in nuclear waste streams. In comparison, low concentrations of nitrate (1 mM) inhibited the reduction of 250 μM Tc(VII) by *E. coli*, possibly by diverting reducing equivalents to denitrifying enzymes or by affecting the regulation of the Tc-reducing formate hydrogenlyase complex (125).

It seems that the ability to reduce the pertechnetate anion may be widespread, an observation that has implications regarding in situ bioremediation of Tc contamination. Other organisms capable of reducing and precipitating the radionuclide include laboratory cultures of *Rhodobacter sphaeroides, Paracoccus denitrificans,* and some pseudomonads (J. R. Lloyd unpublished observations) in addition to the acidophiles *Thiobacillus ferrooxidans* and *T. thiooxidans* (137) and the hyperthermophile *Pyrobaculum islandicum* (K. Kashefi and D. Lovley, submitted for publication). Although such investigations are warranted, the Tc(VII)- and, indeed, U(VI)-reducing capabilities of the microflora indigenous to contaminated sediments remain uncharacterized.

Group VIII: Ruthenium, Cobalt, and Nickel Removal

Ruthenium

Although ^{106}Ru is one of the nuclides present in low-activity wastes (7, 138), it has received very little attention. Unlike the cations Co^{2+} and Ni^{2+}, Ru tends to occur as nitrosyl and other complexes. Very little information is available on the biology of this element, other than a study by Gibson et al. (83). A common use of ruthenium is as the ruthenium red dye, used as a complexing agent for the quantification of microbial exopolymers (173). Thus, its removal by biosorption as a dye complex is assured, but there have been no attempts to evaluate this in terms of bioremediation.

Cobalt

Although ^{60}Co is considered relatively innocuous by virtue of its relatively short half-life (5.27 years), radioactive Co(III) complexed by EDTA is noted as a contaminant at some Department of Energy sites (131). Recent studies have shown that iron-reducing bacteria are able to reduce Co(III) when complexed with EDTA (41, 42, 85). The Co(II) formed does not associate strongly with EDTA [it is over

25 orders of magnitude less thermodynamically stable than Co(III)-EDTA], and absorbs to soils, offering potential for in situ immobilization of the metal in contaminated soils. However, Gorby et al. (85) have demonstrated that reduced Co(II)-EDTA can transfer electrons abiotically to solid Mn(IV) oxides, effectively acting as an electron shuttle between the bacterial cell and the metal oxide. Mn(IV) minerals could therefore play an important role in maintaining concentrations of Co(III)-EDTA in the subsurface.

Several biosorbents have also been tested against Co, including the marine alga *Sargassum natans*, which was able to accumulate the metal to a loading of 156 mg g (dry weight)$^{-1}$ (112), and De Vo-Holbein composition G-1 (138), which was used to remove detectable levels of ^{60}Co and ^{63}Ni from a contaminated solution (starting concentrations of metals, 88 ppm). The yeast *Candida utilis* removed approximately 80% of the ^{60}Co from a solution containing 1,357 nCi/liter, probably attributable at this low concentration to biosorption onto biomass-bound phosphate (249).

Nickel

^{59}Ni is produced as an activation product from fuel assemblies and is considered problematic due to its long half-life (7.6×10^4 years). Microbial interactions with this isotope have been relatively poorly studied, but the uptake of "cold" nickel has been studied by several authors. An *Oscillatoria* sp. accumulated Ni from freshwater to a concentration factor of 9,000 (226). The initial concentration of the Ni was 0.12 ppm. A *Zoogloea* sp. accumulated up to 50% of 500 ppm Ni, with an extracellular gelatinous matrix implicated as a major site of metal biosorption (73). A strain lacking the matrix accumulated only 25% of the Ni supplied. More recently, Tsezos et al. (234) noted poor Ni biosorption by two pseudomonads, which was severely depressed by the presence of Ag and Y. The Co-accumulating alga *S. natans* accumulated Ni poorly (B. Volesky, personal communication) and Ni is generally held to be fairly recalcitrant, for reasons which are still not clear. Tsezos et al. (235), in a study of biosorption of Ni by six strains, considered the hydrolysis behavior, chemical coordination, stereochemical, and redox characteristics and suggested that solubility considerations alone could not explain the observed low biosorption of Ni; stereochemical factors were considered the most likely. Addition of phosphate to a solution of Ni^{2+} released by biodegradation of citrate resulted in poor Ni^{2+} removal compared to Co^{2+} removal by two pseudomonads (212). Removal of Ni^{2+} by phosphate-liberating *Citrobacter* sp. was also poor (29), although Ni^{2+} was removed by a strain of *Alcaligenes eutrophus* (242) that is well known to remove heavy metals as the carbonates via localized alkalinization concomitant with metal efflux (209).

In contrast to biosorption, nickel was removed effectively by ion exchange into preformed biogenic HUP ($HUO_2PO_4 \cdot 4H_2O$) by *Citrobacter* sp. primed with uranyl ion in the presence of glycerol 2-phosphate (MECM; see above). The Ni^{2+} solution (aqueous) was then passed through columns of primed biomass, with quantitative removal of Ni into the HUP, converting this into $Ni(UO_2PO_4)_2$ (NiHUP) (19, 29). The columns behaved as "classical" ion exchangers, functioning in repeated sorption-desorption cycles (18). Using citrate buffer as the eluant, the Ni could be selectively desorbed and fresh uranyl phosphate could be loaded in simultaneously, taking advantage of the different complexation properties of Ni^{2+} and $UO_2{}^{2+}$ with citrate. Such regenerated columns had an enhanced Ni-sorbing capacity, and Ni was confirmed to penetrate within HUP deposits in the center of the polyacrylamide gel immobilizing matrix, using the complementary techniques of energy-dispersive X-ray microanalysis and proton-induced X-ray emission analysis in conjunction with electron microscopy of cryofixed sections (19).

The "loop" for processing of Ni from the desorption phase was closed by the use of citrate-degrading micoorganisms (212) to release the Ni into an available, concentrated form (212). The latter illustrates the application of "clean" technology and the modern requirement of remediation processes to, in themselves, generate minimal waste.

CONCLUSIONS

Microbial interactions with all radionuclides of environmental concern are documented. Many interactions result in the accumulation of the radionuclide on or within the microbial cell, effectively removing it from solution. In most cases, the biochemical factors underlying such processes are now well characterized. An exciting development has been the marriage of microbiology and crystallography to use biochemical activity to promote and steer chemical crystallization processes. It is possible to overcome chemical constraints by using biotransformations to convert a recalcitrant metal species into one which is amenable to remediation. Superimposition of crystallographic aspects upon this introduces a multilevel sophistication and has resulted in recent success against metals for which no other methods have been found applicable. The problems of mixed and chelated wastes, achieving prominence only within the last 5 to 10 years, are becoming less intransigent with the development of new microorganisms and microbial communities. The present challenges are to gain a greater understanding of the effect of process variables on radionuclide accumulation, allowing optimization in complex matrices characteristic of real wastes, and to develop bioprocesses with minimal end-of pipe waste in terms of associated biological and chemical oxygen demand. To implement biotechnology to treat large areas contaminated with historic waste, the challenges are to gain a better understanding of microbial communities at site and devise effective methods of stimulating or augmenting microbial activities required in situ.

Acknowledgments. We thank all colleagues who have allowed access to unpublished information; in particular, the help of R. A. P. Thomas and G. Basnakova in the provision of unpublished data and some of the literature is acknowledged.

Financial support from the NABIR program of the DOE (J.R.L.) and BBSRC (L.E.M.; grant 6/TO6494) is acknowledged.

REFERENCES

1. **Abdelouas, A., Y. L. Lu, W. Lutze, and H. E. Nuttall.** 1998. Reduction of U(VI) to U(IV) by indigenous bacteria in contaminated ground water. *J. Contam. Hydrol.* **35:**217–233.
2. **Anderson, S., and V. D. Apanna.** 1994. Microbial formation of crystalline strontium carbonate. *FEMS Microbiol. Lett.* **116:**42–48.

3. **Andres, Y., H. J. MacCordick, and J.-C. Hubert.** 1993. Adsorption of several actinide (Th, U) and lanthanide (La, Eu, Yb) ions by *Mycobacterium smegmatis*. *Appl. Microbiol. Biotechnol.* **39:** 413–417.
4. **Andres, Y., H. J. MacCordick, and J. C. Hubert.** 1994. Binding sites of sorbed uranyl ion in the cell wall of *Mycobacterium smegmatis*. *FEMS Microbiol. Lett.* **115:**27–32.
5. **Andres, Y., H. J. MacCordick, and J. C. Hubert.** 1995. Selective biosorption of thorium ions by an immobilized mycobacterial biomass. *Appl. Microbiol. Biotechnol.* **44:**271–276.
6. **Appanna, V. D., L. G. Gazso, J. Huang, and M. St. Pierre.** 1996. A microbial model for ceasium containment. *Microbios* **86:**121–126.
7. **Ashley, N. V., and D. J. W. Roach.** 1990. Review of biotechnology applications to nuclear waste treatment. *J. Chem. Technol. Biotechnol.* **49:**381–394.
8. **Avery, S. A., and J. M. Tobin.** 1992. Mechanisms of strontium uptake by laboratory and brewing strains of *Saccharomyces cerevisiae*. *Appl. Environ. Microbiol.* **58:**3883–3889.
9. **Avery, S. V.** 1995. Microbial interactions with caesium—implications for biotechnology. *J. Chem. Technol. Biotechnol.* **62:**3–16.
10. **Avery, S. V., G. A. Codd, and G. M. Gadd.** 1991. Caesium accumulation and interactions with other monovalent cations in the cyanobacterium *Synechocystis* PCC 6803. *J. Gen. Microbiol.* **137:** 405–413.
11. **Avery, S. V., G. A. Codd, and G. M. Gadd.** 1992. Caesium transport in the cyanobacterium *Anabaena variabilis* kinetics and evidence for uptake via ammonium transport system(s). *FEMS Microbiol. Lett.* **95:**253–258.
12. **Avery, S. V., G. A. Codd, and G. M. Gadd.** 1993. Transport kinetics, cation inhibition and intracellular location of accumulated caesium in the green microalga *Chlorella salina*. *J. Gen. Microbiol.* **139:**827–834.
13. **Bailar, J. C., H. J. Emelium, R. Nyholm, and A. F. Trotman-Dickenson.** 1973. *The Actinides*, vol. 5. Pergamon Press, Oxford, United Kingdom.
14. **Barnes, L. J., F. J. Janssen, J. Sherren, J. H. Versteegh, R. O. Koch, and P. J. H. Scheeren.** 1991. A new process for the microbial removal of sulphate and heavy metal from contaminated waters extracted by a geohydrological control system. *Chem. Eng. Res. Des.* **69A:**184–186.
15. **Barnhart, B. J., E. W. Campbell, E. Martinez, D. E. Caldwell, and R. Hallett.** 1980. *Potential Microbial Impact on Transuranic Wastes under Conditions Expected in the Waste Isolation Pilot Plant* (*WIPP*). Document LA-8297-PR. Los Alamos National Laboratory, Los Alamos, N.Mex.
16. **Basnakova, G., and L. E. Macaskie.** 1999. Accumulation of zirconium and nickel by *Citrobacter* sp. *J. Chem. Technol. Biotechnol.* **74:**509–514.
17. **Basnakova, G., and L. E. Macaskie.** 1998. Microbially-enhanced chemisorption of heavy metals: a method for the bioremediation of solutions containing long-lived isotopes of neptunium and plutonium. *Environ. Sci. Technol.* **32:**184–187.
18. **Basnakova, G., and L. E. Macaskie.** 1997. Microbially-enhanced chemisorption of nickel into biologically-synthesized hydrogen uranyl phosphatea novel system for the removal and recovery of metals from aqueous solution. *Biotechnol. Bioeng.* **54:**319–328.
19. **Basnakova, G., A. J. Spencer, E. Palsgard, G. W. Grime, and L. E. Macaskie.** 1998. Identification of the nickel uranyl phosphate deposit on *Citrobacter* sp. cells by electron microscopy with electron probe X-ray microanalysis (EPXMA) and by proton induced X-ray emission analysis (PIXE). *Environ. Sci. Technol.* **32:**760–765.
20. **Basnakova, G., E. Stephens, M. C. Thaller, G. M. Rossolini, and L. E. Macaskie.** 1998. The use of *Escherichia coli* bearing a *phoN* gene for the removal of uranium and nickel from aqueous flows. *Appl. Microbiol. Biotechnol.* **50:**266–272.
21. **Battista, J. R.** 1997. Against all odds: the survival strategies of *Deinococcus radiodurans*. *Annu. Rev. Microbiol.* **51:**203–204.
22. **Belly, R. T., J. J. Lauff, and C. T. Goodhue.** 1975. Degradation of ethylenediaminetetraacetic acid by microbial populations from an aerated lagoon. *Appl. Microbiol.* **29:**787–794.
23. **Bengtsson, L., B. Johansson, T. J. Hackett, L. McHale, and A. P. McHale.** 1995. Studies on the biosorption of uranium by *Talaromyces emersonii* CBS 814.70 biomass. *Appl. Microbiol. Biotechnol.* **42:**807–811.

24. **Beveridge, T. J., M. N. Hughes, H. Lee, K. T. Leung, R. K. Poole, I. Savvaidis, S. Silver, and J. T. Trevors.** 1997. Metal-microbe interactions: contemporary approaches. *Adv. Microb. Physiol.* **38:**177–243.
25. **Binks, P. R.** 1996. Radioresistant bacteria: have they got industrial uses? *J. Chem. Technol. Biotechnol.* **67:**319–322.
26. **Blount, J. G.** 1998. Physicochemical and biogeochemical stabilization of uranium in a low level radioactive waste disposal cell. *Environ. Eng. Geosci.* **4:**491–502.
27. **Boegley, W. J. J., and H. J. Alexander.** 1986. Radioactive wastes. *J. Water Pollut. Control Fed.* **58:**594–600.
28. **Bolton, H. J., S. W. Li, D. J. Workman, and D. C. Girvin.** 1993. Biodegradation of synthetic chelates in subsurface sediments from the southeastern coastal plain. *J. Environ. Qual.* **22:**125–132.
29. **Bonthrone, K. M., G. Basnakova, F. Lin, and L. E. Macaskie.** 1996. Bioaccumulation of nickel by intercalation into polycrystalline hydrogen uranyl phosphate deposited via an enzymatic mechanism. *Nat. Biotechnol.* **14:**635–638.
30. **Borst-Pauwels, G. W. F. H.** 1981. Ion transport in yeast. *Biochim. Biophys. Acta* **650:**88–127.
31. **Bosecker, K.** 1997. Bioleaching: metal solubilisation by microorganisms. *FEMS Microbiol. Rev.* **20:** 591–604.
32. **Bossemeyer, D., A. Schlosser, and E. Bakker.** 1989. Specific cesium transport via the *Escherichia coli* Kup (TrkD) K^+ uptake system. *J. Bacteriol.* **171:**2219–2221.
33. **Brady, D., A. Stoll, and J. R. Buncan.** 1994. Biosorption of heavy metal cations by non-viable yeast biomass. *Environ. Technol.* **15:**429–439.
34. **Brierley, C. L., and J. Brierley.** 1981. *Biological Processes for Concentrating Trace Elements from Uranium Mine Wastes.* Technical Completion Report 140. New Mexico Water Resources Research Institute, Las Cruces.
35. **Brierley, I. A., G. M. Goyak, and C. L. Brierley.** 1986. Considerations for commercial use of natural products for metal recovery, p. 105–120. *In* H. Eccles and S. Hunt (ed.), *Immobilisation of Ions by BioSorption.* Ellis Horwood, Chichester, United Kingdom.
36. **Bryers, J. D., and S. Sanin.** 1994. Resuscitation of starved ultramicrobacteria to improve in-situ bioremediation. *Ann. N.Y. Acad. Sci.* **745:**61–76.
37. **Brynhildsen, L., and B. Allard.** 1994. Influence of metal complexation on the metabolism of citrate by *Klebsiella oxytoca. Biometals* **7:**163–169.
38. **Brynhildsen, L., and T. Rosswall.** 1989. Effects of copper, magnesium and zinc on the decomposition of citrate by a *Klebsiella* sp. *Appl. Environ. Microbiol.* **55:**1375–1379.
39. **Burnett, W. C., J. B. Cowart, and P. A. Chin.** 1987. Polonium in the superficial aquifer of West Central Florida, p. 251–269. *In* B. Graves (ed.), *Radon Radium and Other Radioactivity in Groundwater: Hydrogeologic Impact and Application to Indoor Airborne Contamination.* Lewis Publishers, Boca Raton, Fla.
40. **Bustard, M., A. Donnellan, A. Rollan, L. McHale, and A. P. McHale.** 1996. The effect of pulsed field strength on electric field stimulated biosorption of uranium by *Kluyveromyces marxianus* IMB3. *Biotechnol. Lett.* **18:**479–482.
41. **Caccavo, F., J. D. Coates, R. A. Rossello-Mora, W. Ludwig, K. H. Schleifer, D. R. Lovley, and M. J. McInerney.** 1996. *Geobacter ferrireducens*, a phylogenetically distinct dissimilatory Fe(III)-reducing bacterium. *Arch. Microbiol.* **165:**370–376.
42. **Caccavo, F., Jr., D. J. Lonergan, D. R. Lovley, M. Davis, J. F. Stolz, and M. J. McInerney.** 1994. *Geobacter sulfurreducens* sp. nov., a hydrogen- and acetate-oxidizing dissimilatory metal-reducing microorganism. *Appl. Environ. Microbiol.* **60:**3752–3759.
43. **Cataldo, D. A., T. R. Garland, R. E. Wildung, and R. J. Fellows.** 1989. Comparative metabolic behaviour and interrelationships of Tc and S in soybean plants. *Health Phys.* **57:**281–288.
44. **Cherrier, J., W. C. Burnett, and P. A. LaRock.** 1995. Uptake of polonium and sulfur by bacteria. *Geomicrobiol. J.* **13:**103–115.
45. **Clearfield, A.** 1988. Role of ion exchange in solid-state chemistry. *Chem. Rev.* **88:**125–148.
46. **Cleveland, J. M., and T. F. Rees.** 1981. Characterisation of plutonium in Maxey Flats radioactive trench leachates. *Science* **212:**1506–1509.
47. **Crameri, A., G. Dawes, E. Rodriguez, S. Silver, and W. P. C. Stemmer.** 1997. Molecular evolution of an arsenate detoxification pathway by DNA shuffling. *Nat. Biotechnol.* **15:**436–438.

48. **Cripps, R. E., and A. S. Noble.** 1973. The metabolism of nitrilotriacetate by a pseudomonad. *Biochem. J.* **136:**1059–1068.
49. **Davis, W. J.** 1984. Radiolytic behavior, p. 221–265. *In* W. W. Schulz, J. D. Navratil, and A. E. Talbot (ed.), *Science and Technology of Tributyl Phosphate.* CRC Press Inc, Boca Raton, Fla.
50. **Dehut, J. P., K. Fosny, C. Myttenaere, D. Deprins, and C. M. Vandecasteele.** 1989. Bioavailability of Tc incorporated in plant material. *Health Phys.* **57:**263–267.
51. **Delegard, C. H.** 1987. Solubility of PuO_2 in alkaline high-level waste solution. *Radiochim. Acta* **41:**11–21.
52. **Derks, W. J. G., and G. W. F. H. Borst-Pauwels.** 1979. Apparent three-site kinetics of Cs^+-uptake by yeast. *Physiol. Plant.* **46:**241–246.
53. **deRome, L., and G. M. Gadd.** 1991. Use of pelleted and immobilized yeast and fungal biomass for heavy metal and radionuclide recovery. *J. Ind. Microbiol.* **7:**97–104.
54. **Dhami, P. S., V. Gopalakrishnan, R. Kannan, A. Ramanujam, N. Salvi, and S. I. Udupa.** 1998. Biosorption of radionuclides by *Rhizopus arrhizus. Biotechnol. Lett.* **20:**225–228.
55. **Diels, L., Q. Dong, D. van der Lelie, W. Baeyens, and M. Mergeay.** 1995. The *czc* operon of *Alcaligenes eutrophus* CH34: from resistance mechanism to the removal of heavy metals. *J. Ind. Microbiol.* **14:**142–153.
56. **Dodge, C. J., and A. J. Francis.** 1994. Photodegradation of uranium citrate complex with uranium recovery. *Environ. Sci. Technol.* **28:**1300–1306.
57. **Dorhout, P. K., R. J. Kissane, K. D. Abney, L. R. Avens, P. G. Eller, and A. B. Ellis.** 1989. Intercalation reactions of the neptunyl (VI) dication with hydrogen uranyl phosphate and hydrogen neptunyl host lattices. *Inorg. Chem.* **28:**2926–2930.
58. **Dwivedy, K. K., and A. K. Mathur.** 1995. Bioleaching—our experience. *Hydrometallurgy* **38:**99–109.
59. **Eccles, H.** 1999. Nuclear waste managementa bioremediation approach, p. 187–208. *In* G. R. Choppin and M. K. Khankhasayev (ed.), *Chemical Separation Technologies and Related Methods of Nuclear Waste Management.* Kluwer Academic Publishers, Dordrecht, The Netherlands.
60. **Eccles, H.** 1995. Removal of heavy metals from effluent streams—why select a biological process? *Intl. Biodeterior. Biodegrad.* **35:**5–16.
61. **Ehrlich, H. L.** 1996. *Geomicrobiology*, 3rd ed. Marcel Dekker, Inc, New York, N.Y.
62. **Ellwood, D. C., M. J. Hill, and J. H. P. Watson.** 1992. Pollution control using microorganisms and metal separation, p. 89–112. *In* J. C. Fry, G. M. Gadd, R. A. Herbert, C. W. Jones, and I. A. Watson-Craik (ed.), *46th Symposium of the Society for General Microbiology. Microbial Control of Pollution.* Cambridge University Press, Cambridge, United Kingdom.
63. **Entry, J. A., P. T. Rygiewicz, and W. H. Emmingham.** 1994. Sr-90 uptake by *Pinus ponderosa* and *Pinus radiata* seedlings inoculated with ectomycorrhizal fungi. *Environ. Pollut.* **86:**201–206.
64. **Ferris, F. G., C. M. Fratton, J. P. Gertis, S. Schultzelam, and B. S. Lollar.** 1995. Microbial precipitation of a strontium calcite phase at a groundwater discharge zone near Rock Creek, British Columbia, Canada. *Geomicrobiol. J.* **13:**57–67.
65. **Fisher, N. S., P. Bjerregaard, L. Huynh-Ngoc, and G. R. Harvey.** 1983. Interactions of marine plankton with transuranic elements. Influence of dissolved organic compounds on americium and plutonium accumulation in diatoms. *Mar. Chem.* **13:**45–56.
66. **Fortin, D., D. Davis, and T. J. Beveridge.** 1996. The role of *Thiobacillus* and sulfate-reducing bacteria in iron biocycling in oxic and acidic mine tailings. *FEMS Microbiol. Ecol.* **21:**11–24.
67. **Francis, A. J.** 1994. Microbial transformations of radioactive wastes and environmental restoration through bioremediation. *J. Alloys Compounds* **213/214:**226–231.
68. **Francis, A. J., and C. J. Dodge.** 1998. Remediation of soils and wastes contaminated with uranium and toxic metals. *Environ. Sci. Technol.* **32:**3993–3998.
69. **Francis, A. J., and C. J. Dodge.** March 1994. U.S. patent 5,292,456.
70. **Francis, A. J., C. J. Dodge, J. B. Gillow, and J. E. Cline.** 1991. Microbial transformations of uranium in wastes. *Radiochim. Acta* **52/53:**311–316.
71. **Francis, A. J., C. J. Dodge, F. Lu, G. P. Halada, and C. R. Clayton.** 1994. XPS and XANES studies of uranium reduction by *Clostridium* sp. *Environ. Sci. Technol.* **28:**636–639.
72. **Frausto da Silva, J. J. R., and R. J. P. Williams.** 1993. *The Biological Chemistry of the Elements.* Clarendon Press, Oxford, United Kingdom.

73. **Friedman, B. A., and P. R. Dugan.** 1968. Concentration and accumulation of metallic ions by the bacterium *Zoogloea*. *Dev. Ind. Microbiol.* **9:**381–395.
74. **Friis, N., and P. Myers-Keith.** 1986. Biosorption of uranium and lead by *Streptomyces longwoodensis*. *Biotechnol. Bioeng.* **28:**21–28.
75. **Gadd, G. M.** 1996. Influence of microorganisms on the environmental fate of radionuclides. *Endeavour* **20:**150–156.
76. **Gadd, G. M.** 1997. Roles of microorganisms in the environmental fate of radionuclides. *CIBA Found. Symp.* **203:**94–104.
77. **Gadd, G. M., and C. White.** 1989. Heavy metal and radionuclide accumulation and toxicity in fungi and yeasts, p. 19–38. *In* R. K. Poole and G. M. Gadd (ed.), *Metal-Microbe Interactions*. IRL Press, Oxford, United Kingdom.
78. **Gadd, G. M., and C. White.** 1989. Removal of thorium from simulated acid process streams by fungal biomass. *Biotechnol. Bioeng.* **33:**592–597.
79. **Gale, G. R., and H. H. McLain.** 1963. Effect of ethambutol on cytology of *Mycobacterium smegmatis*. *J. Bacteriol.* **86:**749–756.
80. **Garnham, G. W., G. A. Codd, and G. M. Gadd.** 1993. Accumulation of zirconium by microalgae and cyanobacteria. *Appl. Microbiol. Biotechnol.* **39:**666–672.
81. **Garnham, G. W., G. A. Codd, and G. M. Gadd.** 1992. Uptake of technetium by fresh water green microalgae. *Appl. Microbiol. Biotechnol.* **37:**679–684.
82. **Ghosh, S., A. Sharma, and G. Talukder.** 1992. Zirconium, an abnormal trace element in biology. *Biol. Trace Element Res.* **35:**247–271.
83. **Gibson, J. F., R. K. Poole, M. N. Hughes, and J. R. Rees.** 1986. Ruthenium nitrosyl complexes—toxicity to *Escherichia coli* and yeasts, and uptake by marine bacteria. *Arch. Environ. Contam. Toxicol.* **15:**519–523.
84. **Giesy, J. P. J., and D. Paine.** 1977. Uptake of americium-241 by algae and bacteria. *Prog. Water Technol.* **9:**845–857.
85. **Gorby, Y. A., F. Caccavo, and H. Bolton.** 1998. Microbial reduction of cobalt^{III} EDTA$^-$ in the presence and absence of manganese(IV) oxide. *Environ. Sci. Technol.* **32:**244–250.
86. **Gorby, Y. A., and D. R. Lovley.** 1992. Enzymatic uranium precipitation. *Environ. Sci. Technol.* **26:** 205–207.
87. **Gray, K. R., and A. J. Biddlestone.** 1995. Engineered reed-bed systems for waste-water treatment. *Trends Biotechnol.* **13:**248–252.
88. **Gray, N. F.** 1992. *Biology of Wastewater Treatment*. Oxford University Press, Oxford, United Kingdom.
89. **Greene, B., M. T. Henzl, J. M. Hosea, and D. W. Darnall.** 1986. Elimination of bicarbonate interference in the binding of U(VI) in mill-waters to freeze-dried *Chlorella vulgaris*. *Biotechnol. Bioeng.* **28:**764–772.
90. **Guibal, E., C. Roulph, and P. Le Cloirec.** 1995. Infrared spectroscopic study of uranyl biosorption by fungal biomass and materials of biological origin. *Environ. Sci. Technol.* **29:**2496–2503.
91. **Guibal, E., C. Roulph, and P. Le Cloirec.** 1992. Uranium biosorption by a filamentous fungus *Mucor miehei*: pH effect on mechanisms and performances of uptake. *Water Res.* **26:**1139–1145.
92. **Hafez, N., A. S. Abdel-Razek, and H. M. B.** 1997. Accumulation of some heavy metals on *Aspergillus flavus*. *J. Chem. Technol. Biotechnol.* **68:**19–22.
93. **Hard, B. C., S. Friedrich, and W. Babel.** 1997. Bioremediation of acid mine water using facultatively methylotrophic metal-tolerant sulfate-reducing bacteria. *Microbiol. Res.* **152:**65–73.
94. **Harvey, R. S., and R. Patrick.** 1967. Concentration of ^{137}Cs, ^{65}Zn and ^{85}Sr by freshwater algae. *Biotechnol. Bioeng.* **9:**449–456.
95. **Henrot, J.** 1989. Bioaccumulation and chemical modification of Tc by soil bacteria. *Health Phys.* **57:**239–245.
96. **Higham, D. P., P. J. Sadler, and M. D. Scawen.** 1984. Cadmium-resistant *Pseudomonas putida* synthesizes novel cadmium binding proteins. *Science* **225:**1043–1046.
97. **Hu, M. Z.-C., and M. Reeves.** 1997. Biosorption of uranium by *Pseudomonas aeruginosa* strain CSU immobilized in a novel matrix. *Biotechnol. Prog.* **13:**60–70.
98. **Hughes, M. N., and R. K. Poole.** 1989. *Metals and Micro-Organisms*. Chapman & Hall, London, United Kingdom.

99. **Hunsberger, L. R., and A. B. Ellis.** 1990. Excited-state properties of lamellar solids derived from metal complexes and hydrogen uranyl phosphate. *Coord. Chem. Rev.* **97:**209–224.
100. **Jeffers, T. H., P. G. Bennett, and R. R. Corwin.** 1994. *Biosorption of Metal Contaminents Using Immobilized Biomass-Field Studies.* Document R1 9461. U.S. Bureau of Mines, Salt Lake City, Utah.
101. **Jeong, B. C., C. Hawes, K. M. Bonthrone, and L. E. Macaskie.** 1997. Localization of enzymically enhanced heavy metal accumulation by *Citrobacter* sp. and metal acumulation *in vitro* by liposomes containing entrapped enzyme. *Microbiology* **143:**2497–2507.
102. **Jeong, B. C., P. S. Poole, A. J. Willis, and L. E. Macaskie.** 1998. Purification and characterization of acid-type phosphatases from a heavy metal-accumulating *Citrobacter* sp. *Arch. Microbiol.* **169:** 166–173.
103. **Johnson, D. B.** 1995. Acidophilic microbial communities: candidates for bioremediation of acidic mine effluents. *Int. Biodeterior. Biodegrad.* **35:**41–58.
104. **Johnson, E. E., A. G. O'Donnell, and P. Ineson.** 1991. An autoradiographic technique for selecting Cs-137-sorbing microorganisms from soil. *J. Microbiol. Methods* **13:**293–298.
105. **Joshitope, G., and A. J. Francis.** 1995. Mechanisms of biodegradation of metal-citrate complexes by *Pseudomonas fluorescens. J. Bacteriol.* **177:**1989–1993.
106. **Kapoor, A., and T. Viraraghavan.** 1995. Fungal biosorption—an alternative treatment option for heavy metal bearing wastewatera review. *Bioresource Technol.* **53:**185–206.
107. **Karavaiko, G. I., A. S. Kareva, Z. A. Avakian, V. I. Zakharova, and A. A. Korenevsky.** 1996. Biosorption of scandium and yttrium from solutions. *Biotechnol. Lett.* **18:**1291–1296.
108. **Katz, J. J., G. T. Seaborg, and L. R. Morss.** 1986. *Chemistry of the Actinide Elements.* Chapman & Hall, London, United Kingdom.
109. **Khalid, A. M., S. R. Ashfaq, T. M. Bhatti, and M. A. Anwar.** 1993. The uptake of microbially leached uranium by microbial biomass, p. 299–300. *In* A. E. Torma, M. L. Apel, and C. L. Brierley (ed.), *Biohydrometallurgical Technologies,* vol. 2. The Minerals, Metals and Materials Society, Warrendale, Pa.
110. **Kirby, H. W.** 1986. *The Chemistry of the Actinide Elements.* Chapman & Hall, London, United Kingdom.
111. **Kotegov, K. V., O. N. Pavlov, and V. P. Shvendov.** 1968. Technetium, p. 1–90. *In* H. J. Emelius and A. G. Sharpe (ed.), *Advances in Inorganic Chemistry and Radiochemistry.* Academic Press, Inc., New York, N.Y.
112. **Kuyucak, N., and B. Volesky.** 1989. Accumulation of cobalt by marine algae. *Biotechnol. Bioeng.* **33:**809–814.
113. **Landa, E. R., E. J. P. Phillips, and D. R. Lovley.** 1991. Release of ^{226}Ra from uranium mill tailings by microbial Fe(III) reduction. *Appl. Geochem.* **6:**647–652.
114. **Lange, C. C., L. P. Wackett, K. W. Minton, and M. J. Daly.** 1998. Engineering a recombinant *Deinococcus radiodurans* for organopollutant degradation in radioactive mixed waste environments. *Nat. Biotechnol.* **16:**929–933.
115. **LaRock, P., J.-H. Hyun, S. Boutelle, W. C. Burnett, and C. D. Hull.** 1996. Bacterial mobilization of polonium. *Geochim. Cosmochim. Acta* **60:**4321–328.
116. **Lauff, J. J., D. B. Steel, L. A. Coogan, and J. M. Breitfeller.** 1990. Degradation of the ferric chelate of EDTA by a pure culture of an *Agrobacterium* sp. *Appl. Environ. Microbiol.* **56:**3346–3353.
117. **Lear, D. W., and C. H. Oppenheimer.** 19. Biological removal of radioisotopes ^{90}Sr and ^{90}Y from seawater by marine microrganisms. *Limnol. Oceeanogr.* **7**(Suppl.)**:**44–62.
118. **Lieser, K. H., and A. Muhlenweg.** 1988. Np in the hydrosphere and in the geosphere. *Radiochim. Acta* **43:**27–35.
119. **Lloyd, J. R., J. A. Cole, and L. E. Macaskie.** 1997. Reduction and removal of heptavalent technetium from solution by *Escherichia coli. J. Bacteriol.* **179:**2014–2021.
120. **Lloyd, J. R., C. L. Harding, and L. E. Macaskie.** 1997. Tc(VII) reduction and precipitation by immobilized cells of *Escherichia coli. Biotechnol. Bioeng.* **55:**505–510.
121. **Lloyd, J. R., and L. E. Macaskie.** 1997. Microbially-mediated reduction and removal of technetium from solution. *Res. Microbiol.* **148:**530–532.

122. **Lloyd, J. R., and L. E. Macaskie.** 1996. A novel phosphorImager based technique for monitoring the microbial reduction of technetium. *Appl. Environ. Microbiol.* **62:**578–582.
123. **Lloyd, J. R., H.-F. Nolting, V. A. Solé, K. Bosecker, and L. E. Macaskie.** 1998. Technetium reduction and precipitation by sulphate-reducing bacteria. *Geomicrobiol. J.* **15:**43–56.
124. **Lloyd, J. R., J. Ridley, T. Khizniak, N. N. Lyalikova, and L. E. Macaskie.** 1999. Reduction of technetium by *Desulfovibrio desulfuricans*: biocatalyst characterisation and use in a flowthrough bioreactor. *Appl. Environ. Microbiol.* **65:**2691–2696.
125. **Lloyd, J. R., G. H. Thomas, J. A. Finlay, J. A. Cole, and L. E. Macaskie.** 1999. Microbial reduction of technetium by *Escherichia coli* and *Desulfovibrio desulfuricans*: enhancement via the use of high activity strains and effect of process parameters. *Biotechnol. Bioeng.* **66:**122–130.
125a. **Lloyd, J. R., P. Yong, and L. E. Macaskie.** Biological reduction and removal of Np(V) by two microorganisms. *Environ. Sci. Technol.*, in press.
126. **Loewenschuss, H.** 1982. Metal-ferrocyanide complexes for the decontamination of cesium from aqueous radioactive waste. *Radioact. Waste Manage.* **2:**327–324.
127. **Lonergan, D. J., H. Jenter, J. D. Coates, E. J. P. Phillips, T. Schmidt, and D. R. Lovley.** 1996. Phylogenetic analysis of dissimilatory Fe(III)-reducing bacteria. *J. Bacteriol.* **178:**2402–2408.
128. **Lovley, D., and E. J. Phillips.** 1992. Reduction of uranium by *Desulfovibrio desulfuricans. Appl. Environ. Microbiol.* **58:**850–856.
129. **Lovley, D. R.** 1995. Bioremediation of organic and metal contaminants with dissimilatory metal reduction. *J. Ind. Microbiol.* **14:**85–93.
130. **Lovley, D. R.** 1993. Dissimilatory metal reduction. *Annu. Rev. Microbiol.* **47:**263–290.
131. **Lovley, D. R., and J. D. Coates.** 1997. Bioremediation of metal contamination. *Curr. Opin. Biotechnol.* **8:**285–289.
132. **Lovley, D. R., S. J. Giovannoni, D. C. White, J. E. Champine, E. J. P. Phillips, Y. A. Gorby, and S. Goodwin.** 1993. *Geobacter metallireducens* gen. nov. sp. nov., a microorganism capable of coupling the complete oxidation of organic compounds to the reduction of iron and other metals. *Arch. Microbiol.* **159:**336–344.
133. **Lovley, D. R., and E. J. P. Phillips.** 1992. Bioremediation of uranium contamination with enzymatic uranium reduction. *Environ. Sci. Technol.* **26:**2228–2234.
134. **Lovley, D. R., E. J. P. Phillips, Y. A. Gorby, and E. Landa.** 1991. Microbial reduction of uranium. *Nature* **350:**413–416.
135. **Lovley, D. R., E. E. Roden, E. J. P. Phillips, and J. C. Woodward.** 1993. Enzymatic iron and uranium reduction by sulfate reducing bacteria. *Mar. Geol.* **113:**41–53.
136. **Lovley, D. R., P. K. Widman, J. C. Woodward, and E. J. P. Phillips.** 1993. Reduction of uranium by cytochrome c_3 of *Desulfovibrio vulgaris. Appl. Environ. Microbiol.* **59:**3572–3576.
137. **Lyalikova, N. N., and T. V. Khizhnyak.** 1996. Reduction of heptavalent technetium by acidophilic bacteria of the genus *Thiobacillus. Microbiol.* **65:**468–473.
138. **Macaskie, L. E.** 1991. The application of biotechnology to the treatment of wastes produced from nuclear fuel cycle biodegradation and bioaccumulation as a means of treating radionuclide-containing streams. *Crit. Rev. Biotechnol.* **11:**41–112.
139. **Macaskie, L. E.** 1997. Bioaccumulation of heavy metals and application to the remediation of acid mine drainage water containing uranium. *Res. Microbiol.* **148:**528–530.
140. **Macaskie, L. E.** 1990. An immobilized cell bioprocess for the removal of heavy metals from aqueous flows. *J. Chem. Technol. Biotechnol.* **49:**357–379.
141. **Macaskie, L. E., and K. M. Bonthrone.** 1996. *Modelling of Genetic, Biochemical, Cellular and Microenvironmental Parameters Determining Bacterial Sorption and Mineralization Processes for Recuperation of Heavy or Precious Metals.* Final report on EU contract BE 5350.
142. **Macaskie, L. E., K. M. Bonthrone, P. Yong, and D. Goddard.** Enzymatically-mediated bioprecipitation of uranium by a *Citrobacter* sp.: a concerted role for extracellular lipopolysaccharide and associated phosphatase in biomineral formation. *Microbiology*, in press.
143. **Macaskie, L. E., and A. C. R. Dean.** 1985. Strontium accumulation by immobilized cells of a *Citrobacter* sp. *Biotechnol. Lett.* **7:**627–630.
144. **Macaskie, L. E., R. M. Empson, A. K. Cheetham, C. P. Grey, and A. J. Skarnulis.** 1992. Uranium bioaccumulation by a *Citrobacter* sp. as a result of enzymically-mediated growth of polycrystalline HUO_2 PO_4. *Science* **257:**782–784.

145. **Macaskie, L. E., R. M. Empson, F. Lin, and M. R. Tolley.** 1995. Enzymatically-mediated uranium accumulation and uranum recovery using a *Citrobacter* sp. immobilised as a biofilm within a plug-flow reactor. *J. Chem. Technol. Biotechnol.* **63:**1–16.
146. **Macaskie, L. E., B. C. Jeong, and M. R. Tolley.** 1994. Enzymically-accelerated biomineralization of heavy metalsapplication to the removal of americium and plutonium from aqueous flows. *FEMS Microbiol. Rev.* **14:**351–368.
147. **Macaskie, L. E., J. R. Lloyd, R. A. P. Thomas, and M. R. Tolley.** 1996. The use of microoorganisms for the remediation of solutions contaminated with actinide elements, other radionuclides and organic contaminants generated by nuclear fuel cycle activities. *Nuclear Energy* **35:**257–271.
148. **Macaskie, L. E., P. Yong, T. C. Doyle, M. G. Roig, M. Diaz, and T. Manzano.** 1997. Bioremediation of uranium-bearing wastewaterbiochemical and chemical factors influencing bioprocess application. *Biotechnol. Bioeng.* **53:**100–109.
149. **MacKenzie, A. B., and R. D. Scott.** 1993. Sellafield waste radionuclides in Irish Sea intertidal and salt marsh sediments. *Environ. Geochem. Health* **15:**173–178.
150. **Marques, A. M., R. Bonet, M. D. Simon-Oujol, M. C. Fuste, and F. Congregado.** 1990. Removal of uranium by an exopolysaccharide from *Pseudomonas* sp. *Appl. Microbiol. Biotechnol.* **34:**429–431.
151. **Marques, A. M., X. Roca, M. D. Simon-Pujol, M. C. Fuste, and C. Francisco.** 1991. Uranium acumulation by *Pseudomonas* sp. EPS-5028. *Appl. Microbiol. Biotechnol.* **35:**406–410.
152. **McCready, R. G. L., and H. R. Krouse.** 1980. Sulfur isotope fractionation by *Desulfovibrio vulgaris* during metabolism of $BaSO_4$. *Geomicrobiol. J.* **2:**55–62.
153. **McCready, R. G. L., and V. I. Lakshmanan.** 1986. Review of bioadsorption research to recover uranium from leach solutions in Canada, p. 219–226. *In* H. Eccles and S. Hunt (ed.), *Immobilization of Ions by Bio-Sorption.* Ellis Horwood, Chichester, United Kingdom.
154. **McCullough, J., T. C. Hazen, S. M. Benson, F. B. Metting, and A. C. Palmisano.** 1999. *Bioremediation of Metals and Radionuclides ... What Is It and How It Works.* Lawrence Berkeley National Laboratory, Berkeley, Calif.
155. **McDonald, P., G. T. Cook, M. S. Baxter, and J. C. Thomson.** 1990. Radionuclide transfer from Sellafield to South West Scotland. *J. Environ. Radioact.* **12:**285–298.
156. **McHale, A. P., and S. McHale.** 1994. Microbial biosorption of metals: potential in the treatment of metal pollution. *Biotechnol. Adv.* **12:**647–652.
157. **Means, J. L., D. A. Crerar, and J. O. Duguid.** 1978. Migration of radioactive wastes: radionuclide mobilisation by complexing agents. *Science* **200:**1477–1481.
158. **Mellor, R. B., J. Ronnenberg, W. H. Campbell, and S. Diekmann.** 1992. Reduction of nitrate and nitrite in water by immobilized enzymes. *Nature* **355:**717–719.
159. **Mohegheghi, A., D. M. Updegraff, and M. B. Goldhaber.** 1994. The role of sulfate-reducing bacteria in the deposition of sedimentary uranium ores. *Geomicrobiol. J.* **4:**153–173.
160. **Mudder, T. I., and J. L. Whitlock.** 1984. Biological treatment of cyanidation wastewaters. *Miner. Metall. Process. SME-AIME Trans.* **276:**161–165.
161. **Myers, C. R., and J. M. Myers.** 1992. Localization of cytochromes to the outer membrane of anaerobically grown *Shewanella putrefaciens* MR-1. *J. Bacteriol.* **174:**3429–3438.
162. **Nakajima, A., T. Horikoshi, and T. Sakaguchi.** 1982. Recovery of uranium by immobilized microorganisms. *Eur. J Appl. Microbiol. Biotechnol.* **16:**88–91.
163. **Nealson, K. H., and C. R. Myers.** 1992. Microbial reduction of manganese and iron new approaches to carbon cycling. *Appl. Environ. Microbiol.* **58:**439–443.
164. **Nelson, D. M., and M. B. Lovett.** 1978. Oxidation state of plutonium in the Irish Sea. *Nature* **276:**599–601.
165. **Nemec, P., H. Prochazka, K. Stamberg, J. Katzer, J. Stamberg, R. Jilek, and P. Hulak.** May 1977. U.S. patent 4,021,368.
166. **Nero, A. V.** 1979. *A Guidebook to Nuclear Reactors.* University of California Press, Berkeley.
167. **Niki, T., T. Yagi, I. Inokuchi, and K. Kimura.** 1977. Electrode reaction of cytochrome c_3 of *Desulfovibrio vulgaris* Miyazaki. *J. Electrochem. Soc.* **124:**1889–1892.
168. **Norberg, A. B., and H. Persson.** 1984. Accumulation of heavy metal ions by *Zoogloea ramigera. Biotechnol. Bioeng.* **26:**239–246.

169. **Nortemann, B.** 1992. Total degradation of EDTA by mixed cultures and a bacterial isolate. *Appl. Environ. Microbiol.* **58:**671–676.
170. **O'Boyle, N. C., G. P. Nicholson, T. J. Piper, D. M. Taylor, D. R. Williams, and G. Williams.** 1997. A review of plutonium (IV) selective ligands. *Appl. Radiat. Isot.* **48:**183–200.
171. **Okorov, L. A., L. P. Lichko, V. M. Kodomtseva, V. P. Kholodenko, V. T. Titovsky, and I. S. Kulaev.** 1977. Energy-dependent transport of manganese into yeast cells and distribution of accumulated ions. *Eur. J. Biochem.* **75:**373–377.
172. **Omar, N. B., M. L. Merroun, M. T. Gonzalez-Munoz, and J. M. Arias.** 1996. Brewery yeast as a biosorbent for uranium. *J. Appl. Bacteriol.* **81:**283–287.
173. **Organbide, G., S. Philip-Holingsworth, E. Tola, R. A. Cedergren, A. Squartini, F. B. Dazzo, and R. P. Hollingsworth.** 1996. Glycoconjugate and lipid components of *Rhizobium hedysara* IS123. *Can. J. Microbiol.* **42:**340–345.
174. **Paccard, E., and B. Besnanou.** 1995. French patent 9509563.
175. **Palumbo, A. V., S. Y. Lee, and P. Boerman.** 1994. Effect of media composition of EDTA degradation by *Agrobacterium* sp. *Appl. Biochem. Biotechnol.* **45:**811–822.
176. **Peretrukhin, V. F., N. N. Khizhniak, N. N. Lyalikova, and K. E. German.** 1996. Biosorption of technetium-99 and some actinides by bottom sediments of Lake Belsso Kosino of the Moscow region. *Radiochem.* **38:**440–443.
177. **Perkins, J., and G. M. Gadd.** 1993. Caesium toxicity, accumulation and intracellular location in yeasts. *Mycol. Res.* **97:**712–724.
178. **Pham-Thi, M., and P. Columban.** 1985. Cationic conductivity, water species motions and phase transitions in $H_3OUO_2PO_4 \cdot 3H_2O$ (HUP) and MUP related compounds ($M^+ = Na^+, K^+, Ag^+, Li^+, NH_4^+$). *Solid State Ion* **17:**295–306.
179. **Phillips, E. J. P., E. R. Landa, and D. R. Lovley.** 1995. Remediation of uranium contaminated soils with bicarbonate extraction and microbial U(VI) reduction. *J. Ind. Microbiol.* **14:**203–207.
180. **Pignolet, L., K. Fonsny, F. Capot, and Z. Moureau.** 1989. Role of various microorganisms on Tc behaviour in sediments. *Health Phys.* **57:**791–800.
181. **Pinar, G., E. Duque, A. Haidour, J. M. Oliva, L. Sanchez-Barcero, V. Calvo, and J. L. Ramos.** 1997. Removal of high concentrations of nitrate from industrial wastewater by bacteria. *Appl. Environ. Microbiol.* **63:**2071–2073.
182. **Pitt, W. W., C. W. Hancher, and B. D. Patton.** 1981. Biological reduction of nitrates in wastewater from nuclear fuel reprocessing using a fluidised bed reactor. *Nuclear Chem. Waste Manage.* **2:**57–70.
183. **Plato, P., and J. T. Denovan.** 1974. The influence of potassium on the removal of ^{137}Cs by live *Chlorella* from low level radioactive wastes. *Radiat. Bot.* **14:**37–41.
184. **Pons, M. P., and M. C. Fuste.** 1993. Uranium uptake by immbilized cells of *Pseudomonas* strain EPS 5028. *Appl. Microbiol. Biotechnol.* **39:**661–665.
185. **Postgate, J. R.** 1979. *The Sulphate Reducing Bacteria.* Cambridge University Press, Cambridge, United Kingdom.
186. **Pozas-Tormo, R., L. Moreno-Real, Martinez-Lara, and S. Bruque-Gamez.** 1987. Intercalation of lanthanides into $H_3OUO_2PO_4 \cdot 3H_2O$ and $C_4H_9NH_3UO_2PO_4 \cdot 3H_2O$. *Inorg. Chem.* **26:**1442–1445.
187. **Pozas-Tormo, R., L. Moreno-Real, M. Martinez-Lara, and S. Bruque-Gamez.** 1986. Layered metal uranyl phosphates. Retention of divalent ions by amine intercalates of uranyl phosphates. *Can. J. Chem.* **64:**30–34.
188. **Premuzic, E. T., A. J. Francis, M. Lin, and J. Schubert.** 1985. Induced formation of chelating agents by *Pseudomonas aeruginosa* grown in the presence of thorium and uranium. *Arch. Environ. Contam. Toxicol.* **14:**759–768.
189. **Premuzic, E. T., M. S. Lin, J.-Z. Jin, and K. Hamilton.** 1997. Geothermal waste treatment biotechnology. *Energy Sources* **19:**9–17.
190. **Pumpel, T.** 1997. Metal biosorption: a structured data space? *Res. Microbiol.* **148:**514–515.
191. **Reid, G. W., P. Lassovszky, and S. Hathaway.** 1985. Treatment, waste management and cost for removal of radioactivity from drinking water. *Health Phys.* **48:**671–694.
192. **Riordan, C., M. Bustard, R. Putt, and A. P. McHale.** 1997. Removal of uranium from solution using residual brewery yeast: combined biosorption and bioprecipitation. *Biotechnol. Lett.* **19:**385–387.

193. **Roig, M. G., J. F. Kennedy, and L. E. Macaskie.** 1995. *Biological Rehabilitation of Metal Bearing Wastewaters.* Final report. EC contract EV5V-CT93-0251. European Commission, Brussels, Belgium.
194. **Rosenberg, A., and M. Alexander.** 1979. Microbial cleavage of various organophosphorus insecticides. *Appl. Environ. Microbiol.* **37:**886–891.
195. **Ross, I. S., and C. C. Townsley.** 1986. The uptake of heavy metals by filamentous fungi, p. 49–58. *In* H. Eccles and S. Hunt (ed.), *Immobilisation of Ions by Bio-Sorption.* Ellis Horwood, Chichester, United Kingdom.
196. **Rusin, P. A., Q. L., J. R. Brainard, B. A. Strietelmeier, C. D. Tait, S. A. Ekberg, P. D. Palmer, T. W. Newton, and D. L. Clark.** 1994. Solubilization of plutonium hydrous oxide by iron reducing bacteria. *Environ. Sci. Technol.* **28:**1686–1690.
197. **Salt, D. E., M. Blaylock, N. P. B. A. Kumar, V. Dushenkov, B. D. Ensley, I. Chet, and I. Raskin.** 1995. Phytoremediation: a novel strategy for the removal of toxic metals from the environment using plants. *Bio/Technol.* **13:**468–474.
198. **Scott, C. D.** 1992. Removal of dissolved metals by plant tissue. *Biotechnol. Bioeng.* **39:**1064–1068.
199. **Shumate, S. E. III, and G. W. Strandberg.** 1985. Accumulation of metals by microbial cells, p. 235–249. *In* M. Moo-Young (ed.), *Comprehensive Biotechnology,* vol. 4. Pergamon Press, New York, N.Y.
200. **Simmons, P., J. M. Tobin, and I. Singleton.** 1995. Considerations on the use of commercially available yeast biomass for the treatment of metal-containing effluents. *J. Ind. Microbiol.* **14:**240–246.
201. **Singh, S., S. Negi, S. Barati, and H. N. Singh.** 1994. Common nitrogen control of caesium (Cs^+) uptake, caesium (Cs^+) toxicity and ammonium (methylammonium) uptake in the cyanobacterium *Nostoc muscorum. FEMS Microbiol. Lett.* **117:**243–247.
202. **Spear, J. R., L. A. Fugueroa, and B. D. Honeyman.** 1999. Modeling the removal of uranium U(VI) from aqueous solutions in the presence of sulfate reducing bacteria. *Environ. Sci. Technol.* **33:**2667–2675.
203. **Spooner, G. M.** 1949. Observation of the absorption of radioactive strontium and yttrium by marine algae. *J. Mar Biol. Assoc.* **28:**587–625.
204. **Stoner, D. L., and A. J. Tien.** September 1995. U.S. patent 5,453,375.
205. **Strandberg, G., and W. D. Arnold.** 1988. Microbial accumulation of neptunium. *J. Ind. Microbiol.* **3:**329–331.
206. **Strandberg, G. W., S. E. Shumate II, and J. R. Parrott.** 1981. Microbial cells as biosorbents for heavy metals: accumulation of uranium by *Saccharomyces cerevisiae* and *Pseudomonas aeruginosa. Appl. Environ. Microbiol.* **41:**237–245.
207. **Sun, H., X. R. Wang, L. S. Wang, L. M. Dai, Z. Li, and Y. J. Cheng.** 1997. Bioconcentration of rare earth elements lanthanum, gadolinium and yttrium in algae (*Chlorella vulgaris* Beijerinck): influence of chemical species. *Chemosphere* **34:**1753–1760.
208. **Swanson, J. L.** 1990. Purex process flowsheets, p. 55–79. *In* W. W. Schulz and J. D. Navratil (ed.), *Science and Technology of Tributyl Phosphate,* vol. 3. CRC Press Inc., Boca Raton, Fla.
209. **Taghavi, S., M. Mergeay, D. Nies, and D. Van der Lelie.** 1997. *Alcaligenes eutrophus* as a model system for bacterial interaction with heavy metals in the environment. *Res. Microbiol.* **148:**536–551.
210. **Tebo, B. M., and A. Y. Obraztsova.** 1998. Sulfate-reducing bacterium grows with Cr(VI), U(VI), Mn(IV), and Fe(III) as electron acceptors. *FEMS Microbiol. Lett.* **162:**193–198.
211. **Tengerdy, R. P., J. E. Johnson, J. Hollo, and J. Toth.** 1981. Denitrification and removal of heavy metals from waste water by immobilized microorganisms. *Appl. Biochem. Biotechnol.* **6:**3–13.
212. **Thomas, R. A. P., A. J. Beswick, G. Basnakova, R. Moller, and L. E. Macaskie.** Growth of naturally-occurring microbial isolates in metal-citrate medium and bioremediation of metal-citrate wastes. *J. Chem. Technol. Biotechnol.*, in press.
213. **Thomas, R. A. P., K. Lawlor, M. Bailey, and L. E. Macaskie.** 1998. The biodegradation of metal-EDTA complexes by an enriched microbial population. *Appl. Environ. Microbiol.* **64:**1319–1322.

214. **Thomas, R. A. P., and L. E. Macaskie.** 1996. Biodegradation of tributyl phosphate by naturally occuring microbial isolates and coupling to the removal of uranium from aqueous solution. *Environ. Sci. Technol.* **30:**2371–2375.
215. **Thomas, R. A. P., and L. E. Macaskie.** 1998. The effect of growth conditions on the biodegradation of tributyl phosphate and potential for the remediation of acid mine drainage waters by a naturally-occurring mixed microbial culture. *Appl. Microbiol. Biotechnol.* **49:**202–209.
216. **Thomas, R. A. P., A. P. Morby, and L. E. Macaskie.** 1997. The biodegradation of tributyl phosphate by naturally-occurring microbial isolates. *FEMS Microbiol. Lett.* **155:**155–159.
217. **Tiedje, J. M.** 1975. Microbial biodegradation of ethylenediaminetetraacetic acid in soils and sediments. *Appl. Microbiol.* **30:**327–329.
218. **Tobin, J. M., D. G. Cooper, and R. J. Neufeld.** 1984. Uptake of metal ions by *Rhizopus arrhizus* biomass. *Appl. Environ. Microbiol.* **47:**821–824.
219. **Tobin, J. M., C. White, and G. M. Gadd.** 1994. Metal accumulation by fungi: applications in environmental biotechnology. *J. Ind. Microbiol.* **13:**126–130.
220. **Tolley, M. R., and L. E. Macaskie.** 1993. Bioaccumulation of heavy metals: aplication to the decontamination of solutions containing americium, plutonium and neptunium, p. 89–96. *In* A. E. Torma, M. L. Apel, and C. L. Brierley (ed.), *Biohydrometallurgical Technologies.* The Minerals, Metals and Materials Society, Nepean, Ontario, Canada.
221. **Tolley, M. R., and L. E. Macaskie.** 1994. United Kingdom patent GB94/00626.
222. **Tolley, M. R., L. F. Strachan, and L. E. Macaskie.** 1995. Lanthanum accumulation from acidic solutions using *Citrobacter* sp. immobilized in a flow-through bioreactor. *J. Ind. Microbiol.* **14:** 271–280.
223. **Tomioka, N., H. Uchiyama, and O. Yagi.** 1992. Isolation and characterization of cesium-accumulating bacteria. *Appl. Environ. Microbiol.* **58:**1019–1023.
224. **Trabalka, J. R., and C. T. Garten, Jr.** 1983. Behaviour of the long-lived synthetic elements and their natural analogues in food chains. *Adv. Radiat. Biol.* **10:**68–73.
225. **Treen-Sears, M. E., B. Volesky, and R. J. Neufeld.** 1984. Ion exchange/complexation of the uranyl ion by *Rhizopus* biosorbent. *Biotechnol. Bioeng.* **26:**1323–1329.
226. **Trollope, D. R., and B. Evans.** 1976. Concentrations of copper, iron, lead, nickel and zinc in fresh water algal blooms. *Environ. Pollut.* **11:**109–116.
227. **Truex, M. J., B. M. Peyton, N. B. Valentine, and Y. A. Gorby.** 1997. Kinetics of U(VI) reduction by a dissimilatory Fe(III)-reducing bacterium under non-growth conditions. *Biotechnol. Bioeng.* **55:** 490–496.
228. **Tsezos, M.** 1983. The role of chitin in uranium adsorption by *Rhizopus arrhizus. Biotechnol. Bioeng.* **25:**2025–2040.
229. **Tsezos, M., M. H. I. Baird, and L. W. Shemilt.** 1987. The elution of radium adsorbed by microbial biomass. *Chem. Eng. J.* **34:**B57–B64.
230. **Tsezos, M., M. H. I. Baird, and L. W. Shemilt.** 1986. The kinetics of radium biosorption. *Chem. Eng. J.* **33:**B35–B41.
231. **Tsezos, M., M. H. I. Baird, and L. W. Shemilt.** 1987. The use of immobilised biomass to remove and recover radium from Elliot Lake Uranium Tailing Streams. *Hydrometallurgy* **17:**357–368.
232. **Tsezos, M., and D. M. Keller.** 1983. Adsorption of radium-226 by biological origin absorbents. *Biotechnol. Bioeng.* **25:**201–215.
233. **Tsezos, M., R. G. L. McCready, and J. P. Bell.** 1989. The continuous recovery of uranium from biologically leached solutions using immobilized biomass. *Biotechnol. Bioeng.* **34:**10–17.
234. **Tsezos, M., E. Remoudaki, and V. Angelatou.** 1996. A study of the effects of competing ions on the biosorption of metals. *Int. Biodeterior. Biodegrad.* **38:**19–29.
235. **Tsezos, M., E. Remoudaki, and V. Angelatou.** 1996. A systematic study on the equilibrium and kinetics of biosorptive accumulation. The case of Ag and Ni. *Int. Biodeterior. Biodegrad.* **35:**129–154.
236. **Tsezos, M., and B. Volesky.** 1981. Biosorption of uranium and thorium by *Rhizopus arrhizus. Biotechnol. Bioeng.* **23:**583–604.
237. **Tsezos, M., and B. Volesky.** 1982. The mechanism of thorium biosorption. *Biotechnol. Bioeng.* **24:**955–969.

238. **Tsezos, M., and B. Volesky.** 1982. The mechanism of uranium biosorption by *Rhizopus arrhizus*. *Biotechnol. Bioeng.* **24:**385–401.
239. **Tucker, M. D., L. L. Barton, and B. M. Thompson.** 1998. Removal of U and Mo from water by immobilized *Desulfovibrio desulfuricans* in column reactors. *Biotechnol. Bioeng.* **60 (1):**90–96.
240. **Tucker, M. D., L. L. Barton, and B. M. Thomson.** 1996. Kinetic coefficients for simultaneous reduction of sulfate and uranium by *Desulfovibrio desulfuricans. Appl. Microbiol. Biotechnol.* **46:** 74–77.
241. **Turner, J. S., and N. J. Robinson.** 1995. Cyanobacterial metallothioneins: Biochemistry and molecular genetics. *J. Ind. Microbiol.* **14:**119–125.
242. **Van Roy, S., K. Peys, T. Dresselaers, and L. Diels.** 1997. The use of an *Alcaligenes eutrophus* biofilm in a membrane bioreactor for heavy metal recovery. *Res. Microbiol.* **148:**526–528.
243. **Vesely, V., and V. Pekarek.** 1972. Synthetic inorganic ion exchangers. 1. Hydrous oxides and acidic salts of multivalent metals. *Talanta* **19:**219–262.
244. **Volesky, B.** 1994. Advances in biosorption of metals: selection of biomass types. *FEMS Microbiol. Rev.* **14:**291–302.
245. **Volesky, B.** 1990. *Biosorption of Heavy Metals.* CRC Press Inc., Boca Raton, Fla.
246. **Volesky, B., and Z. R. Holan.** 1995. Biosorption of heavy metals. *Biotechnol. Prog.* **11:**235–250.
247. **Volesky, B., and H. A. May-Phillips.** 1995. Biosorption of heavy metals by *Saccharomyces cerevisiae. Appl. Microbiol. Biotechnol.* **42:**797–806.
248. **Volesky, B., and M. Tsezos.** March 1981. U.S. patent 4,320,093.
249. **Watson, J. H. P., and D. C. Ellwood.** 1994. Biomagnetic separation and extraction process for heavy metals from solution. *Miner. Eng.* **7:**1017–1028.
250. **Watson, J. S., C. D. Scott, and B. D. Faison.** 1989. Adsorption of Sr by immobilized microorganisms. *Appl. Biochem. Biotechnol.* **20/21:**699.
251. **Watson, J. S., C. D. Scott, and B. D. Faison.** 1989. Adsorption of Sr by soil microorganisms. *Appl. Biochem. Biotechnol.* **21:**201–209.
252. **Weidemann, D. P., R. D. Tanner, G. W. Strandberg, and S. E. Shumate II.** 1981. Modelling the rate of transfer of uranyl ions onto microbial cells. *Enzyme Microb. Technol.* **3:**33–40.
253. **Wersin, P., M. F. Hochella, Jr., P. Persson, G. Redden, J. O. Leckie, and D. W. Harris.** 1994. Interaction between aqueous uranium (VI) and sulfide minerals: spectoscopic evidence for sorption and reduction. *Geochim. Cosmochim. Acta* **58:**2829–2843.
254. **White, C., and G. M. Gadd.** 1990. Biosorption of radionuclides by fungal biomass. *J. Chem. Technol. Biotechnol.* **49:**331–343.
255. **White, C., and G. M. Gadd.** 1987. Inhibition of H^+ efflux and K^+ uptake and induction of K^+ efflux in yeast by heavy metals. *Tox. Assess.* **2:**437–447.
256. **White, C., and G. M. Gadd.** 1996. Mixed sulphate-reducing bacterial cultures for bioprecipitation of toxic metals: factorial and response-surface analysis of the effects of dilution rate, sulphate and substrate concentration. *Microbiology* **142:**2197–2205.
257. **White, C., A. K. Sharman, and G. M. Gadd.** 1998. An integrated microbial process for the bioremediation of soil contaminated with toxic metals. *Nat. Biotechnol.* **16:**572–575.
258. **Wildung, R. E., K. M. McFadden, and T. R. Garland.** 1979. Technetium sources and behaviour in the environment. *J. Environ. Qual.* **8:**156–161.
259. **Woolfolk, C. A., and H. R. Whiteley.** 1962. Reduction of inorganic compounds with molecular hydrogen by *Micrococcus lactilyticus.* I. Stoichiometry with compounds of arsenic, selenium, tellurium, transition and other elements. *J. Bacteriol.* **84:**647–658.
260. **Wurtz, E. A., T. H. Sibley, and W. R. Schell.** 1986. Interactions of *Escherichia coli* and marine bacteria with ^{241}Am in laboratory cultures. *Health Phys.* **50:**79–88.
261. **Yakubu, N. A., and A. W. L. Dudeney.** 1986. Biosorption of uranium with *Aspergillus niger*, p. 183–200. *In* H. Eccles and S. Hunt (ed.), *Immobilization of Ions by Biosorption.* Ellis Horwood, Chichester, United Kingdom.
262. **Yong, P.** 1996. PhD. thesis. University of Birmingham, Birmingham, United Kingdom.
263. **Yong, P., and L. E. Macaskie.** 1998. Bioaccumulation of lanthanum, uranium and thorium, and use of a model system to develop a method for the biologically-mediated removal of plutonium from solution. *J. Chem. Technol. Biotechnol.* **71:**15–26.

264. **Yong, P., and L. E. Macaskie.** 1997. Effect of substrate concentration and nitrate inhibition on product release and heavy metal removal by a *Citrobacter* sp. *Biotechnol. Bioeng.* **55:**821–830.
265. **Yong, P., and L. E. Macaskie.** 1995. Enhancement of uranium bioaccumulation by a *Citrobacter* sp. via enzymatically-mediated growth of polycrystalline NH_4 UO_2 PO_4. *J. Chem. Technol. Biotechnol.* **63:**101–108.
266. **Yong, P., and L. E. Macaskie.** 1997. Removal of lanthanum, uranium and thorium from the citrate complexes by immobilized cells of *Citrobacter* sp. in a flow-through reactor: implications for the decontamination of solutions containing plutonium. *Biotechnol. Lett.* **19:**251–255.
267. **Yong, P., and L. E. Macaskie.** 1995. Removal of the tetravalent actinide thorium from solution by a biocatalytic system. *J. Chem. Technol. Biotechnol.* **64:**87–95.
268. **Yurkova, N. A., and N. N. Lyalikova.** 1991. New vanadate-reducing facultative chemolithotrophic bacteria. *Microbiology* **59:**672–677.
269. **Zajic, J. E., and Y. S. Chiu.** 1972. Recovery of heavy metals by microbes. *Dev. Ind. Microbiol.* **13:**91–100.

Environmental Microbe-Metal Interactions
Edited by Derek R. Lovley

Chapter 14

Biosorption Processes for Heavy Metal Removal

Silke Schiewer and Bohumil Volesky

INTRODUCTION

Biosorption: Metal Removal by Passive Sequestration

Metal accumulation in natural systems can involve living organisms as well as nonliving biomass such as particulate or dissolved organic matter in the water column and organic fraction of sediments. The metal binding properties, e.g., of the cell walls of many microorganisms probably play a role in the flow of ions into the cell. Essential elements can be concentrated from very dilute solutions for future use, and toxic heavy metals, on the other hand, can be rendered innocuous through the formation of insoluble metal aggregates.

Metal binding in natural systems does not contribute much to alleviating the problems of heavy metal pollution we are facing today. Whether toxic heavy metals are bound to biomass or whether they occur free in solution, they remain bioavailable for the most part. Although sediments can act as a temporary sink for heavy metals, resuspension and liberation of heavy metals due to changes in the chemical environment (pH, redox potential) may occur. Burrowing organisms feeding on sediment organic matter can constitute a vector carrying metals back into the food chain of the aquatic environment.

For the removal of heavy metals from the food cycle, we can therefore not rely on nature. However, we can make natural processes work for us: the same biomolecules that bind metals in natural systems can make certain types of biomass suitable for metal sequestration in industrial biosorption processes which are described in this chapter. Research is in progress to establish biosorption as a commercially viable technique to trap and accumulate metals. Biosorption can serve as a tool for the recovery of precious metals (e.g., from processing solutions or the seawater) and for the elimination of toxic metals (particularly from industrial wastewaters).

Silke Schiewer and Bohumil Volesky • Department of Chemical Engineering, McGill University, M. H. Wong Building, 3610 University St., Montreal, Quebec H3A 2B2, Canada.

The term "biosorption" is used to describe the passive (i.e., not metabolically mediated) accumulation of metals or radioactive elements by biological materials. Usually, dead biomass serves as a basis for a family of biosorbents. Biosorption must be distinguished from bioaccumulation, which is an active process relying on metabolic activity and therefore occurs only in living organisms.

There are distinct advantages in using either living or dead biomass (14, 32, 74). The main advantages of using living biomass are that it is self-renewing, that active transport into the cell may lead to higher metal uptake levels, and that excreted metabolic products can contribute to metal removal. However, it was observed that in terms of the quantity of metal accumulated, dead and living biomass may even be comparable. The use of dead biomass avoids problems with toxicity (the toxicity of metals is often the very reason for using biosorption). Costs of cultivation of live biomass (e.g., nutrient supply) far exceed the costs of obtaining nonliving biomass. Waste products from fermentation processes or naturally abundant biomass can be used. Biosorption by dead biomass is often faster, since only the cell wall-based binding, not active transport into the cell, occurs. Another advantage of using dead biomass is the easier and nondestructive recovery of bound metals, which allows regeneration of the biosorbent material. Metals accumulated intracellularly by living biomass can often be recovered only when the cell is destroyed. In most cases, working with dead biomass offers more advantages and is therefore the object of the majority of more practically oriented biosorption studies.

Applications of Biosorption

Biosorption can concentrate metals several thousandfold: the concentration factor by which the metal concentration in algae exceeds that in the surrounding seawater was as high as a few thousand (10). Therefore, biosorption could serve as a means of obtaining precious metals from the seawater or from processing solutions. Its main target is, however, to remove heavy metals which can be quite toxic even at low concentrations. Biosorption is particularly suited as a polishing step whereby wastewater with a low to medium initial metal concentration (from a few to about 100 ppm) is purified such that drinking-water quality is obtained. Biosorptive treatment of wastewaters with very high initial metal concentrations would lead to rapid exhaustion of the biosorbent material and thus require large amounts of biosorbent. Therefore, pretreatment of such effluents using other techniques, such as precipitation (which is currently used for 90% of heavy metal removal from industrial wastewater) or electrolytic recovery, may be more economical. On the other hand, advantages of biosorption (compared to precipitation) are that it offers high effluent quality and avoids the generation of toxic sludges (15).

As a polishing technique, biosorption has to compete with other processes such as ion exchange, activated-carbon sorption, or membrane technologies (electrodialysis, reverse osmosis) that are currently used for this purpose. The advantages of biosorption are that it can be used under a broad range of operating conditions (pH, temperature, metal concentration, other ions in solution) and, especially, that it is cost-effective (Table 8 in reference 15) (64). Cheap raw materials can be used like suitable naturally abundant biomass types (e.g., certain seaweeds) or industrial waste by-product biomass (e.g., bacteria and fungi from the fermentation industry, apple cores, and tree bark). In the latter case, an additional benefit is that waste from one industry can be used to clean the waste from other processes. Moreover, very high cost-effectiveness can be achieved through the regeneration of biosorbents with the possibility of recovering the metal.

Reactor Types for Industrial Biosorption Processes

Several types of reactors could be used for biosorption, and each has certain advantages and drawbacks (13, 15, 116). The most frequently recommended contactor is a packed-bed column, operating in a mode similar to the one where synthetic ion-exchange resins are employed. The wastewater flows (usually downward) through a column packed with biosorbent (Fig. 1; also see Fig. 5). This type of a reactor offers the advantage of very high effluent quality because the stream exiting the column is in contact with fresh sorbent material. There is no need for an additional solid-liquid separation step. Potential disadvantages are that the column could be clogged if the wastewater contains significant concentrations of suspended solids are that excessive channeling (i.e., axial mixing) may occur.

In fluidized or expanded-bed reactors, the wastewater passes upward through the reactor. The flow rate has to be balanced with the biomass size and density such that the flow suspends the biomass particles without carrying them out of the reactor. Similarly to fixed-bed columns, the water leaving the reactor is in contact with relatively fresh biomass. However, the metal concentration profile in fluidized-bed reactors is more blurred than in fixed-bed columns. Fluidized-bed reactors avoid the danger of clogging. The metal-loaded particles tend to exhibit a higher density, so that they sink to the bottom of the reactor, where they can be periodically collected and replaced with fresh biosorbent (about 1/10 of the biomass in the reactor is replaced at a time).

Stirred tanks contain biomass dispersed throughout the reactor. Since the exiting solution is in contact with particles of average loading (not with fresh biomass, as in the other reactor types), more biomass has to be used to achieve the same effluent quality as in the other techniques. If the biosorbent particles are fragile, such that they would be damaged by impellers, the tank can be air mixed, which requires higher energy input. The use of this type of reactor requires a solid/liquid separation step, where the metal-laden biomass is separated from the treated effluent. Gravity settling, flotation, centrifugation, filtration, or magnetic separation (if the biomass is magnetic) can be used for this purpose.

Regeneration of Metal-Laden Biomass

There are generally two avenues of dealing with the metal-laden biosorbent: either it must be disposed of (e.g., by incineration or deposition in landfills) or, preferably, it is regenerated. In the first case, biosorption serves to reduce the volume of waste: a large liquid waste stream is converted to a much smaller amount of solid waste, which can perhaps be more easily disposed of. In the second case, it is possible to avoid waste generation altogether and to regenerate the biosorbent as well as to recover the metal. The metal is desorbed from the biomass, yielding

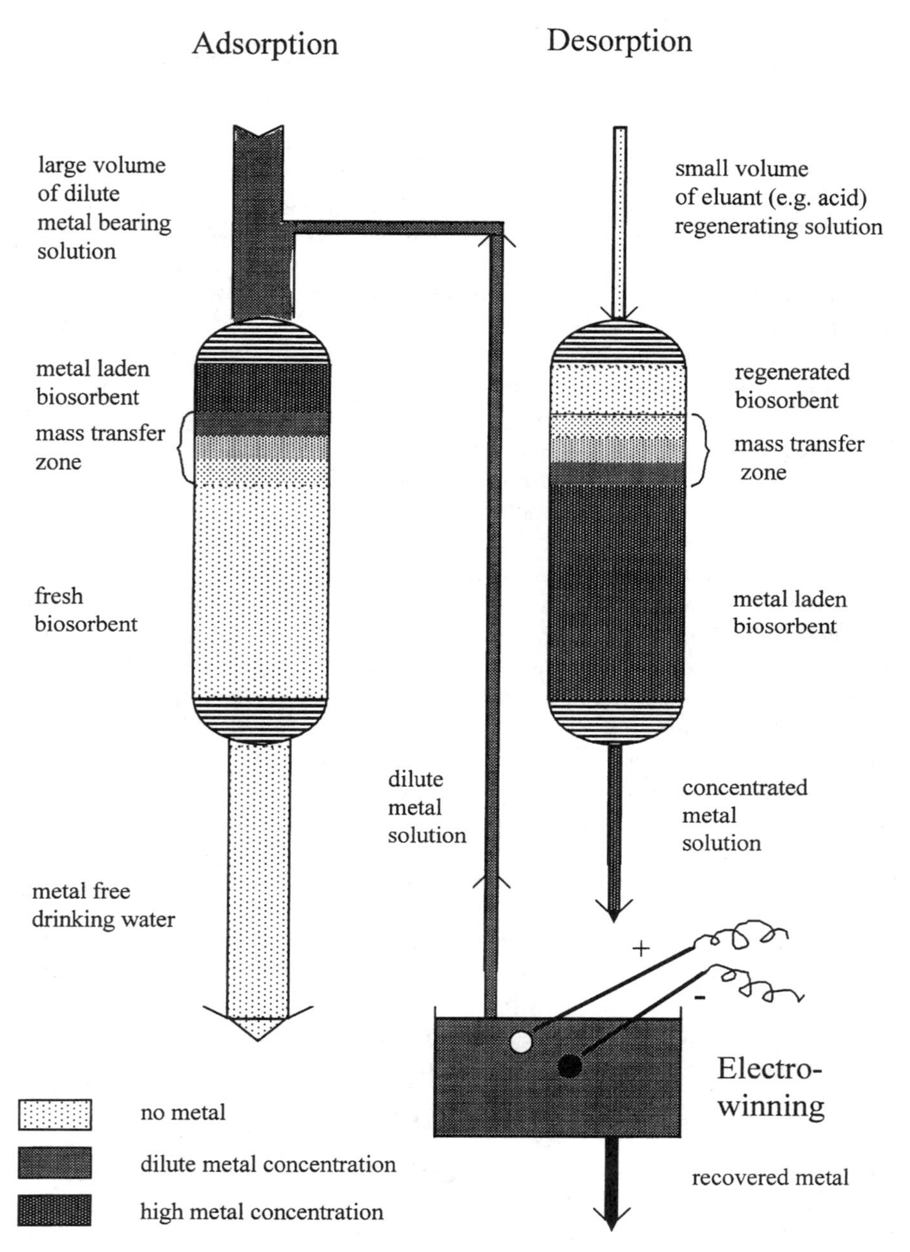

Figure 1. Flow scheme of a possible biosorption application using packed-bed columns for adsorption and desorption.

reusable biosorbent and a highly concentrated metal solution (of a concentration of at least 100 times that of the original solution). This metal solution can then be treated by precipitation or electrowinning. While the former process removes the metal but generates a toxic sludge, the latter technique can serve to recuperate the metal. A possible biosorption process involving fixed-bed columns, employed alternatively for adsorption and desorption, is shown in Fig. 1.

Many types of eluants can be used to desorb the metal from the biomass (2, 33, 67, 109). Desorption can be based on two general principles, which may also act in combination: certain desorbing agents such as acids or metal salts provide cations that compete with the bound metal ions for binding sites, thereby "pushing them away" from the sites. The effectiveness of desorption depends in this case on the binding strength of the cation added to the biosorbent. The alternative is to employ ligands which complex the metal in solution (e.g., EDTA), such that the concentration of free metal in solution is lowered. In this way, the metal is "drawn" into the solution and thus the metal uptake is reversed.

Criteria for the choice of desorbing agents are as follows (67): (i) a high ratio of regenerated sorbent to eluant volume; (ii) a high metal concentration factor (after desorption and before adsorption); (iii) efficient desorption (a minimum of ca. 95% of the metal has to be desorbed); (iv) fast kinetics; (v) selectivity; (vi) no structural damage of sorbent material; (vii) high uptake in the next cycle; (viii) low cost; and (ix) environmentally benign. Some of these criteria are conflicting: if a smaller volume of eluant is used such that a high concentration factor is achieved (criteria i and ii), the efficiency may drop (criterion iii). Mineral acids (0.1 to 1 M) have proven to be efficient low cost desorption agents (2). Alhough higher concentrations of eluant yield higher efficiency (criterion iii and concentration factors (criterion ii), extreme pH values can, for example, cause damage to the biosorbent (criterion vi) (67).

MECHANISM OF BIOSORPTION

Understanding the mechanism of biosorption is not merely a question of academic interest; practical benefits are also gained. The main objective of studying biosorption is obviously to optimize its application. Rather than establishing optimal conditions by a lengthy and expensive trial-and-error process, one should aim for conceptual understanding that allows predictions to be made. Knowing, for example, that the mechanism of biosorption is based largely on ion exchange implies that changes in the ionic strength of the solution will affect metal uptake. To predict quantitatively how much a given factor (such as ionic strength) influences metal uptake, we must use mathematical models. These will be more reliable if they are not arbitrary mathematical correlations but, rather, are based on the actual mechanism. Furthermore, the choice of desorption technique also depends on the mechanism involved. Metal binding to acidic groups can, for example, be reversed by lowering the pH and thereby protonating these groups.

Unfortunately, knowledge of the biosorption mechanism is not easily obtained, since we are not dealing with simple, clearly defined chemical compounds. Biosorbents comprise different types of cells with a highly complex structure whose various building blocks consist of a multitude of different molecules which in turn can display several binding sites. Moreover, even one binding site can participate in different binding mechanisms: carboxyl groups can, for example, engage in both complexation and electrostatic attraction of metal cations. Consequently, several mechanisms often act in combination. The mechanism may also vary considerably with external conditions such as pH.

As for overall metal binding mechanisms, we can distinguish between ion exchange, sorption of electrically neutral material (soluble metal-ligand complexes) to specific binding sites, and microprecipitation. These main mechanisms are based on sorbate-sorbent or solute-solvent interactions, which in turn rely on some combination of covalent, electrostatic, and van der Waals forces.

Overall Mechanisms: Ion Exchange, Adsorption, and Microprecipitation

While some authors consider only an exchange of electrostatically bound ions to be ion exchange, in this chapter we adopt a broader definition of this term. The term "ion exchange" is used when the charge of ions taken up equals the charge of ions released (so that the charge neutrality of the particle is maintained), regardless of whether these ions are bound electrostatically or by complexation. The driving force of ion exchange is mostly the attraction of the biosorbent for the sorbate (metal). Metals can be bound electrostatically or by complexation. Interactions between the solute (metal) and the solvent (usually water) play a role in so far as less hydrophilic (and consequently less hydrated) molecules have a lower affinity for the liquid phase and are therefore sorbed more easily (102). The importance of ion exchange in biosorption has frequently been reported. The amounts of ions from the natural environment (Na^+, K^+, Ca^{2+}, Mg^{2+}, and H^+) and from biosorbent pretreatment (such as protonation) which are released during biosorption balance the heavy metal uptake by algae (24, 68, 94), bacteria (A. C. C. Plette, L. Haanstra, and W. V. Van Riensdijk, Metals-Microorg. Relat. App., FEMS Symp. Abstr., p. 8, 1993), fungi (31, 107), and peat (20, 25).

Desorption can, in many cases, also be interpreted in terms of ion exchange. Such "competitive" desorption can be achieved by acids (e.g., HCl and H_2SO_4) and/or salt solutions (e.g., $CaCl_2$) (2, 67). In each case, the cation (H^+, Ca^{2+}) competes with the bound metal ion for the binding sites and replaces it if the concentration of the desorption agent is high enough.

We use the terms "adsorption" and "microprecipitation" to describe the accumulation of electrically neutral material which does not involve the release of a stoichiometric amount of previously bound ions. The difference between adsorption and microprecipitation is that in the former case, affinity between sorbent and sorbate (metal complex) and in the latter case limited solubility (i.e., an interaction between the solute and solvent) represents the main driving force. In microprecipitation, the metal cation and an anion (e.g., SO_4^{2-}, S^{2-}, oxalate, or HPO_4^{2-}), itself often a metabolic product of certain biomass types, form insoluble aggregates (salts or complexes) such as sulfides, carbonates, oxides, oxalates, and phosphonates (9, 93). Changed local pH or redox potential can also influence the occurrence of precipitation. Microprecipitation does not necessarily involve a bond between biomass and metal. The process may, however, be nucleated by metal initially bound to active sites in the biomass (82). This means that a two-stage process takes place where binding to specific sites is followed by microprecipitation. The latter process is not limited by the number of binding sites but can occur in multiple layers (76, 77). Sorption of neutral complexes is thought to be responsible for Cu binding to peat at high concentrations (20).

Contribution of Electrostatic Attraction and Complexation

Ligands in the biomass (such as carboxyl groups) can form complexes (or coordination compounds) with metal ions. Chelation, i.e., binding of one metal ion to two coordinating atoms in the same biomolecule, may also occur. Complex formation involves both covalent and electrostatic components, whose relative contribution can be estimated by investigating how specific the binding is. When purely electrostatic attraction occurs, the binding strength should correlate with the charge density (z^2/r_{hyd}). Ions of same charge (z) and hydrated radius (r_{hyd}) should therefore be bound with equal strength. Major deviations of the binding strength from the z^2/r_{hyd} correlation indicate a tendency toward a covalent bond character.

The nature of the ions released provides information about the bond type. Electrostatically bound ions cannot displace covalently bound ions. It was observed that proton release occurred only during heavy metal uptake, not during light metal uptake (26, 40). Since protons are mainly bound covalently, the binding of heavy metals must have been more covalent than that of light metal ions. Similarly, the more Na^+ (which binds only electrostatically) reduces the uptake of other ions, the higher is the contribution of electrostatic attraction in the binding of those ions (96).

The bond character in biosorption can partially be explained by Pearson's (90) concept of hard and soft acids and bases. So-called hard (or type a) ions (such as the alkaline earth ions and Mn^{2+}) participate in ionic bonds. Easily polarizable soft (or type b) ions (such as Ag^+, Au^+, Hg^+, Hg^{2+}, and Cu^+) tend to form covalent bonds. Many transition metals (e.g., Fe^{2+}, Co^{2+}, Ni^{2+}, Cu^{2+}, Zn^{2+}, Pb^{2+}, and Cd^{2+}) as well as protons have intermediate characteristics, whereby Zn^{2+} tends toward hard and Pb^{2+} tends toward soft (1, 87, 90). Among ligand atoms, O and F are considered hard; S, P, and As are considered soft; while N is classified as intermediate.

Different correlations have been proposed to describe the increase of ionic-bond character with an increasing difference in the electronegativity between the two bonding atoms (16, 28, 89). For typical elements in biological ligands (O, N, and S), the ionic-bond character therefore increases with the electronegativity of the metal. It follows from the concept of hard and soft acids and bases that Pb^{2+} and Cu^{2+} are expected to display more covalent bond character and consequently stronger binding than are the hard ions Na^+ or Ca^{2+}.

Binding Sites

Numerous chemical groups have been proposed to contribute to biosorptive metal binding by, e.g., algae (26, 38), bacteria (13), or biopolymers (50), including hydroxyl, carbonyl, carboxyl, sulfhydryl, thioether, sulfonate, amine, imine, amide, imidazole, phosphonate, and phosphodiester groups. Whether any given group is important for biosorption of a certain metal by a certain biomass depends on factors such as the quantity of sites in the biosorbent material, the accessibility of the sites, the chemical state of the site (i.e., its availability), and the affinity between the site and the metal (i.e., the binding strength). For covalent metal binding, even an already occupied site is theoretically available. To what extent the site can be used by the metal in question depends on its binding strength and concentration compared to the metal already occupying the site. For electrostatic metal binding, a site is available only if it is ionized.

The major binding sites in biosorption are acidic. Many groups (hydroxyl, carboxyl, sulfhydryl, sulfonate, and phosphonate) are neutral when protonated and negatively charged when deprotonated. When the pH of the solution exceeds its pK_a, these groups become mostly available for the attraction of cations. Amine, imine, amide, and imidazole groups, on the other hand, are neutral when deprotonated and positively charged when protonated. Therefore, they attract anions if the pH is lowered such that the groups are protonated. The structural formulae and pK_a values of binding groups are summarized in Table 1. The occurrence of these groups in different types of biomass is discussed in the next section.

For the freshwater alga *Chlorella*, the charge was positive (probably due to amine groups), favoring anion binding, at pH < 3 and negative (mostly carboxyl groups probably) at higher pHs, so that electrostatic attraction of cations occurred (38). The charge of the biosorbent does not depend exclusively on the pH value. Covalent binding of metals can "consume" negatively charged groups. Groups become charge neutral that would otherwise have been negatively charged in metal-free solution of the same pH (94).

BIOSORBENT MATERIALS

A multitude of different biomaterials have been examined for their biosorptive properties, and different types of biomass have shown levels of metal uptake high enough (an the order of $\sim$100 mg/g or 1 mmol/g) to warrant further research (119). The three classes of biosorbents most frequently used are based on algal, bacterial, and fungal biomass. Other types of biosorbents include peat, pectic substances, and cellulosic materials (such as wood chips). Carboxylate and phenolic groups of peat may bind metals through ion exchange and complexation (20, 23). Pectin, a polymer of galacturonic acid, contains carboxyl groups that can chelate divalent cations. It is present in apple, citrus, sunflower, and sugar beet (60). Since cellulose itself contains only hydroxyl groups and displays very low metal uptake when untreated, it has been used in modified (e.g., phosphorylated) forms for metal uptake (45). The main properties to consider in choosing appropriate biosorbents are low cost, availability, high metal uptake, favorable kinetics, and easy recovery and regeneration.

Algae

The abundance of algae can hardly be overestimated. A total of 90% of all photosynthesis occurs through algae, and the mass of algal cell wall polymers alone

Table 1. Binding groups

Group	Structural formula[a]	pK_a[b]	HSAB classifications[c]	Ligand atom	Occurrence in selected biomolecules[d]
Hydroxyl	—OH	9.5–13	Hard	O	PS, UA, SPS, AA
Carbonyl (ketone)	>C=O		Hard	O	Peptide bond
Carboxyl	—C(=O)OH	1.7–4.7	Hard	O	UA, AA
Sulfhydryl (thiol)	—SH	8.3–10.8	Soft	S	AA
Thioether	>S		Soft	S	AA
Sulfonate	—S(=O)(=O)OH	1.3	Hard	O	SPS
Amine	$—NH_2$	8–11	Int.	N	Cto, AA
Secondary amine	>NH	13	Int.	N	Cti, PG, peptide bond
Imine	=NH	11.6–12.6	Int.	N	AA
Amide	$—C(=O)NH_2$		Int.	N	AA
Imidazole	—C—N—H, ‖ \ CH, H—C—N//	6.0	Soft	N	AA
Phosphonate	—P(=O)(OH)OH	0.9–2.1, 6.1–6.8	Hard	O	PL
Phosphodiester	>P(=O)OH	1.5	Hard	O	TA, LPS

[a] From references 17, 50, and 80.
[b] From references 5, 16, and 50.
[c] From reference 87. Int., intermediate.
[d] PS, polysaccharides; UA, uronic acids, SPS, sulfated PS; Cto, chitosan; Cti, chitin; PG, peptidoglycan; AA, amino acids (e.g., in proteins and PG); TA, teichoic acid; PL, phospholipids; LPS, lipopolysaccharides.

equals about half of the total nonaqueous biomass (98). The world harvest of seaweeds for food and algal products (e.g., agar, carrageenan, and alginate, which can be employed as thickeners) already exceeds 3 million tons annually (73) and potential harvests are estimated as 2.6 million tons for red algae and 16 million tons for brown algae (18). Apart from being readily available, the advantages of algal biosorbents are that the particles can be large enough to eliminate the need for immobilization and that the material is better defined: conditions in natural waters do not change as much as those of the culture media. The latter aspect concerns both the available biomolecules (substrate dependent formation) and impurities that may be present.

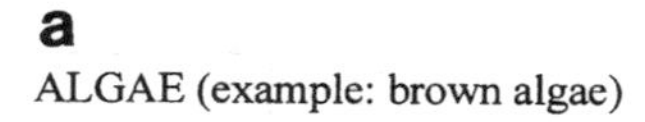

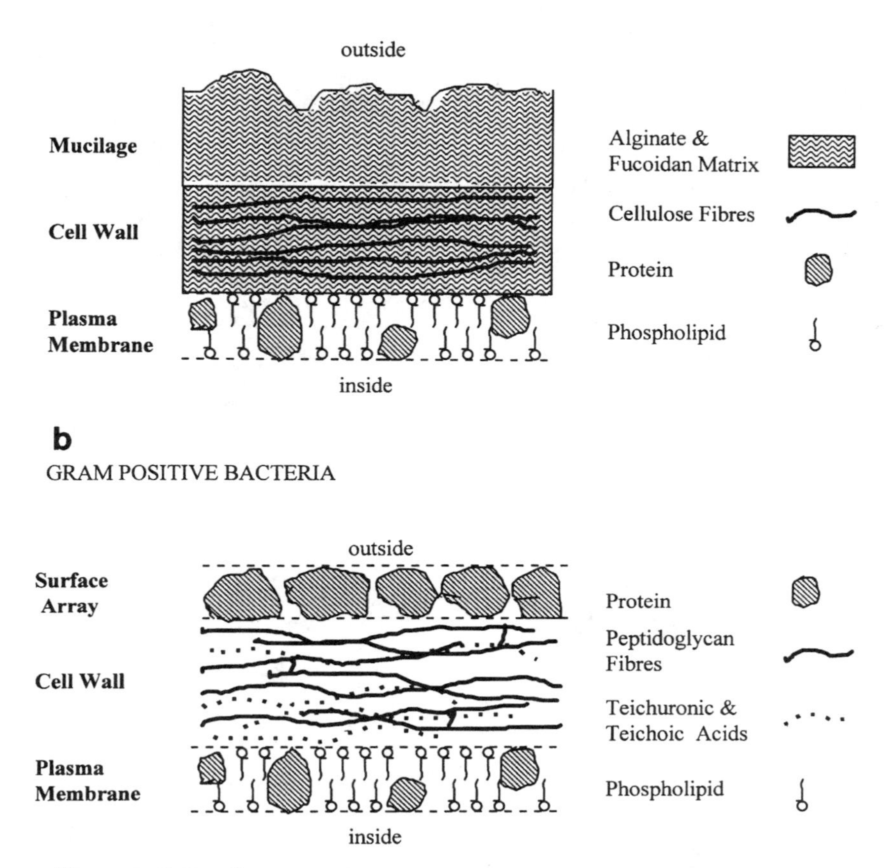

Figure 2. Cell wall structure in algae (example, brown algae) (a), gram-positive bacteria (modified from references 8 and 93) (b), gram-negative bacteria (modified from references 8 and 93) (c), and fungi (example, type V, e.g., Euascomycetes) (modified from reference 84) (d).

Biosorption in algae has been attributed mainly to the cell wall, where both electrostatic attraction and complexation can play a role. The most common types of algae that possess complete cell walls are Rhodophyta (red algae), Chlorophyta (green algae), and Phaeophyta (brown algae) (11). A typical algal cell wall consists of a fibrillar skeleton and an amorphous embedding matrix (Fig. 2a). The most common fibrillar material is cellulose, but it can be replaced by mannan or xylan in some siphonaceous green algae and in some genera of red algae. Brown and red algae contain the largest amounts of amorphous matrix polysaccharides. While alginic acid is the dominant constituent, followed by the sulfated polysaccharides such as fucoidan, in the brown algal matrix, a number of sulfated galactans (agar, carrageenan, porphyran, etc.) play a major role in red algae. The green algal matrix

c

GRAM NEGATIVE BACTERIA

outside

Surface Array

Outer Membrane

Cell Wall

Plasma Membrane

inside

Lipopolysaccharide

Lipoprotein

Peptidoglycan fibres

Phospholipid

Protein

d

FUNGI (example: type V, e.g. Euascomycetes)

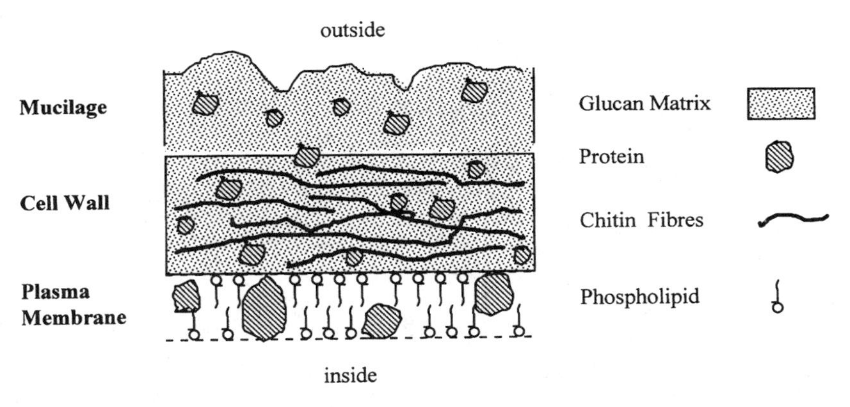

Figure 2. *Continued*

consists of complex heteropolysaccharides (involving galactose, arabinose, xylose, rhamnose, and glucuronic acid) that may be sulfated (70). Cell wall characteristics of different algae are summarized in Table 2, and the compositions of different biopolymers are given in Table 3.

The main acidic groups responsible for metal uptake are the carboxyl groups of uronic acids (guluronic, mannuronic, and glucuronic) as well as sulfonate groups. Hydroxyl groups are present in all polysaccharides, but they become negatively charged only at pH $>$ 10, at lower pHs, they play only a secondary role. Carboxyl groups can be the dominating binding groups, especially in brown algae. This is the case because alginic acid, which is composed of mannuronic and guluronic acids (Fig. 3a), can constitute up to 40% of the dry weight of brown algae (91). Carboxyl groups are found to a lesser extent in the glucuronic acid of green algal matrix polysaccharides and red algal agaropectin (98). Sulfonate groups are abun-

Table 2 Biomolecules in different types of biomass[a]

Biomass	Polysaccharide				With carboxyl			Sulfonate		(Second) amine			Phosphonate			Comment
	Cell	Glu	Man	Xyl	Muc	Alg	TuA	Sgal	Fuc	Cto	Cti	PG	TA	LPS	PLi	
Algae																
Rhodophyta	f		(x)	(f)		(x)		x								(PC), (Ca)
Chlorophyta	f		(f)	(f)	x											(no), (PC), (Ca)
Charophyta	f															(Ca)
Euglenophyta																no, PC, (lor)
Phaeophyta	f					x			x							
Pyrrophyta	(x)				x											(no), theca
Chrysophyta	(f)				(x)						(x)					(no), (lor), (sca), (Si), (Ca)
Cryptophyta																no, periplast plates
Bacteria																
Gram positive								x				x	x			
Gram negative												x		OM	OM	
Fungi[b]																
Myxomycetes	f															
Acrasiomycetes (I)	f	x														
Plasmodiophoromycetes										f						
Hyphochytridiomycetes (III)	f									f						
Chytridiomycetes (V)		x									f					
Oomycetes (II)	f	x														
Trichomycetes (VIII)	Galactan									Galactosamine						
Zygomycetes (IV)										x	f					
Euascomycetes (V)		x									f					
Hemiascomycetes (VI)		f	x													
Deuteromycetes (V)		x									f					
Homobasidiomycetes (V)		x									f					
Heterobasidiomycetes (VII)			x								f					

[a]This table was compiled from literature information as follows: algae, references 12 and 100; bacteria, references 50 and 83; fungi, references 85 and 93. Cell, cellulose; Glu, noncellulosic glucan; Man, mannan; Xyl, Xylan; Muc, mucilage, also containing uronic acids and sulfonate gorups; Alg, alginic acid; TuA, teichuronic acid; Sgal, sulfated galactans; Fuc, fucoidan; Cto, chitosan; Cti, chitin; PG, peptidoglycan; TA, teichoic acid; LPS, lipopolysaccharide; PLi, phospholipid; x, present; f, in inner fibrillar skeleton; OM, in outer membrane; no, no cell wall; PC, proteinaceous cuticle; lor, lorica; theca, thecal plates; sca, scales; Ca, calcified; Si, silicified. Parentheses indicate that the molecule is present only in some genera or in small quantities.

[b]Fungal wall types are given according to the widely used classification of Bartnicki-Garcia (6) and are indicated by roman numerals.

Table 3. Composition of biopolymers

Polymer[a]	Link	Monomer	Group
Polysaccharides			
Cellulose	β-1,4	D-Glucose	
Fungal α glucans	α-1.3; α-1,4	D-Glucose	
Fungal β glucans	β-1,3; β-1,4; β-1, 6	D-Glucose	
Mannan (algal)	β-1,4	D-Mannose	
Mannan (fungal)	α-1,2; α-1,6; α-1,3	D-Mannose	
Xylan	β-1,3, β-1,4	D-Xylose	
Uronic acids			
Alginic acid	β-1,4	D-Mannuronic acid	Carboxyl
	α-1,4	L-Guluronic acid	Carboxyl
Teichuronic acid	Linear hexuronic	Glucuronic acid	Carboxyl
Sulfated polysaccharides			
Agar, porphyran[alt]	β-1,3	D-Galactose	Sulfonate
	α-1,4	L-Anhydrogalactose	Sulfonate
Carrageenan[alt]	β-1,3	D-Galactose	Sulfonate
	α-1,4	D-(Anhydro)galactose	Sulfonate
Fucoidan	α-1,2 (α-1,3; α-1,4)	L-Fucose	Sulfonate
Nitrogenated biopolymers			
Chitosan	β-1,4	D-Glucosamine	Amine
Chitin	β-1,4	*N*-Acetyl-D-glucosamine	Second amine, acetyl
Peptidoglycan[alt]	β-1,4	*N*-Acetyl-D-glucosamine	Second amine, acetyl
	β-1.4	Muramic acid with peptide side chain	Second amine, acetyl, carboxyl, etc.
Phosphated biomolecules			
Teichoic acid	Ribitol or glycerol phosphate		Phosphodiester
Lipopolysaccharides	Phosphorylated lipid, oligosaccharide core, and polysaccharide side chain		Phosphodiester
Phospholipids	Phosphatidyl ethanolamine and fatty acids		Phosphonate

[a] alt, alternating monomers.

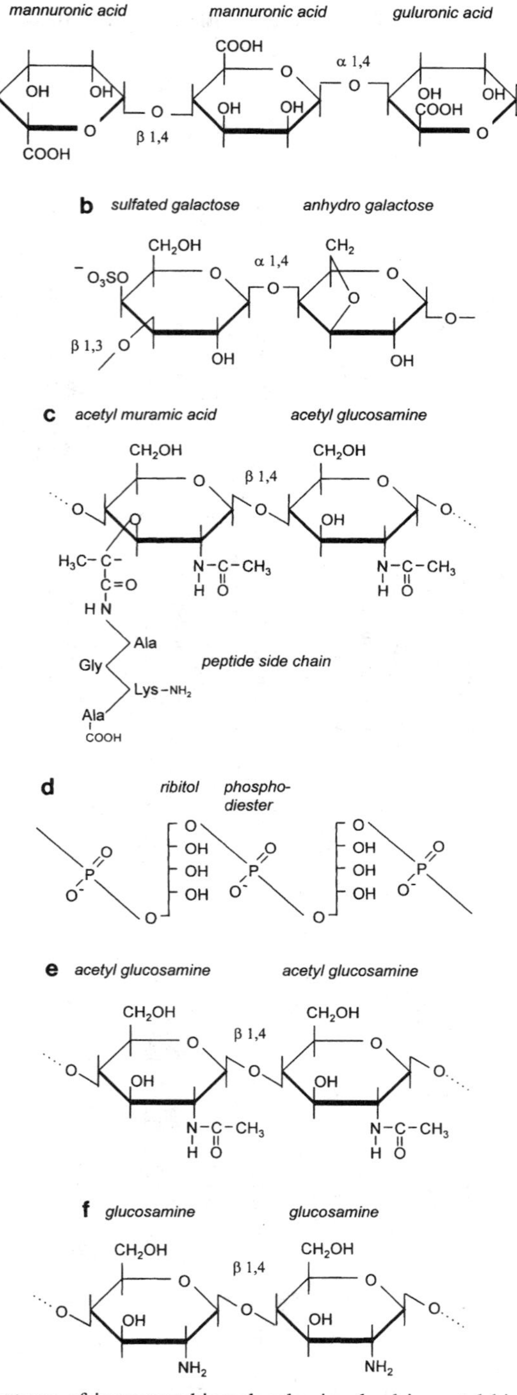

Figure 3. Structures of important biomolecules involved in metal binding. (a) Alginic acid of brown algae (modified from reference (1); (b) κ carrageenan of red algae (modified from reference 18); (c) peptidoglycan of bacteria (modified from references 50 and 114); (d) teichoic acid of bacteria (modified from reference 50); (e) chitin of fungi; (f) chitosan of fungi.

dant in red algae due to their high content of carrageenan (Fig. 3b), porphyran, and agar (up to 70% of the cell wall [12]), but they also occur in fucoidan of brown algae and sulfated heteropolysaccharides of green algae. Particularly for green algae, the contribution of functional groups of amino acids (hydroxyl, carboxyl, sulfhydryl, amine, imine, amide, and imidazole [5]) should not be neglected: protein can constitute 10 to 70% of the (outer cuticular layer [12] of the) cell wall (98).

Bacteria

Bacterial biomass (e.g., *Bacillus*, *Streptomyces*, and *Citrobacter*) can be obtained as waste products from fermentation industries, which makes it a cheap raw material. However, the raw biomass may contain residual chemicals that affect metal binding, and the product may be of variable quality due to variations in the fermentation conditions. It may also be necessary to immobilize the biomass before application in reactors, which adds to the cost.

Microprecipitation is a common phenomenon in metal binding by bacteria (83), but complexation by extracellular substances or by N and O ligands in the cell wall, as well as electrostatic attraction to charged groups in the cell wall, may also occur (13). Microprecipitation is often preceded by binding to specific sites, which provides nucleation points (83).

The higher metal binding capacity of gram-positive than of gram-negative bacteria can be traced back to their cell wall makeup (Fig. 2b and c, Tables 2 and 3): gram-positive bacteria possess a thicker peptidoglycan (PG) layer (13).

The gram-positive bacterial cell wall features a ca. 20 to 30-nm-thick layer (ca. 25 molecules) of PG into which teichoic acids (TA) and teichuronic acids (TUA) are embedded (8, 83). The total cell wall can be 50 to 150 nm thick (93). PG is reported to represent 40 to 90% of the cell wall (93); other sources mention that TA and TUA can constitute up to 80% of the wall (8). PG is a linear polymer of alternating glucosamine and muramic acid with peptide side chains (Fig. 3c). These side chains bear one carboxyl group at the terminal amino acid and additional functional groups on certain intermediate amino acids like asparagine, lysine, cysteine, or aspartic acid (83). TA contains phosphodiesters (Fig. 3d), and TUA feature carboxyl groups; both of these contribute to the negative charge of the biomass and enable ion exchange (13). For *Bacillus subtilis*, the major importance of the carboxyl groups of the PG as well as the minor contribution of phosphate groups in metal uptake was demonstrated by blocking experiments, amine groups did not appear to be relevant (83).

Gram-negative bacteria have a much thinner PG layer are to three molecules thick [8]), which makes up about 10% of the weight of the total cell wall, which can be 30 to 80 nm thick (93). The PG layer of gram negative bacteria does not contain TA or TUA. Therefore, they offer less negatively charged carboxyl groups which is a reason for their lower biosorptive capacity (13, 83). On the other hand, a characteristic of these bacteria is an outer membrane which contains lipopolysaccharides and phospholipids. Their phosphonate groups, which create a negative surface charge conducive to cation binding, were confirmed to be the primary metal binding site in *Escherichia coli* (8, 83).

Proteinaceous surface arrays, or S-layers, are present in many bacteria (8, 83). Extracellular polymers of the capsule or slime layer contain carboxyl and occasionally phosphonate or sulfonate groups (83). Certain bacteria produce SO_4^{2-} or S^{2-} or enzymes that liberate HPO_4^{2-}. These ligands can form microprecipitates with metal cations (13, 83).

Fungi

Fungi can be cheaply available as industrial waste products. *Aspergillus niger* is used in the production of citric acid and of the enzyme glucamylase, *Saccharomyces cerevisiae* is used in the food and beverage industry, and *Rhizopus arrhizus* produces the enzyme lipase, just to name a few examples of fungi and yeasts that have been employed in biosorption studies (27, 32, 81). Some filamentous fungi such as *A. niger* grow as pellets, which aids the recovery of the metal-laden biosorbent (32). Other types of fungi create problems of solid/liquid separation and are not easily filterable (115). Another potential disadvantage of the use of fungi is that of impurities due to adhering fermentation broth residue that may affect metal uptake.

Similarly to algae and bacteria, the cell wall is the main site of metal deposition in fungi (115). Polysaccharides constitute up to 90% of the fungal cell wall (93).

Figure 2d shows the fungal cell wall architecture. The inner microfibrillar layer of the wall usually consists of chitin (Fig. 3e), but cellulose or, in rarer cases, noncellulosic β-glucan (in Hemiascomycetes, e.g., *Saccharomyces*) can take its place, depending on the taxonomic group (85, 93). The outer, more amorphous layer is made up of mostly α-glucans but can also contain mannan, galactans, chitosan (Fig. 3f) (Zygomycetes, e.g., *Mucor* and *Rhizopus*), or glycogen (93). As seen by a comparison of Fig. 3e and f, chitin is acetylated chitosan. The main components of cell walls of different groups of fungi are listed in Table 2, and the compositions of macromolecules are given in Table 3.

Phosphated polysaccharides may occur (32); the phosphate content in *Mucor* can exceed 20% of the cell wall dry weight (115). The phosphate and carboxyl groups (of glucuronic acid) are thought to be responsible for the negative charge in the fungal wall, whereas the amine groups of the chitosan create a positive charge (93, 115). Apart from electrostatic attraction to these charged groups, complexation with N or O donors (e.g., of chitin) may occur (81). Since the protein content of the fungal cell wall is only about 10% (27, 115), the importance of amino acid functional groups in metal uptake is small.

As in the case of bacteria, released metabolites can lead to microprecipitation (oxalates due to oxalic acid, sulfides due to H_2S) or chelation (citric acid, siderophores).

OPERATING CONDITIONS

The efficiency of biosorption is usually reported in terms of the metal uptake (metal binding), ^{M}q, in milligrams or millimoles of metal bound per gram of biomass. Typical metal uptake values range from a few to several hundred milligrams per gram, depending on the type of metal and biomass (117). Biosorption may be used under a wide range of operating conditions whereby the metal uptake is affected by the parameters discussed below. Other quantitative parameters characterizing biosorption, such as the percentage of metal removed, may not be very appropriate and can serve for crude comparison at best. The biosorption equilibrium should, of course, be characterized in terms of the relevant parameters under equilibrium conditions. Therefore, it should be reported for which equilibrium concentration (not initial concentration) certain metal uptake values were obtained. This also pertains to graphical depictions. Correspondingly, the final metal concentration is chosen as the horizontal axis in Fig. 4a and c. In this section, the influence of the most important parameters on the biosorption equilibrium is described in qualitative terms. The next section deals with quantitative modeling of the key phenomena.

Metal Concentration

One of the most important factors influencing metal uptake is, not surprisingly, the concentration of the metal itself. Biosorption can be applied in a broad range of metal concentrations. With increasing equilibrium metal concentration [M], the metal uptake ${}^{M}q$ increases too. Whereas metal accumulation can occur in multiple layers in microprecipitation, leading to very high uptake values, a plateau value, ${}^{M}q_{max}$, of the metal uptake is usually reached at high metal concentrations (usually several hundred parts per million) in sorption or ion exchange. This is illustrated in Fig. 4a. Such a plot of ${}^{M}q$ versus [M] is called the sorption isotherm. The sorption capacity is limited by the number of binding sites in the biomass. At low metal concentrations, as occur in effluents from mining industry (which may contain less than 5 ppm of metal), the capacity of the biosorbent is not fully used.

Temperature

Sorption is usually temperature dependent. Therefore, uptake values for different conditions are comparable only if they are obtained at similar temperatures (as implied in the term "sorption isotherm"). Simple physical sorption is generally exothermic, i.e., the equilibrium constants decrease with increasing temperature (99).

For two reasons, this cannot be extrapolated to biosorption. First, in an exchange of two metals, the overall reaction can be either endo- or exothermic even if the binding of both ions is exothermic: while the binding of one ion liberates energy (exothermic reaction), the release of the other ion is endothermic. Second, biosorption not only is physisorption but also can involve complex formation. As a result, both endo- and exothermic exchange reactions have been reported (40, 119).

Since biosorption usually operates in a narrow temperature range (5 to 40°C), temperature effects are only of secondary importance. When the temperature was increased from about 5 to about 50°C, the equilibrium metal uptake increased only by a factor of about 2 (65) or less (35, 71).

pH Value

Biosorption can be used at a large range of pH values, from the low pH of acid mine drainage to the high pH of wastewaters pretreated by precipitation. Since

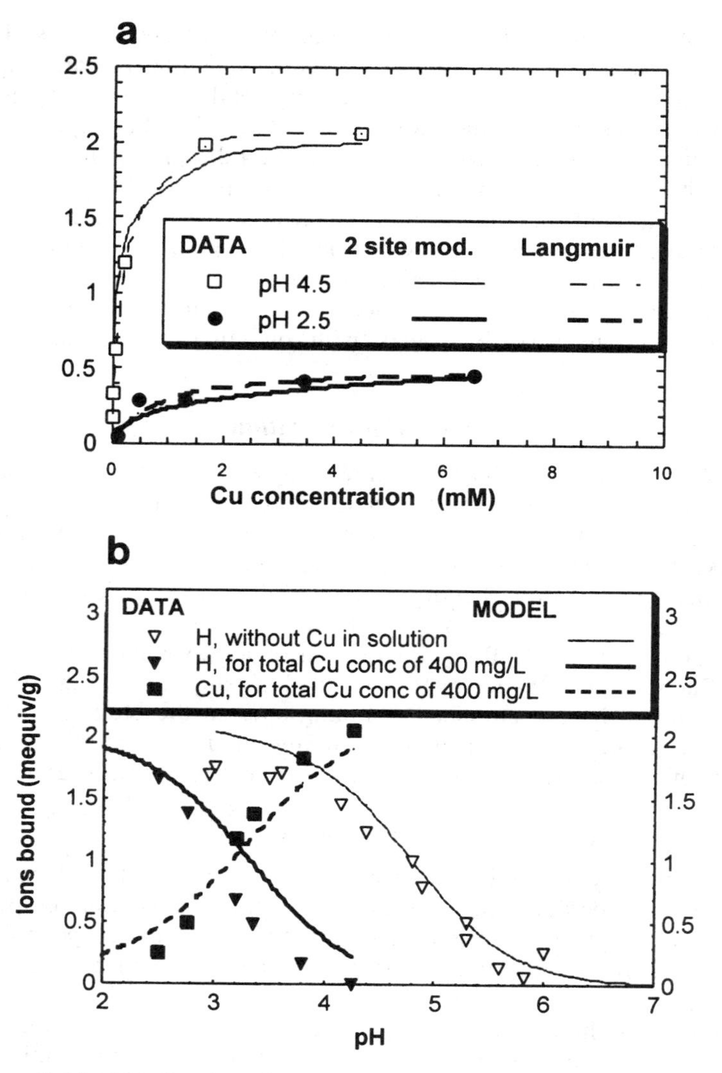

Figure 4. Metal binding (experimental data and model predictions). (a) Langmuir isotherm: Cu^{2+} binding at pH 4.5 and 2.5. (b) Influence of pH on binding of Cu^{2+} and protons. reprinted from reference 94 with permission of the publisher. (c) Three-dimensional plot of total binding of Cd^{2+} and Zn^{2+} as a function of both metal concentrations. Reprinted from reference 95 with permission of the publisher.

sorption can dramatically change with pH, this key parameter should always be reported. Data are comparable only if they are obtained at the same pH (e.g., all points on a sorption isotherm should be at the same pH).

There are several ways in which pH may influence sorption. First, the speciation of the metal in solution is pH dependent. While many metals occur as free hydrated species at lower pH, hydroxides are formed with increasing pH and eventually

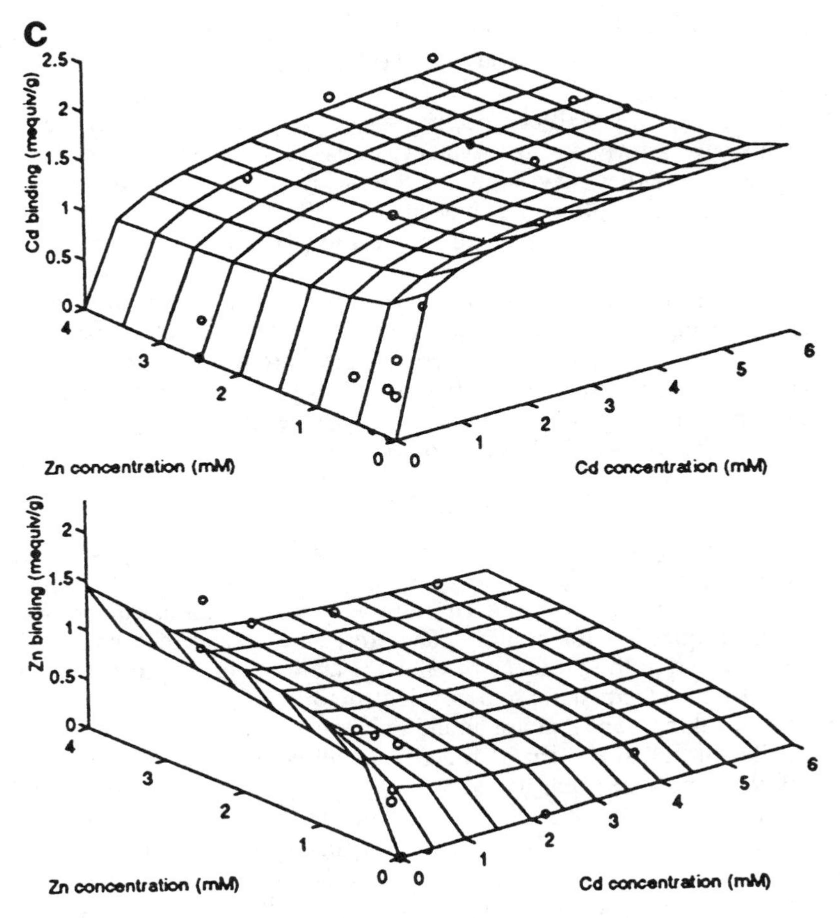

Figure 4. *Continued*

precipitation may occur. If several species of one metal occur in solution, each of them should be considered an individual sorbate. Sorption increases with decreasing solubility of the sorbate. Consequently, hydrolyzed metal ions may sorb better than the free metal ion, because they are less hydrophilic, so that less energy is necessary for removal or reorientation of the hydrated water molecules of hydrolyzed metals (22, 88, 102).

Second, extreme pH values can damage the structure of the sorbent, as could be seen in some electron micrographs of distorted cells reported in the literature (67).

The third and often most important factor is that the pH value can change the state of the active binding sites, which in biosorption are usually acidic. Their protonation and consequently their availability can change dramatically if the pH is varied by 1 or 2 units. Cation binding usually increases with increasing pH, as shown in Figs. 4a and b. This occurs because negatively charged free sites allow electrostatic attraction of cations and, if the metal is covalently bound, high pH means less competition with protons for the binding sites (few sites are occupied by protons).

Only the metal ions which can occur as negatively charged complexes or have a strong tendency to form strong covalent bonds, such as Ag, Hg, or Au (e.g., as tetrachloroaureate), either may show a decrease in uptake with increasing pH or may not be sensitive to pH changes at all (37, 38, 92). However, the biosorbent uptake of most metals increases with increasing pH. An example is depicted in Figs. 4a and b.

Nevertheless, it may not be appropriate to choose higher pH values for biosorption, because precipitation of the metal may occur. On the laboratory scale, this complicates the determination of the actual metal removal by binding to the biomass. On an industrial scale, precipitation may be desirable because it increases the overall removal of metal from the solution. However, it is possible that precipitates will clog the fixed-bed biosorption column.

Other Cations

Light metals (e.g., Na and Ca) frequently occur in wastewaters, especially after pretreatment by precipitation. Moreover, industrial wastewaters may contain more than one type of heavy metal. Similarly to protons, other cations can compete with the metal ion in question for binding. Therefore, the uptake of each metal cation is generally reduced in the presence of others (Fig. 4c). This competition can occur at the level of both electrostatic and covalent binding.

Generally speaking, ions that are bound only weakly through mostly electrostatic attraction, like the alkaline (earth) metals, are effective in competing only with other weakly bound ions such as Zn^{2+} (30). To a lesser extent, they are able to reduce the uptake of ions like Cu^{2+}, Pb^{2+}, and Ag^{+}, where the bond character is more covalent (30, 105). This is because light metal ions such as Na^{+} are able to compete only with the electrostatically bound fraction of a heavy metal ion such as Cd^{2+}. To the degree that Cd^{2+} is covalently bound, it cannot be replaced by Na^{+} (96).

Among heavy metal ions, the effect of a more strongly binding ion such as Cu^{2+} on the uptake of the weaker competitor such as Zn^{2+} is obviously more pronounced than vice versa, as illustrated in Fig. 4c (21, 95). Radioactive elements (such as uranium and radium) can be strongly binding and therefore radioactive, and heavy metal ions mutually influence each other's uptake (105, 111). For biosorption of anions such as tetrachloroaureate, there would of course be electrostatic competition by other anions, not by cations. However, since Au is a soft metal with a covalent binding tendency, the competition by electrostatic binding of other anions should not be severe for this metal. For competition in covalent binding, the charge of the metal is irrelevant. The level of competition depends mainly on whether two metals use the same binding sites and how strongly they are bound to any given site.

Anions

Necessarily, anions of the metal salt that balance the positive charge of the metal ion occur in any metal-bearing solution. Additionally, chelating agents may be added in electroplating processes (13). These anions may affect the metal uptake. Anions could theoretically bind to the active sites of the biomass, thereby changing their state such that metal uptake is affected. However, there is little evidence of such phenomena. The more important effect of anions is that they may form complexes with the metals in solution. Whether this has a positive or negative effect on overall metal sorption depends on whether the metal-ligand complex has a higher or lower affinity for the biomass than the free cation does. In most cases of biosorption, metal uptake is reduced at increased ligand concentrations (37, 66, 106), which indicates that many complexes bind less strongly than the free metal ions. This effect is most pronounced for strong complexing agents such as EDTA (92, 106). Most anions (e.g., sulfonate, phosphonate, chloride, and carbonate) show only a weak influence (106). Consequently, the influence of anions is usually minor unless complex formation occurs.

MODELING THE BIOSORPTION EQUILIBRIUM

After establishing the general phenomena and trends in biosorption, the next step before applying it on an industrial scale is to develop mathematical models that describe the process quantitatively and aid in optimizing its operating conditions. Even for bench scale experiments, mathematical modeling can assist in reducing the number of experiments performed: the initial conditions of the experiment can be chosen more judiciously if one knows what outcome may be expected. Instead of systematically trying all possible variations of all parameters, one can limit the experiments to a number of spot-checks under relevant conditions in order to verify the validity of the model. Modeling is all the more important for applications on an industrial scale since any trials at this level are rather expensive. Modeling can aid the reactor design and help discover bottlenecks and optimize the operating conditions.

In this section, equilibrium models are presented. These are the basis for modeling of dynamic processes, e.g., in columns, that are of greater industrial relevance and are described in the following section. In both cases, the objective is to obtain models which are not just arbitrary statistical correlations but which are compatible with the observed binding mechanisms and which can be interpreted in a meaningful physicochemical way.

Isotherm Models of Langmuir and Freundlich

Simple isotherm models such as the Langmuir (25, 30, 44, 68) and Freundlich (20, 112) sorption isotherms, which predict metal uptake as a function of the concentration of that one metal, are commonly used in biosorption. In their basic form, both models describe the binding of one sorbate; i.e., they are suited for monometal systems without pH effects. The Freundlich isotherm can be interpreted as sorption to sites with an affinity distribution whereby the sites with higher affinity for the metal become occupied first (99, 101). The metal uptake, ${}^{M}q$ (in milliequivalents per gram), is then given by

$$^{M}q = c[M]^{1/n} \quad (1)$$

where c is related to the maximum binding capacity and n is related to the affinity or binding strength. The Langmuir isotherm (69), on the other hand, assumes that all sites (total number of binding sites, ^{t}B) have the same affinity and that one sorbate molecule binds to one binding site B. Secondary effects are neglected. The equilibrium constant ^{BM}K expresses the affinity between metal and biomass.

$$^{M}q = \frac{^{t}B\ ^{BM}K[M]}{1 + ^{BM}K[M]} \quad (2)$$

From equation 2, it follows that the maximum value of the metal uptake $^{M}q_{max}$, which is reached at high concentrations, corresponds to the total number of binding sites, ^{t}B. The slope in the origin is proportional to $^{t}B\ ^{BM}K$. It is desirable that both a high metal binding capacity, ^{t}B or $^{M}q_{max}$, and a high affinity, ^{BM}K (i.e., a steep slope in the origin), occur (Fig. 4a).

Both the Langmuir and Freundlich isotherms have been applied successfully to model biosorption. Their drawback is, however, that these simple isotherms cannot predict the effect of pH or other ions in solution. They are not appropriate for ion exchange where the metal that was initially occupying the binding sites is released and has to be regarded as a competitor. When ion exchange occurs, there is always more than one sorbate to consider.

Modeling Competition among Metal Cations

There are often several metals present in industrial wastewaters, and experiments have shown that these usually reduce each other's uptake. As pointed out in the preceding section, ion exchange among cations is a common phenomenon in biosorption. One can describe perfect ion exchange, where the total number of occupied binding sites remains constant, by ion exchange constants. For two monovalent ions, M and N, the ion exchange constant is

$$K = \frac{^{M}q[N]}{^{N}q[M]} \quad (3)$$

Ion exchange constants do not, however, allow the direct calculation of metal uptake, ^{M}q from known concentrations of the metals in solution; it is necessary to perform iterations. To avoid these iterations, it is useful to rather use an explicit isotherm equation that allows the calculation of ^{M}q as a function of the concentrations of all ions in solution. There are numerous possible multicomponent isotherm equations (123). In biosorption, a multicomponent Langmuir isotherm has been used (21, 108) which assumes a 1:1 stoichiometry between metal ions and binding sites whereby all metals make use of the same sites and compete for them. Apart from an easier calculation of the metal uptake, a further advantage over the use of ion exchange constants is that the isotherm accounts not only for ion exchange but also for free binding sites that become occupied with increasing metal concentrations. When metal uptake is modeled only in terms of chemical binding constants, free sites may occur at low metal concentrations. Once electrostatic attraction is also taken into account (see below), all sites are somehow occupied, if not with covalently bound ions then with electrostatically bound cations such as Na^+ in their vicinity.

Divalent ions often bind to two monovalent binding sites each (1:2 stoichiometry). This should also be accounted for in the model. Both for ion exchange and for isotherm equations, one can assume two different reaction mechanisms, (i) $2B + M \rightleftharpoons B_2M$ or (ii) $2B + M \rightleftharpoons 2\ BM_{0.5}$.

$$ {}^{BM}K = \frac{[BM_{0.5}]^2}{[B]^2[M]} \tag{4} $$

A modified Langmuir model based on the formation of B_2M complexes has been proposed (54); however, it required iterative calculations for ${}^{M}q$. Assuming the formation of $BM_{0.5}$ complexes allowed the formulation of an explicit multicomponent isotherm that rendered direct calculation of ${}^{M}q$ without iteration possible (94). An extended version of this isotherm that also includes pH effects is presented in equation 5 (see below). Predictions of this model for competition between divalent heavy metal ions are depicted in Fig. 4c.

Models That Include pH Effects

In many biosorption systems, metal uptake increases with pH. Due to the complexity of biosorbent materials, which may contain several functional groups with different unknown quantities and pK_a values, it is not easy to account for the pH effect in modeling. When pH-insensitive models (e.g., classical Langmuir isotherms) are used, it is therefore necessary to determine a new set of parameters (${}^{BM}K$ and ${}^{t}C$ for the simple Langmuir model) for each pH value (43, 48, 120). To be able to use the same constants for different pH values and also to predict the influence of pH on metal uptake (as well as the change of pH during sorption or the necessary amount of acid or base to be added to prevent pH changes), it is necessary to incorporate the pH as a variable into the isotherm equation. As pointed out in the preceding section, the strong effect of pH in biosorption of many metals is often due to competition of protons for the same binding sites that metal cations use. Therefore, the pH effect can be incorporated by treating protons as one of the competing cations (48, 49, 53, 94, 97). Preferably, the derived isotherm should allow direct calculation of the metal uptake, ${}^{M}q$, without iteration. Further advantages are gained if the isotherm can be used for more than one type of binding site and multimetal systems. Moreover, the observed binding stoichiometry (e.g., whether one divalent metal ion binds to one or two sites) should be taken into account. An isotherm model which fulfills these criteria has recently been derived from the definition of the equilibrium constants of reactions like equation 4 and the equations for conservation of mass for the binding sites (95). The case of a 1:2 stoichiometry for binding of two divalent metals, ${}^{1}M$ and ${}^{2}M$, as well as a 1:1

stoichiometry for proton uptake whereby two different types of binding sites (C and S) can be used by all three ions, is described by

$$
{}^{1}q = {}^{t}C \frac{\sqrt{{}^{C1}K[{}^{1}M]}}{1 + {}^{CH}K[H] + \sqrt{{}^{C1}K[{}^{1}M]} + \sqrt{{}^{C2}K[{}^{2}M]}} + {}^{t}S \frac{\sqrt{{}^{S1}K[{}^{1}M]}}{1 + {}^{SH}K[H] + \sqrt{{}^{S1}K[{}^{1}M]} + \sqrt{{}^{S2}K[{}^{2}M]}} \quad (5)
$$

The resulting predictions of this model regarding pH effects and competition between two metal ions are shown in Fig. 4b and c.

Modeling the Effects of Ionic Strength and Electrostatic Attraction

Light metals are frequently present in industrial wastewaters and are bound only weakly, mostly through electrostatic attraction. Nevertheless, a high Na^{+} concentration (or, more generally, a high ionic strength) can reduce the uptake of heavy metals. Due to electrostatic attraction, the concentration of all cations is higher in the vicinity of the negatively charged binding sites than it is in the bulk solution. That means that even heavy metals that tend to form complexes with the biomass are partially bound through electrostatic attraction. The total number of accumulated cations is limited, it has to balance the number of negatively charged sites (the biomass particle has to be overall charge neutral). If the concentration of one cation (e.g., Na^{+}) in the bulk solution rises, it will displace other electrostatically bound ions such that their concentration near the binding sites is lowered. That means that Na^{+} can reduce the amount of electrostatically bound heavy metals. Moreover, since the amount of covalently bound metals is determined by their concentrations near the binding site, their covalent binding is consequently reduced as well.

According to the Donnan theory (29), the polyelectrolyte (e.g., the charged biomass particle) constitutes a separate phase with homogenous concentrations throughout. The concentration factors (concentration $[X_p]$ near the interface/concentration [X] in the bulk) for ions of different valence are related through the equation

$$
\frac{[H_p^{+}]}{[H^{+}]} = \frac{[Na_p^{+}]}{[Na^{+}]} = \sqrt{\frac{[M_p^{2+}]}{[M^{2+}]}} = \left(\frac{[X_p^{zx+}]}{[X^{zx+}]}\right)^{1/zx} \quad (6)
$$

This means that the concentration factor for divalent cations, M^{2+}, is the square of the concentration factor for H^{+} or Na^{+}. The Donnan model has been applied to different polyelectrolytes (79), among others also to alginic acid (52, 57, 72), and recently to marine algal biomass (96). Other models for electrostatic effects imply a concentration gradient between the charged sites and the bulk solution (16). A Helmholtz model that assumes a constant capacitance was used for metal uptake by algae (121).

KINETICS

Batch Kinetics: General Observtions

Passive metal binding to cell surfaces is generally considered a rapid process. A large percentage of the equilibrium uptake is already reached in a few minutes, and complete equilibration is attained after a few hours (36, 71, 118). Desorption requires similar reaction times (122). Metal accumulation takes longer to reach equilibrium when transport into the cells of living organisms is involved (34) or in the case of microprecipitation, which can continue at a constant rate for hours or days (39, 74).

Theoretically, the rate of biosorption is determined by three steps: first, film diffusion from the bulk of the solution to the surface of the biosorbent particle has to occur, followed by diffusion through the particle, eventually followed by the chemical reaction of metal complexation.

The chemical reaction of metal complexation in weak-acid ion exchangers (47) or in natural materials (42, 103) is very fast. The rate of biosorption is therefore determined by mass transfer. Consequently, any claims of determining the reaction order of biosorption (65, 118) are to be regarded with caution. The reaction order of the chemical reaction can be determined only when mass transfer resistance is eliminated or can be reliably minimized or accounted for.

To what extent film or pore diffusion determines the reaction rate depends on the circumstances. Film diffusion is likely to play a limiting role for low agitation rates (71), at the initial instant, or when metal binding is confined to the outer parts of the particle (i.e., if the initial metal concentration is low and the amount of biosorbent per solution volume [S/L ratio] is high [3]). Under these conditions, an influence of the agitation rate can be observed (65). Pore or gel diffusion is rate limiting under the opposite conditions.

Batch Kinetics: Quantitative Modeling

Theoretically, a kinetic model should include the three terms for external film diffusion, internal pore (gel) diffusion, and chemical reaction. It is, however, possible to assume that the last step is fast enough as to be neglected, which was done by all the models mentioned below.

Simple lumped-parameter models assume that the rate is proportional to the displacement from equilibrium ($q_t - q_{eqm}$) or ($[M]_t - [M]_{eqm}$) and to the external surface area or the mass of the sorbent (108, 118). Such empirical models cannot be extrapolated to other conditions.

Other models are based exclusively on external film diffusion and therefore have a limited validity. The film diffusion coefficient is either calculated from literature correlations (for no relative motion, k_{film} = diffusivity/r_{hyd}) (58) or fitted to the experimental data using the microcolumn or the initial-slope method (3, 71). A model for conditions, where pore diffusion is rate limiting, yielded a diffusivity of a reasonable magnitude, which was about 40% of the molecular diffusivity. Other kinetic models (4, 19, 51), describing metal binding in alginate, yielded unreasonable values of the diffusivity.

A more comprehensive model was used for encapsulated biomass, comprising terms for external film diffusion, diffusion through the layer of encapsulating material, and diffusion through the biomass. Negligible metal accumulation in the pores was assumed (113). However, only very few data were given to support the model.

Necessity and Techniques for Immobilization of Biosorbent Particles

Particles to be used in biosorption reactors have to fulfill a number of requirements (14, 46, 113). First, they have to possess a high sorption capacity under equilibrium conditions. That implies that the particles should contain as little as possible of inert material that does not offer binding sites. Second, the biosorption kinetics should be favorable. For that purpose, it is desirable that the particles be hydrophilic, possess a high porosity or a gel-like structure, and have a particle size of 0.5 to 1.5 mm (110). Third, the flow dynamics in the reactor has to be reasonably smooth. This prohibits the use of very small or strongly swelling particles in packed-bed columns, since they would build up an excessive pressure drop, i.e., clog the column. A particle size of about 1 mm is often suitable. For the operation of fluidized beds, the required particle size and density depend on the flow velocity (13). Fourth, regeneration and frequent reuse of the particles must be possible. Consequently, the separation of metal-laden particles from the treated solution must be efficient (e.g., particles have to be easily filterable or magnetic [13]). Metal elution must be possible with little desorbing agent (and without damaging the biosorbent), and the particles have to withstand numerous sorption-desorption cycles (i.e., they must possess mechanical strength, be temperature stable, and be resistant to chemicals). Finally, cost-effectiveness is an important criterion.

To obtain particles that are stable and suited for use in reactors, it may be necessary to reinforce or immobilize the biomass, especially when microorganisms of small particle size are to be used. "Macro algae," such as, for example, *Sargassum*, can be used in columns without pretreatment (61). Reinforcement can be achieved by chemical cross-linking (e.g., with formaldehyde or glutaraldehyde), or immobilization can be achieved by entrapment, encapsulation, or attachment to a solid surface (46). Agents used for immobilization of biosorbents include alginate, silica, and polyacrylamide (7). Stainless steel wire, pan scourers, cotton webbing, alumina, coal, reticulated foams, or polyvinyl chloride can be used as solid supports (13, 75). The trade-off of these treatments is, however, that the sorption kinetics may be slowed and that a reduction of the sorption capacity (per total mass) by addition of inert materials or by blocking or consumption of binding sites may occur (13, 112).

Choice of Ionic Form of Sorbent

An important consideration is the choice of the ionic form of biosorbent, i.e., the ion with which its active sites are initially saturated before heavy metal uptake. Ideally, this should be the same ion as the one used for desorption, otherwise an intermediate conditioning or activation step must be added between desorption and adsorption, which increases the overall process costs.

When easily replaceable ions such as Ca^{2+} are used for initial saturation, metal sorption is facilitated but desorption with these weakly bound ions can consume large amounts of regenerating solution (and require a long time) and/or be ineffective (78). This means that it may yield low metal concentrations in the eluted solution or fail to replace all of the bound metal. Generally, desorption is more efficient if high concentrations of the desorbing agent are used. However, the solubility of some light metal ions such as Ca^{2+} is limited. Additional constraints are that ions such as Na^{+} and Mg^{2+} may solubilize alginate, which is the active ingredient of some biosorbents based on seaweed biomass.

Another alternative is to choose strongly binding ions such as protons. Desorption with acids is often fast and complete, and high concentration of metals in the regenerating solution can be achieved. However, the metal sorption to protonated biomass is impeded: release of protons leads to a lower pH (proton competition) in the column, the metal uptake is therefore lower, and breakthrough occurs earlier (63). Therefore, an intermediate conditioning step may be used after desorption and before adsorption, whereby the protons binding to the active sites (after acidic desorption) are replaced with light metals (63). This avoids an uncontrollable decrease of pH in the adsorption column and increases the metal uptake so that longer column operating times can be achieved (31, 56, 108).

Packed-Bed Column: Advantages and General Principles

The major advantage of choosing packed-bed columns as a reactor for biosorption is that they combine a reasonably high exploitation of the sorption capacity with very low effluent concentrations. As illustrated in Fig. 1 and 5a, this is achieved because at the influent (usually at the top of the column), the biosorbent is equilibrated at the relatively high concentration of the metal-bearing solution, so that high uptake values are achieved, whereas the effluent, whose metal concentration is very low, is in contact with fresh sorbent material (i.e., is operating near the origin of the sorption isotherm) (Fig. 4a).

In a biosorption column, three zones can be distinguished: one in which the sorbent is saturated and where the concentration equals that in the influent, one in which the sorbent is fresh and where the metal is virtually eliminated from solution, and an intermediate mass transfer zone which exhibits a concentration gradient. This mass transfer zone travels gradually down through the column until it hits the end of the column. At that time, the so-called breakthrough occurs, where the effluent concentration of one component exceeds its maximum allowed concentration and the column operation has to be stopped. A typical breakthrough curve that shows the change of the effluent concentration with time until it eventually reaches the influent concentration is depicted in Fig. 5b. Note that in industrial applications the column operation is already stopped at the earlier time when breakthrough occurs.

For optimal column performance, i.e., long operating times and high column utilization, it is desirable that the mass transfer zone be short. The length of the mass transfer zone increases with increasing mass transfer resistance, i.e., when the sorption kinetics is slow. Moreover, a shortening (sharpening) of the mass

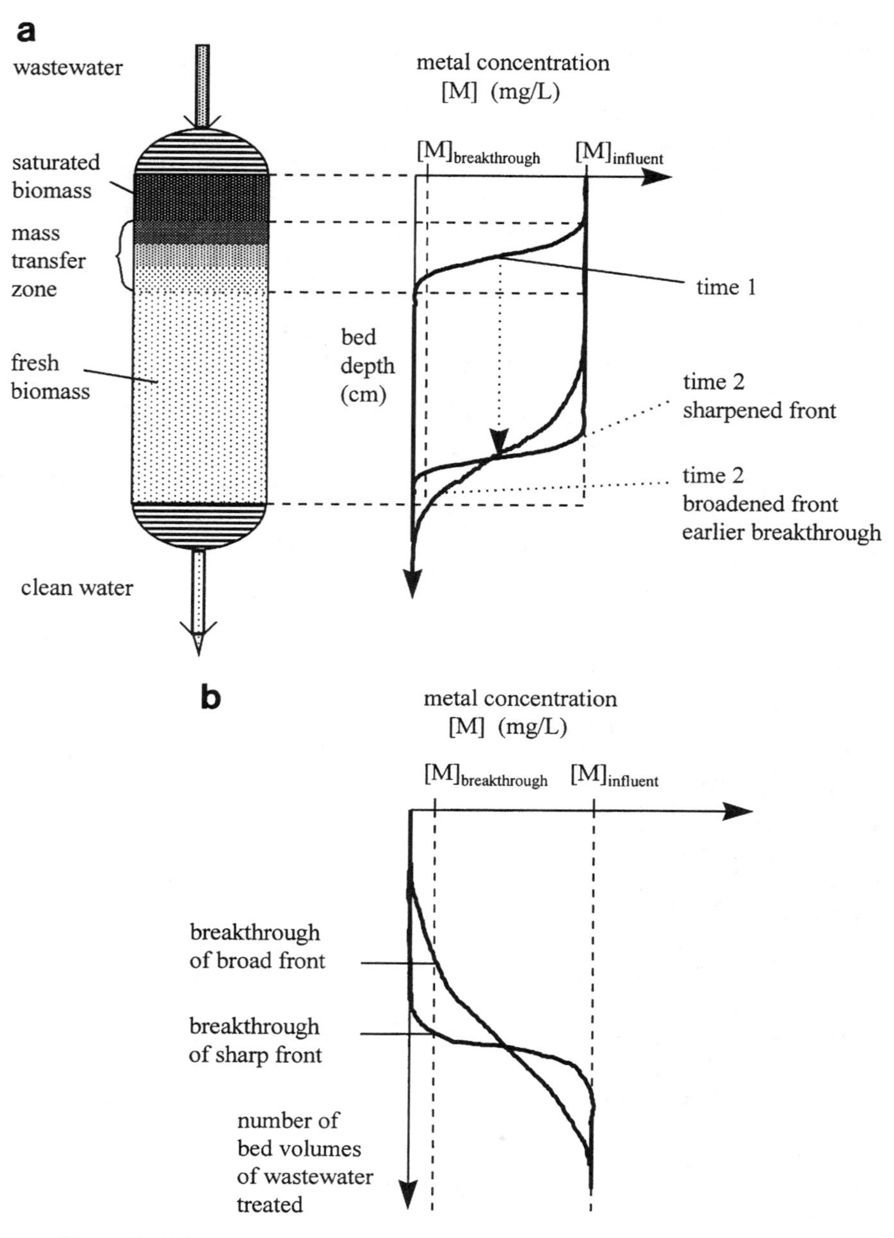

Figure 5. Column operation. (a) Metal concentration profiles in the column. (b) Breakthrough curve of metal concentration exiting the column.

transfer zone can occur when the sorbed species is bound more strongly than the released one (41, 63). Broadening occurs under the opposite conditions.

It is evident that the breakthrough occurs faster at high influent concentrations or for metals that bind weakly (78), as well as at higher flow rates and in shorter columns (118). To avoid giving undue importance to the effects of flow rate and column length and to make the data more comparable, it is advisable to plot the breakthrough curve versus the number of bed volumes treated (volume of solution treated until time t divided by column volume) rather than versus time.

Packed-Bed Column: Modeling

A model that is easy to use is the one developed by Bohart and Adams for activated carbon columns, which has also been applied to biosorption (55, 86, 118). This model assumes that the chemical reaction is the rate-limiting step. In biosorption, however, kinetics is likely to be determined by mass transfer. Moreover, the model does not take into account the fact that biosorption is often an ion exchange process and is unable to predict the effect of pH and competing ions. Therefore, it is only of limited use.

The equilibrium column model, which was developed for synthetic ion-exchange columns (59), assumes negligible mass transfer resistance and no axial dispersion, as well as constant feed composition, uniform sorbent presaturation, and a homogeneous bed. Requiring only the equilibrium constants, it can predict the order of breakthrough, the highest effluent concentrations of all components, and the minimum amount of sorbent required per volume of wastewater treated. This model has been applied to biosorption. For an influent containing mainly the strongly binding Cu^{2+} and small amounts of weakly binding Zn^{2+} it successfully predicted that the Zn^{2+} breakthrough occurs before the one for Cu^{2+} and that the Zn^{2+} concentration overshoots, exceeding its influent concentrations (62). Moreover, the equilibrium column model can indicate which species must be considered in more complete column models including mass transfer.

A model originally developed for ion exchangers that includes an overall mass transfer resistance (104) was successfully used to predict the breakthrough curves in biosorption (63).

OUTLOOK

Screening of biosorbents under different conditions has identified a number of potent biosorbents and established how different parameters affect the metal uptake. Important progress has been made in understanding the mechanism of biosorption and in quantitative modeling of this process under controlled laboratory conditions. To confirm the ability of biosorption to effectively treat industrial wastewaters, trials with actual wastewater must be performed, both on the bench scale and eventually on the pilot plant scale. Companies that generate metal-bearing effluents will only be convinced to jump ahead and go for biosorption if the potential of this process is demonstrated. To provide ready-to-use biosorbents, immobilization procedures, especially for microbial biomass, must be further optimized.

REFERENCES

1. **Ahrland, S., J. Chatt, and N. R. Davis.** 1958. The relative affinities of ligand atoms for acceptor molecules and ions. *Q. Rev. Chem.* **12:**265–276.
2. **Aldor, I., E. Fourest, and B. Volesky.** 1995. Desorption of cadmium from algal biosorbent. *Can. J. Chem. Eng.* **73:**516–522.
3. **Allen, S., P. Brown, G. McKay, and O. Flynn.** 1992. An evaluation of single resistance transfer models in the sorption of metal ions by peat. *J. Chem. Technol. Biotechnol.* **54:**271–276.
4. **Apel, M. L., and A. E. Torma.** 1993. Determination of kinetics and diffusion coefficients of metal sorption on Ca-alginate beads. *Can. J. Chem. Eng.* **71:**652–656.
5. **Bailey, J. E., and D. F. Ollis.** 1986. *Biochemical Engineering Fundamentals*, 2nd ed., p. 58–59, 120. McGraw-Hill Book Co., New York, N.Y.
6. **Bartnicki-Garcia, S.** 1973. Fungal cell wall composition, p. 201. *In* A. I. Laskin and H. A. Lechevalier (ed.), *CRC Handbook of Microbiology*, vol. 2. *Microbial Composition.* CRC Press, Inc., Boca Raton, Fla.
7. **Bedell, G. W., and D. W. Darnall.** 1990. Immobilization of nonviable, biosorbent, algal biomass for the recovery of metal ions, p. 313–326. *In* B. Volesky (ed.), *Biosorption of Heavy Metals.* CRC Press, Inc., Boca Raton, Fla.
8. **Beveridge, T. J.** 1986. The immobilization of soluble metals by bacterial walls, p. 127–140. *In* H. L. Ehrlich and D. S. Holmes (ed.), *Biotechnology and Bioengineering Symposium no. 16. Biotechnology for the Mining, Metal-Refining, and Fossil Fuel Processing Industries.* Wiley Interscience, New York, N.Y.
9. **Beveridge, T. J.** 1990. Interactions of metal ions with components of bacterial cell walls and their biomineralization, p. 65–83. *In* R. K. Poole and G. M. Gadd (ed.), *Metal-Microbe Interactions.* IRL Press, Oxford, United Kingdom.
10. **Black, W. A. P., and R. L. Mitchell.** 1952. Trace elements in the common brown algae and in sea water. *J. Mar. Biol. Assoc.* **30:**575–584.
11. **Bold, H. C., C. J. Alexopoulos, and T. Delevoryas.** 1987. *Morphology of Plants and Fungi*, p. 42–48. Harper & Row, New York, N.Y.
12. **Bold, H. C., and M. J. Wynne.** 1985. *Introduction to the Algae*, p. 20–22, 70–75, 288–289, 301, 478–479, and 516. Prentice-Hall, Englewood Cliffs, N.J.
13. **Brierley, C. L.** 1990. Metal immobilization using bacteria, p. 303–324. *In* H. L. Ehrlich and C. L. Brierley (ed.), *Microbial Mineral Recovery.* McGraw-Hill Book Co., New York, N.Y.
14. **Brierley, J. A.** 1990. Production and application of a *Bacillus*-based product for use in metals biosorption, p. 305–312. *In* B. Volesky (ed.), *Biosorption of Heavy Metals.* CRC Press, Inc., Boca Raton, Fla.
15. **Brierley, J. A., C. L. Brierley, and G. M. Goyak.** 1986. AMT-BIOCLAIM: a new wastewater treatment and metal recovery technology, p. 291–304. *In* R. W. Lawrence, R. M. R. Branion, and H. G. Ebner (ed.), *Fundamental and Applied Biohydrometallurgy.* Elsevier Science Publishing, Amsterdam, The Netherlands.
16. **Buffle, J.** 1988. *Complexation Reactions in Aquatic Systems: an Analytical Approach*, p. 47–72, 195–303, 306–330. Ellis Horwood Ltd., Chichester, United Kingdom.
17. **Cahn, R. S., and O. L. Dermer.** 1979. *Introduction to Chemical Nomenclature*, p. 17–18. Butterworths, London, United Kingdom.
18. **Chapman, V. J.** 1980. *Seaweeds and Their Uses.* p. 122, 253–278. Chapman & Hall, London, United Kingdom.
19. **Chen, D., Z. Lewandowski, F. Roe, and P. Surapaneni.** 1993. Diffusivity of Cu^{2+} in calcium alginate beads. *Biotechnol. Bioeng.* **41:**755–760.
20. **Chen, X. H., T. Gosset, and D. R. Thevenot.** 1990. Batch copper ion binding and exchange properties of peat. *Water Res.* **24:**1463–1471.
21. **Chong, K. H., and B. Volesky.** 1995. Description of two-metal biosorption equilibria by Langmuir-type models. *Biotechnol. Bioeng.* **47:**451–460.
22. **Collins, Y. E., and G. Stotzky.** 1992. Heavy metals alter the electrokinetic properties of bacteria, yeasts and clay minerals. *Appl. Environ. Microbiol.* **58:**1592–1600.

23. **Crist, R. H., J. R. Martin, J. Chonko, and D. R. Crist.** 1996. Uptake of metals on peat moss: an ion-exchange process. *Environ. Sci. Technol.* **30:**2456–2461.
24. **Crist, R. H., J. R. Martin, P. W. Guptill, J. M. Eslinger, and D. R. Crist.** 1990. Interactions of metals and protons with algae. 2. Ion exchange in adsorption and metal displacement by protons. *Environ. Sci. Technol.* **24:**337–342.
25. **Crist, R. H., K. Oberholser, J. McGarrity, D. R. Crist, J. K. Johnson, and J. M. Brittsan.** 1992. Interaction of metals and protons with algae. 3. Marine algae, with emphasis on lead and aluminum. *Environ. Sci. Technol.* **26:**496–502.
26. **Crist, R. H., K. Oberholser, N. Shank, and M. Nguyen.** 1981. Nature of bonding between metallic ions and algal cell walls. *Environ. Sci. Technol.* **15:**1212–1217.
27. **Deacon, J. W.** 1984. *Introduction to Modern Mycology*, p. 4–7. Blackwell Scientific Publications Ltd., Oxford, United Kingdom.
28. **Dean, J. A.** 1985. *Lange's Handbook of Chemistry*, p. 3.11–3.12. McGraw-Hill Book Co., New York, N.Y.
29. **Donnan, F. G.** 1911. Theorie der Membrangleichgewichte und Membranpotentiale bei Vorhandensein von nicht dialysierenden Elektrolyten. *Z. Elektrochem.* **17:**572–581.
30. **Ferguson, J., and B. Bubela.** 1974. The concentration of Cu (II), Pb (II), and Zn (II) from aqueous solutions by particulate algal matter. *Chem. Geol.* **13:**163–186.
31. **Fourest, E., and J. C. Roux.** 1994. Improvement of heavy metal biosorption by mycelial dead biomass (*Rhizopus arrhizus, Mucor miehei* and *Penicillium chrysogenum*): pH control and cationic activation. *FEMS Microbiol. Rev.* **14:**325–332.
32. **Gadd, G. M.** 1990. Fungi and yeasts for metal accumulation, p. 249–276. *In* H. L. Ehrlich and C. L. Brierley (ed.), *Microbial Mineral Recovery.* McGraw-Hill Book Co., New York, N.Y.
33. **Gadd, G. M., and C. White.** 1992. Removal of thorium from simulated acid process stream by fungal biomass: potential for thorium desorption and reuse of biomass and desorbent. *J. Chem. Technol. Biotechnol.* **55:**39.
34. **Garnham, G. W., G. A. Codd, and G. M. Gadd.** 1992. Kinetics of uptake and intracellular location of cobalt, manganese and zinc in the estuarine green alga *Chlorella salina. Appl. Microbiol. Biotechnol.* **37:**270–276.
35. **Greene, B., and D. W. Darnall.** 1988. Temperature dependence of metal ion sorption by *Spirulina.* Biorecovery **1:**27–41.
36. **Greene, B., M. T. Henzl, J. M. Hosea, and D. W. Darnall.** 1986. Elimination of bicarbonate interference in the binding of U(VI) in mill-waters to freeze-dried *Chlorella vulgaris.* Biotechnol. Bioeng. **28:**764.
37. **Greene, B., M. Hosea, R. McPherson, M. Henzl, M. D. Alexander, and D. W. Darnall.** 1986. Interaction of gold(I) and gold(III) complexes with algal biomass. *Environ. Sci. Technol.* **20:**627–632.
38. **Greene, B., R. McPherson, and D. Darnall.** 1987. Algal sorbents for selective metal ion recovery, p. 315–338. *In* J. W. Patterson and R. Pasino (ed.), *Metals Speciation, Separation and Recovery.* Lewis, Chelsea, Mich.
39. **Hatch, R. T., and A. Menawat.** 1978. Biological removal and recovery of trace heavy metals. *Biotechnol. Bioeng. Symp.* **8:**191–203.
40. **Haug, A., and O. Smidsrod.** 1970. Selectivity of some anionic polymers for divalent metal ions. *Acta Chem. Scand.* **24:**843–854.
41. **Helfferich, F.** 1962. *Ion Exchange*, p. 72–94. McGraw-Hill Book Co., New York, N.Y.
42. **Hering, J. G., and F. M. M. Morel.** 1990. The kinetics of trace metal complexation: implications for metal reactivity in natural waters, p. 145–171. *In* W. Stumm (ed.), *Aquatic Chemical Kinetics.* Wiley Interscience, New York, N.Y.
43. **Ho, Y. S., D. A. J. Wase, and C. F. Forster.** 1995. Batch nickel removal from aqueous solution by sphagnum moss peat. Water. Res. **29:**1327–1332.
44. **Holan, Z. R., and B. Volesky.** 1994. Biosorption of lead and nickel by biomass of marine algae. *Biotechnol. Bioeng.* **43:**1001–1009.
45. **Holan, Z. R., and B. Volesky.** 1995. Accumulation of cadmium, lead and nickel by fungal and wood biosorbents. *Appl. Biochem. Biotechnol.* **53:**133–142.

46. **Holbein, B. E.** 1990. Immobilization of metal-binding compounds, p. 327–340. *In* B. Volesky (ed.), *Biosorption of Heavy Metals.* CRC Press, Inc., Boca Raton, Fla.
47. **Holl, W., and H. Sontheimer.** 1977. Ion exchange kinetics of the protonation of weak acid ion exchange resins. *Chem. Eng. Sci.* **32:**755–762.
48. **Huang, C., C. P. Huang, and A. L. Morehart.** 1991. Proton competition in Cu (II) adsorption by fungal mycelia. *Water Res.* **25:**1365–1375.
49. **Huang, J.-P., C. P. Huang, and A. L. Morehart.** 1991. Removal of heavy metals by fungal (*Aspergillus oryzae*) adsorption, p. 329–349. *In* J. P. Vernet (ed.), *Heavy Metals in the Environment.* Elsevier Science Publishers, Amsterdam, The Netherlands.
50. **Hunt, S.** 1986. Diversity of biopolymer structure and its potential for ion-binding applications, p. 15–45. *In* H. Eccles and S. Hunt (ed.), *Immobilisation of Ions by Bio-Sorption.* Ellis Horwood, Chichester, United Kingdom.
51. **Jang, L. K., W. Brand, M. Resong, W. Mainieri, and G. G. Geesey.** 1990. Feasibility of using alginate to absorb dissolved copper from aqueous media. *Environ. Prog.* **9:**269–274.
52. **Jang, L. K., N. Harpt, D. Grasmick, L. N. Vuong, and G. Geesey.** 1990. A two-phase model for determining the stability constants for interactions between copper and alginic acid. *J. Phys. Chem.* **94:**482–488.
53. **Jang, L. K., D. Nguyen, and G. G. Geesey.** 1995. Effect of pH on the absorption of Cu (II) by alginate gel. *Water Res.* **29:**315–321.
54. **Jang, L. K., D. Nguyen, and G. G. Geesey.** 1995. Selectivity of alginate gel for Cu vs. Co. *Water Res.* **29:**307–313.
55. **Jansson-Charrier, M., E. Guibal, J. Roussy, R. Surjous, and P. LeCloirec.** 1996. Dynamic removal of uranium by chitosan: influence of operating parameters. *Water Sci. Technol.* **34:**169–177.
56. **Jeffers, T. H., C. R. Ferguson, and P. G. Bennett.** 1991. *Biosorption of Metal Contaminants Using Immobilized Biomass—a Laboratory Study.* Report of investigations. U.S. Bureau of Mines, Salt Lake City, Utah.
57. **Katchalsky, A., R. E. Cooper, J. Upadhyay, and A. Wassermann.** 1961. Counter-ion fixation in alginates. *J. Am. Chem. Soc.* **83:**5198–5204.
58. **Khummongkol, D., G. S. Canterford, and C. Fryer.** 1982. Accumulation of heavy metals in unicellular alga. *Biotechnol. Bioeng.* **24:**2643–2660.
59. **Klein, G., D. Tondeur, and T. Vermeulen.** 1967. Multicomponent ion-exchange in fixed beds. *Ind. Eng. Chem. Fundam.* **6:**339–350.
60. **Kohn, R.** 1975. Ion binding on polyuronates—alginate and pectin. *Pure Appl. Chem.* **42:**371–397.
61. **Kratochvil, D., E. Fourest, and B. Volesky.** 1995. Biosorption of copper by *Sargassum fluitans* biomass in fixed-bed column. *Biotechnol. Lett.* **17:**777–782.
62. **Kratochvil, D., and B. Volesky.** Multicomponent biosorption in fixed beds. *Water Res.*, in press.
63. **Kratochvil, D., B. Volesky, and G. Demopoulos.** 1997. Optimizing Cu removal/recovery in a biosorption column. *Water Res.* **31:**2327–2339.
64. **Kuyucak, N.** 1990. Feasibility of biosorbents application, p. 371–378. *In* B. Volesky (ed.), *Biosorption of Heavy Metals.* CRC Press, Inc., Boca Raton, Fla.
65. **Kuyucak, N., and B. Volesky.** 1989. Accumulation of cobalt by marine alga. *Biotechnol. Bioeng.* **33:**809–814.
66. **Kuyucak, N., and B. Volesky.** 1989. Accumulation of gold by algal biosorbent. *Biorecovery* **1:**189–204.
67. **Kuyucak, N., and B. Volesky.** 1989. Desorption of cobalt-laden algal biosorbent. *Biotechnol. Bioeng.* **33:**815–822.
68. **Kuyucak, N., and B. Volesky.** 1989. The mechanism of cobalt biosorption. *Biotechnol. Bioeng.* **33:** 823–831.
69. **Langmuir, I.** 1918. The adsorption of gases on plane surfaces of glass, mica and platinum. *J. Am. Chem. Soc.* **40:**1361–1403.
70. **Lee, R. E.** 1989. *Phycology*, p. 10–13, 34–37, 534–539, 584–599. Cambridge University Press, Cambridge, United Kingdom.
71. **Leusch, A., and B. Volesky.** 1995. The influence of film diffusion on cadmium biosorption by marine biomass. *J. Biotechnol.* **43:**1–10.

72. **Lin, F. G., and J. A. Marinsky.** 1993. A Gibbs-Donnan-based interpretation of the effect of medium counterion concentration levels on the acid dissociation properties of alginic acid and chondroitin sulfate. *React. Polymers* **19:**27–45.
73. **Lobban, C. S., and P. J. Harrison.** 1994. *Seaweed Ecology and Physiology*, p. 283–298. Cambridge University Press, Cambridge, United Kingdom.
74. **Macaskie, L. E.** 1990. An immobilized cell bioprocess for the removal of heavy metals from aqueous flows. *J. Chem. Technol. Biotechnol.* **49:**357–379.
75. **Macaskie, L. E., and A. C. R. Dean.** 1989. Microbial metabolism, desolubilization and deposition of heavy metals: metal uptake by immobilized cells and application to the detoxification of liquid wastes, p. 159. *In* A. Mizrahi (ed.), *Advances in Biotechnological Processes.* Alan R. Liss, Inc., New York, N.Y.
76. **Macaskie, L. E., A. C. R. Dean, A. K. Cheetham, R. J. B. Jakeman, and A. J. Skarnulis.** 1987. Cadmium accumulation by a *Citrobacter* sp.: the chemical nature of the accumulated metal precipitate and its location on the bacterial cells. *J. Gen. Microbiol.* **133:**539–544.
77. **Macaskie, L. E., R. M. Empson, A. K. Cheetham, C. P. Grey, and A. J. Skarnulis.** 1992. Uranium bioaccumulation by a *Citrobacter* sp. as a result of enzymatically mediated growth of polycrystalline HUO_2PO_4. *Science* **257:**782–784.
78. **Maranon, E., and E. Sastre.** 1991. Heavy metal removal in packed beds using apple wastes. *Bioresource Technol.* **38:**39–43.
79. **Marinsky, J. A.** 1987. A two-phase model for the interpretation of proton and metal ion interaction with charged polyelectrolyte gels and their linear analogs, p. 49–81. *In* W. Stumm (ed.), *Aquatic Surface Chemistry.* Wiley Interscience, New York, N.Y.
80. **Martell, A. E., and M. Calvin.** 1952. *Chemistry of Metal Chelate Compounds,* p. 168–169. Prentice-Hall, Inc., Englewood Cliffs, N.J.
81. **May, H.** 1984. *Biosorption by Industrial Microbial Biomass.* M.Eng. thesis. McGill University, Montreal, Canada.
82. **Mayers, I. T., and T. J. Beveridge.** 1989. The sorption of metals to *Bacillus subtilis* walls from dilute solutions and simulated Hamilton Harbour (Lake Ontario) water. *Can. J. Microbiol.* **35:**764–770.
83. **McLean, R. J. C., and T. J. Beveridge.** 1990. Metal-binding capacity of bacterial surfaces and their ability to form mineralized aggregates, p. 185–222. *In* H. L. Ehrlich and C. L. Brierley (ed.), *Microbial Mineral Recovery.* McGraw-Hill Book Co., New York, N.Y.
84. **Moore-Landecker, E.** 1996. *Fundamentals of the Fungi*, p. 15–18. Prentice-Hall, Inc., Upper Saddle River, N.J.
85. **Mueller, E., and W. Loeffler.** 1976. *Mycology*, p. 59–64. Thieme, Stuttgart, Germany.
86. **Muraleedharan, T. R., L. Lyengar, and C. Venkobachar.** 1994. Further insight into the mechanism of biosorption of heavy metals by *Ganoderma lucidum. Environ. Technol.* **15:**1015–1027.
87. **Nieboer, E., and D. H. S. Richardson.** 1980. The replacement of the nondescript term 'heavy metals' by a biologically and chemically significant classification of metal ions. *Environ. Pollut.* **1B:** 11–13.
88. **Pagenkopf, G. K.** 1978. *Introduction to Natural Water Chemistry*, p. 161–167, 214–216, 220–230. Marcel Dekker, Inc., New York, N.Y.
89. **Pauling, L.** 1967. *Nature of the Chemical Bond*, p. 55–73. Cornell University Press, Ithaca, NY.
90. **Pearson, R. G.** 1967. Hard and soft acids and bases. *Chem. Bri.* **3:**103–107.
91. **Percival, E., and R. H. McDowell.** 1967. *Chemistry and Enzymology of Marine Algal Polysaccharides*, p. 99–126, 127–156. Academic Press, Ltd., London, United Kingdom.
92. **Ramelow, G. J., D. Fralick, and Y. Zhao.** 1992. Factors affecting the uptake of aqueous metal ions by dried seaweed biomass. *Microbios* **72:**81–93.
93. **Remacle, J.** 1990. The cell wall and metal binding, p. 83–92. *In* B. Volesky (ed.), *Biosorption of Heavy Metals.* CRC Press, Inc., Boca Raton, Fla.
94. **Schiewer, S., and B. Volesky.** 1995. Modeling of the proton-metal ion exchange in biosorption. *Environ. Sci. Technol.* **29:**3049–3058.
95. **Schiewer, S., and B. Volesky.** 1996. Modeling of multi-metal ion exchange in biosorption. *Environ. Sci. Technol.* **30:**2921–2927.

96. **Schiewer, S., and B. Volesky.** 1997. Ionic strength and electrostatic effects in biosorption of divalent metal ions and protons. *Environ. Sci. Technol.* **31:**2478–2485.
97. **Seki, H., A. Suzuki, and I. Kashiki.** 1990. Adsorption of lead ions on immobilized humic acid. *J. Colloid Interface Sci.* **134:**59–65.
98. **Siegel, B. Z., and S. M. Siegel.** 1973. The chemical composition of algal cell walls. *Crit. Rev. Microbiol.* **3:**1–26.
99. **Smith, J. M.** 1981. *Chemical Engineering Kinetics*, p. 310–322. McGraw-Hill Book Co., New York, N.Y.
100. **South, G. R., and A. Whittick.** 1987. *Introduction to Phycology*, p. 27–28, 46–63, 104–115, 176–177, 268–271. Blackwell Scientific Publications Ltd., Oxford, United Kingdom.
101. **Stumm, W.** 1992. *Chemistry of the Solid-Water Interface*, p. 87–97. John Wiley & Sons, Inc., New York, N.Y.
102. **Stumm, W., and J. J. Morgan.** 1970. *Aquatic Chemistry*, p. 445–513. John Wiley & Sons, Inc., New York, N.Y.
103. **Stumm, W., L. Sigg, and B. Sulzberger.** 1994. The role of coordination at the surface of aquatic particles, p. 45–89. *In* J. Buffle and R. R. De Vitre (ed.), *Chemical and Biological Regulation of Aquatic Systems.* Lewis Publishers, Boca Raton, Fla.
104. **Tan, H. K. S., and I. H. Spinner.** 1994. Multicomponent ion exchange column dynamics. *Can. J. Chem. Eng.* **72:**330–341.
105. **Tobin, J. M., D. G. Cooper, and R. J. Neufeld.** 1988. The effects of cation competition on metal adsorption by *Rhizopus arrhizus* biomass. *Biotechnol. Bioeng.* **31:**282–286.
106. **Tobin, J. M., D. G. Cooper, and R. J. Neufeld.** 1987. Influence of anions on metal adsorption by *Rhizopus arrhizus* biomass. *Biotechnol. Bioeng.* **30:**882–886.
107. **Treen-Sears, M. E., B. Volesky, and R. J. Neufeld.** 1984. Ion exchange/complexation of the uranyl ion by *Rhizopus* biosorbent. *Biotechnol. Bioeng.* **26:**1323–1329.
108. **Trujillo, E. M., T. H. Jeffers, C. Ferguson, and H. Q. Stevenson.** 1991. Mathematically modeling the removal of heavy metals from wastewater using immobilized biomass. *Environ. Sci. Technol.* **25:**1559–1565.
109. **Tsezos, M.** 1984. Recovery of Uranium from biological adsorbents—desorption equilibrium. *Biotechnol. Bioeng.* **26:**973–981.
110. **Tsezos, M.** 1986. Adsorption by microbial biomass as a process for removal of ions from process or waste solutions, p. 200–209. *In* H. H. Eccles and S. Hunt (ed.), *Immobilisation of Ions by Biosorption.* Ellis Horwood, Chichester, United Kingdom.
111. **Tsezos, M.** 1990. Engineering aspects of metal binding by biomass, p. 325–340. *In* H. L. Ehrlich and C. L. Brierley (ed.), *Microbial Mineral Recovery.* McGraw-Hill Book Co., New York, N.Y.
112. **Tsezos, M., and A. A. Deutschmann.** 1990. An investigation of engineering parameters for the use of immobilised biomass particles in biosorption. *J. Chem. Technol. Biotechnol.* **48:**29–39.
113. **Tsezos, M., S. H. Noh, and M. H. I. Baird.** 1988. A batch reactor mass transfer kinetic model. *Biotechnol. Bioeng.* **32:**545–553.
114. **Vogel, G., and H. Angermann.** 1984. Atlas zur Biologie, p. 60. DTV, Munich, Germany.
115. **Volesky, B.** 1990. Biosorption by fungal biomass, p. 139–172. *In* B. Volesky (ed.), *Biosorption of Heavy Metals.* CRC Press, Inc., Boca Raton, Fla.
116. **Volesky, B.** 1990. Removal and recovery of heavy metals by biosorption, p. 7–43. *In* B. Volesky (ed.), *Biosorption of Heavy Metals.* CRC Press, Inc., Boca Raton, Fla.
117. **Volesky, B., and Z. R. Holan.** 1995. Biosorption of heavy metals. *Biotechnol. Prog.* **11:**235–250.
118. **Volesky, B., and I. Prasetyo.** 1994. Cadmium removal in a biosorption column. *Biotechnol. Bioeng.* **43:**1010–1015.
119. **Weppen, P., and A. Hornburg.** 1995. Calorimetric studies on interactions of divalent cations and microorganisms or microbial envelopes. *Thermochim. Acta* **269/270:**393–404.
120. **Xue, H. B., and L. Sigg.** 1990. Binding Cu(II) to algae in a metal buffer. *Water Res.* **24:**1129–1136.
121. **Xue, H.-B., W. Stumm, and L. Sigg.** 1988. The binding of heavy metals to algal surfaces. *Water Res.* **22:**917–926.
122. **Yang, J., and B. Volesky.** 1996. Intraparticle diffusivity of Cd ions in a new biosorbent material. *J. Chem. Technol. Biotechnol.* **66:**355–364.
123. **Yu, J.-W., and I. Neretnieks.** 1990. Single-component and multicomponent adsorption equilibria on activated carbon of methylcyclohexane, toluene and isobutyl methyl ketone. *Ind. Eng. Chem. Res.* **29:**220–231.

Environmental Microbe-Metal Interactions
Edited by Derek R. Lovley

Chapter 15

Biodegradation of Synthetic Chelating Agents

Harvey Bolton, Jr., Luying Xun, and Don C. Girvin

Multidentate aminopolycarboxylic synthetic chelating agents such as EDTA and nitrilotriacetate (NTA) form stable water-soluble complexes with a wide range of metals and radionuclides (4, 53, 71). Synthetic chelating agents have many useful applications in various industries including the cosmetic, food, leather, metal, pulp and paper, pharmaceutical, photographic, rubber, and textile industries (70). Synthetic chelating agents have been used to decontaminate nuclear reactors and other materials (6, 64) and to process nuclear waste (55).

The codisposal of synthetic chelating agents with radionuclides at U.S. Department of Energy (DOE) sites has lead to the enhancement of the transport of radionuclides and metals in groundwater (18, 57, 69), because the chelating agents substantially alter the adsorptive properties of the radionuclides and metals (31, 36, 56, 97).

Ion-exchange studies and analyses of groundwater from Oak Ridge National Laboratory indicated that codisposed EDTA influenced the mobility of ^{60}Co and to some extent Pu (57). At Chalk River, Canada, sediment porewater from a chemical pit that received radioactive waste and EDTA contained mobile ^{60}Co, Pu, and Am in anionic organic complexes (16, 42). At Maxey Flats, Kentucky, Cleveland and Rees (18) detected soluble ^{60}Co and Pu in trench leachates as EDTA complexes with 0.1 to 0.5 mg of EDTA/liter. Later studies (86) found 1.3 mg of EDTA/liter in experimental wells at Maxey Flats. Radionuclide-EDTA mixtures have also been codisposed at the DOE Hanford Site's 300 and 200 Areas, the Radioactive Waste Management Complex at the Idaho National Engineering and Environmental Laboratory, and the Bear Creek burial grounds and low-level waste trenches at the Oak Ridge National Laboratory (69). As an example, approximately 2.2×10^5 kg of EDTA (1, 43) and 240 to 381 kg of Pu (1, 30) were discharged to single-shell storage tanks at the Hanford Site from 1944 to 1975. Sixty-seven of the tanks have leaked or are suspected of leaking (30) Pu and EDTA into the vadose zone beneath

Harvey Bolton, Jr., and Don C. Girvin • Pacific Northwest National Laboratory, Richland, WA 99352. ***Luying Xun*** • Washington State University, Pullman, WA 99164.

the tanks. No monitoring of the vadose zone beneath the tanks has been conducted, and so the amount of Pu or EDTA released and the volume of sediment impacted are unknown. Although some of the EDTA is expected to have been radiolytically degraded before release, the tanks still contain a significant quantity of EDTA. As an example, approximately 25% of the original EDTA was degraded by radiolysis in tank SY-101 at the Hanford Site when assayed in 1993 (14). No Pu has been detected yet in groundwater sampled from monitoring wells near the tanks, but the potential exists for enhanced Pu transport as the PuEDTA complex in the vadose zone to the groundwater.

Microorganisms can influence the mobility of chelated radionuclides and metals through several processes, which can either remove the radionuclide or metal from the chelating agents (e.g., bioaccumulation, biosorption, oxidation, reduction, and production of competing ligands) or remove the chelating agent (e.g., biodegradation) (11). Biodegradation of the chelating agent is an important process and the focus of this chapter because the organic complexant can no longer influence radionuclide and metal mobility once it is biodegraded. However, the radionuclide contamination of groundwater facilitated by chelating agents is unique to DOE sites and may be a long-term problem because of the recalcitrance of EDTA to biodegradation in near-surface (12, 58, 82, 83) and subsurface (12) sediments. As an example, the maximum amount of 10^{-5} M EDTA mineralized to CO_2 in subsurface sediments from the DOE Savannah River Plant at 36 to 376 m deep under optimal laboratory conditions was 15% after 115 days of incubation (12). Biodegradation of the EDTA-complexing radionuclides can significantly alter the mobility of the initially chelated radionuclides. However, little is known about the biodegradation of radionuclide-EDTA complexes. Bacteria able to degrade both NTA and EDTA have been isolated and offer an opportunity to study the mechanisms of biodegradation of radionuclide-chelate complexes. Bacteria able to degrade NTA have been isolated from freshwaters (24, 29), soils and sediments (20, 22, 23, 65, 85), and sewage effluents (28, 39, 49, 93). Bacteria able to degrade EDTA have been isolated by selective enrichment from EDTA-containing environments, they include an *Agrobacterium* sp. from a waste treatment facility (47), the gram-negative bacterium BNC1 from a soil enrichment (60), and the gram-negative bacterium DSM 9103 from an enrichment culture (92).

The objectives of this chapter are to discuss the current state of knowledge of the biodegradation of the synthetic chelating agents NTA and EDTA. The understanding of the biodegradation of chelating agents requires a mechanistic coupled understanding of geochemistry and microbiology, including the way chelated metal influences biodegradation, the way metal in the metal-chelate complex can exchange with other metals, the way adsorption of the chelating agent influences biodegradation, the biodegradation of chelating agents by whole cells, the cellular transport of chelating agents into the cell, its enzymatic biodegradation, and, finally, the genetics of biodegradation.

INFLUENCE OF AQUEOUS SPECIATION AND LABILITY OF THE COMPLEX ON THE BIODEGRADATION OF CHELATING AGENTS

Studies of the biodegradation of radionuclide- and metal-chelate complexes in aqueous environments are complicated because several physicochemical factors affect chelating-agent biodegradation. These factors include the aqueous speciation of the chelating agent or the form of the chelating agent in solution [e.g., Pu(IV)EDTA, $MgEDTA^{2-}$, and $HEDTA^{3-}$), the solubility and oxidation state of the chelated metal, and the pH of the environment. The dominance of a specific metal-chelate complex in solution depends on the solution pH, the ratio of metal to ligand concentration (e.g., [Pu]/[EDTA]), and aqueous composition including competing metals cations (e.g., Ca^{2+}, Fe^{3+}, and Al^{3+}), and competing inorganic ligands (e.g., hydroxyl and carbonate) and organic ligands (e.g., siderophores and exopolysaccharides). Metals and protons compete for the chelating agent at low pH, while hydroxyl and carbonate ions and organic ligands compete with the chelating agent for metals at higher pH. Thus, these multiple equilibria will dictate the form of the chelate in solution. The lability of the metal-chelate complex will also influence the aqueous speciation of the chelating agent. Lability is a kinetic property of a metal-chelate complex and is defined as the ability and frequency at which carboxyl oxygen and nitrogen moieties of the chelating agent make and break their bonds with the chelated metal and metal ions in solution. Lability therefore represents the kinetic tendency with which metals in the chelate complex exchange with metals in solution.

The formulas of NTA and EDTA and the structures of $CoNTA^-$ and $CoEDTA^{2-}$ are shown in Fig. 1. The NTA forms a tetradentate complex with various metals, using the one nitrogen and three carboxylate oxygens. $CoNTA^-$ also has two water of hydration molecules present. EDTA forms a hexidentate or sexidentate complex in which the metal is coordinated with two nitrogens and up to four carboxylate oxygens. The Pu ion in solution has seven to eight water of hydration molecules

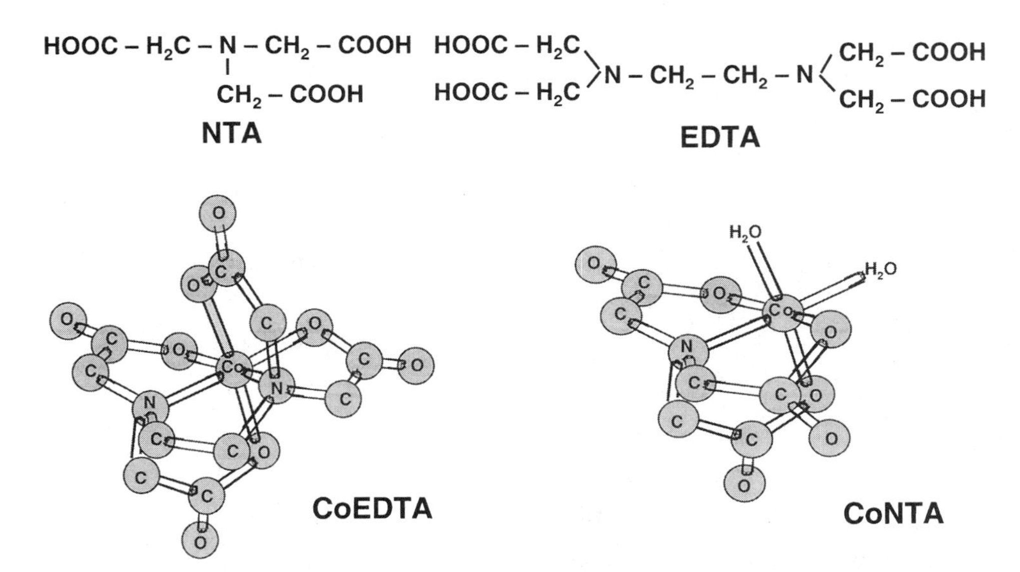

Figure 1. Structures of NTA, EDTA, CoNTA, and CoEDTA.

in the first coordination sphere (17), and so it retains two water of hydration molecules when complexed by EDTA.

NTA Biodegradation

The chelated metal is known to influence the biodegradation of NTA in soils (84), sediments (12), waters (76), sewage (49, 75), and microbial cultures (25, 49). More recent studies have explicitly examined the influence of aqueous speciation on the biodegradation of NTA by *Chelatobacter heintzii* (9, 10). In solutions containing micromolar concentrations of NTA and Al^{3+} ions, where the Al concentration was controlled by the solubility of the common Al mineral gibbsite [$Al(OH)_3$; log K_{sp} = 8.7)], the distribution of NTA species was highly pH dependent (Fig. 2). The first-order biodegradation rates at pH 6, 7, and 8 were 0.033, 0.14, and 2.1 h^{-1}, respectively (9). The change in the biodegradation rate for NTA as a function of pH was attributed to the change in solution species from $HNTA^{2-}$ (14%), AlNTA (14%), and $AlOHNTA^{-}$ (70%) at pH 6 to $HNTA^{2-}$ (87%), $Al(OH)_2NTA^{2-}$ (7%), and $AlOHNTA^{-}$ (4%) at pH 8 (9). In complementary studies, we reported that the relative biodegradation rates of NTA by *C. heintzii* followed the series $HNTA^{2-}$ > $CoNTA^{-}$ = $FeOHNTA^{-}$ = $ZnNTA^{-}$ > $AlOHNTA^{-}$ > $CuNTA^{-}$ > $NiNTA^{-}$ (10). The concentration of metals used in the NTA biodegradation experiments did not inhibit glucose biodegradation, indicating that the toxicity of the chelated metal did not influence the rates of NTA biodegradation. The biodegradability of the various metal-NTA complexes in this series was not related to the thermodynamic stability constants for the various metal-NTA complexes but was related to differences in the labilities of the various metal-NTA complexes. As stated above, lability represents the kinetic tendency of the metals in the chelate complex to exchange with

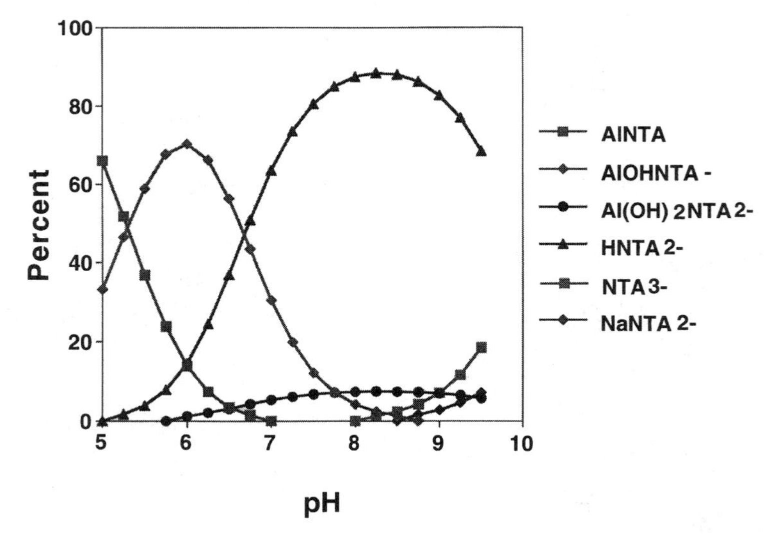

Figure 2. Speciation of 1 μM NTA in equilibrium with gibbsite ($Al[OH]_3$; log K_{sp} = 8.77) in a 0.01 M $NaClO_4$ solution.

metals or protons in solution. The labilities of the metal-NTA complexes investigated above vary significantly for the forward rate of the exchange reaction, $MeNTA^{z-3} + H^+ \rightleftharpoons Me^z + HNTA^{2-}$. The relative rates of this forward exchange reaction have been summarized for the various metal-NTA complexes (10) as Co = Fe = Zn > Al = Cu > Ni. Both $CoNTA^-$ and FeNTA are labile complexes, while $CuNTA^-$ and $NiNTA^-$ are nonlabile and inert complexes, respectively (52). The lability of metal-NTA complexes could influence biodegradation by affecting transport or uptake of the NTA into the cell and/or the subsequent enzymatic biodegradation of the NTA in the cytoplasm. These processes are addressed in later sections of this chapter.

EDTA Biodegradation

There has been some recent work on the influence of the complexed metal or the aqueous speciation of the EDTA on its biodegradation. Several systems have been studied, including an *Agrobacterium* sp. (47, 62), the gram-negative bacterium BNC1 (44, 60; H. Bolton, Jr., unpublished data), the gram-negative bacterium DSM 9103 (92), and a mixed culture (81). Little work on the influence of aqueous speciation on the degradation of EDTA by whole cells of strain DSM 9103 has been reported (92). However, Nörtemann (59) reported that taxonomic experiments have shown that DSM 9103 will degrade Mg-, Ca-, and Mn-EDTA.

Lauff et al. (47) investigated whether various metal-EDTA complexes would support the growth of an *Agrobacterium* sp. Only Fe(III)EDTA supported growth of this bacterium and degradation of the EDTA at initial concentrations of 100 and 35 mM. The degradation of 35 mM Fe(III)EDTA resulted in 2.4 mM EDTA remaining in solution, and so not all the Fe(III)EDTA was biodegraded. The *Agrobacterium* sp. was unable to grow with uncomplexed EDTA, NiEDTA, or CuEDTA. Subsequent work by Palumbo et al. (62) using this *Agrobacterium* sp. found that CoEDTA was not biodegraded. As stated above, ^{60}CoEDTA is a problem at several waste sites (16, 18, 42, 57). Palumbo et al. (62) suggested that displacing the Co from the CoEDTA complex via exchange reactions with Fe(III) and the production of Fe(III)EDTA might be one way for the *Agrobacterium* sp. to degrade the EDTA.

Klüner et al. (44) found that BNC1 degraded 0.6 mM metal-EDTA complexes with thermodynamic stability constants below 10^{12} (e.g., Ba-, Ca-, Mg-, and Mn-EDTA). Metal-EDTA complexes with stability constants greater than 10^{12} were not degraded (Cd-, Co-, Cu-, Fe(III)-, Ni-, and PbEDTA) except for ZnEDTA. This suggests that the degradability of various metal-EDTA complexes can be predicted based on their thermodynamic stability constants. Recent results with BNC1 using a much lower initial concentration of EDTA (e.g., 1 μM) showed very different results. Speciation and lability influence the biodegradation of 1 μM EDTA by BNC1 (Bolton, unpublished). Most 1 μM metal-EDTA complexes, including Al-, Ca-, Cu-, Fe(III)-, Mg-, and $ZnEDTA^{2-}$, were biodegraded by BNC1. However, $Co(II)EDTA^{2-}$, $Co(III)EDTA^-$, and $NiEDTA^{2-}$ were not biodegraded (Fig. 3). The Co or Ni concentrations used were not toxic to BNC1, because acetate mineralization was not inhibited by 1 μM metal or by the 1 μM metal-EDTA complex. The $Co(II)EDTA^{2-}$ was not oxidized to $Co(III)EDTA^-$ during our experiments, as

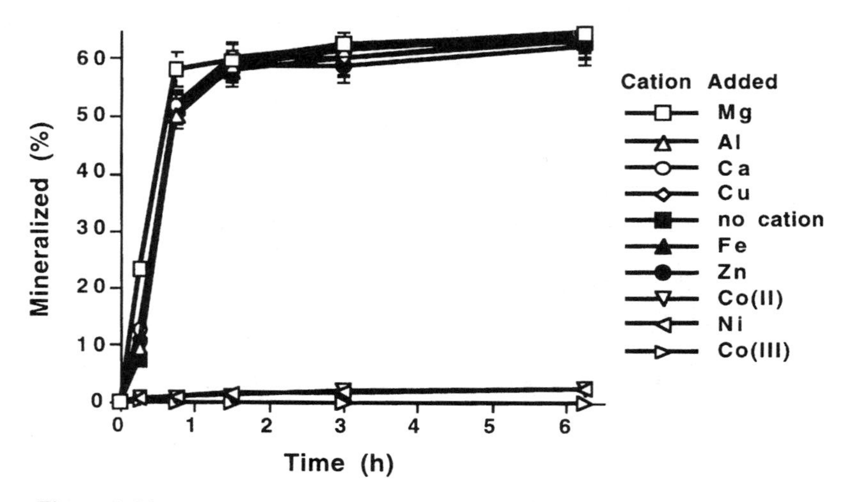

Figure 3. Biodegradation of 1 μM EDTA to CO_2 at pH 7 by BNC1 with different metals added. Error bars indicate ± 1 standard deviation of the mean.

determined by ion chromatographic analysis of Co(III)EDTA (78). The $Co(II)EDTA^{2-}$, $Co(III)EDTA^{-}$, and $NiEDTA^{2-}$ were not biodegraded by BNC1 when they were present at concentrations from 0.1 to 100 μM or when they were incubated for extended times of 48 h (Bolton, unpublished). The relative rates of biodegradation of the metal-EDTA chelates by BNC1 were Mg > H > Ca = Cu = Zn = Fe(III) = Al > Co(II) = Ni = Co(III) (Fig. 3) (Bolton, unpublished). The biodegradability of the various metal-EDTA complexes was not related to their thermodynamic stability constants, similar to the biodegradation of NTA by *C. heintzii* (10). The $Co(III)EDTA^{-}$ and $NiEDTA^{2-}$ complexes are nonlabile (19, 46, 50, 51), and do not dissociate rapidly to form other metal-EDTA complexes or protonated EDTA. The relative rates of biodegradation of metal-EDTA complexes may be due to their variable labilities, as was found for the various metal-NTA complexes (10). The relative exchange rates for metals chelated by EDTA with Mg^{2+} follow the order Ca >> Cu = Zn > Fe(III) > Co(II) >> Ni >> Co(III) (Bolton, unpublished). Whether lability influenced EDTA biodegradation at the cellular transport or enzymatic catalysis step is addressed below.

These results and conclusions for the degradation of various metal-EDTA complexes are somewhat different from those of Klüner et al. (44). Fe(III)- and Cu-EDTA were degraded at 1 μM in the experiments of Bolton et al. (Bolton, unpublished) but not at 600 μM in the experiments of Klüner et al. (44). This suggests that concentration as well as speciation influences the degradation of specific metal-EDTA complexes.

Thomas et al. (81) investigated the degradation of metal-EDTA complexes using an enriched microbial population from a river and liquid effluent sludge. The consortium was enriched on Fe(III)EDTA. The consortium slowly degraded 5 mM metal-EDTA complexes in the order Fe(III)EDTA > CuEDTA > CoEDTA > NiEDTA > CdEDTA, with 60, 30, 25, 23, and 19%, respectively, being degraded during a 28-day incubation. Therefore, all treatments were associated with a significant residual amount of EDTA left in solution after 28 days, with FeEDTA having the lowest concentration (2 mM) and the other metal-EDTA complexes having higher concentrations (3.5 mM or greater) remaining in solution. Thus, the extent of degradation reported here and the rate of EDTA biodegradation by this consortium were much less than for BNC1 (59). However, this is the first reported instance where Cu-, Co-, Ni-, and Cd-EDTA complexes were biodegraded.

Fate of the Metal during Biodegradation

The biodegradation of either $^{60}CoNTA^-$ or $^{63}NiNTA^-$ by *C. heintzii* resulted in very low concentrations of ^{57}Co and ^{63}Ni (i.e., 1 to 3%) becoming associated with viable cells; almost all the initially chelated radionuclide was soluble as biodegradation approached completion (10). Therefore, after the NTA is biodegraded, the initially chelated radionuclide or metal would behave as the free metal cation. If the released radionuclides were associated with the cell biomass, their fate and transport could be substantially altered and would reflect the movement and fate of the bacterial population. The localization of radionuclides following biodegradation of the chelating agent has a profound impact on the transport of radionuclides and metals initially complexed by NTA, where biodegradation of NTA and radionuclide-NTA complexes may be occurring. The location of the metal during and after metal-EDTA biodegradation has not been investigated. Therefore, it is not known if the chelated metal would become associated with bacterial cells or behave as the free metal cation. This would be of particular interest at DOE waste sites, where the presence of EDTA can increase the mobility of the highly insoluble $Pu(IV)O_2$ oxide (log K_{sp} = 54) (67) via the formation of the highly soluble and stable Pu(IV)EDTA (formation constant, log K = 27) (15). Biodegradation of this highly mobile Pu(IV)EDTA and the extracellular release of the Pu(IV) ion would probably result in the immediate formation of the insoluble oxide $Pu(IV)O_2$. If the formation of $Pu(IV)O_2$ occurred either inside the cell or in association with the cell wall, the transport of the Pu would mimic the transport of the bacterial population. Thus, the location of radionuclides (e.g., Pu) and metals during and after the biodegradation of radionuclide/metal-EDTA complexes must be determined to understand the potential transport properties of the initially chelated radionuclide in the environment.

INFLUENCE OF SORPTION ON CHELATING-AGENT BIODEGRADATION

The studies discussed so far on the biodegradation of chelating agents were conducted in aqueous systems with bacteria being the only potential sorptive surfaces present. In the natural environment, sediment and soil solids would also be present. These solids, consisting of mineral and organic particles, can potentially interact with both the chelating agents and the radionuclide/metal-chelate complexes and can influence the solution concentration and aqueous speciation of the chelating agent. The naturally occurring solid phases can function as adsorbents (e.g., Al, Fe, and Mn oxides and clays), increasing the difficulty of studying and understanding chelating-agent biodegradation by adding adsorption to the suite of chemical reactions among metals, sorbent cations, and chelating agents. The reactive sites on the oxide and clay adsorbents can be considered to be ligands that react with solution constituents. These adsorbate ligands will compete with the chelate ligand for the metal ions in metal-chelate complexes. For example, with the CoNTA complex, the commonly occurring Al and Fe oxides such as gibbsite and goethite can outcompete CoNTA for the Co^{2+} ion (32; D. C. Girvin, unpublished data). The competition of the reactive adsorption sites on these solids with the NTA results in the dissociation of the CoNTA complex, with Co and NTA adsorbing independently as a function of pH. Therefore, knowledge of the initial sorbing metal-NTA complex may not be critical for interpreting biodegradation studies. Rather, the identity of the metal-NTA complex present in solution after desorption, that is, the "desorbed metal-NTA complex," is important for understanding and predicting NTA biodegradation when sorbents are present.

Because the thermodynamic stability constants for the formation of metal-EDTA complexes are significantly greater than for NTA complexes with the same metals [e.g., $\log K\ (NiNTA^-) = 11.5 << \log K\ (NiEDTA^{2-}) = 18.5$] and because of the insolubility of Fe and Al hydroxides at high pH, the adsorption of metal-EDTA complexes does not generally result in the dissociation of the original metal-EDTA complex at pH 7 (31, 98). However, at pH 7 the original chelated metal may exchange with structural Fe or Al, forming Fe- or Al-EDTA complexes. This exchange occurs both in solution and on the surface of the oxide and is represented by the reaction $Fe/Al_{(aq\ or\ surface)} + MeEDTA_{(aq\ or\ surface)} \rightleftharpoons FeEDTA_{(aq)}$ or $AlEDTA_{(aq)} + Me_{(aq)}$, where aq designates an aqueous ion (98). At an equilibrium pH of 7, the amounts of Co and EDTA adsorbed are different when equal molar concentrations of Co and EDTA are present. Some of the $Co(II)EDTA^{2-}$ complex dissociates, and Fe^{3+} and Al^{3+} ions are introduced into solution as EDTA complexes. Thus, EDTA enhances the solubility of the Fe and Al oxides, which in turn affects the aqueous speciation of the EDTA. NTA can also enhance the solubility of Al and Fe oxides, but to a lesser extent than EDTA, because NTA has lower stability constants than EDTA for Al and Fe. Thus, for pH 7, the "desorbed metal-NTA complex" referred to in the preceding paragraph is predominantly $AlOHNTA^-$ and AlNTA, with the relative proportion of these species being dependent on the pH (Fig. 2). Clearly, chelating-agent adsorption and chelate-enhanced dissolution of Fe and Al sorbents will alter the speciation and thus the composition of soluble metal-chelate complexes.

The biodegradation of $CoNTA^-$ by *C. heintzii* was studied in the presence of the Al oxide sorbent gibbsite to investigate how sorption of a synthetic chelating agent influenced biodegradation (9). It was hypothesized that adsorbed NTA would not be biodegraded and that the rate of NTA desorption would limit the rate of biodegradation. The experimental system included 1 μM NTA at pH 6 and 8 and CoNTA at pH 7, high cell concentrations of *C. heintzii*, and gibbsite. The experimental parameters investigated included biodegradation of NTA in solutions containing concentrations of Al that would be present in equilibrium with the gibbsite suspension, desorption of NTA and Co from gibbsite, coupled sorption-biodegradation with NTA or CoNTA initially all sorbed to gibbsite, and determination of whether the ^{60}Co during and after NTA biodegradation was in solution or associated with the cell biomass or gibbsite. Biodegradation depended upon the aqueous speciation of the NTA in the various systems, with different rates of biodegradation depending upon the complexed metal, as found previously (10). At pH 7, the NTA and Co desorbed from gibbsite independently, with NTA desorption being much slower than Co desorption (9). The desorption half-lives of NTA and Co at this pH were 16 and 2.5 h, respectively, while the desorption half-lives of NTA at pH 8 and 6 were 1 and 80 h, respectively. The biodegradation half-lives for NTA in solution at pH 8, 7, and 6 were 0.8, 5, and 21 h, respectively. These solutions contained dissolved Al at concentrations that would be present in equilibrium with the gibbsite suspension and would therefore represent the aqueous speciation of desorbing NTA in such suspensions. The desorption rate of NTA was much lower than the biodegradation rate, so that desorption should limit biodegradation. This was confirmed in experimental systems, where all the NTA was initially sorbed to gibbsite prior to the addition of *C. heintzii*. A coupled sorption-biodegradation process model successfully modeled the results of experiments at all three pHs and showed that sorbed NTA was not biodegraded. The implications of these results are that in sediments containing gibbsite, adsorption may have a major impact on the biodegradation of synthetic chelating agents such as NTA.

It was essential to determine the location or distribution of ^{60}Co among the aqueous and solid phases during and after NTA biodegradation, because the distribution of ^{60}Co will influence its fate and mobility. It was unclear how the gibbsite and *C. heintzii* cells would influence ^{60}Co sorption and mobility during biodegradation of the $^{60}CoNTA^-$. Gram-negative bacteria such as *C. heintzii* can sorb metals (8, 66) and compete with the gibbsite for sorption of ^{60}Co. However, neither desorption nor adsorption of Co to gibbsite was substantially altered by the presence of *C. heintzii* or by NTA biodegradation. Although similar investigations of metal-EDTA desorption rates and the influence of these rates on EDTA biodegradation by the EDTA-degrading bacterium BNC1 have not been conducted, it is likely that adsorption-desorption processes would also limit EDTA biodegradation. These issues must be addressed for EDTA-biodegrading microorganisms to better understand the potential effects of EDTA biodegradation on radionuclide mobility at sites where radionuclides and metals have been codisposed with EDTA.

OXIDATION STATE CHANGE OF THE CHELATED METAL

Multivalent metal-EDTA complexes can sorb to the surface of redox-active Fe or Mn oxide solid phases, e.g., Fe(III)OOH or Mn(IV)O_2, resulting in a change in the valance of the chelated metal, which in turn will result in a change in the structure, stability, and reactivity of the metal-chelate complex. A change in the valance of the chelated radionuclide or metal could directly or indirectly affect the microbial biodegradation of the chelating agent. Naturally occurring manganese oxides, such as todorokite and rancieite in subsurface sediments (97), and the pure mineral-phase pyrolusite (β-MnO_2) (37) rapidly and completely oxidized Co(II)EDTA to Co(III)EDTA in batch and flow systems, respectively. This oxidation also occurred under anoxic conditions, but more slowly and to a lesser extent, suggesting that Mn(IV) was the dominant oxidant (37). Similarly, goethite (FeOOH) (95) and ferrihydrite [$Fe(OH)_3$] (13) oxidized Co(II)EDTA to Co(III)EDTA under oxic conditions. The oxidation of Co(II)EDTA by ferrihydrite decreased under anoxic conditions but was not eliminated. This suggests that the structural Fe(III) acted as the oxidant (13). Co(III)EDTA adsorbs to a much lesser extent than does Co(II)EDTA to either Al oxides (31) or Fe oxides (13, 97); therefore, Co(III)EDTA is more mobile in surface waters and groundwaters than is Co(II)EDTA (13, 31, 37, 97). Although neither Co(II)EDTA nor Co(III)EDTA is biodegraded, the formation of Co(III)EDTA prevents the dissociation of $CoEDTA^{2-}$ and the formation of other metal-EDTA complexes that can be biodegraded [e.g., $CaEDTA^{2-}$ and Fe(III)EDTA]. As an example, EDTA-complexed Co(II) can exchange with Fe(III) upon sorption of Co(II)EDTA to Fe(III) oxides to form Fe(III)EDTA (98). The decrease in the total concentration of EDTA through the biodegradation of Fe(III)EDTA or other biodegradable complexes and the resultant respeciation of EDTA would decrease the concentration of $^{60}Co(II)EDTA^{2-}$ and result in uncomplexed ^{60}Co.

BIOCHEMISTRY AND GENETICS OF CHELATING-AGENT BIODEGRADATION

NTA Monooxygenase

The biochemistry of NTA biodegradation by *C. heintzii* has been extensively studied (45, 89, 94, 96). The oxidation of NTA to iminodiacetic acid (IDA) and glyoxylate in cell extracts of *C. heintzii* was first reported by Firestone and Tiedje in 1978 (26) (Fig. 4). The enzyme responsible for the oxidation of NTA to IDA is a two-component monooxygenase, NTA monooxygenase; each component has been purified, and the reconstituted enzyme has been characterized (89). NTA monooxygenase oxidized NTA to IDA with the coconsumption of oxygen and NADH in the presence of flavin mononucleotide (FMN) and Mg^{2+}. FMN was proposed to be a coenzyme. Of 26 compounds assayed, NTA was the only substrate for the enzyme (89). The Mg^{2+} in the enzyme assay mixture resulted in the formation of $MgNTA^-$, suggesting that $MgNTA^-$ was a substrate for NTA monooxygenase. Further proof that $MgNTA^-$ and other metal-NTA complexes were substrates came from kinetic analysis combined with aqueous speciation studies of various metal-NTA complexes (96) as discussed below.

NTA monooxygenase did not degrade the protonated form of NTA ($HNTA^{2-}$) or $CaNTA^-$ (89, 96). The previous discussion indicated that whole cells of *C. heintzii* degraded $HNTA^{2-}$ much faster than they degraded all the metal-NTA complexes assayed (10). If $HNTA^{2-}$ was transported into the cell, $MgNTA^-$ would form, given the 1 to 4 mM concentrations of Mg in the bacterial cytoplasm (35). NTA monooxygenase degraded $MgNTA^-$, $CoNTA^-$, $NiNTA^-$, and $ZnNTA^-$ at the same rate (96). This was in contrast to biodegradation of metal-NTA complexes by

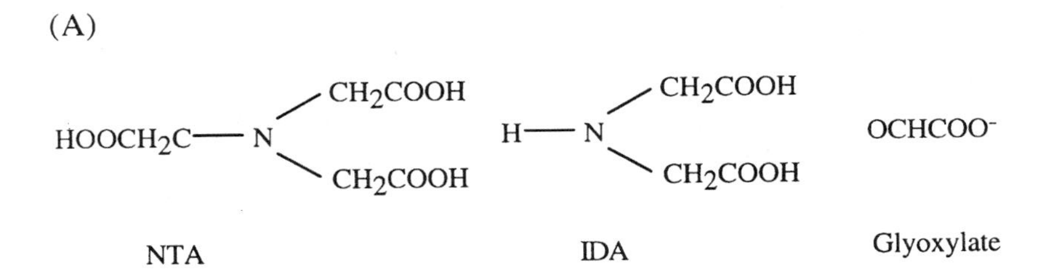

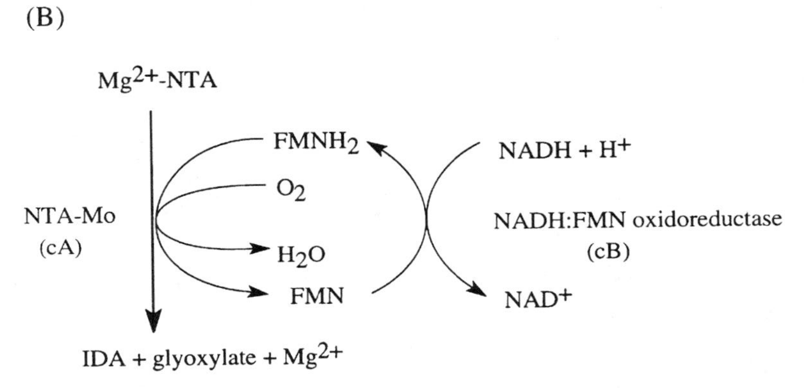

Figure 4. (A) Structures of NTA, IDA, and glyoxylate. (B) Reactions catalyzed by NADH:FMNB oxidoreductase (cB) and NTA monooxygenase (cA).

whole cells described above, where $CoNTA^-$ and $ZnNTA^-$ were degraded at a similar rate while $NiNTA^-$ was degraded more slowly (10). The variation in NTA biodegradation by whole cells was proposed to be influenced by the lability of the metal-NTA complex or the frequency of making and breaking the metal-carboxyl oxygen and the metal-nitrogen bonds of the complex (10). The specificity of NTA monooxygenase indicated that an intact metal-NTA complex was necessary for biodegradation (96) and that the lability of the metal-NTA complex did not influence its biodegradation by this enzyme. Lability may still influence the biodegradation of metal-NTA complexes by whole cells if lability influences the transport of NTA into the cell.

The genes encoding the two components of NTA monooxygenase have been cloned with oligonucleotide probes designed from N-terminal sequences of purified proteins, and the sequences have determined (45, 94). A cluster containing genes encoding both components (NmoA and NmoB), a hypothetical regulatory protein (NmoR), and a transposase (NmoT) has been reported (Fig. 5). The two genes *nmoA* and *nmoB* were transcribed in opposite directions (Fig. 5). Since only 308 bp separated the two translation starting codons, it was likely that a single regulatory protein controls the expression of both genes. The expression of the two genes required induction by NTA (45).

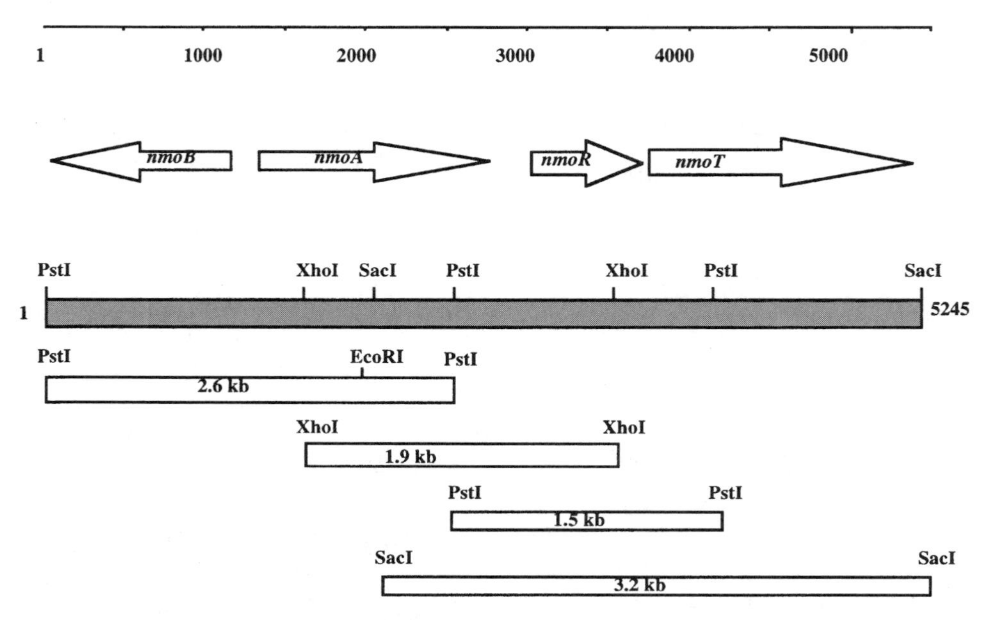

Figure 5. Overview of the gene cluster encoding NTA monooxygenase of *C. heintzii*. The genes *nmoA*, *nmoB*, *nmoR*, and *nmoT* and their directions of transcription are indicated by arrows. Cloned DNA fragments are indicated by open bars.

Sequence analysis showed a strong sequence similarity between NmoA and a monooxygenase that uses $FMNH_2$ as the reducing agent instead of NADH and between NmoB and an FMN reductase (94). Biochemical analysis demonstrated that NmoB effectively reduced FMN to $FMNH_2$ with NADH as the reducing agent (94). When NmoA was assayed with $FMNH_2$ as the reducing agent, it oxidized NTA to IDA. Therefore, the two components were recharacterized as two separate enzymes. NmoB is an NADH:FMN oxidoreductase that generates FMN_2 and NmoA is NTA monooxygenase, which uses $FMNH_2$ to oxidize $MgNTA^-$ to IDA, glyoxylate, and Mg^{2+} (Fig. 4) (94).

IDA Dehydrogenase

The IDA produced by NTA monooxygenase was converted to glyoxylate and glycine by the membrane-bound IDA dehydrogenase in *C. heintzii* (88). This enzyme was also induced by NTA. It has not been purified, but evidence suggests that it passes the electrons to O_2 through the electron transport chain. Membrane proteins extracted with detergents and reconstituted into soybean phospholipid vesicles oxidized IDA with ubiquinone Q_1 as the intermediate electron carrier and acetate iodonitrotetrazolium as the final electron acceptor. These results suggested that the electrons derived from IDA enter the electron transport chain via a ubiquinone. IDA is the only known substrate for the enzyme. IDA oxidation by the enzyme was stimulated by $CaCl_2$, $BaCl_2$, and $MgCl_2$. $MnCl_2$ and $FeCl_3$ inhibited the oxidation slightly, while $ZnCl_2$, $CuCl_2$, and $NiCl_2$ inhibited the oxidation completely. The nature of the metal-IDA complexes probably affected the enzyme activity, but the effects of aqueous speciation on enzyme activity are unknown. The IDA dehydrogenase appears to be ubiquitous in NTA-degrading organisms and has been detected in 10 different NTA-degrading strains belonging to three different genera (88).

In summary, *C. heintzii* produces three enzymes that together degrade NTA to common metabolic intermediates, which can enter other metabolic pathways. The first step involves both NADH:FMN oxidoreductase and NTA monooxygenase to oxidize NTA to IDA and glyoxylate (Fig. 4). The second step involves the conversion of IDA to glycine and glyoxylate by IDA dehydrogenase. Both glycine and glyoxylate are common metabolic intermediates and can be further processed by common catabolic and anabolic pathways. Future areas of research on the biochemistry and genetics of NTA biodegradation include the purification of IDA dehydrogenase and the cloning, sequencing, and characterization of the corresponding gene(s). Detailed regulation of all the genes involved in NTA biodegradation should also be investigated.

EDTA Monooxygenase

The biochemistry of EDTA biodegradation has not been extensively characterized; however, some progress has been made. The EDTA-degrading activity has been detected in cell extracts of two EDTA-degrading bacterial strains, DSM 9103 (92) and BNC1 (44, 63). A cytoplasmic cell extract of BNC1 degraded all metal-EDTA complexes assayed, including Al-, Ca-, Co(II)-, Co(III)-, Cu-, Fe(III)-, Mg-, Ni-, and $ZnEDTA^{2-}$, as well as free EDTA, to form glyoxylate (Bolton, unpublished). Another report on BNC1 showed that the cell extracts degraded EDTA when it was chelated with Ba, Co(II), Mg, Mn(II), or Zn but not when it formed complexes with Ca, Cd, Cu, Fe(III), Pb, or Sn(II) (44). The discrepancy is probably caused by different assay conditions and the aqueous speciation of the EDTA in the various experimental systems. The $MgEDTA^{2-}$ complex was degraded most rapidly by cell extracts. $MgEDTA^{2-}$ is probably a substrate for the cytoplasmic enzyme, given the high $[Mg^{2+}]$ in the cytoplasm (35). $Co(II)EDTA^{2-}$, $Co(III)EDTA^{-}$, and $NiEDTA^{2-}$, which were not biodegraded by whole cells, were rapidly degraded by cell extracts, suggesting that transport or uptake into the cell is the rate-limiting step in the biodegradation of these recalcitrant complexes (Bolton, unpublished).

In strain DSM 9103, cell extracts converted 1 mmol of EDTA to 1 mmol of ethylenediaminetriacetate (ED3A) and then to 1 mmol of ethylenediaminediacetate (EDDA) with the concomitant release of 2 mmol of glyoxylate. The EDDA was further degraded by an NADH-independent reaction (92). Similarly, BNC1 cell extracts converted EDTA to ED3A and then to EDDA, but the sum of ED3A and EDDA was smaller than the amount of EDTA consumed and the level of glyoxylate produced was higher than expected (44). The consumption of EDDA by BNC1 cell extracts was not detected. Therefore, Klüner et al. (44) proposed an alternative pathway in which ED3A was cleaved into IDA and iminoacetaldehydeacetate, which could be further oxidized to IDA. Since IDA can be readily oxidized by strain BNC1 resting cells but EDDA cannot (44), the alternative pathway is plausible for strain BNC1.

An EDTA monooxygenase was purified from strain DSM 9103 (92), and another was purified from strain BNC1 (63). Both enzymes are $FMNH_2$-utilizing monooxygenases, similar to NTA monooxygenase (94). They both oxidized EDTA, NTA, and DTPA. Minor differences have been reported for these two enzymes. The DSM 9103 enzyme cannot use free EDTA as a substrate and can use only several selected EDTA metal complexes as the substrate (92), while the BNC1 enzyme can use both free EDTA and EDTA complexed with metals (63). The enzyme from BNC1 has a higher affinity for MgEDTA than for free EDTA (63). The DSM 9103 enzyme oxidized EDTA to ED3A and then to EDDA (92), whereas the BNC1 enzyme oxidized EDTA primarily to ED3A, but EDDA was not detected in the assay mixture (63). The end product for the BNC1 enzyme is probably ED3A, but the possibility exists that ED3A is further oxidized to EDDA. Further experiments to clarify the reaction end products are needed. The two EDTA monooxygenases and NTA monooxygenase (formerly component A) of *C. heintzii* (89, 94) have some similarities. All enzymes require $FMNH_2$ and O_2 as cosubstrates and are single polypeptides of about 45 kDa. The EDTA monooxygenases oxidize EDTA to ED3A and glyoxylate and oxidize NTA to iminodiacetate and glyoxylate. NTA monooxygenase oxidizes NTA to iminodiacetate and glyoxylate (89), but does not oxidize EDTA or DTPA (J. W. Payne, H. Bolton, Jr., and L. Xun, Abstr. 96th Gen. Meet. Am. Soc. Microbiol. 1996, abstr. Q-275, 1996). The NTA monooxygenase degrades only specific metal-NTA complexes (89, 96).

Flavin-dependent monooxygenases are ubiquitous. Flavin adenine dinucleotide (FAD) and FMN are normally prosthetic groups, not cosubstrates, which are reduced by the monooxygenases with NADH or NADPH as the reductant (27, 79). EDTA monooxygenase uses $FMNH_2$ directly. Since the enzyme does not appear to contain any chromophores and does not require any specific transition metal cofactors, the activated oxygen species is probably a C(4a)-flavin hydroperoxide, as proposed for other flavin-dependent monooxygenases (41, 54). Thus, $FMNH_2$ functions as both the reductant and the prosthetic group for EDTA monooxygenase. An endogenous FMN reductase supplied EDTA monooxygenase with $FMNH_2$. Since the reductase was replaced by two other FMN reductases, it is unlikely that there is any direct protein-protein interaction between the reductases and the oxygenase. These data suggest that EDTA monooxygenase belongs to a small group of monooxygenases, which utilize $FMNH_2$ as both the reductant and the prosthetic group. Bacterial luciferase of *Photobacterium fischeri* was the first $FMNH_2$-utilizing monooxygenase studied (87). Recently, pristinamycin IIA synthase of *Streptomyces pristinaespiralis* (80), NTA monooxygenase of *C. heintzii* (94), and two monooxygenases involved in desulfurization of dibenzothiophene by *Rhodococcus* sp. strain IGTS8 (33, 48) have been characterized and shown to use $FMNH_2$ as the cosubstrate. EDTA monooxygenase appears to be a sixth member of this group. The substrate range of $FMNH_2$-dependent monooxygenases now includes carbon-nitrogen, carbon-sulfur, and carbon-carbon double bonds.

CELLULAR TRANSPORT OF CHELATING AGENTS

Cellular Transport of NTA

Catalysis of NTA by NTA monooxygenase is not influenced by the lability of the metal-NTA complex. Therefore, the specificity of the degradation of the various metal-NTA complexes observed with whole cells is probably conferred by cellular transport proteins. The transport of NTA into *C. heintzii* cells was studied to test the hypothesis that $HNTA^{2-}$ is the transported NTA species (Bolton, unpublished). As stated above, the lability of the various metal-NTA complexes and the relative rates of formation of $HNTA^{2-}$ (10) appeared to limit the rate of NTA biodegradation by whole cells. The uptake of various metal-NTA complexes by *C. heintzii* was studied by varying the concentrations of NTA and metals to alter the aqueous speciation (2) of the various metal-NTA complexes. The presence of 1 mM Ca^{2+} significantly enhanced the uptake of 5 μM NTA by *C. heintzii*, in comparison to 1 mM Na^{+}, K^{+}, or Mg^{2+}. Uptake of NTA in the presence of Ca displayed saturation kinetics at 1 μM NTA, while uptake in the absence of Ca was much slower and appeared to be diffusion limited. There was an equimolar uptake of ^{45}Ca and ^{14}C-NTA as a function of concentration, consistent with $CaNTA^{-}$ being transported into the cell as an intact complex. Uptake of NTA with Ca was inhibited by sodium azide, dinitrophenol, and carbonyl cyanide *m*-chlorophenylhydrazone, suggesting active uptake requiring either ATP or the proton motive force. This uptake of NTA was an inducible process, because uptake did not occur in acetate-grown cells (Bolton, unpublished). These results indicated that $CaNTA^{-}$ and perhaps other metal-NTA complexes were taken up by *C. heintzii*. This is somewhat surprising given that $CaNTA^{-}$ was not a substrate for NTA monooxygenase (96). However, if $CaNTA^{-}$ is the complex transported into the cell, the excess intracellular Mg^{2+} should displace the Ca from the NTA complex to form $MgNTA^{-}$, which is a substrate for NTA monooxygenase.

Cellular Transport of EDTA

The cellular transport of EDTA by BNC1 also depends on the speciation of the EDTA in solution, analogous to NTA transport by *C. heintzii*. The recalcitrance of $Co(II)EDTA^{2-}$, $Co(III)EDTA^{-}$, and $NiEDTA^{2-}$ to biodegradation by whole cells of BNC1 (described above) was probably due to the inability of cells to transport these complexes into the cell (Bolton, unpublished). The uptake of EDTA by BNC1 cells did not increase as a function of time (up to 48 h) or at higher concentrations (10 and 100 μM) when EDTA was present as $Co(II)EDTA^{2-}$, $Co(III)EDTA^{-}$, and $NiEDTA^{2-}$, while uptake did increase for other metal-EDTA complexes. The relative rates of 1 μM EDTA uptake for the metal-EDTA complexes followed the order Mg > Ca = Cu = Zn > Fe(III) >> Co(II) = Ni = Co(III) (Bolton, unpublished), indicating a specificity in the uptake of EDTA when present as various metal-EDTA complexes. This selectivity may be due to the uptake of a specific metal-EDTA complex by the cell. The rate of formation of this specific metal-EDTA complex from the other metal-EDTA complexes would then limit the uptake and biodegradation of the EDTA. The maximum uptake rate of EDTA as well as the highest rate of biodegradation by whole cells (Fig. 3) was with the $MgEDTA^{2-}$ complex (Bolton, unpublished). This suggests that $MgEDTA^{2-}$ is the complex transported into the cell, given the relatively high concentrations of Mg^{2+} associated with bacterial cell walls (35). When 1 μM EDTA was added to iced cell suspensions of BNC1 (to inhibit EDTA biodegradation and uptake), the EDTA complexed primarily Ca and Mg from the cell wall (aqueous speciation of the resulting solution was $CaEDTA^{2-}$ at 84% and $MgEDTA^{2-}$ at 13%) (Bolton, unpublished). If $MgEDTA^{2-}$ is the complex transported into the cell, Mg^{2+} would have to exchange with the other chelated metals before EDTA uptake could occur. Estimates of the relative exchange rates for $Mg^{2+} + \text{Me-EDTA}^{a-4} \rightleftharpoons Me^{a+} + MgEDTA^{2-}$ were made using the exchange model of Margerum et al. (56). The calculated rates, based on several approximations and assumptions, followed the order Ca >> Cu ≈ Zn > Fe(III) > Co(II) >> Ni >> Co(III). The series of relative uptake and exchange rates for the various metal-EDTA complexes were similar, although the exchange rates varied to a greater extent than did the uptake rates. This suggests that a biochemical selectivity for a specific form of EDTA (e.g., $MgEDTA^{2-}$) may control EDTA uptake into BNC1. However, the validity of the calculated rates is not clear, and the origin and mechanism of the biological selectivity in metal-EDTA uptake by BNC1 are currently unknown.

The uptake of EDTA by BNC1 was an active inducible process because EDTA-grown cells with metabolic inhibitors or acetate-grown cells are unable to take up EDTA into the cell (Bolton, unpublished). Active uptake typically requires either an electrochemical ionic gradient or ATP (3). No work has been conducted to determine the mechanisms of chelating-agent transport into microorganisms. However, work has been conducted on the transport into bacteria of citrate, an effective metal complexant. The uptake of citrate by several genera of bacteria involves an electrochemical ionic gradient utilizing a single hydrophobic membrane protein (5, 7, 21, 38, 40, 61, 68, 72, 90, 91). Uptake requiring ATP hydrolysis involves an ATP binding cassette-type (ABC-type) transporter (34, 77). These transporters contain transmembrane proteins, an ATP binding protein on the cytoplasmic side of the membrane, and a high-affinity extracytoplasmic solute-binding protein in the periplasm (77), the region between the inner and outer membranes in gram-negative bacteria. These solute binding proteins confer specificity and high affinity for the uptake of the specific substrate (77). Several bacteria transport citrate via ABC-type transporters with a periplasmic binding protein (73, 74). It is presently unclear whether the NTA uptake system of *C. heintzii* or the EDTA uptake system of BNC1 utilizes an electrochemical ionic gradient or an ABC-type transporter with a periplasmic binding protein. The presence of a binding protein would suggest that specificity for uptake is conferred by the binding protein. The absence of a binding protein suggests that specificity is conferred by the transport protein in the cytoplasmic membrane. Understanding the mechanism of uptake would provide insights into the selectivity of NTA and EDTA uptake and into the way this process might be altered to enhance the biodegradation of recalcitrant radionuclide/metal-EDTA complexes.

CONCLUSIONS

Understanding the biodegradation of synthetic chelating agents is complex because several coupled geochemical and microbiological processes interact to influence biodegradation. First, the chelating agent can complex a wide variety of metals and can have different rates of biodegradation depending upon the complexed metal. Second, the chelating agent can interact with sorbents in the environment and decrease the aqueous concentration of the chelating agent, impact the speciation of the chelating agent, and alter the oxidation state of the complexed metal. Third, metal-chelate complexes have different thermodynamic stability constants, which allow predictions of the percentage of a specific metal-chelate complex in solution at equilibrium, given that all the stability and acid dissociation constants are known and accurate. Fourth, the kinetics of metal exchange can be different depending upon the complexed metal. The lability of metal-chelate complexes is probably a major factor controlling the biodegradation of both NTA and EDTA, because the nonlabile metal-chelate complexes are not transported into the cell. Thus, the rate-limiting step for biodegradation of various metal-NTA and metal-EDTA complexes appears to be the transport or uptake of the complex into the cell. Understanding these coupled interactions of chelate chemistry and microbial biodegradation is critical to the understanding of the fate and transport of chelated radionuclides in the environment. As an example, the current understanding of CoEDTA biodegradation is that the EDTA degrader BNC1 will not degrade ^{60}Co(II,III)EDTA at nuclear waste sites because these complexes are not transported into the cell. However, BNC1 will degrade other metal-EDTA complexes expected at these waste sites [e.g., $CaEDTA^{2-}$ and $Fe(III)EDTA^{-}$], resulting in a respeciation of the EDTA, which would decrease the concentration of $^{60}Co(II)EDTA^{2-}$ and result in the release of ^{60}Co from the complex. The net decrease in the $^{60}Co(II)EDTA^{2-}$ concentration would reduce the formation of highly mobile $^{60}Co(III)EDTA^{-}$ by Mn- and Fe-oxide minerals (36, 97). The $^{60}Co(III)EDTA^{-}$ is weakly adsorbed and enhances ^{60}Co transport at waste sites because of its nonlability and high stability constant (31, 36, 97). Cell extracts of BNC1 degraded metal-EDTA complexes recalcitrant to whole cells [e.g., ^{60}Co(II,III)EDTA]. This suggests that modifying cellular transport to include these recalcitrant cation-EDTA complexes could, if successful, provide a method to directly biodegrade ^{60}Co(II, III)EDTA at nuclear-waste sites.

Acknowledgments. This research was supported by the Natural and Accelerated Bioremediation Research Program, Office of Biological and Environmental Research, U.S. Department of Energy (DOE). Pacific Northwest National Laboratory is operated for the DOE by Battelle Memorial Institute under Contract DE-AC06-76RLO 1830.

The support of Anna Palmisano is greatly appreciated.

REFERENCES

1. **Allen, G. K.** 1976. *Estimated Inventories of Chemicals Added to Underground Waste Tanks, 1944 through 1975.* Report ARH-CD-601B. Atlantic Richfield Hanford Co., Richland, Wash.
2. **Allison, J. D., D. B. Brown, and K. J. Novo-Gradac.** 1991. *MINTEQA2/PRODEFA2. A Geochemical Assessment Model for Environmental Systems: Version 3.0 User's Manual.* Document EPA/600/3-91/021. U.S. Environmental Protection Agency, Office of Research and Development, Washington, D.C.

3. **Ames, G. F. L., and A. K. Joshi.** 1990. Energy coupling in bacterial periplasmic permeases. *J. Bacteriol.* **172:**4133–4137.
4. **Anderegg, G.** 1982. Critical survey of stability constants of NTA complexes. *Pure Appl. Chem.* **54:** 2693–2758.
5. **Ashton, D. M., G. D. Sweet, J. M. Somers, and W. W. Kay.** 1980. Citrate transport in *Salmonella typhimurium*: studies with 2-fluoro-L-erythro-citrate as a substrate. *Can. J. Biochem.* **58:**797–803.
6. **Ayers, J. A.** 1970. *Decontamination of Nuclear Reactors and Equipment.* The Ronald Press Co., New York, N.Y.
7. **Bergsma, J., and W. N. Konings.** 1983. The properties of citrate transport in membrane vesicles from *Bacillus subtilis. Eur. J. Biochem.* **134:**151–156.
8. **Beveridge, T. J.** 1989. *Metal Ions and Bacteria.* John Wiley & Sons, Inc., New York, N.Y.
9. **Bolton, H., Jr., and D. C. Girvin.** 1996. Effect of adsorption on the biodegradation of nitrilotriacetate by *Chelatobacter heintzii. Environ. Sci. Technol.* **30:**2057–2065.
10. **Bolton, H., Jr., D. C. Girvin, A. E. Plymale, S. D. Harvey, and D. J. Workman.** 1996. Degradation of metal-nitrilotriacetate (NTA) complexes by *Chelatobacter heintzii. Environ. Sci. Technol.* **30:**931–938.
11. **Bolton, H., Jr., and Y. A. Gorby.** 1995. An overview of the bioremediation of inorganic contaminants, p. 1–16. *In* R. E. Hinchee and J. L. Means (ed.), *Bioremediation of Inorganics.* Battelle Press, Columbus, Ohio.
12. **Bolton, H., Jr., S. W. Li, D. J. Workman, and D. C. Girvin.** 1993. Biodegradation of synthetic chelates in subsurface sediments from the southeast coastal plain. *J. Environ. Qual.* **22:**125–132.
13. **Brooks, S. C., D. L. Taylor, and P. M. Jardine.** 1996. Reactive transport of EDTA-complexed cobalt in the presence of ferrihydrite. *Geochim. Cosmochim. Acta* **60:**1899–1908.
14. **Campbell, J. A., S. A. Clauss, K. E. Grant, V. Hoopes, G. M. Mong, J. Rau, R. Steele, and K. L. Wahl.** 1996. *Flammable Gas Safety Program.* Actual Waste Organic Analysis FY 1996 Progress Report PNNL-11307/UC-601. Pacific Northwest National Laboratory, Richland, Wash.
15. **Caucheiter, P., and C. Guichard.** 1973. Electrochemical and spectrophotometric study of the complexes of plutonium ions with EDTA. I. Plutonium(III) and (IV). *Radiochim. Acta* **19:**137–146.
16. **Champ, D. R., and D. E. Robertson.** 1986. Chemical speciation of radionuclides in contaminant plumes at the Chalk River Nuclear Laboratories, p. 114–120. *In* R. A. Bulman and J. R. Cooper (ed.), *Speciation of Fission and Activation Products in the Environment.* Elsevier Science Publishing Co., New York, N.Y.
17. **Choppin, G. R.** 1983. Aspects of plutonium solution chemistry, p. 484. *In* W. T. Carnall and G. R. Choppin (ed.), *Plutonium Chemistry*, vol. 216. American Chemical Society, Washington, D.C.
18. **Cleveland, J. M., and T. F. Rees.** 1981. Characterization of plutonium in Maxey Flats radioactive trench leachates. *Science* **212:**1506–1509.
19. **Cook, J., and F. A. Long.** 1958. Kinetics of the exchange of nickel ethylenediaminetetraacetate ion with nickelous ion. *J. Am. Chem. Soc.* **80:**33–37.
20. **Cripps, R. E., and A. S. Noble.** 1973. The metabolism of nitrilotriacetate by a pseudomonad. *Biochem. J.* **136:**1059–1068.
21. **Dimroth, P., and A. Thomer.** 1986. Citrate transport in *Klebsiella pneumoniae. J. Biol. Chem.* **367:** 813–823.
22. **Egli, T., and H. U. Weilenmann.** 1989. Isolation, characterization, and physiology of bacteria able to degrade nitrilotriacetate. *Toxic. Assess.* **4:**23–34.
23. **Egli, T., H. U. Weilenmann, T. El-Banna, and G. Auling.** 1988. Gram-negative, aerobic, nitrilotriacetate-utilizing bacteria from wastewater and soil. *Syst. Appl. Microbiol.* **10:**297–305.
24. **Enfors, S. O., and N. Molin.** 1973. Biodegradation of nitrilotriacetic acid (NTA) by bacteria. I. Isolation of bacteria able to grow anaerobically with NTA as a sole carbon source. *Water Res.* **7:** 881–888.
25. **Firestone, M. K., and J. M. Tiedje.** 1975. Biodegradation of metal-nitrilotriacetate complexes by a *Pseudomonas* species: mechanism of reaction. *Appl. Microbiol.* **29:**758–764.
26. **Firestone, M. K., and J. M. Tiedje.** 1978. Pathway of degradation of nitrilotriacetate by a *Pseudomonas* species. *Appl. Environ. Microbiol.* **35:**955–961.
27. **Flashner, M. S., and V. Massey.** 1974. Flavoprotein oxygenases, p. 245–283. *In* O. Hayaishi (ed.), *Molecular Mechanisms of Oxygen Activation.* Academic Press, Inc., New York, N.Y.

28. **Focht, D. D., and H. A. Joseph.** 1971. Bacterial degradation of nitrilotriacetic acid (NTA). *Can J. Microbiol.* **17:**1553–1556.
29. **Forsberg, C., and G. Lindqvist.** 1967. Experimental studies on bacterial degradation of nitrilotriacetate, NTA. *Vatten* **23:**265–277.
30. **Gephart, R. E., and R. E. Lundgren.** 1996. *Hanford Tank Clean Up: A Guide to Understanding the Technical Issues.* Report PNNL-10773. Pacific Northwest National Laboratory, Richland, Wash.
31. **Girvin, D. C., P. L. Gassman, and H. Bolton, Jr.** 1993. Adsorption of aqueous cobalt ethylenediaminetetraacetate by γ-Al_2O_3. *Soil Sci. Soc. Am. J.* **57:**47–57.
32. **Girvin, D. C., P. L. Gassman, and H. Bolton, Jr.** 1996. Adsorption of cobalt, nitrilotriacetate (NTA), and Co-NTA by gibbsite. *Clay Clay Miner.* **44:**757–768.
33. **Gray, K. A., O. S. Pogrebinsky, G. T. Mrachko, L. Xi, D. J. Monticello, and C. H. Squires.** 1996. Molecular mechanisms of biocatalytic desulfurization of fossil fuels. *Nat. Biotechnol.* **14:** 1705–1709.
34. **Higgins, C. F.** 1992. ABC transporters: from microorganisms to man. *Annu. Rev. Cell Biol.* **8:**67–113.
35. **Hughes, M. N., and R. K. Poole.** 1989. *Metals and Microorganisms.* Chapman & Hall, New York, N.Y.
36. **Jardine, P. M., G. K. Jacobs, and J. D. O'Dell.** 1993. Unsaturated transport processes in undisturbed heterogeneous porous media. II. Co-contaminants. *Soil Sci. Soc. Am. J.* **57:**954–962.
37. **Jardine, P. M., and D. L. Taylor.** 1995. Kinetics and mechanisms of Co(II)EDTA oxidation by pyrolusite. *Geochim. Cosmochim. Acta* **59:**4193–4203.
38. **Johnson, C. L., Y. A. Cha, and J. R. Stern.** 1975. Citrate uptake in membrane vesicles of *Klebsiella aerogenes. J. Bacteriol.* **121:**682–687.
39. **Kakii, K., H. Yamaguchi, Y. Iguchi, M. Teshima, T. Shirakashi, and M. Kuriyama.** 1986. Isolation and growth characteristics of nitrilotriacetate-degrading bacteria. *J. Ferment. Technol.* **64:** 103–108.
40. **Kay, W. W., and M. Cameron.** 1978. Citrate transport in *Salmonella typhimurium. Arch. Biochem. Biophys.* **190:**270–280.
41. **Kemal, C., T. W. Chan, and T. C. Bruice.** 1977. Reaction of $^3O_2$ with dihydroflavins. 1. *N*3,5-Dimethyl-1,5-dihydrolumiflavin and 1,5-dihydroisoalloxazines. *J. Am. Chem. Soc.* **99:**7272–7286.
42. **Killey, R. W. D., J. O. McHugh, D. R. Champ, E. L. Cooper, and J. L. Young.** 1984. Subsurface cobalt-60 migration from a low level waste disposal site. *Environ. Sci. Technol.* **18:**148–157.
43. **Klem, M. J.** 1988. *Inventory of Chemicals Used at Hanford Production Plants and Support Operations (1944–1980).* Westinghouse Hanford Co., Richland, Wash.
44. **Klüner, T., D. C. Hempel, and B. Nörtemann.** 1998. Metabolism of EDTA and its metal chelates by whole cells and cell-free extracts of strain BNC1. *Appl. Microbiol. Biotechnol.* **49:**194–201.
45. **Knobel, H. R., T. Egli, and J. R. van der Meer.** 1997. Cloning and characterization of the genes encoding nitrilotriacetate monooxygenase of *Chelatobacter heintzii* ATCC 29600. *J. Bacteriol.* **178:** 6123–6132.
46. **Krishnan, S. S., and R. E. Jervis.** 1967. Kinetic isotope exchange studies of metal ion substitution in EDTA chelates. I. Co(II), cobalt(II)-EDTA exchange. *J. Inorg. Chem.* **29L:**87–95.
47. **Lauff, J. J., D. B. Steele, L. A. Coogan, and J. M. Breitfeller.** 1990. Degradation of the ferric chelate of EDTA by a pure culture of an *Agrobacterium* sp. *Appl. Environ. Microbiol.* **56:**3346–3353.
48. **Lei, B., and S.-C. Tu.** 1996. Gene overexpression, purification, and identification of a desulfurization enzyme from *Rhodococcus* sp. strain IGTS8 as a sulfide/sulfoxide monooxygenase. *J. Bacteriol.* **178:**5699–5705.
49. **Madsen, E. L., and M. Alexander.** 1985. Effects of chemical speciation on the mineralization of organic compounds by microorganisms. *Appl. Environ. Microbiol.* **50:**342–349.
50. **Margerum, D. W.** 1959. Coordination kinetics of ethylenediaminetetraacetate complexes. *J. Phys. Chem.* **63:**336–339.
51. **Margerum, D. W.** 1963. Exchange reactions of multidentate ligand complexes. *Rec. Chem. Prog.* **24:**237–251.

52. **Margerum, D. W., G. R. Cayley, D. C. Weatherburn, and G. K. Pagenkopf.** 1978. Kinetics and mechanisms of complex formation and ligand exchange. *In* A. E. Martell (ed.), *Coordination Chemistry*, vol. 2. American Chemical Society, Washington, D.C.
53. **Martell, A. E., and R. M. Smith.** 1974. *Critical Stability Constants*, vol. 1. *Amino Acids*. Plenum Press, New York, N.Y.
54. **Massey, V.** 1994. Activation of molecular oxygen by flavins and flavoproteins. *J. Biol. Chem.* **269:** 22549–22562.
55. **McFadden, K. M.** 1980. *Organic Components of Nuclear Wastes and Their Potential for Altering Radionuclide Distribution when Released to Soil.* Report PNL-2563. National Technical Information Service, Springfield, Va.
56. **Means, J. L., and C. A. Alexander.** 1981. The environmental biogeochemistry of chelating agents and recommendations for the disposal of chelated radioactive wastes. *Nucl. Chem. Waste Manage.* **2:**183–196.
57. **Means, J. L., D. A. Crerar, and J. O. Duguid.** 1978. Migration of radioactive wastes: radionuclide mobilization by complexing agents. *Science* **200:**1477–1486.
58. **Means, J. L., T. Kucak, and D. A. Crerar.** 1980. Relative degradation rates of NTA, EDTA, and DTPA and environmental implications. *Environ. Pollut Ser. B* **1:**45–60.
59. **Nörtemann, B.** 1999. Biodegradation of EDTA. *Appl. Microbiol. Biotechnol.* **51:**751–759.
60. **Nörtemann, B.** 1992. Total degradation of EDTA by mixed cultures and a bacterial isolate. *Appl. Environ. Microbiol.* **58:**671–676.
61. **O'Brien, R. W., and J. R. Stern.** 1969. Requirement for sodium in the anaerobic growth of *Aerobacter aerogenes* on citrate. *J. Bacteriol.* **98:**388–393.
62. **Palumbo, A. V., S. Y. Lee, and P. Boerman.** 1994. The effect of media composition on EDTA degradation by *Agrobacterium* sp. *Appl. Biochem. Biotechnol.* **45/46:**811–822.
63. **Payne, J. W., H. Bolton, Jr., J. A. Campbell, and L. Xun.** 1998. Purification and characterization of EDTA monooxygenase from the EDTA-degrading bacterium BNC1. *J. Bacteriol.* **180:**3823–3827.
64. **Piciulo, P. L., J. W. Adams, M. S. Davis, L. W. Milian, and C. I. Anderson.** 1986. *Release of Organic Chelating Agents from Solidified Decontamination Wastes.* Report NUREG/CR-4709, BNL-NUREG-52014. National Technical Information Service, Springfield, Va.
65. **Pickaver, A. H.** 1976. The production of N-nitrosoiminodiacetate from nitrilotriacetate and nitrate by microorganisms growing in mixed culture. *Soil Biol. Biochem.* **8:**13–17.
66. **Poole, R. K., and G. M. Gadd (ed.).** 1989. *Metal-Microbe Interactions.* OIRL Press at Oxford University Press, Oxford, United Kingdom.
67. **Rai, D.** 1984. Solubility product of Pu(IV) hydrous oxide and equilibrium constants of Pu(IV)/Pu(V), Pu(IV)/Pu(VI) and Pu(V)/Pu(IV) couples. *Radiochim. Acta* **35:**97–106.
68. **Reynolds, C. H., and S. Silver.** 1983. Citrate utilization by *Escherichia coli*: plasmid- and chromosome-encoded systems. *J. Bacteriol.* **156:**1019–1024.
69. **Riley, R. G., and J. M. Zachara.** 1992. *Chemical Contaminants on DOE Lands and Selection of Contaminant Mixtures for Subsurface Science Research.* Report DOE/ER-0547T. National Technical Information Service, Springfield, Va.
70. **Sillanpaa, M.** 1997. Environmental fate of EDTA and DTPA. *Rev. Environ. Contam. Toxicol.* **152:** 85–111.
71. **Smith, R. M., and A. E. Martell.** 1987. Critical stability constants, enthalpies and entropies for the formation of metal complexes of aminocarboxylic acids and carboxylic acids. *Sci. Total Environ.* **64:**125–147.
72. **Somers, J. M., G. D. Sweet, and W. W. Kay.** 1981. Fluorocitrate resistant tricarboxylate transport mutants of *Salmonella typhimurium*. *Mol. Gen. Genet.* **181:**338–345.
73. **Staudenmaier, H., B. V. Hove, Z. Yaraghi, and V. Braun.** 1989. Nucleotide sequences of the *fecBCDE* genes and locations of the proteins suggest a periplasmic-binding-protein-dependent transport mechanism for Fe(III)dicitrate in *Escherichia coli*. *J. Bacteriol.* **171:**2626–2633.
74. **Sweet, G. D., C. M. Kay, and W. W. Kay.** 1984. Tricarboxylate-binding proteins of *Salmonella typhimurium*. *J. Biol. Chem.* **259:**1586–1592.
75. **Swisher, R. D., M. M. Crutchfield, and D. W. Caldwell.** 1967. Biodegradation of nitrilotriacetate in activated sludge. *Environ. Sci. Technol.* **1:**820–827.

76. **Swisher, R. D., T. A. Taulli, and E. J. Malec.** 1974. Biodegradation of NTA metal chelates in river water, p. 237–263. *In* P. C. Singer (ed.), *Trace Metals and Metal-Organic Interactions in Natural Waters.* Ann Arbor Science Publishers, Ann Arbor, Mich.
77. **Tam, R., and M. H. Saier, Jr.** 1993. Structural, functional, and evolutionary relationships among extracellular solute-binding receptors of bacteria. *Microbiol. Rev.* **57:**320–346.
78. **Taylor, D. L., and P. M. Jardine.** 1995. Analysis of cobalt(II)EDTA and cobalt(III)EDTA in pore water by ion chromatography. *J. Environ. Qual.* **24:**789–792.
79. **Testa, B.** 1995. The nature and functioning of cytochromes P450 and flavin-containing-monooxygenases, p. 70–121. *In* B. Testa and J. Caldwell (ed.), *The Metabolism of Drugs and Other Xenobiotics: Biochemistry of Redox Reactions.* Academic Press, Inc., San Diego, Calif.
80. **Thibaut, D., H. Ratet, D. Bisch, D. Faucher, L. Debussche, and F. Blanche.** 1995. Purification of the two-enzyme system catalyzing the oxidation of the D-proline residue of pristinamycin II_A biosynthesis. *J. Bacteriol.* **177:**5199–5205.
81. **Thomas, R. A. P., K. Lawlor, M. Bailey, and L. E. Macaskie.** 1998. Biodegradation of metal-EDTA complexes by an enriched microbial population. *Appl. Environ. Microbiol.* **64:**1319–1322.
82. **Tiedje, J. M.** 1977. Influence of environmental parameters on EDTA biodegradation in soils and sediments. *J. Environ. Qual.* **6:**21–26.
83. **Tiedje, J. M.** 1975. Microbial degradation of ethylenediaminetetraacetic acid in soils and sediments. *Appl. Environ. Microbiol.* **30:**327–329.
84. **Tiedje, J. M., and B. B. Mason.** 1974. Biodegradation of nitrilotriacetic acid (NTA) in soils. *Soil Sci. Soc. Am. Proc.* **38:**278–283.
85. **Tiedje, J. M., B. B. Mason, C. B. Warren, and E. J. Malec.** 1973. Metabolism of nitrilotriacetate by cells of *Pseudomonas* species. *Appl. Microbiol.* **25:**811–818.
86. **Toste, A. P., L. J. Kirby, D. E. Robertson, K. H. Abel, and R. W. Perkins.** 1983. Characterization of radionuclide behavior in low level waste sites. *IEEE Trans. Nucl. Sci.* **NS-30:**580–585.
87. **Tu, S.-C., and H. I. X. Mager.** 1995. Biochemistry of bacterial bioluminescence. *Photochem. Photobiol.* **62:**615–624.
88. **Uetz, T., R., and T. Egli.** 1993. Characterization of an inducible, membrane-bound iminodiacetate dehydrogenase from *Chelatobacter heintzii* ATCC 29600. *Biodegradation* **3:**423–434.
89. **Uetz, T., R. Schneider, M. Snozzi, and T. Egli.** 1992. Purification and characterization of a two-component monooxygenase that hydroxylates nitrilotriacetate from "*Chelatobacter*" strain ATCC 29600. *J. Bacteriol.* **174:**1179–1188.
90. **Van der Rest, H. E., E. Schwarz, D. Oesterhelt, and W. N. Konings.** 1990. DNA sequence of a citrate carrier of *Klebsiella pneumoniae*. *Eur. J. Biochem.* **189:**401–407.
91. **Willecke, K., E. M. Gries, and P. Oehr.** 1973. Coupled transport of citrate and magnesium in *Bacillus subtilis*. *J. Biol. Chem.* **248:**807–814.
92. **Witschel, M., S. Nagel, and T. Egli.** 1997. Identification and characterization of the two-enzyme system catalyzing oxidation of EDTA in the EDTA-degrading bacterial strain DSM 9103. *J. Bacteriol.* **179:**6937–6943.
93. **Wong, P. T. S., D. Liu, and B. J. Dutka.** 1972. Rapid biodegradation of NTA by a novel bacterial mutant. *Water Res.* **6:**1577–1584.
94. **Xu, Y., M. W. Mortimer, T. S. Fisher, M. L. Kahn, F. J. Brockman, and L. Xun.** 1997. Cloning, sequencing and analysis of a gene cluster from *Chelatobacter heintzii* ATCC 29600 encoding nitrilotriacetate monooxygenase and NADH:flavin mononucleotide oxidoreductase. *J. Bacteriol.* **179:** 1112–1116.
95. **Xue, Y., and S. J. Traina.** 1996. Oxidation kinetics of Co(II)-EDTA in aqueous and semi-aqueous goethite suspensions. *Environ. Sci. Technol.* **30:**1975–1981.
96. **Xun, L., R. B. Reeder, A. E. Plymale, D. C. Girvin, and J. H. Bolton.** 1996. Degradation of metal-nitrilotriacetate complexes by nitrilotriacetate monooxygenase. *Environ. Sci. Technol.* **30:** 1752–1755.
97. **Zachara, J. M., P. L. Gassman, S. C. Smith, and D. Taylor.** 1995. Oxidation and adsorption of $Co(II)EDTA^{2-}$ complexes in subsurface materials with iron and manganese oxide grain coatings. *Geochim. Cosmochim. Acta* **59:**4449–4463.
98. **Zachara, J. M., S. C. Smith, and L. S. Kuzel.** 1995. Adsorption and dissociation of Co-EDTA complexes in iron oxide-containing subsurface sands. *Geochim. Cosmochim. Acta* **59:**4825–4844.

INDEX